The Facts on File
DICTIONARY of
ENVIRONMENTAL SCIENCE

New Edition

Bruce Wyman
L. Harold Stevenson

Checkmark Books®

An imprint of Facts On File, Inc.

To Mark and Kate
B.C.W.

To Sheila, Eric, and April
L.H.S.

The Facts On File Dictionary of Environmental Science
New Edition

Copyright © 2001, 1991 by Bruce C. Wyman and L. Harold Stevenson

Checkmark Books
An imprint of Facts On File, Inc.
11 Penn Plaza
New York NY 10001

Library of Congress Cataloging-in-Publication Data

Wyman, Bruce
The Facts on File dictionary of environmental science/
    edited by Bruce Wyman, L. Harold Stevenson — New ed.
        p. cm. — (Facts on File science library)
    ISBN 0-8160-4233-0 (alk.paper)—ISBN 0-8160-4234-9 (pbk.; alk.paper)
    1. Environmental engineering—Dictionaries. 2. Environmental protection—Dictionaries. 3. Environmental sciences—Dictionaries. I. Wyman, Bruce C.
II. Stevenson, L. Harold, 1940– . IV. Facts on File, Inc. V. Title: Dictionary of environmental science. VI. Series.

TD9.F33   2000
363.7'003—dc21   00-055554

Text and cover design by Cathy Rincon
Illustrations by Jeremy Eagle

Printed in the United States of America.

MP Hermitage 10 9 8 7 6 5 4 3 2 1
        (pbk) 10 9 8 7 6 5 4 3 2 1

This book is printed on acid-free paper.

CONTENTS

ACKNOWLEDGMENTS

We would like to thank the following individuals for suggesting words for inclusion in this dictionary and/or contributing information used to prepare definitions: James Beck, Tricia Byler, George Fister, Gary Gaston, Gale Haigh, Mary Gay Heagler, Dorothy Jones, James Lane, Robert Maples, Barbara Monroe, Patty Palmer, Davis Parker, Ronald Reasonover, Sandra Stephens, William Taylor, Aubrey Thompson, and Reid Trekell.

For invaluable technical help we are indebted to Chad Thibodeaux, Dennis Stutes, and George Mead. Efficient and cheerful library research support came from R. Brantley Cagle, Jr., Ellen Robinson, and Kenneth Awagain.

Encouraging words and understanding were generously supplied by Elizabeth Stevenson and Barbara Wyman.

Finally, we would like to thank Frank K. Darmstadt, Senior Editor at Facts On File, for his insight and patience in our preparation of this manuscript.

PREFACE

Our primary objective in writing the first edition of *The Facts On File Dictionary of Environmental Science* was to put something useful into the hands of students, both those in the sciences and engineering and those nonscience students enrolled in courses intended to fulfill general science requirements. That motivation is no less important in this new edition; however, the wide acceptance and application of the original version by industry environmental management departments, governmental regulatory agencies, and the consulting community have provided an additional dimension.

With this diverse readership in mind, the new edition was constructed as a comprehensive dictionary that would reflect the variety of disciplines represented in the grab bag that is environmental science. In addition to terms from the basic sciences that are applied to environmental problems, a lexicon has grown up within the field, especially from the federal, state, and local environmental regulations. So terms from agriculture, biology, chemistry, ecology, engineering, medicine, microbiology, social science, soil science, management, general administrative law, toxicology, and physics are here, plus regulatory words and phrases like *bubble, opt in,* and the *God committee.*

Nearly a decade has passed since the first edition was published in 1991, and we have added terms from U.S. environmental legislation and the accompanying regulatory programs begun during the 1990s. The dictionary continues to emphasize American terms, but the emerging global problems of species extinction, sustainable development, ozone layer depletion, and global warming are included, along with organized responses to these issues, such as the Earth Summit, the Kyoto and Montreal Protocols, and the multiple groups within the United Nations Environment Program. Special additions were made in the areas of solid waste management and recycling, along with the terminology associated with volcanic activity. Selected individuals important to environmental health and conservation, past and present, have been added, as have a few of the places made infamous by various environmental misdeeds, such as Chernobyl, Love Canal, the Cuyahoga River, and the Valley of the Drums. The growing protest movement spawned by the rise of genetic engineering in agriculture and medicine has been recognized through the incorporation of terminology associated with biotechnology. Also included are the Responsible Care® initiative developed over the last 10 years by the American Chemical Council, community right-to-know programs such as the Toxics Release Inventory and risk management plans, and the new focus on pollution prevention.

Each definition is considered a short lesson constructed to be understandable as an independent entry. [Entries are alphabetized word-by-word.] Undefined jargon and acronyms have been avoided as much as possible, and terms used in a definition or related to the dictionary entry are cross-referenced. We have used "U.S." as an abbreviation for *United States* and "USEPA" as an acronym for *United States Environmental Protection Agency.*

We hope that we have produced a handy and succinct reference. This is not an encyclopedia, much less a specialized treatise, and brevity and clarity are old enemies. As Dr. Johnson said, "Dictionaries are like watches; the worst is better than none, and the best cannot be expected to go quite true."

Bruce C. Wyman
L. Harold Stevenson
Lake Charles, Louisiana

aa flow Hawaiian term (pronounced "ah'ah") describing a lava flow characterized by the presence of relatively little gas. Consequently, the flow tends to move very slowly, a few meters per hour, and to be very thick, with a depth of 3 to 10 meters. The surface of the lava flow cools to form a crust while the interior remains molten.

abandoned site A closed hazardous waste storage or disposal site, the original owner or operator of which is normally no longer in business.

abandoned well A well that is no longer used for the intended purpose. These wells represent an environmental hazard because they provide a conduit for the unimpeded migration of pollutants from the surface directly into aquifers used for drinking water.

abatement Reducing or eliminating pollution.

Abbey, Edward (1927–89) American Radical activist and writer of *Desert Solitaire* (1968), set in the American Southwest, and *The Monkey-Wrench Gang* (1974), which has inspired vandalism and violent attacks against industry, loggers, and developers.

abiotic Nonliving components of the environment (or ECOSYSTEM), including chemicals in the air, water, and soil; the level and variability of solar radiation; and other aspects of the CLIMATE.

abrasive blasting The removal of coatings, rust, or SCALE from surfaces with a pressurized airstream containing silica sand, aluminum oxide, steel grit, nut shells, or other coarse material.

abscission The dropping of leaves from a plant. Excessive exposure to the air contaminants hydrogen fluoride, chlorine, and ethylene, among others, is associated with premature abscission in certain plant species.

absolute error The difference between a measured value and the true value, indicating the accuracy of a measurement.

absolute humidity The amount of water vapor in a unit mass of air, usually expressed in grams of water per kilogram of air.

absolute pressure The pressure exerted by a gas relative to zero pressure. Common units are millimeters of mercury (mm Hg) and POUNDS PER SQUARE INCH (ABSOLUTE). Compare GAUGE PRESSURE.

absolute temperature (T) A temperature expressed on the thermodynamic scale, measured from absolute zero. If θ is the temperature in degrees Celsius, then
$$T = \theta + 273.15$$

absolute zero The zero value of thermodynamic temperature; 0 Kelvin or $-273.15°C$. Theoretically, the volume of an ideal gas at this temperature becomes zero, according to CHARLES'S LAW. In practice, however, substances become solids.

absorbance An expression of the amount of light that is absorbed by a solution; the measure is used to determine the concentration of certain ions or molecules in the solution. Specifically, the logarithm of the radiant power of light striking the solution minus the logarithm of the amount of light transmitted through the solution. See BEER-LAMBERT LAW.

absorbed dose The amount of a chemical substance that the body absorbs through the skin, gastrointestinal tract, or respiratory system. One can be exposed to a substance without absorption taking place. The substance may not penetrate the skin; it may pass out with body wastes unabsorbed; or it may be exhaled without absorption in the lungs.

absorber A material capable of taking in a substance (such as oil) or radiation, as a sponge takes up water.

absorption The uptake of water or chemicals into the cells or circulatory system of an organism. For example, the passing of a nutrient from the intestinal tract into the bloodstream of an individual, the movement of a toxic chemical through the skin and into the circulatory system of an individual, or the uptake of a gas from the lungs into the bloodstream of an individual exposed to the agent from the atmosphere.

absorption band 1. The wavelengths of ELECTROMAGNETIC RADIATION that are absorbed by a substance; chemical-specific absorption is applied by many instruments used for environmental analysis. 2. In atmospheric science, the wavelengths of the ELECTROMAGNETIC SPECTRUM absorbed by a particular atmospheric gas, such as the absorption wavelengths for INFRARED RADIATION by GREENHOUSE GASES.

absorption coefficient See SABIN ABSORPTION COEFFICIENT.

absorption factor The amount (expressed as a fraction) of a chemical that is absorbed into body tissues upon exposure.

absorption tower See PACKED TOWER.

absorptivity The ratio of the amount of radiation absorbed by a body to the maximum amount it can absorb. A surface that is a poor reflector is a good absorber. If no radiation is reflected, the surface acts as a BLACKBODY and has an absorptivity and EMISSIVITY of 1, equal to the emissivity of a body, according to KIRCHHOFF'S LAW.

abyssal zone The bottom of a deep ocean. Compare to BATHYAL ZONE and EUPHOTIC ZONE.

acceleration (a) The rate of change of the velocity of a moving body with respect to time. Units of acceleration are velocity (e.g., ft s^{-1}, m s^{-1}) per time (e.g., ft s^{-2}, m s^{-2}).

accelerator An additive that increases the rate of a chemical reaction.

acceptable daily intake (ADI) The daily dose of a chemical substance determined to be without adverse effect on the general human population after long-term exposure. The dose is expressed in milligrams of the substance per kilogram body weight per day and is often used for the intake of PESTICIDES and PESTICIDE RESIDUES or for the amount of a food additive that can be consumed on a daily basis without the development of a health hazard. The USEPA has replaced acceptable daily intake with the REFERENCE DOSE for protection against noncarcinogenic health effects.

accident rate In an industrial facility, the number of injuries that require treatment beyond first aid multiplied by 200,000 (the number of hours 100 employees would normally work in one year) and divided by the number of hours actually worked by the employees of the facility in one year. Also known as the total injury incident rate (TIIR), injury incident rate (IIR), incident rate (IR), and recordable injury rate (RIR). Used to compare one facility or industry with another. An accident rate of 0.2 would be considered good, whereas a rate of 8.0 would be considered excessive.

acclimation Synonym for ACCLIMATIZATION.

acclimatization The physiological adjustment by an organism to new, usually physical environmental conditions. Often used in reference to the greater ability to tolerate increased air or water temperature after exposure to elevated temperatures.

accuracy The degree to which a reading given by an instrument or the calculation of a statistic agrees with the true value of what is being measured. Compare PRECISION.

acetylcholine A chemical transmitter of impulses across a SYNAPSE between nerve cells or between nerve cells and muscles. After the impulse is delivered, the enzyme ACETYLCHOLINESTERASE catalyzes the reaction that destroys the acetylcholine. Excessive exposure to ORGANOPHOSPHATE insecticides inhibits the action of acetylcholinesterase, and the resulting buildup of acetylcholine can overstimulate the nervous system and muscles, causing tremors, convulsions, or death.

acetylcholinesterase An enzyme that catalyzes the HYDROLYSIS of ACETYLCHOLINE. See ORGANOPHOSPHATE.

acid Chemicals that release hydrogen ions (H^+) in solution and produce hydronium ions (H_3O^+). Such solutions have a sour taste, neutralize bases, and conduct electricity.

acid aerosol Airborne particles composed of SULFATES, sulfuric acid, nitrates, and/or nitric acid. Moderate air concentrations have been shown to cause constriction of the respiratory passages in exercising asthmatics, and mush higher levels have been linked to excess mortality rates in past air pollution EPISODES. Dry particle diameters are typically less than 1–2 micrometers. See ACID FOG.

acid bright dip The cleaning of metal surfaces of corrosion and scale by immersion in acid.

acid deposition The introduction of acidic material to the ground or to surface waters. Includes WET DEPOSITION, from precipitation; dry deposition, from particle fallout; and ACID FOG. See ACID RAIN.

acid dissociation constant (K_a) See DISSOCIATION CONSTANT.

acid fog Airborne water droplets containing sulfuric acid and/or nitric acid; typical diameters are 5–30 micrometers. See ACID AEROSOL.

acid gas Air pollutant with a pH value of less than 5 released as a by-product of incomplete combustion of solid waste and fossil fuels.

acid mine drainage Acidic water that flows into streams from abandoned coal mines or piles of coal mining waste. The acid arises from the OXIDATION of iron sulfide compounds in the mines by air, DISSOLVED OXYGEN in the water, and CHEMOAUTOTROPHS, which are bacteria that can use the iron sulfide as an energy source. Iron sulfide oxidation products include sulfuric acid, the presence of which has reduced or eliminated aquatic life in many streams in coal mining areas.

acid-neutralizing capacity Measure of the ability of soil, water, or some other medium to resist a lowering of pH when exposed to an acid. The greater the neutralizing capacity, the more acid that can be added without damaging the system.

acid rain Rainfall with a greater acidity than normal. Normal rainfall absorbs CARBON DIOXIDE from the air and forms CARBONIC ACID, which has a pH of about 5.6. Greater rainfall acidity is observed downwind (often at significant distances) from areas with many large sources of the air contaminants SULFUR DIOXIDE and NITROGEN DIOXIDE. These oxides react with water to form acids. Long-term deposition of these acids is linked to adverse effects on aquatic organisms and vegetation in geographical areas with poor BUFFERING CAPACITY within the sediments.

Exposure to acid rain also causes deterioration of marble and limestone buildings and statuary. Because acidic materials also can be delivered to the Earth by acidic particles, acid rain is a form of acid deposition. See also WET DEPOSITION, DRY DEPOSITION.

acidification Raising the acidity (lowering the pH) of a fluid by adding an acid.

acidity The level of hydrogen ion concentration in a solution. On the pH scale, used to measure acidity, a solution with a concentration of hydrogen ions per mole greater than 1×10^{-7}, or a pH less than 7.

acoustic Related or pertaining to sound or hearing. Excessive noise can affect hearing or otherwise impact individuals negatively.

acoustical lining A sound-absorbing lining inside air-conditioning ducts or electrical conduits that reduces the transmission of noise through walls of otherwise-soundproof enclosures.

acoustical materials Sound-absorbing materials, generally soft and porous in nature, such as acoustic tile or carpet.

acre-foot A unit of water use of consumption; one acre-foot is the amount (volume) of water covering one acre to a depth of one foot. Equal to 43,560 cubic feet or 1232.75 cubic meters.

acrolein An aldehyde (C_3H_4O) used widely in the manufacture of organic chemicals. The compound is a very strong contact irritant to humans.

actinic range See ULTRAVIOLET RADIATION, ACTINIC RANGE.

actinides The chemical elements for ATOMIC NUMBER 89 (actinium) or atomic number 90 (thorium) through atomic number 103 (lawrencium). The elements are grouped separately on the PERIODIC TABLE of the elements.

actinolite A calcium-iron-magnesium AMPHIBOLE; a type of ASBESTOS.

actinometer An instrument that measures radiant (solar) energy. See also PYRGEOMETER, PYRHELIOMETER, PYRANOMETER.

actinomycetes A group of soil bacteria that are moldlike in appearance and that are important in the ecology of soil. These organisms are involved in the decomposition and conversion of complex plant polymers into HUMUS. The characteristic musty smell of soil is derived from compounds released by this group of bacteria. Streptomycin and related antibiotics are produced by members of the actinomycetes group.

action level (AL) Under restrictions of the Occupational Safety and Health Administration, a measured concentration of certain airborne contaminants or a noise level in the workplace that requires medical testing of exposed workers. Action levels are usually set at one-half of the permissible exposure level.

activated carbon See ACTIVATED CHARCOAL.

activated charcoal A material produced by burning a carbon material such as wood or coconut shells and then increasing the adsorptive capacity by steam treatment, which creates a very large surface area. The product is used to adsorb organic materials, either as an air or water pollution control technique or as an air-sampling device. See ADSORPTION.

activated sludge The suspended solids, mostly microorganisms, present in the AERATION tank of a sewage treatment plant. See ACTIVATED SLUDGE PROCESS.

activated sludge process A form of SECONDARY TREATMENT of sewage, this process uses microorganisms suspended in well-aerated wastewater to degrade the organic material. A sedimentation tank placed after the activated sludge/aeration tank removes the SLUDGE, routing some of it back to the inflow side of the aeration tank. The rest is pumped outside the system for treatment and disposal.

activation In nuclear physics, the neutron bombardment of stable atoms, which absorb the neutrons and become RADIOACTIVE.

activation energy The energy that must be put into a system in order to start some reaction or process. The activation energy is later recovered during the reaction or process. The simplest example is a match that is used to start a forest fire.

active Describing chemicals that readily combine or react with other chemicals. For example, oxygen gas is an active chemical because it reacts with other substances during the combustion process; metallic sodium is an active material because it reacts spontaneously with water to release hydrogen gas. Compare INERT.

active exchange program An organized effort utilized by a laboratory to redistribute surplus chemicals through an intermediary to other facilities or laboratories that need them. The program is designed to reduce the amount of waste chemicals requiring disposal as well as to reduce costs of operation.

active ingredient That part of a commercial preparation that accomplishes the purpose of the formulation. For example, the active ingredient of a children's medicine may be formulated with coloring agents, sugar or other taste enhancers, and solvents to dissolve the active component; or a commercial insecticide preparation may include an inert carrier that allows for the distribution of the active ingredient that actually kills the insects.

activity plans Required by the Asbestos Hazard Emergency Response Act as applied to a school containing asbestos. Written procedures in an asbestos-management plan that present the steps the local education agency will follow in performing the initial and later cleaning, operation, and maintenance-program tasks; periodic surveillance; and reinspection.

active solar system A system used to capture solar energy as heat and to distribute that heat by a system of pumps and pipes. A typical system consists of solar collectors that use sunlight to heat water, which is stored in an insulated tank. The heated water is then pumped through a series of heat exchangers to provide heat for a building. Compare PASSIVE SOLAR SYSTEMS.

activity Another term for radioactivity, expressed in CURIE units. One Curie equals 3.7×10^{10} atomic decays per second.

actual cubic feet per minute (ACFM) The volumetric rate of air, uncorrected for STANDARD CONDITIONS of temperature and pressure. See STANDARD CUBIC FEET PER MINUTE.

acute Describing a sudden exposure to a significant dose of a dangerous chemical or radiation that usually results in a severe reaction.

acute exposure In toxicology, doses administered or received over a period of 24 hours or less. Compare CHRONIC EXPOSURE.

acute health effect A circumstance in which exposure to a chemical or physical agent results in the rapid development of severe symptoms in exposed humans or animals.

acute toxicity Toxicity resulting from an acute exposure; adverse effects closely spaced in time to the absorbed dose of a toxic material.

adaptation The changes in the metabolism, structure, or habits that allow for survival of an organism in a new environment. The changes must be within the genetic potential of the organism, and the organism will revert to the previous status once conditions return to normal. For example, an individual living in the coastal plain just a few hundred feet above sea level will undergo adaptive changes in body physiological mechanisms upon moving to a 6000-foot elevation in Colorado. The body will shift back upon returning to the coastal plain. However, adaptation to living permanently at 26,000 feet on Mount Everest by adjusting physiological processes is not possible within the realities of the human body.

additive effect The interaction of two or more chemical substances on a biological system that has a combined effect equal to the sum of the effects of each substance acting alone. Compare SYNERGISTIC EFFECT, ANTAGONISTIC EFFECT.

additive reagent A chemical element or compound that is included with the normal reagents employed in a pollution abatement procedure to facilitate the operation. A substance added in trace amounts to improve some standard process.

additives Chemicals added in small amounts to a product such as food in an effort to preserve freshness or quality.

add-on control device An apparatus such as carbon absorber or incinerator added to the exhaust system to prevent air pollution. The device usually does not reduce pollutants through a modification of the process; rather, potential pollutants are removed from the exhaust gases.

adenoma A benign tumor of the epithelial (protective, membranous) tissue of a gland. An adenoma observed in a rat that has been administered a dose of a test chemical is taken to imply that the chemical may pose a human cancer risk.

adenosine triphosphate (ATP) An energy-storage compound common to all biological systems. The high-energy intermediate is formed during photosynthesis or by the breakdown of energy-containing material, such as glucose. ATP supplies the energy for cell reactions and functions.

adequately wet In an asbestos removal project, the mixing of asbestos-containing

Adenosine triphosphate

material with liquid sufficiently to prevent the release of particles into the atmosphere.

adiabatic cooling lines (ABC lines) Plots of temperature versus ABSOLUTE HUMIDITY found on a PSYCHROMETRIC CHART.

adiabatic lapse rate The constant decline in temperature of an AIR PARCEL as it rises in the atmosphere. The temperature change is due to the pressure drop and gas expansion only, and no heat is considered to be exchanged with the surrounding air through convection or mixing. The DRY ADIABATIC LAPSE RATE for air not saturated with water is 0.98°C per 100-meter rise (0.54°F per 100 feet). The WET ADIABATIC LAPSE RATE is about 0.6°C per 100 meters (0.33°F per 100 feet).

adiabatic process A change involving no gain or loss of heat.

adjusted rate In EPIDEMIOLOGY, a rate that has been statistically modified to allow a proper comparison with other (adjusted) rates. For example, the death rate for a retirement community will be greater than the death rate for a typical suburb. The difference in the age distributions of the two areas is statistically eliminated when an adjusted death rate is computed. See AGE ADJUSTMENT, DIRECT METHOD; AGE ADJUSTMENT, INDIRECT METHOD.

administered dose In the testing of substances to determine the relationship between dose of an agent and the response of the test species, the amount of a substance given to a test subject. The amount of the chemical absorbed by the test subject may not be equal to the amount given.

administrative controls A system that reduces or controls exposure of workers to chemical or physical agents. The time workers spend in higher-exposure areas may be limited, and those workers with greater ABSORBED DOSES are rotated to lower-exposure areas. Activities with potentially high exposures are scheduled at times or in places that reduce the number of workers present in the area.

administrative law The legal process by which STATUTORY LAW is implemented. For environmental matters, the process pertains to administrative agency decisions and procedures, judicial review of agency actions, issuance of regulations, permission to establish facilities, and citizen suits, among other areas. See CITIZEN SUIT PROVISION.

administrative law judge A government officer who presides during a formal HEARING held by an administrative body, such as the USEPA.

administrative order Directive issued by the USEPA to an individual, business, or facility in violation of the enforceable provisions of environmental regulations. The order can direct compliance, require corrective action, order monitoring and analysis, identify an imminent hazard, or require the payment of a penalty for the violation.

administrative order on consent An agreement between the USEPA and an individual, business, or facility in violation of environmental regulations committing the offending party to refrain from the inappropriate activity or to take corrective or cleanup action. The agreement describes the actions to be taken, applies to civil actions, and can be enforced by the courts.

Administrative Procedure Act (APA) A federal law enacted in 1946 that requires regulatory agencies to follow certain procedures when they make decisions, such as issuing regulations or standards and granting permits to facilities. Judicial review of environmental regulatory agency actions follows the provisions of the APA.

administrative record The studies, reports, hearings, letters, internal memoranda, or other material used by an administrative agency to write a regulation or standard, make a permit decision, and so on. For example, the information used by the USEPA to select a response action under the COMPREHENSIVE ENVIRONMENTAL RESPONSE, COMPENSATION, AND LIABILITY ACT (Superfund); the information includes the REMEDIAL INVESTIGATION/FEASIBILITY STUDY, the RECORD OF DECISION, and any public comments.

Administrator, the The chief officer of the USEPA.

adsorbate The chemical(s) being adsorbed or taken up in a pollution control device used for the ADSORPTION of a chemical substance. See ADSORBENT.

adsorbent The material to which a pollutant is being adsorbed in a pollution control device operating by ADSORPTION. See ADSORBATE.

adsorber A solid or liquid that can hold molecules of another substance on its surface.

adsorption The collection of solids, liquids, or condensed gases on the surface of another substance or body; for example, the adsorption of volatile organic compounds by activated charcoal.

adsorption isotherm test (AI test) See SOIL AND SEDIMENT ADSORPTION ISOTHERM TEST.

adulterant 1. A chemical impurity or substance that by law should not be in either a food or a pesticide. 2. Any substance present in food or drugs at concentrations exceeding those allowed by the federal FOOD AND DRUG ADMINISTRATION.

adulterated 1. Any food or feed that contains illegal pesticide residues or other dangerous substances. 2. A pesticide whose strength and purity are below the quality stipulated on the label.

advance notice of proposed rule making (ANPR) An announcement in the *federal register*, usually with an extended discussion, of the regulatory approach an agency (such as the USEPA) is considering in an area of concern; the notice is not an actual proposed rule.

advanced wastewater treatment Any phase in the treatment of wastewater that goes beyond the usual biological processing to degrade dissolved or suspended organic material (see SECONDARY TREATMENT). Common additional treatments include the removal of inorganic compounds of nitrogen and phosphorus that serve as plant nutrients or the removal of low concentrations of toxic organic compounds (see TERTIARY TREATMENT).

advection Transport by a moving fluid (liquid or gas). Differs from CONVECTION in being related mainly to the horizontal movement of airborne materials (pollutants) by wind.

adverse effects data Information required of pesticide manufacturers under provisions of the FEDERAL INSECTICIDE, FUNGICIDE, AND RODENTICIDE ACT. Those who register a pesticide with the USEPA must submit to the agency information from any study regarding unreasonable adverse effects of the agent obtained at any time after the initial registration.

advisory Information intended to communicate potential risk so that those affected can take appropriate action. In environmental health, nonbinding guidance issued by a government agency that allows the public to manage voluntarily health risks, such as a suggested dietary limit of three meals per month of fish caught in an estuary contaminated by CHLORINATED HYDROCARBONS.

aeration lagoon A pond used to process organic wastes by ensuring that sufficient oxygen is dissolved in the water to promote biological decomposition. Often the devices used agitate the contents of the pond to promote contact between the contents and the air. The presence of oxygen stimulates the activity of microorganisms capable of degrading organic materials. See LAGOON

aerobic treatment Any processing of waste material in the presence of oxygen. See AERATION LAGOON, SECONDARY TREATMENT, ACTIVATED SLUDGE PROCESS.

Aerometric Information Retrieval System (AIRS) A USEPA computer database containing air quality monitoring data, point source emissions levels, and mobile source (vehicle) emissions data. Ambient air quality data from STATE AND LOCAL AIR MONITORING SYSTEM (SLAMS) and NATIONAL AIR MONITORING SYSTEM (NAMS) stations are in the AIR QUALITY SUBSYSTEM (AQS), and point source emissions are found in the AIRS FACILITY SUBSYSTEM (AFS). AIRS data can be accessed online from www.epa.gov/airsdata/.

aeolian soil Soil transported from one area to another by wind.

aeration In wastewater treatment, the addition of air (oxygen) to wastewater. This prevents the DISSOLVED OXYGEN content of the water from falling to levels insufficient for rapid degradation of the organic material in sludge or other wastewater.

aeroallergens Airborne particles that can cause allergic responses (such as hay fever) in susceptible persons.

aerobic Requiring oxygen; the usual application is to bacteria that require oxygen for growth. An aquatic system characterized by the presence of DISSOLVED OXYGEN. Compare ANAEROBIC.

aerobic bacteria Single-celled microscopic organisms that require oxygen to live. These organisms are responsible, in part, for the AEROBIC DECOMPOSITION of organic wastes. Also called DECOMPOSERS.

aerobic decomposition The biodegradation of materials by AEROBIC microorganisms; the process produces CARBON DIOXIDE, water, and other mineral products. Generally a faster breakdown process than ANAEROBIC DECOMPOSITION.

aerodynamic diameter A way to express the diameter of a (usually airborne) particle by using the diameter of a perfect sphere of UNIT DENSITY with the same SETTLING VELOCITY as the particle. For example, a nonspherical particle roughly six micrometers across with a density of two grams per cubic centimeter may have an aerodynamic diameter of eight micrometers meaning that the nonspherical particle has the same settling velocity as an eight-micrometer sphere having a density of one gram per cubic centimeter. Compare STOKES DIAMETER.

aerosols A suspension of liquid or solid particles in air or gas.

aesthetic degradation Corruption of the natural environment in a way that is offensive to a sense of good taste. This type of damage does not pose a threat to health or represent a peril faced by the natural environment. This type of defilement just looks bad. The classic example is the installation of scores of billboards along roads that would otherwise offer a view of a beautiful natural landscape.

affected landfill Under the provisions of the Clean Air Act, any facility used to bury solid waste meeting the criteria of capacity and age indicating the potential to produce significant amounts of methane and other gases. Such a facility is required to collect and combust gas emissions.

affected public 1. Citizens living or working near a hazardous waste disposal site. 2. Individuals adversely impacted after exposure to a toxic pollutant in food, water, or air.

afforestation The establishment of a forest by human planting in an area where trees have not grown previously.

aflatoxin Toxin that is produced during the growth of fungi or molds in grains or grain meals stored under moist conditions. Some of the aflatoxins are thought to be responsible for the development of cancer in animals that consume the contaminated grain or meal.

afterburner An air pollution control device that incinerates organic compounds in an airstream to carbon dioxide and water. Also called a vapor incinerator.

age adjustment, direct method The application of AGE-SPECIFIC RATES for area or community populations to a STANDARD POPULATION to allow a proper comparison of two or more different populations. The method eliminates the effect of different age distributions in the populations being compared.

age adjustment, indirect method The application of a standard (i.e., nationwide) MORBIDITY RATE or MORTALITY RATE to an area or community population to obtain an expected number of illnesses or deaths. This number is compared with the actual number of illnesses or deaths in the population.

age pyramid See AGE-STRUCTURE DIAGRAM.

Agency for Toxic Substances and Disease Registry (ATSDR) An agency in the United States Public Health Service that performs public health assessments in areas near hazardous waste sites, including health surveillance and epidemiological studies. The agency also sponsors research on the toxicological characteristics of specific chemical materials and develops health information databases for use in future health assessments. Website: www.atsdr.cdc.gov

Agenda 21 A planning document adopted by the representatives attending the United Nations Conference on the Environment and Development at Rio de Janeiro (the EARTH SUMMIT) in 1992; promotes SUSTAINABLE DEVELOPMENT. Website: www.un.org/esa/sustdev/agevda21.htm

agent Used to designate any biological, chemical, or physical entity that can cause harm to individuals or the environment.

Agent Orange The herbicide mixture used as a defoliant during the Vietnam War. The primary agent in the mixture was 2,4,5-trichlorophenoxyacetic acid (2,4,5-T). The chemical synthesis of this herbicide results in slight contamination of the agent with DIOXIN. Consequently, persons exposed to Agent Orange in Vietnam also were exposed to a small amount of dioxins. Dioxins are known to cause cancer and birth defects in experimental animals, but the hazard they pose to human health at small doses is uncertain.

age-specific rate In EPIDEMIOLOGY, a MORTALITY RATE that is calculated for a specific age group; for example, the death rate for persons 70 to 75 years old in the United States in 1988.

age-structure diagram A bar graph showing the number or percentage of individuals in a population within various age ranges: 0–5 years, 6–10 years, and so on. The graph is often divided horizontally to show the relative number of males and females within the age ranges.

agglomeration In wastewater treatment, the grouping of small suspended particles into larger particles that are easier to remove. See COAGULATION.

aggregate risk The sum of the estimated individual excess risks of disease or death in a population caused by an exposure to a chemical or physical agent.

aggregation The collection of smaller units into a larger single mass.

agitator/mixer Blades or paddles that slowly rotate within a tank to facilitate the mixing of reagents or suspended materials.

agricultural pollution General term used to indicate adverse consequences associated with the operation of a farm. Runoff and leaching of pesticides and fertilizer from cultivated fields, dust and erosion problems resulting from cultivation practices, improper disposal of animal manure and carcasses, and accumulated crop residues and other debris can potentially cause environmental problems.

agricultural solid waste Manures, plant residues, or discarded stalks and hulls produced during the operation of a farm. Many of these materials are disposed of on the farm by plowing them into the fields.

Agrobacterium tumefaciens Genus and species of a bacterium that can infect certain plants, inducing the development of tumorlike growths commonly called galls. In addition to the potential to act as a plant pathogen, this organism is useful in the introduction of foreign genes into plants to produce new strains of crop plants. See GENETIC ENGINEERING, Ti PLASMID and TRANSGENIC PLANTS.

agroecosystems The farm and surrounding area impacted by agricultural practices, including physical structures; land used for crops, pasture, and livestock; adjacent uncultivated land; and the associated atmosphere, soils, surface drainage water, and groundwater.

AHERA designated person (ADP) An individual identified by a local education agency or school to ensure that the provisions of the Asbestos Hazard Emergency Response Act for asbestos management and abatement programs are properly implemented. The local person responsible for control of asbestos contamination in schools.

air The gaseous mixture that makes up the ATMOSPHERE of the Earth. Four gases account for 99.997% (by volume) of clean, dry air: nitrogen (78.084%), oxygen (20.946%), argon (0.934%), and carbon dioxide (0.033%). The remaining components are neon, helium, methane, krypton, nitrous oxide, hydrogen, xenon, and various organic vapors. Under actual conditions, air contains up to about 3% water vapor (by volume) and many solid, liquid, or gaseous contaminants introduced by human activities and natural processes such as wind erosion.

Air and Waste Management Association (AWMA) A professional organization based in Pittsburgh, Pennsylvania, dedicated to research and management issues involving air quality and hazardous waste. Founded in 1907, it now has 10,000 members. Formerly known as the Air Pollution Control Association. Website: www.awma.org

air changes per hour (ACH) The rate at which a volume of air inside a room or building is replaced by outside air, calculated as the flow in/out divided by the volume of the enclosure. For example, a flow in/out of 2000 cubic feet per hour through a room containing 1000 cubic feet has 2 air changes per hour. Also called the air exchange rate.

air classification A process that uses an airstream to separate shredded municipal solid waste into light and heavy components to facilitate recycling or incineration. The light materials consist of items like paper; glass would be a heavier item.

air classifier A device used in solid waste recovery operations that separates paper and other low-density materials from mixed waste by pumping air up through the waste.

air contaminant An airborne solid, liquid, or gas present in a time-averaged CONCENTRATION that may cause adverse effects on humans, wildlife, or vegetation.

air curtain The flow of air bubbles upward through water from a submerged perforated pipe. Used to contain oil spills.

air exchange rate See AIR CHANGES PER HOUR.

air injection 1. In groundwater management, the pumping of compressed air into the soil to move water in the unsaturated zone down to the saturated zone, or water table. 2. In automobile emission control, the addition of air into the exhaust gas to complete the oxidation of CARBON MONOXIDE to CARBON DIOXIDE, now accomplished by the CATALYTIC CONVERTER.

air knife A blower employing an airstream to remove selected items from a conveyer. Used in the separation of the components of municipal solid waste.

air mass A relatively homogeneous body of air. On a regional scale, usually described by its meteorological features: air temperature, pressure, wind speed, and direction. On an urban scale the term may refer to its pollution levels.

air mover A fan, pump, or other device that induces the movement of air in a duct in workplace ventilation or air pollution control equipment or through hoses in the case of an air-sampling apparatus.

air parcel A theoretical volume of air considered in air pollution dispersion predictions to rise (or fall) without exchange of heat with surrounding air. An air parcel will rise if its temperature is greater than that of the air around it. See ADIABATIC PROCESS, ADIABATIC LAPSE RATE, MIXING HEIGHT.

air plenum See PLENUM.

air pollutant See AIR POLLUTION, AIR CONTAMINANT.

air pollution The presence of gases or AEROSOLS in the AMBIENT air at levels considered to be detrimental to human health, wildlife, visibility, or materials. The air contaminants can have a human origin (smokestacks, tailpipes) or a natural origin (dust storms, volcanoes).

Air Pollution Control Association (APCA) Former name of the AIR AND WASTE MANAGEMENT ASSOCIATION.

air pollution control device Any of many technologies designed to capture air contaminants before they are released to the atmosphere. See CYCLONE COLLECTOR, CATALYTIC CONVERTER, ELECTROSTATIC PRECIPITATOR, FABRIC FILTER, SCRUBBER, BAGHOUSE, TRAY TOWER, SPRAY TOWER, PACKED TOWER.

air pollution episode See EPISODE.

air preheater In industrial and utility boilers, a device in which hot flue gas exhausting from the boiler flows past cool air flowing into the boiler burners and warms the incoming air. The use of preheated combustion air increases boiler fuel efficiency by saving the natural gas, coal, or fuel oil energy otherwise used to raise the incoming air temperature.

air-purifying respirator (APR) A device worn by workers when the air contains, or may contain, harmful levels of contaminants, for example, during certain hazardous workplace tasks or during the cleanup of chemical spills outside facility property. A filter or cartridge removes specific air contaminants from the incoming air breathed by the wearer. The purifying filters or cartridges can become saturated or exhausted and must be replaced. Each APR is effective only against a certain chemical or chemical type; it must be chosen to match the airborne chemical in each situation. Compare SELF-CONTAINED BREATHING APPARATUS (SCBA). See also SUPPLIED-AIR RESPIRATOR.

air quality control region (AQCR) One of 247 air management areas in the United States designated by the USEPA. Each region contains at least two urban areas, often in adjoining states, that share actual or potential air quality problems.

air quality criteria The adverse effects on human health, human welfare, wildlife, or the environment used to set ambient AIR QUALITY STANDARDS. See CRITERIA DOCUMENT.

air quality dispersion model Any of a variety of mathematical simulations of the downwind diffusion and/or physical/chemical removal of air contaminants emitted by one or more sources. The inputs to the model include the type of emission, emission rate, release height, and appropriate weather conditions. The model output is an estimated air concentration at specified points in the vicinity of the source.

air quality maintenance area (AQMA) Administrative area defined by the USEPA for the production of STATE IMPLEMENTATION PLANS. Ambient air quality data, emissions data, air pollution control strategies, and so forth that are part of a state implementation plan are organized by AQMA.

air quality-related value (AQRV) For areas designated as Class I under the PREVENTION OF SIGNIFICANT DETERIORATION of air quality program, criteria other than the allowable INCREMENT CONSUMPTION that may be used to assess the impact of a new emission source. The value mainly refers to any reduction in the visibility that may be caused by the new air emissions. Class I areas include major national parks and wilderness areas, mostly in the American West.

air quality standard (AQS) The allowable amount of a material in the AMBIENT air, for a certain AVERAGING TIME or, less often, the allowable emissions of air contaminants into the atmosphere. See NATIONAL AMBIENT AIR QUALITY STANDARD; STANDARDS, ENVIRONMENTAL. Compare EMISSION STANDARD.

air sparging See SPARGING.

air stripping Removing a dissolved contaminant from water to transfer it to the atmosphere by pumping water over a large surface area in order to provide as much contact as possible between the liquid and the air. A common method for the removal of VOLATILE ORGANIC COMPOUNDS from contaminated groundwater.

air-to-cloth ratio (A/C) In computations involving FABRIC FILTERS for air pollution control, the ratio of the volumetric flow rate of the gas to be cleaned to the fabric area. The air-to-cloth ratio is typically expressed as a FILTERING VELOCITY and is measured in meters per minute.

air-to-fuel ratio (AFR) In internal combustion engines, the ratio of the mass of air available for fuel combustion to the mass of the fuel. High AFRs are described as lean; low ratios are said to be rich. Exhaust emissions are affected by AFRs. A rich mixture will emit higher levels of CARBON MONOXIDE and VOLATILE ORGANIC COMPOUNDS (unburned fuel); a lean (air-rich) mixture will cause poor engine performance. The theoretical AFR, called the STOICHIOMETRIC RATIO, is neither rich nor lean. When an engine operates with an almost ideal mixture, carbon monoxide and volatile organic compound emissions are low, but emissions of NITROGEN OXIDES greatly increase. The THREE-WAY CATALYST emission control device must operate within a narrow AFR to be effective at controlling carbon monoxide, volatile organic compounds, and nitrogen oxides.

air toxics Air pollutants that may pose chronic health risks to human populations if their ambient levels are excessively high for a prolonged period. The ambient level and the length of time required to cause harm vary by pollutant. The health risks of concern include cancer; neurotoxic, mutagenic, and teratogenic effects; and reproductive disorders. Most air toxics are metals, VOLATILE ORGANIC COMPOUNDS, or PRODUCTS OF INCOMPLETE COMBUSTION. EXTREMELY HAZARDOUS SUBSTANCES—those posing acute health risks—are sometimes included among air toxics.

airborne Suspended in the atmosphere.

airborne particles See PARTICULATE MATTER; PARTICULATE MATTER, 10-MICROMETER DIAMETER; DUST; AEROSOL; FUME; SMOKE.

airshed An area with fairly homogeneous air mixing and affected by uniform air pollution sources, thus experiencing roughly the same air quality. Large-scale air movements commonly void the geographical boundaries of an airshed.

airway resistance (R_{aw}) The resistance to airflow in the respiratory system, measured by lung function tests, including SPIROMETRY.

Aitken counter A device for counting the number of CONDENSATION NUCLEI in an air sample. Sample air is cooled by rapid expansion, and the droplets that form on suspended dust particles (nuclei) are counted with the aid of a microscope.

Alachlor A herbicide used to control broadleaf weeds, marketed as Lasso.

Alar Trade name for daminozide, pesticide used on apples and other fruits linked to tumors in mice, but not rats. The NATURAL RESOURCES DEFENSE COUNCIL led a media campaign to ban daminozide; the manufacturer withdrew it from the market voluntarily in 1989.

albedo The fraction of electromagnetic radiation reflected after striking a surface. The average albedo of the Earth to sunlight is about 30%, which is a combination of many different surface reflectivities: water (at noon), about 5%; fresh snow, about 80%; deciduous forest, about 20%.

aldehyde A type of organic molecule containing a carbonyl group in which a carbon atom is bonded to an oxygen atom with a double covalent bond and to a hydrogen atom with a single covalent bond. Aldehydes can be reduced to form alcohols and oxidized to acids. Two common aldehydes are formaldehyde and acetylaldehyde. The basic chemical formula for aldehydes is as follows. The R-group represents any type of organic side chain. Excessive exposure to aldehydes can result in central nervous system depression, skin or mucous membrane

irritation, or allergic reactions in sensitive groups.

$$R\text{-}CH_2\text{-}CH \overset{O}{\underset{\|}{}}$$

Aldicarb A synthetic CARBAMATE pesticide used to control chewing and sucking insects (aphids, whiteflies, leaf miners), spider mites, and nematodes that damage crops such as sugar beets, strawberries, potatoes, onions, coffee, pecans, cotton, and sugarcane. The agent is applied to soil and is taken up by the plant. Aldicarb is a CHOLINESTERASE INHIBITOR that is considered sufficiently toxic to be placed on the dirty dozen list by the Pesticide Action Network.

Aldrin An insecticide of the chlorinated hydrocarbon class. The compound is a cyclodiene, meaning that the parent molecule is cyclic and contains many double bonds between the carbon atoms. The insecticide is persistent in the environment and considered dangerous. The use of aldrin is strictly controlled.

algae A group of one-celled, free-floating green plants often found in aquatic ecosystems. The singular form is *alga*.

algal bloom Rapid growth of ALGAE caused by the addition of a LIMITING FACTOR, such as the nutrient phosphorus, to water. The extent of growth may result in a coloration of the water or the production of toxins at levels that present problems for humans or wildlife.

algicide Chemical, usually added to water, to kill or control algae. Inorganic copper salts like copper sulfate are common examples.

aliphatic Organic compounds with a straight-chain structure. One of the two major groups of HYDROCARBONS; the other are the AROMATIC hydrocarbons (closed-ring structures). The aliphatic hydrocarbons are classified as alkanes (carbon atoms connected by single bonds), alkenes (carbon atoms connected by double bonds), and alkynes (carbon atoms connected by triple bonds). See also STRAIGHT-CHAIN HYDROCARBONS.

aliquot The amount of a sample used for analysis.

alkali A chemical substance that can neutralize an ACID. Also refers to soluble salts in soil, surface water, or groundwater.

alkaline fly ash scrubbing A system for the removal of sulfur oxides from the flue gas of a coal-burning boiler by using as an absorbing medium the alkaline constituents of ash that remain after the burning of coal.

alkalinity Generally, a measure of the ability of water to neutralize acids. Measured by determining the amount of acid required to lower the pH of water to 4.5. In natural waters, the alkalinity is effectively the BICARBONATE ion concentration plus twice the CARBONATE ion concentration, expressed as milligrams per liter calcium carbonate.

alkylbenzene sulfonate (ABS) An early synthetic detergent that, because of its branched molecular structure, was not decomposed effectively by bacteria in wastewater treatment plants and therefore produced foaming discharges. The detergent molecule was reconfigured as LINEAR ALKYL SULFONATE (LAS) in the 1960s; the result was a chemical form of detergent readily decomposed by bacteria.

alkyl benzenes Any member of a family of compounds consisting of a BENZENE ring (a single-ring aromatic compound, C_6H_6) containing one or more aliphatic (straight) side chains (e.g., methyl or ethyl groups) attached to the ring. The common solvents toluene and xylene are included in this group. The usual problems associated with inhalation of these substances in excessive doses are caused by irritation of mucous membranes in the respiratory system. Acute exposure leads to depression of the central nervous system. The adverse

reactions experienced by "glue sniffers" are usually related to this class of compound.

alkyl mercury A group of compounds consisting of aliphatic organic compounds (e.g., methyl and ethyl groups) bonded to mercury atoms. The most well-known member of this group is METHYL MERCURY, the most toxic form of mercury to people, which affects a variety of systems in humans, including the central nervous system and the reproductive process.

allelopathy The secretion by plants of chemical substances that inhibit the growth of competing species; a significant influence on the rate and type of plant SUCCESSION.

allergen A substance that acts as an antigen, causing the formation of antibodies that react with the antigenic substance. This particular form of allergic antigen-antibody reaction manifests itself only in certain persons for certain substances and can be accompanied by skin or mucous membrane irritation, respiratory function impairment, and eye irritation, among other effects. Allergens can cause a reaction at extremely low doses to sensitive persons.

allergic sensitizers Chemical substances that act as antigens to produce an allergic reaction after repeated (sensitizing) exposures to the skin or respiratory system. The agents that can act as sensitizers to susceptible persons include the epoxy resins, toluene diisocyanate, formaldehyde, chromium compounds, and the plant resins in poison ivy.

allochthonous Describing organisms that are nonnative or transient members of a community in a specific habitat. They do not carry out a metabolic function in that location, and they do not reproduce and occupy the habitat permanently. The term is most frequently used to describe bacteria native to one environment (e.g., the human body) that are transported to a completely different habitat (e.g., the deep ocean) where they might be able to survive but cannot perform metabolic functions or colonize. Compare AUTOCHTHONOUS.

allogenic Exogenous, caused by external factors, such as a change in a habitat caused by flooding; the opposite of AUTOGENIC.

allogenic succession Predictable changes in the plant and animal communities in which alterations are caused by events external to the community, such as fire. Opposite of AUTOGENIC SUCCESSION.

alluvial Referring to ALLUVIUM, the type of soil found in floodplains.

alluvium The fertile sediment deposited in the floodplain of a river.

alopecia The loss of hair. The condition can result from exposure to toxic chemicals in the environment or chemicals administered as anticancer agents.

alpha counter A type of radiation detector that has a selective sensitivity to the ALPHA PARTICLE.

alpha decay The spontaneous decomposition of the nuclei of atoms that results in the emission of ALPHA PARTICLES.

alpha particle A form of IONIZING RADIATION ejected from the nucleus of a radioactive material. The particle consists of two protons and two neutrons: the equivalent of a helium ion. Although alpha particles only travel at most inches in the air, the ingestion or inhalation and absorption of radioactive materials that emit these particles are significant health hazards because the ionization occurs in intimate contact with body tissues.

alpha radiation See ALPHA PARTICLE.

alpine tundra The grassland area found above the TREE LINE on mountain ranges.

alternate-case scenario Based on the history and operations of a specific facility, the most likely accidental release of toxic substances into the environment. RISK MANAGEMENT PLANS are required of certain industrial, utility, military, water treatment, and small business facilities that either manufacture or use any of 140 specific chemicals. The plan must document prevention and emergency response programs in the event of an accidental release that affects the community outside the facility. Also termed creditable case scenario. Compare WORST-CASE SCENARIO.

alternate concentration limits (ACLs) Three types of standards that may be applied when a leak is detected at a TREATMENT, STORAGE, OR DISPOSAL facility and groundwater compliance monitoring is required. ACLs are set by the USEPA for specific hazardous waste constituents at levels that are designed to prevent a substantial hazard to human health or the environment. Groundwater compliance monitoring can use the following standards: (1) background concentrations, or the levels found in the area naturally; (2) specific values set by federal regulations in Title 40, Part 264.94, of the *code of federal regulations* for eight metals and six pesticides and herbicides; or (3) alternate concentration limits.

alternate dispute resolution (ADR) The use of techniques other than litigation, especially third-party mediation, to solve disagreements. In environmental law, ADR is often used to negotiate liability (cost) allocations during waste site cleanups required by the COMPREHENSIVE ENVIRONMENTAL RESPONSE, COMPENSATION, AND LIABILITY ACT or the RESOURCE CONSERVATION AND RECOVERY ACT. NEGOTIATED RULE MAKING by federal administrative agencies is another application of ADR. See REG-NEG.

alternate hypothesis In a statistical test, the interpretation of the data that indicates an association between two variables. For example, in a laboratory test one group of test subjects may be treated or exposed in some way, and a second group serving as the control group may receive no experimental treatment. In this case, the primary statement or NULL HYPOTHESIS will be that there is no difference between the two groups relative to some measured outcome. The alternate hypothesis will be that there is a difference, and therefore the treatment or exposure had a significant effect. Compare NULL HYPOTHESIS.

alternative compliance A policy of the USEPA that allows facilities to choose from among various methods for achieving emission or risk reductions. The company is allowed to choose where and how within the facility compliance with pollution emission standards is achieved. Opposite of a command and control system, wherein the agency specifies standards and ways to meet them.

alternative fuels Substitutes for traditional liquid, oil-based fuels like gasoline and diesel fuel for the operation of motor vehicles. Common possibilities include mixtures of gasoline with ethanol or methanol, ethanol-based fuel, and compressed natural gas. The primary objective of using alternative fuels is to lessen dependence on oil.

alternative method Any method that is not an official approved or registered method or equivalent method sanctioned by the USEPA for the sampling or analyzing for an air or water pollutant.

alternative remedial contract strategy contractor Private company that provides project management and technical services to support cleanup activities at abandoned or uncontrolled hazardous waste disposal sites.

alternative technology See APPROPRIATE TECHNOLOGY.

altruistic preservation Conservation philosophy based on the proposition that nature deserves to exist for its own sake without regard to its usefulness to humans. The protection of nature based on aesthetic and spiritual value, emphasizing the fundamental right of other organisms to exist. John Muir, naturalist and first president of the Sierra Club, is credited with establishing this philosophy in the early part of the 20th century. Compare UTILITARIAN CONSERVATION.

alum Aluminum sulfate, used as a coagulant in wastewater treatment. See COAGULATION.

aluminosis A disorder of the lung resulting from excessive inhalation of aluminum-containing particulate matter. In some cases, pulmonary FIBROSIS is present.

alveolar macrophage A scavenger cell that can engulf (phagocytize) inhaled microbes or particulate matter deposited in the ALVEOLAR REGION of the lungs. These migratory cells constitute a major part of the natural resistance of the body to diseases related to bacteria and viruses that gain entry through the lungs.

alveolar region The terminal parts of the respiratory system containing many millions of small air sacs, or alveoli; the area of gas exchange and, therefore, the entrance point for some air contaminants to the bloodstream. The alveolar region itself can be scarred by certain PARTICULATE MATTER air contaminants. See FIBROSIS.

alveoli The primary gas exchange structures of the lungs. Each alveolus has an extremely thin wall that facilitates close contact with blood capillaries of the circulatory system. The structure allows for very effective exchange of carbon dioxide and oxygen between the air space and the blood. The millions of alveoli in the lungs are arranged like clusters of grapes at the end of each bronchiole, or small airway. See ALVEOLAR REGION.

amalgam A mixture (alloy) of the metal mercury and one or more other metals.

ambient Describing a natural outside environment. The term *ambient air,* for example, commonly excludes indoor and workplace air environments.

ambient air See AMBIENT.

ambient air quality standards See NATIONAL AMBIENT AIR QUALITY STANDARDS.

ambient measurement Quantification of the concentration of a normal atmospheric constituent or of a pollutant within the immediate surroundings of an organism.

ambient medium Air, water, or soil surrounding an organism through which a pollutant must move before impacting the organism.

ambient noise In sound surveys, the general noise level arising from all sources within the confines of a specific area. Compare SOURCE MEASUREMENT.

ambient quality standard (AQS) See AMBIENT STANDARD.

ambient standard The allowable amount of material, as a CONCENTRATION, in air, water, or soil. The standard is set to protect against anticipated adverse effects on human health or welfare, wildlife, or the environment, with a MARGIN OF SAFETY in the case of human health. See STANDARDS, ENVIRONMENTAL. Compare EMISSION STANDARD.

ambient temperature Natural temperature of the surrounding air. Normal temperature.

amebiasis See AMEBIC DYSENTERY.

amebic dysentery A disorder of the gastrointestinal tract caused by a protozoan parasite belonging to the genus and species *Entamoeba histolytica*. The disorder is commonly found in communities with poor sanitary conditions and is transmitted by contaminated water supplies, contaminated food, flies, and person-to-person contact. Infected individuals experience abdominal cramps, diarrhea, and blood and mucus in the feces. The parasite invades the liver in some cases.

amensalism A two-species interaction in which one population is adversely affected and the other is not, such as the secretion of chemicals by certain aromatic shrubs, inhibiting the growth of herbaceous plants in their vicinity. Compare MUTUALISM.

American Chemistry Council See CHEMICAL MANUFACTURERS ASSOCIATION.

American Conference of Governmental Industrial Hygienists (ACGIH) A professional organization of individuals involved in occupational health activities. Based in Cincinnati, Ohio, the organization publishes recommended occupational exposure limits for chemicals and physical agents in the workplace. See THRESHOLD LIMIT VALUE. Website: www.acgih.org

American National Standards Institute (ANSI) An organization that coordinates nationwide consensus methods of testing, rating, and performance standards, including safety and health standards. The ANSI represents the United States in the INTERNATIONAL ORGANIZATION FOR STANDARDIZATION (IOS). Website: www.ansi.org

American Petroleum Institute (API) A trade association of large companies involved in the production, refining, and distribution of petroleum, natural gas, and their associated products. Based in Washington, D.C., the organization promotes the petroleum industry, supports research in all aspects of the oil and gas industry, and collects and publishes worldwide oil industry statistics. Many of its publications concern environmental matters. Website: www.api.org

American Public Health Association (APHA) Founded in 1872, the largest organization of public health professionals in the world. Membership includes researchers, health service providers, administrators, and teachers. The membership of the organization has a broad interest in issues related to personal and environmental health, including federal and state funding of health issues, pollution control, education, and programs related to chronic and infectious diseases. The mission of the organization is to improve public health. Website: www.apha.org

American Public Works Association (APWA) A national organization of individuals and organizations involved in the management of municipal solid waste and in the design and operation of wastewater treatment plants. Founded in 1894, the organization has 25,000 members and is based in Chicago, Illinois. Website: www.pubworks.org

American Society for Testing and Materials (ASTM) The largest organization setting standards for materials, products, and services. ASTM standards include methods for the sampling and testing of many physical and chemical agents that may be of environmental concern. Based in Philadelphia, Pennsylvania. Website: www.astm.org

American Society of Civil Engineers (ASCE) A professional organization in New York City that supports the practice of, and research in, environmental engineering. Founded in 1852, it now has 108,000 members. Website: www.asce.org

American Water Works Association (AWWA) The national organization, based in Denver, Colorado, of individuals involved in the design and operation of public water supplies. Founded in 1881, it now includes 45,000 members. Website: www.awwa.org

Ames, Bruce (1918–) American Biochemist/molecular biologist, developed the AMES TEST, a petri dish screening test for MUTAGENS. Conducted path-breaking research on the causes of cancer and aging; former board member, NATIONAL CANCER INSTITUTE.

Ames test A test of the ability of a chemical to cause mutations and thereby act as a CARCINOGEN. The process involves the use of a histidine-dependent strain of the bacterial species *Salmonella typhimurium*. The organism used is actually a mutant strain that requires the presence of the amino acid histidine for growth. The chemical being screened is usually mixed with an extract of rat liver, which provides enzymes that convert the chemical being tested from an inactive to an active form. The suspension of bacteria and the suspected mutagen is allowed to stand for an appropriate period. The mixture is then spread on the surface of an agar medium that contains all the ingredients needed for growth of the test strain of *Salmonella* species except histidine. If a mutation occurs that converts the test strain back to the wild type, which does not need to find histidine in the environment in order to grow, colonies will develop on the surface of the agar medium. If mutations do not occur, nothing will grow on the medium, and the test result will be considered negative. Because carcinogens are also mutagens, the test is used as a screening technique to test for carcinogenicity. The test is inexpensive and produces overnight results but misses perhaps 40% of carcinogens. Also, the appearance of mutations does not prove that the chemical tested is a carcinogen. The procedure was developed by Bruce Ames at the University of California, Berkeley.

amictic lake A lake that does not experience mixing or turnover on a seasonal basis. See DIMICTIC LAKE.

amines A family of compounds related to ammonia (NH_3). Subclasses of amines are named for the number of groups attached to the nitrogen atom: primary amines (i.e., methylamine [CH_3NH_2]); secondary amines (i.e., dimethylamine [$(CH_3)_2NH$]); and tertiary amines (i.e., trimethylamine [$(CH_3)_3N$]). Hundreds of nitrogen-containing ring compounds are related to the amines, some of which are toxic to humans.

amino acid An organic compound constituting the basic unit from which all proteins are made. All amino acids share a common characteristic of an alpha amino and an alpha carboxy group. The 20+ amino acids found in all proteins differ from one another on the basis of the chemical properties of the R-group shown in the diagram. Proteins differ from one another on the basis of the kinds of amino acids present and the linear sequence of the amino acids. Eight amino acids are essential nutrients for humans, meaning that they must be included in the diet.

ammonia stripping A method for removing ammonia from wastewater. The pH of the water is raised (made more basic) to convert the dissolved ammonium ions (NH_4^+) to (dissolved) ammonia gas (NH_3). The water is then allowed to cascade down a tower and the ammonia is lost to the atmosphere. A form of TERTIARY TREATMENT of wastewater.

ammonification A process carried out by bacteria in which organic compounds are degraded or mineralized with the release of ammonia (NH_3) into the environment. The simplest example is the bacterial decomposition of urea with the release of ammonia, which occurs in cloth diapers before they are washed.

Amino acid

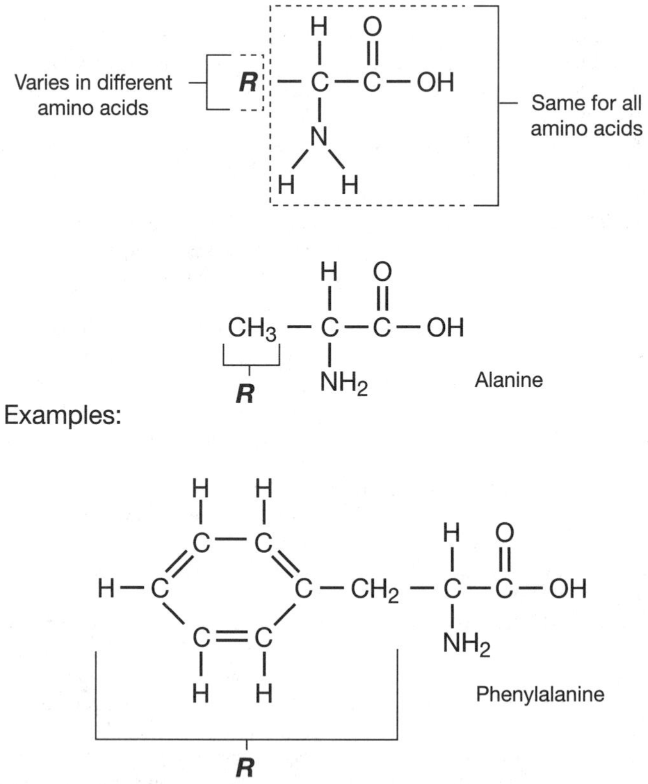

Examples:

Alanine

Phenylalanine

amorphous Noncrystalline, as in amorphous silicates, that do not cause FIBROSIS in the respiratory system as crystalline silicates do.

amosite A type of ASBESTOS, also known as brown asbestos.

amphibole A metasilicate, a type of ASBESTOS.

amphoteric Describing a substance capable of exhibiting properties of an acid or a base, depending on its environment. For example, tin oxide dissolved in water displays properties of a base when placed in an acidic solution but properties of an acid when added to a basic solution.

amplitude For an electromagnetic wave or a sound wave, the distance from the crest (or trough) of the wave to its equilibrium value; the distance from the crest to the trough of a wave is twice the amplitude.

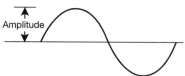

anabolism The metabolic process involving the conversion of simpler substances to more complex substances for the storage of energy. Also called assimilation, biosynthesis, or constructive metabolism. Compare CATABOLISM.

anadromous fish A type of fish that spends its adult life at sea but returns to the upper reaches of a river to spawn (e.g., salmon).

anaerobic An adjective that describes an organism that is able to live without oxygen. The term is frequently used to refer to microorganisms that possess this ability. Also used to describe environments that are devoid of gaseous or dissolved molecular oxygen.

anaerobic decomposition The degradation of materials by ANAEROBIC microorganisms living beneath the ground or in oxygen-depleted water to form reduced compounds such as METHANE or HYDROGEN SULFIDE. Generally slower than AEROBIC DECOMPOSITION.

anaerobic digester An airtight tank in which ANAEROBIC microorganisms decompose organic material and produce BIOGAS, mainly METHANE. SEWAGE TREATMENT PLANTS often use anaerobic digesters to reduce the volume of SLUDGE produced in PRIMARY and SECONDARY TREATMENT, and they sometimes use the methane as a heating fuel. In rural China and India, digesters are used to convert animal manure to methane.

anaerobic respiration A type of metabolism carried out by some bacteria in the absence of oxygen. Some other inorganic compound must be supplied by the environment as a substitute for molecular oxygen. The end products produced when these substitute compounds are reduced have important environmental consequences. For example, sulfate is reduced to hydrogen sulfide, carbon dioxide is reduced to methane, and nitrate is reduced to nitrous oxide or nitrogen gas.

analog Derivative of a naturally occurring compound that an organism cannot distinguish from the natural product; however, the uptake of this derivative results in the formation of a biological molecule that does not carry out its proper function. A common example is *para*-fluorophenylalanine, an analog to the amino acid phenylalanine. An organism may utilize the analog in place of the amino acid in the synthesis of protein; however, the protein containing the analog is not functional.

analysis In environmental testing of air, water, and soil, the separation of a mixed substance into its components and their quantification by type and amount.

analysis of variance (ANOVA) A statistical test that allows an investigator to determine whether mean values observed for several REPLICATES of the same measurement are due to random chance or to real differences. For example, the mean concentration of hexachlorobenzene in the blood of three groups of 30 men each is determined to be 20.0 micrograms per 100 milliliter for group A, 17.8 for group B, and 26.9 for group C. The analysis of variance will allow the testing of the NULL HYPOTHESIS, which states that there are no differences among the groups, and to make that determination with some specific degree of confidence—say 95% sure that there are no differences. The ALTERNATIVE HYPOTHESIS in this case would be that there is a difference among the groups.

analyte 1. The chemical of interest or concern for which a sample is collected and examined. If a program is operated to determine the air pollution caused by the presence of ozone, then samples would be collected to monitor for the presence of ozone. Under that circumstance, ozone would be considered the analyte. 2. The object of a chemical analysis; the chemical for which a CONCENTRATION is to be determined in samples of water, air, soil, or food.

analytical balance A highly sensitive device for determining the mass of a sample of material. The balance can measure mass to the nearest tenth of a milligram or even smaller units.

Andersen sampler An air-sampling device that separates airborne particles into size categories using stacked collection plates.

anechoic space A room in which the surfaces absorb all sound waves hitting them. Also called a free-field space or free-field conditions.

anemometer An instrument that measures air velocity. The cup-type anemometers often used at air-monitoring stations operate by measuring the rotation rate of three or four cups mounted on a vertical shaft. Hot-wire anemometers, used for measurements of indoor air velocities, operate by sending the rate of heat loss from a heated wire to a moving airstream.

aneroid gauge An instrument that does not contain a liquid. An aneroid barometer, for example, contains a metal surface that reacts to differences in air pressure.

aneuploid In humans, having a chromosome number other than 23 or 46 (the normal complement in humans). An irregular number of chromosomes produces physical and chemical abnormalities, many of which can be severe. Opposite of EUPLOID.

angstrom (Å) A unit of length measurement, equal to one ten-billionth. (1×10^{-10}) of a meter. Wavelengths in the electromagnetic spectrum are often expressed in angstroms or nanometers (1×10^{-9} meter).

anhydride Describing a chemical that produces an acid with the addition of water (e.g., CARBON DIOXIDE is the anhydride of CARBONIC ACID and sulfur trioxide [SO_3] is the anhydride of SULFURIC ACID).

anhydrous Describing a chemical compound not containing water.

animal bedding Material used in animal stalls on facilities such as farms and dairies. Straw is a common component; however, shredded paper can be employed, providing a market for recycled paper from the municipal solid waste stream.

animal dander Scales of dead animal skin shed in the normal course of activity. This material is a common constituent of indoor air pollution in those homes with pet dogs. One of the allergens that contribute to asthma.

animal studies Laboratory studies using animals as the test subjects to determine the effects of pollutants, toxins, medicines, and vaccines. The animals are considered to be surrogates for determining the effects of these agents on humans. Animal rights groups object to the use of animals in this way.

anion A negatively charged ION.

anion exchange process See ION EXCHANGE.

annual dose The amount of a substance or radiation to which an individual is exposed over a 12-month period.

annual outage A preset period set aside each year for the shutting down of a boiler or other piece of equipment to allow for routine maintenance and repair.

annulus For a well, the space between the pipe and the outer wall (casing) of the borehole, which may also be a pipe (the well casing).

annulus pressure The positive pressure maintained by a fluid introduced between the well piping and the outer wall (casing) of the bore hole of an underground INJECTION WELL; an indication of the MECHANICAL INTEGRITY of the well.

anode The negative pole of an ELECTROLYTIC cell or a battery. Electrons flow from the anode to the CATHODE.

anomaly The difference between a short-term measurement and the longer-term mean value of a phenomenon, for example, surface air-temperature readings trending upward for several years.

anoxic Without oxygen. Water with no DISSOLVED OXYGEN is said to be anoxic.

antagonistic effect The interaction of two or more chemical substances on a biological system that results in a combined effect less than the sum of the effects of each substance alone. Compare SYNERGISTIC EFFECT and ADDITIVE EFFECT.

Antarctic ice sheet See WEST ANTARCTIC ICE SHEET.

anthophyllite A type of ASBESTOS.

anthracite The type of coal with the highest heat content or energy released per unit burned. Anthracite is the least common and usually most expensive type of coal and is used mainly in the steel industry. The other coal types, BITUMINOUS, SUBBITUMINOUS, and LIGNITE, are used for electric power production and vastly exceed anthracite in tonnage used annually.

anthracosilicosis A chronic lung disease first observed in anthracite coal miners associated with long-term overexposure to dust containing significant concentrations of FREE SILICA. Another term for BLACK LUNG DISEASE. Compare COAL WORKERS' PNEUMOCONIOSIS

anthropocentrism A belief that considers the human being the most important entity. All questions, including those related to environmental management, are resolved on the basis of value to humans.

anthropogenic Arising from human activities, as opposed to a natural origin.

antibacksliding See BACKSLIDING.

antibody A protein produced by the immune system in response to the presence of an ANTIGEN that interacts with that specific antigen to eliminate or inactivate it.

anticyclone A clockwise air circulation associated with a high-pressure center. Anticyclones have descending air in their centers, which warms as it subsides and compresses. This can cause a SUBSIDENCE INVERSION under which air contaminant levels can increase.

antidegradation clause That part of federal water quality and air quality requirements prohibiting the addition of new sources of pollutants in areas where the air and water pollution levels are already above the legal limits.

antidegradation policy Rules or guidelines, required of each state by federal regulations implementing the Clean Water Act, stating that existing water quality be maintained to protect existing uses of the water, even if the water quality in an area is higher than the minimum necessary, as defined by federal ambient water quality standards. Some (controlled) degradation is allowed for economic development. The comparable policy for air is stated in the PREVENTION OF SIGNIFICANT DETERIORATION regulations.

antigen Any substance that causes the formation of an ANTIBODY. See ALLERGEN.

antiknock additive Chemical compound added to gasoline to prevent premature combustion in the engine cylinders, which creates a knocking sound. Tetraethyl lead was used as the primary antiknock additive from the 1920s until the mid-1970s, when USEPA regulations began a phaseout of the lead content of leaded gasoline. Oil refineries now use other antiknock additives in gasoline, such as METHYL TERTIARY BUTYL ETHER (MTBE). See OCTANE NUMBER.

antiknock agent See ANTIKNOCK ADDITIVE.

antinatalist Describing public policies that attempt to discourage births. Opposite of PRONATALIST.

antioxidant A chemical substance that prevents or slows the OXIDATION of another material. The term is applied in several areas related to the environment. **1.** Antioxidants are added to rubber products, such as automobile tires, to protect against the degradation caused by PHOTOCHEMICAL OXIDANT air pollutants. **2.** Antioxidants are also used to prevent rancidity of fats in foods, for example, the food additives butylated hydroxyanisole (BHA) and butylated hydroxytoluene (BHT). **3.** Vitamins C and E, together with a variety of other nutrients, function as antioxidants in the diet because they detoxify free radicals (thereby protecting the body) produced during the metabolism of a variety of toxicants by humans.

antiscavenge ordinance A governmental regulation prohibiting the unauthorized collection of recyclable materials set out in response to a community recycling program. Such prohibitions were necessitated when private individuals removed items such as aluminum beverage cans from materials set out by residents in response to curbside recycling programs. Such scavenging deprives the community of the economic return on valuable recyclables and frequently generates significant litter in residential areas.

any credible evidence rule Established under provisions of the 1990 amendments to the CLEAN AIR ACT (Section 113a), the USEPA regulatory policy that allows the agency to use data from a variety of sources to demonstrate that a facility is out of compliance with the emission limitations in its PERMIT. Previously, enforcement actions were based only on data collected using a federally approved REFERENCE METHOD.

aphotic Without light. The deep waters of large lakes and the ocean are aphotic since they do not receive sufficient light to promote photosynthesis.

API gravity An expression of the density of liquid petroleum products; the arbitrary scale, as defined by the AMERICAN PETROLEUM INSTITUTE (API), is calibrated in degrees. Light crude oil is about 40 degrees API; heavy crude, about 20 degrees API.

API separator A device used to separate oil and oily sludges from water. A primary step in the wastewater treatment system of an oil refinery.

Appendix VII A listing of chemical substances, found in Title 40, Part 261, of the *code of federal regulations*, that are considered to be hazardous constituents and therefore cause certain waste streams to be listed by the USEPA as hazardous wastes. For example, spent halogenated solvents used in degreasing are hazardous waste; Appendix VII gives the basis for this definition as the presence, in this waste, of tetrachloroethylene, methylene chloride, trichloroethylene, 1,1,1-trichloroethane, carbon tetrachloride, and chlorinated fluorocarbons.

Appendix VIII A listing of chemical substances that, if present in waste, may cause it to be defined as hazardous waste. The listing is found in Title 40, Part 261, of the *code of federal regulations*.

Appendix IX list A list of more than 200 chemicals for which groundwater from all monitoring wells must be analyzed if measurements of INDICATOR PARAMETERS/CONSTITUENTS at a hazardous waste TREATMENT, STORAGE, OR DISPOSAL facility show that the facility may be adversely affecting groundwater quality. The list is found in Title 40, Part 264, of the CODE OF FEDERAL REGULATIONS.

applicable or relevant and appropriate requirement (ARAR) Under the SUPERFUND AMENDMENTS AND REAUTHORIZATION ACT, the standard used to define the minimum level of cleanup required at a contaminated site; the standard used to determine "how clean is clean."

applied dose In attempting to assess the effect of a toxic material, the amount of the agent in contact with the skin, lungs, or gastrointestinal tract of an individual or experimental animal. Contact with one of these surfaces does not constitute absorption of the agent into the body. For example, some agents that are consumed by mouth pass through the intestinal tract intact without any absorption into the bloodstream.

appropriate technology The application of scientific knowledge in ways that fit the prevailing economic, social, and cultural conditions and practices in a country. The term also implies low-technology design, simplicity of use, and decentralized provision and maintenance. Usually used in assessing the transfer of agricultural technology from more- to less-developed countries. For example, food crops requiring heavy application of pesticides may be judged inappropriate for an area without sufficient resources to purchase, apply, and manage the pesticides. Not all less-developed countries agree on the virtues of appropriate technology.

aquaculture The managed production of fish or shellfish in ponds or lagoons.

aquatic Related to environments that contain liquid water.

aqueous solubility The amount of a material that will dissolve in water at a specified temperature. See SOLUBILITY.

aquiclude A low-permeability underground formation located above and/or below an AQUIFER.

aquifer An underground formation, usually composed of sand, gravel, or permeable rock, capable of storing and yielding significant quantities of water.

aquifer transmissivity See TRANSMISSIVITY.

aquitard A low-permeability underground formation.

arable land Land that can be farmed.

arbitrary and capricious Shorthand for a standard used in judicial review of informal rule making by a regulatory agency. The actions of the agency should not be "arbitrary, capricious, an abuse of discretion, or otherwise not in accordance with the law," according to the federal ADMINISTRATIVE PROCEDURES ACT (APA). The "arbitrary and capricious" standard causes judges generally to defer to informal agency actions. The APA holds formal rule making to a stricter standard; these agency actions must be supported by "substantial evidence." Environmental regulatory agencies maintain an ADMINISTRATIVE RECORD to support their decisions in case of subsequent judicial review.

arborescent Resembling a tree in appearance or growth pattern.

arctic tundra The grassland BIOME characterized by permafrost (subsurface soil that remains frozen throughout the year). Found in Alaska, Canada, Russia, and other regions near the Arctic Circle.

area method, landfill A technique for depositing municipal solid waste in a landfill. The waste is spread out in narrow strips two to three feet deep and compacted; then more waste is added until a wider compacted layer is six to 10 feet thick, after which at least six inches of cover material is applied. The width of the area receiving the cover material is usually 10 to 20 feet. The volume from one layer of waste and cover material to the next (adjacent or higher) layer is called a cell. Used in areas where the excavation required by the TRENCH METHOD is not practical.

area mining The type of mining used to extract mineral resources close to the surface in relatively flat terrain. Overburden is removed in a series of parallel trenches to allow the extraction of the resource, and the overburden removed from one trench is used to fill in the adjacent trench after the removal of the resource. Also called area strip mining.

area of review The area around an underground injection well that may be influenced adversely by fluid injection. The extent of the area of concern may be calculated by using the specific gravity and rate of introduction of the injected fluids; the size, storage capability, and hydraulic conductivity of the injection zone; and certain underground formation pressures. In other cases, the extent of the area is determined by a fixed radius around the well, not less than one-quarter mile in length.

area source In air pollution, a geographic unit that combines many small sources of an air contaminant or contaminants. For example, part of an urban area could be treated as an area source for VOLATILE ORGANIC COMPOUNDS, which are emitted by hundreds of individual households, service stations, dry cleaners, and similar sources.

arithmetic growth An increase in the number of individuals within a specific area by a constant amount of time per period: for example, a population growth of 10 persons per year, year after year. Compare EXPONENTIAL GROWTH, SIGMOID GROWTH.

arithmetic mean The sum of a set of observations divided by the number of observations.

Aroclor Commercial preparation that contains a mixture of POLYCHLORINATED BIPHENYLS (PCBS) as well as traces of a variety of other dangerous compounds such as DIBENZO-*para*-DIOXIN.

aromatic One of the two main types of hydrocarbon compounds, containing one or more benzene rings or nuclei. Aromatic compounds are usually more difficult to decompose than are straight-chain hydrocarbons, and they present greater hazards to humans and the environment. Compare ALIPHATIC.

arranger liability Under the COMPREHENSIVE ENVIRONMENTAL RESPONSE, COMPENSATION, AND LIABILITY ACT, any person who "arranges" for the transport, treatment, or disposal of HAZARDOUS WASTE can be named a POTENTIALLY RESPONSIBLE PARTY. Arranging includes hiring a transporter or choosing a treatment or disposal facility.

Arrhenius equation The equation, $K = A \exp[^N(E/RT)]$, describing the relationship between the RATE CONSTANT (K) for a chemical reaction and ABSOLUTE TEMPERATURE *(T)*, where *A* is a constant for a given reaction, *R* is the UNIVERSAL GAS CONSTANT, and *E* is the ACTIVATION ENERGY.

arrowhead chart A graphical display of ambient air quality data that plots the frequency of various levels of measured pollutant concentrations against different AVERAGING TIMES, usually for the observations made over a one-year period. The plotted frequencies are the percentages of the total number of measurements, for a given averaging time, that fall below a particular concentration. Longer averaging times produce fewer plotted points, until only one figure represents the annual average. The arrangement of the points on the graph, as they narrow down to the annual average, resembles an arrowhead. The arrowhead chart is used to display the variability and distribution of measured air concentrations at long-term sampling sites.

arsenic (As) A chemical element belonging to the nitrogen family. In pure form, arsenic is a gray crystal. Arsenic reacts readily with a variety of other chemicals, including oxygen in moist air, producing an oxide. Arsenic is released into the environment as a by-product of the smelting of copper ore and through its use as an insecticide. The element undergoes bioconcentration and is considered to be carcinogenic.

artesian aquifer See CONFINED AQUIFER.

artesian water Water drawn from an ARTESIAN WELL.

artesian well A well drilled into a confined geological stratum holding water that is draining from an area higher than the wellhead. The resulting pressure forces the water to the surface, eliminating the need for a pump. A free-flowing well.

artifactual An adjective that describes some compound, material, or process that is made by humans or influenced by human activities.

artificial recharge The deliberate addition by humans of water to an AQUIFER. Two common recharge methods are injecting water through drilled wells and pumping water over the land surface and allowing infiltration. The technology is employed to restore water to aquifers that have been depleted by excessive withdrawal.

asbestos A family of heterogeneous silicate minerals that have a variety of commercial applications. The mineral is fibrous, chemically inert, and nonconductive for heat and electricity. The fiber has been used extensively in insulating materials. A significant danger to individuals is present when airborne fibers of the mineral are taken into the lungs. The fibers are not subject to BIOTRANSFORMATION, nor are they excreted; rather, they remain in the body. Chronic overexposure can lead to ASBESTOSIS. There appears to be a strong association with the development of certain types of lung cancer in individuals who have both an occupational exposure to asbestos and a history of smoking cigarettes. The Asbestos Health Emergency Response Act of 1985 required that all elementary and secondary schools be inspected for the presence of ASBESTOS-CONTAINING MATERIAL and that an asbestos management plan be prepared to include any necessary repair or removal. The school guidelines are now generally used for all types of buildings.

asbestos abatement Procedures to control the release of fibers from asbestos-containing materials in a building or to remove such materials. Control technologies used in lieu of complete removal include encapsulation, repair, enclosure, encasement, and routine maintenance programs. The most common objectives are to reduce indoor air pollution associated with the presence of asbestos fibers suspended in the indoor air and to prevent exposures during renovation and demolition work.

asbestos assessment Evaluation of the physical condition and potential for damage of all asbestos-containing materials and thermal insulation systems that are easily disturbed, causing the production of asbestos dust. This type of evaluation is done in schools and other public buildings as well as before renovation and demolition work in old structures. See FRIABLE.

Asbestos Hazard Emergency Response Act (AHERA) A 1986 federal statute requiring the inspection of U.S. schools for ASBESTOS-CONTAINING MATERIAL and, as appropriate, either in-place management or removal. See ASBESTOS.

asbestosis A disorder resulting in the accumulation of asbestos fibers in the lungs. The particles stimulate the aggregation of lymphoid tissue and the stimulation of collagen synthesis, leading to the accumulation of fibrous tissue in the lungs. The elasticity of the lung is reduced, causing decreased gas exchange efficiency during breathing. The condition is chronic and progressive.

asbestos-containing material (ACM) Construction material containing more than 1% asbestos. Federal regulations require the inspection of public and private elementary and secondary schools for such materials.

as low as reasonably achievable (ALARA) The principle adopted by the INTERNATIONAL COMMISSION ON RADIOLOGICAL PROTECTION and the United States NUCLEAR REGULATORY COMMISSION of limiting exposures to IONIZING RADIATION to less than the allowable levels, if feasible.

ash The residue that remains after the burning of solid waste, wood products, or fossil fuels. Also, the relatively fine powder produced during volcanic eruptions. Ash accumulation in incinerators designed to burn municipal solid waste or in boilers heated by coal represents a significant waste disposal issue.

ash flow A very dangerous type of volcanic eruption. Trapped magma containing large amounts of gas explodes to the surface, ejecting lava, small pieces of rock, ash, crystallized fragments, and gas. The hot mixture expands rapidly and moves downslope at speeds reaching 250 km per hour. The dense cloud of hot ash is denser than air and flows across the surface or the ground. This type of eruption is often referred to as pyroclastic flow.

aspect ratio 1. The ratio of the length to the width of an object. 2. In an ELECTROSTATIC PRECIPITATOR, the ratio of the COLLECTING SURFACE length to its height; it should be large enough to allow dislodged particles to fall into collecting hoppers before they are carried out of the precipitator.

asphyxiants Gases that cause or tend to cause loss of consciousness resembling suffocation resulting from a lack of oxygen uptake. The gases, such as nitrogen and halomethanes, may be nontoxic but simply displace air so that an individual suffocates as a result of oxygen deficiency. See SIMPLE ASPHYXIANT, CHEMICAL ASPHYXIANT.

assessment end point During the process of cleaning up a contaminated environment, the explicit expression of the part of the natural environment that will be monitored to determine whether the cleanup activities have reduced risks of adverse ecological effects that would result from exposure to the contaminants. For example, at a location contaminated with POLYCHLORINATED BIPHENYLS, the protection of the reproductive capacity of predatory birds may represent such an end point. The method of determining whether a formerly contaminated area is clean enough. See ECOLOGICAL RISK ASSESSMENT, MEASUREMENT END POINT.

assimilation 1. In biology, when an organism absorbs a nutrient and incorporates the material into mass. 2. In ecology, the process in a stream or lake whereby added nutrients or other chemicals are incorporated into biota and other compartments, frequently in such a way that the normal community structure and physical environment are not altered. **3.** In ecology, the rate at which CONSUMER organisms gain BIOMASS.

assimilative capacity **1.** The potential of a lake, stream, or other body of water to receive wastewater, drainage, industrial effluents, or other pollutants without apparent adverse effects, damage to aquatic organisms, disruption of the natural ecological balance, or prevention of the use of the resource by humans. **2.** The BIOCHEMICAL OXYGEN DEMAND input rate that a stream can sustain without an unacceptable drop in DISSOLVED OXYGEN.

association In EPIDEMIOLOGY, a relationship between two variables in which as one changes there is also a change in the other. For example, increasing age is associated with increased human cancer incidence. If both variables increase, there is a positive association. Association does not necessarily imply a causal link between the variables.

Association of Boards of Certification An organization dedicated to facilitating communication and cooperation among organizations involved in the certification of environmental occupations and laboratories. The emphasis is on groups involved with water quality issues. The group works to improve and strengthen certification laws, promotion of uniformity of standards and practices, and facilitation of the transfer of certification between certifying authorities. Website: www.abccert.org

Association of State and Territorial Solid Waste Management Officials (ASTSWMO) An organization supporting environmental agencies of the states and trust territories charged with the management of solid waste, hazardous waste, and leaking underground storage tanks. The missions of the group are to enhance and promote effective state and territorial waste management programs and to influence national waste management policies. The organization seeks to assist states in learning from and working with sister states. Website: www.astswmo.org

asthenosphere That part of Earth lying beneath the lithosphere, or more solid crustal material. The MAGMA associated with volcanic eruptions originates in this region.

asthma A respiratory disorder characterized by shortness of breath, wheezing, and difficulty in breathing. Asthma can be triggered by a variety of agents including chemical air pollutants.

at-source prevention See POLLUTION PREVENTION.

at-the-desk separation A system used to facilitate the recycling of office paper. Recyclable office paper is sorted into grades and stored by individual employees at their desk. The process is analogous to a resident's sorting and holding for pickup items from the household solid waste.

atmosphere **1.** The gaseous layer covering the Earth. The regions of the atmosphere are the TROPOSPHERE, STRATOSPHERE, MESOSPHERE, CHEMOSPHERE, thermosphere, IONOSPHERE, and exosphere. The atmosphere is one of the four components, together with the LITHOSPHERE, HYDROSPHERE, and BIOSPHERE, of the Earth's ecosystem. **2.** An expression of ATMOSPHEREIC PRESSURE. One atmosphere is equal to about 14.7 pounds per square inch.

atmospheric deposition Settling of solids or liquids from the atmosphere. Snow, rain, and dust are natural examples, whereas, acids, metallic dust, rock dust, and toxic organic compounds are deposits caused by human activities.

atmospheric pressure The force exerted by the atmosphere per unit area on the surface of the Earth. Sea level atmospheric pressure is equal to about

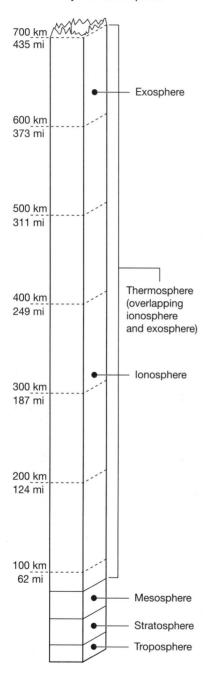

The layered atmosphere

700 km
435 mi

Exosphere

600 km
373 mi

500 km
311 mi

Thermosphere
(overlapping
ionosphere
and exosphere)

400 km
249 mi

Ionosphere

300 km
187 mi

200 km
124 mi

100 km
62 mi

Mesosphere

Stratosphere

Troposphere

14.7 pounds per square inch, or 101.33×10^3 newtons per square meter (pascals), or, in terms of the height of mercury in a barometer, 760 millimeters of mercury.

atmospheric stability An expression of the resistance of the atmosphere to vertical air motion, or dispersion. Stable air resists movement of air upward; unstable conditions result in good vertical dispersion. Atmospheric stability is important to the dispersion and dilution of air contaminants.

atmospheric window See INFRARED WINDOW.

atom A unit of matter consisting of a nucleus and surrounding electrons; the smallest unit of an ELEMENT that retains the chemical characteristics of the element. See MOLECULE.

atomic absorption spectrophotometer (AA) An analytical instrument typically used to determine the concentration of various metals in an environmental sample. The instrument vaporizes the sample and measures the absorbance of the vapor at a wavelength of light that is specific for each element. The sample concentration is obtained by comparing the measured absorbance to an absorbance CALIBRATION CURVE prepared from samples of known concentration.

atomic emission spectroscopy (AES) An analytical technique used to quantify various elements of interest in samples of biota and environmental media. Chemicals in solution are aspirated into the instrument, where they are vaporized and atomized by a flame, discharge, or plasma. The high temperatures provide sufficient energy to promote the atoms to high energy levels. As the atoms of different elements decay back to lower energy levels, light is emitted in narrow energy levels characteristic of each element. The detection of the light emitted by the various elements is used to assay an array of elements simultaneously.

Atomic Energy Commission The predecessor agency to the NUCLEAR REGULATORY COMMISSION.

atomic mass unit (amu) A unit of mass equal to 1.6604×10^{-24} gram. One atomic mass unit is defined as one-twelfth the mass of carbon-12, a carbon ATOM consisting of six PROTONS and six NEUTRONS.

atomic number The number of protons in the nucleus of an atom, representing the atomic mass units in the nucleus accounted for by protons. Atoms of each individual element have a specific number of protons and therefore a unique atomic number. There are approximately 103 different elements, with atomic numbers ranging from 1 (hydrogen, the smallest atom) to 103 (lawrencium, the largest atom).

atomic pile A NUCLEAR REACTOR.

atomic power The generation of electricity using the heat energy produced in nuclear reactors.

atomic waste Radioactive materials created as by-products of activities such as the mining and processing of radioactive minerals, fabrication of nuclear weapons, and operation of nuclear reactors.

atomic weight For each ELEMENT, the weighted sum of the masses of the PROTONS and NEUTRONS composing the ISOTOPE of that element. Approximately equal to the sum of the number of protons and neutrons found in the most abundant isotope.

atomizer An instrument used to produce a fine spray or mist from a liquid.

Atrazine A restricted use pesticide that may be purchased and used only by certified applicators. A selective triazine herbicide used to control broadleaf and grassy weeds in corn, sorghum, sugarcane, and pineapple. The agent is also used in conifer reforestation programs and on nonagricultural sites. Atrazine is slightly toxic to humans and wildlife. The agent does not strongly adsorb to soils and has a long half-life once applied; consequently, Atrazine has a high potential for groundwater contamination when applied to soils of low clay or organic content.

attainment area A geographical area designated by the USEPA that meets the NATIONAL AMBIENT AIR QUALITY STANDARD for a particular pollutant The same area can be in attainment for one or more of the six national ambient air quality standards and be designated as nonattainment for the others. See NONATTAINMENT AREA.

attenuation Weakening; commonly applied to the reduction in noise level or level of IONIZING RADIATION with distance from a source or after passing through barriers or shields. See MONITORED NATURAL ATTENUATION.

attractant A food, light, or sound that lures an insect to a poison, or a chemical sex attractant (PHEROMONE) that disrupts insect propagation by sending false location signals to potential mates.

attributable risk The incremental increase in the risk that an individual will experience an adverse health effect that can be assigned to a particular source or activity.

audible range The sound FREQUENCY range accessible to human hearing, ideally about 20 to 20,000 HERTZ for young adults. The ability to hear higher-frequency sounds deteriorates normally with age or can be lessened by excessive noise exposure. See PRESBYCUSIS, NOISE-INDUCED HEARING LOSS.

audiogram A display of the HEARING THRESHOLD LEVEL, in DECIBELS, for an individual exposed to sound over a range of different frequencies; used to monitor hearing loss.

audiometer An instrument used for testing human hearing. With the person being tested isolated from other sounds, the device generates a pure tone at selected frequencies, and the SOUND PRESSURE LEVEL is increased until the subject can just hear that sound, at a level called AUDIOMETRIC ZERO, or the hearing threshold, for that frequency. The test result is an AUDIOGRAM.

audiometric zero The threshold of normal human hearing, usually defined as a SOUND PRESSURE of 2×10^{-5} newton per square meter, or zero DECIBEL.

audit See ENVIRONMENTAL AUDIT.

Audubon Society See NATIONAL AUDUBON SOCIETY.

autecology The study of a single species within an ecosystem. Compare SYNECOLOGY.

autochthonous Describing an organism that is native, or indigenous, to a locale. The opposite of ALLOCHTHONOUS.

autoclave A sterilizing device that uses high-pressure steam to kill microorganisms.

autogenic Endogenous, arising from within an organism or a community. The opposite of ALLOGENIC.

autogenic succession The orderly and predictable changes of species composition within a community that are caused mainly by actions of the species present in the area. Compare ALLOGENIC SUCCESSION.

autoignition temperature The lowest temperature at which a gas or vapor that is mixed with air will ignite without the presence of a flame or spark.

autoradiography A technique employed to locate RADIOACTIVE substances in small, usually microscopic samples. For example, the part of a cell that incorporates a particular substance can be determined by first incubating the cell in the presence of a radioactive form of the substance in question, then fixing and sectioning the cell. The sections are covered with a photographic emulsion and allowed to stand. The presence of the radioactive material is determined by developing the emulsion and examining the specimen for dark grains of silver.

autotroph An organism that has the capacity to satisfy its requirements for carbon-containing nutrients by the fixation, reduction, and incorporation of carbon dioxide from the environment. Green plants, which derive their energy from sunlight through the process of PHOTOSYNTHESIS, represent a major group of autotrophic organisms. Compare HETEROTROPH.

auxin A plant growth hormone. Synthetic auxins are used as weed killers.

auxotrophic A term employed in MICROBIOLOGY to describe certain mutant strains of bacteria that require for growth the presence of some organic compound or metabolic intermediate that is not required by the parent strain. The requirement generally reflects the loss by the mutant of its ability to produce one or more of the enzymes needed in the metabolic synthesis of the compound or INTERMEDIATE in question.

availability session Informal meeting at a public location where interested citizens can talk with officials from the USEPA and state regulatory agencies on a one-on-one basis.

available chlorine A measure of the amount of chlorine usable for the disinfection of water after the addition of some chlorination agent. For example, when a public water supply is protected with a chlorination agent, chlorine should be chemically active or available when the water reaches a residence in order to ensure protection of the citizens from microbial diseases of the gastrointestinal tract.

available water storage capacity The amount of water that soil can hold and that plants can use. A measure of water available to plants.

averaging time The length of time used for the time-weighted average of a measurement. For example, if the concentration of an air pollutant during an hour is 20 parts per million for 30 minutes and 10 parts per million for 30 minutes, the reported value for a one-hour averaging time would be 15 parts per million.

Avogadro's number (NA) The number of molecules in one mole of a substance, equal to 6.02×10^{23}. Based on the assumption that equal volumes of different gases at the same pressure and temperature contain equal numbers of molecules. Named for the Italian physicist Amedeo Avogadro.

avoided costs Part of the indirect savings realized through a community recycling program. Those expenses that will not be incurred in a solid waste management plan because of the removal of recyclable items from the normal waste stream. Examples are reduced tipping fees at the point of disposal, savings on employee salaries, increasing truck life, and extension of the life of a landfill through the diversion of waste that would otherwise be added to the community discards.

A-weighted scale See DECIBELS, A-WEIGHTING NETWORK.

axenic Describing a culture of a microbe that contains only one kind of organism (i.e., a pure culture). The adjective is frequently used to describe cultures of algae or protozoa that consist of only one type of alga but that may also contain bacteria.

axial flow Fluid flow in the same direction as the AXIS OF SYMMETRY of the duct, vessel, or tank.

axial flow fan A mechanical device for moving air or other gases. The device consists of propellerlike blades rotating in a plane perpendicular to the flow of the gas stream. A fan inside a duct.

axis of symmetry An imaginary line passing through a figure or body around which the figure or body is symmetrically arranged.

B

back corona The buildup of a negative charge on the plates of an ELECTROSTATIC PRECIPITATOR, which reduces the collection efficiency of the plate for negatively charged particles. The condition develops if the collected dust has a high resistivity (resistance to the conduction of electricity). In this case, the poorly conducting, negatively charged particles caught on the positively charged plates do not drain their charge, thus normally allowing the efficient collection of additional dust. The retention of electrostatic attraction between the dust and the plate also makes the periodic removal of the dust more difficult.

back end system The stage in a solid waste resource recovery operation in which recyclable materials are extracted from incinerator ash.

back pressure A pressure that can cause water to back flow into a water supply when the wastewater system of a user is operating at a pressure higher than that of the public water system.

backfilling In the reclamation of land on which contour STRIP MINING has occurred, the placement of spoils (waste soil and rock) in the notches cut in the hills and the restoration of the original slope. This process reduces soil erosion and allows for the reestablishment of vegetation.

backflow/back siphonage A reverse flow condition created by a difference in water pressure that caused water to flow back into the distribution pipes of a drinking water supply from an unintended source.

Background Air Pollution Monitoring Network (BAPMon) Over 200 air-monitoring sites located in remote areas around the world; established by the WORLD METEOROLOGICAL ORGANIZATION in 1968 to record BACKGROUND CONCENTRATION PARTICULATE MATTER, precipitation acidity (ACID RAIN), and GREENHOUSE GASES such as CARBON DIOXIDE, CHLOROFLUOROCARBONS, and METHANE.

background concentration For a chemical substance, the level, or CONCENTRATION found in the air, water, or soil of a geographical area due to natural processes alone; the starting point for the determination of enhanced chemical concentrations caused by human activities.

background level See BACKGROUND CONCENTRATION.

background radiation The IONIZING RADIATION level for a given geographical area produced by the natural radioactive materials in the soil and the cosmic radiation from outer space.

backsliding Prohibited by the CLEAN WATER ACT, the relaxation of the allowable discharge rates when a PERMIT is renewed. See ANTIDEGRADATION POLICY.

backwash The residue that is removed from a RAPID SAND FILTER when clean water is employed to flush away accumulated particles.

backwashing In a wastewater or drinking water treatment facility, the movement of clean water in a direction opposite (upward) to the normal flow of raw water through a RAPID SAND FILTER to clean it.

backyard composting Diversion of organic kitchen waste, yard litter, and grass clippings from the municipal solid waste stream by preparing a pile of this material in the yard of a private residence to encourage decomposition and conversion into a soil conditioner. See COMPOSTING.

bacteria Ubiquitous, one-celled PROCARYOTIC organisms that, together with the fungi, serve as decomposers in aquatic and terrestrial ecosystems. A small number of them cause plant, animal, and human diseases. See PATHOGEN.

bacterial plate count A system employed to quantify the number of bacteria in a sample of solid or liquid material. The sample is mixed and/or diluted with a suitable sterile diluent, and a measured small portion of the mixture is placed on the surface of a solid substrate suitable for the growth of bacteria (a MEDIUM). After incubation, the number of bacterial colonies that have developed on the medium is counted, and the necessary mathematical calculations are made on the assumption that each colony represents one bacterium that was present in the original sample.

bacteriostatic A substance that is inhibitory to bacterial growth but not necessarily lethal. If a bacteriostatic material dissipates or lowers in concentration, bacterial growth may resume.

baffle A wall, barrier, or screen placed in the path of a gas, liquid, light, or sound for the purpose of dispersion or regulation of passage.

bagassosis A lung disease linked to the inhalation of dust from sugarcane residues.

baghouse An air pollution control device that removes particulate matter from an exhaust gas by forcing air through large filter bags; the device is similar to a vacuum cleaner bag. Baghouses vary according to filter material and the method for dislodging the collected dust. See BAGHOUSE, PULSE JET; BAGHOUSE, REVERSE AIR; BAGHOUSE, SHAKER.

baghouse, pulse jet A BAGHOUSE in which the air to be filtered flows through the bags from the outside to the inside. The bags are supported by internal cages and are cleaned by knocking the dust off with short pulses of high-pressure air.

baghouse, reverse air A compartmentalized BAGHOUSE using reverse flows of clean air through the bags to dislodge the collected dust, allowing the dust to fall into a collection hopper. Each compartment is taken off-line before the bags are cleaned.

baghouse, shaker Similar to a REVERSE AIR BAGHOUSE but using a mechanical shaking device to dislodge dust collected on the bags.

bailer An instrument used to extract a water sample from a groundwater well or PIEZOMETER.

baler A mechanical device used to process solid waste into a large strapped bundle for convenient handling and efficient use of landfill space. The waste is mechanically shredded into smaller fragments, compacted, and tied to form a bale.

ball mill A device consisting of a rotating drum containing iron or steel balls and used to grind solid waste before compaction and disposal or as part of a recycling operation.

ballast Heavy material, often seawater, placed in the hold of a ship to gain stability. Routine ballast discharges from oil tankers account for a large portion of the total oil (including spills) introduced to the oceans each year.

ballistic separator A mechanical device that separates organic from inorganic components of municipal solid waste. The process prepares the organic residue for COMPOSTING.

band application The application of chemical pesticides along each row in a row-planted crop.

banking The placement of EMISSION REDUCTION CREDITS (not a monetary sum) in a USEPA-approved account, called a bank.

bar In meteorology, a unit of atmospheric pressure equal to 0.9869 atmosphere or 1×10^5 pascals. Sea-level pressure (one atmosphere) is equal to 1.013 bars or 1013 millibars.

bar rack The closely spaced rods (screen) that remove large solids from the wastewater entering a sewage treatment plant.

bar screen Device used in wastewater treatment facilities to remove gross debris from wastewater before the water enters the treatment process. The debris (commonly rags and plastic bags) would hinder the operation of pumps and other equipment. See BAR RACK.

baritosis The presence in the lungs of small-diameter, chemically inert barium sulfate particles. Baritosis causes no impairment of lung function.

barometric pressure The force exerted on a surface by the acceleration of gravity acting on a column of air above that surface. The pressure can be measured by a liquid barometer, which consists of a vertical tube sealed at the top and a mercury reservoir at the bottom, open to the atmosphere. Atmospheric pressure forces the mercury upward in the tube until the weight of the mercury column equals the weight of the air column above the mercury at the open end of the tube. The barometric pressure measured in this fashion is expressed as the height of the mercury, in millimeters or inches. See ATMOSPHERIC PRESSURE.

barrel A liquid volume, equal to 42 U.S. gallons, frequently used to describe the production or consumption of petroleum or petroleum products. Petroleum is also measured in tons. One ton of crude oil equals about seven barrels, whereas one ton of (less dense) petroleum products is greater than seven barrels.

barrel sampler An open-ended steel tube used to collect soil samples.

barrier island Low, narrow sandy deposits that develop along and just offshore coastlines, mostly along the Atlantic and Gulf coasts of the United States. These islands are unstable, migrating in response to the accretion and erosion of sand under the influence of wind, waves, and currents. Barrier islands are important to the protection of the coastline in storm events. Because of their inherent instability, these areas are unusually sensitive to the deleterious effects of building of homes and roads, operation of motor vehicles, and destruction of natural vegetation.

basal application Application of a pesticide to the plant stem or tree trunk just above ground level to intersect pests as they move up the plant from the soil.

basal diet In a feeding study to determine the toxicity of a chemical or chemical mixture, food that does not contain the test chemical(s); in other words, an unaltered feed available from a commercial source.

basal metabolism The energy used by an organism while at rest.

basalt Rock of volcanic origin. Because of its strength and heat dispersion capability, a desirable type of rock for long-term storage of HIGH-LEVEL RADIOACTIVE WASTE.

basalt aquifers Aquifers found in basalt rock in areas of past volcanic activity, such as the Pacific Northwest region of the United States and in Hawaii.

basaltic eruption The most common type of volcanic activity. Lava at very high temperatures and rich in gas is extruded from fissures and fractures in the crust of the Earth. The lava in this type of eruption is relatively fluid and flows freely downslope until it loses gases, cools, and becomes more viscous. See AA FLOW and PAHOEHOE FLOW.

base Chemicals that release hydroxide ions (OH⁻) in solution. Such solutions have a soapy feel, neutralize adds, and conduct electricity.

base dissociation constant (K_b) See DISSOCIATION CONSTANT.

base flow The flow in a stream arising from groundwater SEEPS alone, excluding surface runoff into the stream.

base load facility A generating facility that operates at a constant output to supply the minimum electric power demand of a system. Such a system is supplemented by a CYCLING LOAD FACILITY as power demand increases.

Basel Convention The Basel Convention on the Control of the Transboundary Movement of Hazardous Wastes and Their Disposal is a 1989 international agreement to control the movement of hazardous wastes, including a 1995 amendment to ban the export of hazardous waste from industrialized countries to less-developed countries. Now, under provisions of the RESOURCE CONSERVATION AND RECOVERY ACT, U.S. waste exporters must communicate to the USEPA a hazardous waste shipment's quantity and chemical composition. The USEPA must get informed consent from the country to which the waste is to be shipped; nonhazardous waste, including medical waste, municipal solid waste, and ash from municipal incinerators, is not covered by this requirement. Over 120 countries have ratified the Basel Conven-

tion, but the United States has not passed the requisite enabling legislation. See also PRIOR INFORMED CONSENT PROCEDURE.

baseline concentration Under the PREVENTION OF SIGNIFICANT DETERIORATION program, the existing level of air quality in an area when the first permit application for a major source of certain air pollutants is submitted.

baseline data Information accumulated concerning the biological, chemical, and physical properties of an ecosystem prior to the initiation of some activity that may result in the pollution of that ecosystem.

baseline emissions For a particular pollutant, the level of air emissions below which the reduction in emission from a facility will be counted as EMISSION REDUCTION CREDITS. For example, if emissions of pollutant A, with a baseline of 1000 pounds per year, are reduced from 1100 pounds per year to 900 pounds per year, emission reduction credits of 100 pounds per year are earned.

basic Describing a solution, sediment, or other material that has a pH greater than 7. See BASE.

basic oxygen process Method of producing steel from molten pig iron. Pure oxygen instead of air is injected into the molten material to oxidize impurities. The process is faster and produces fewer pollutants than other methods. About 28% of the starting material consists of recycled scrap iron. The steel produced by the process can be rolled into thin sheets for use in cans, automobiles, and appliances.

batch method Any industrial or laboratory procedure that involves adding all raw materials or reagents and allowing the process to be completed before the products are removed. For example, a batch method for composting involves adding all leaves, organic materials, or other solid waste and allowing the mixture to sit until the humic substances are removed from the holding vessel. Compare with CONTINUOUS ANALYZER and CONTINUOUS-FEED REACTOR.

batch process An industrial production method in which all or nearly all of the raw materials or reactants are introduced to a process at one time. The materials/reactants are then combined or used; removal or extraction of a product follows. The release of pollutants from a batch process is likely to be intermittent, when the process is opened to be charged or when it is opened and the products are removed. Compare CONTINUOUS PROCESS.

bathing water Water in swimming pools or natural fresh or marine waters used for swimming.

batholiths The largest rock formations in the crust of the Earth. These large expanses of granite are found almost exclusively in the continents. The batholiths consist of huge granitic forms that cover tens of thousands of square kilometers and extend downward for up to 60 kilometers.

bathtub effect The accumulation of LEACHATE in a landfill containing a good LINER but not equipped with a leachate collection and removal system.

bathyal zone The ocean stratum beneath the EUPHOTIC ZONE and above the ABYSSAL ZONE.

Bay of Fundy A bay on the east coast of Canada located between New Brunswick and Nova Scotia. The bay experiences some of the highest natural tides in the world and is considered a likely location for the construction of facilities to utilize tidal energy for the generation of electrical energy.

Bayesian analysis A technique that combines subjective probability estimates of a condition or event with uncertain test results to estimate the conditional probability of the condition or event, given a particular test result. Used in decision analysis involving the risks of introducing

chemical agents to the environment, for example, the interpretation of animal toxicity data.

Beaufort scale A scale equating wind velocities and their effects. A Beaufort number, from 0 to 17 is assigned to each of 13 categories; the largest number represents a hurricane (numbers 12–17 describe six subcategories of hurricane winds). For example, a wind of Beaufort number 2, velocity 4–7 miles per hour, can be felt on the face; Beaufort number 4, 15–18 miles per hour, causes small branches to move and raises dust. The scale was first devised in 1806 by Sir Francis Beaufort, who described the wind effects in terms of those on a fully rigged man-of-war. The inland and terrestrial effects are later additions.

becquerel (Bq) The SI unit of radioactivity, signifying one nuclear disintegration per second.

Becquerel rays A general term for rays emitted by radioactive substances.

bed load The sand, gravel, and other debris that move along the bottom of a stream in response to the current.

bedrock Unbroken and unweathered solid rock that is overlain by broken rock, sand gravel, soil, or other deposits. Usually consisting of granite, the primary crustal material of the continents.

Beer-Lambert law Law allowing the quantitative determination of the concentration of chemicals in solution by measuring the light absorbed upon passage through the solution. It is expressed as $A = ebc$, where A is the absorbency, e is the MOLAR ABORPTIVITY, b is the length of the light path in centimeters, and c is the chemical concentration in moles per liter. The absorbance A is expressed as $A = \ln(P0/P)$, where ln is the natural logarithm, P is the radiant power of the light transmitted through the solution, and $P0$ is the power of the incident light. See SPECTROPHOTOMETER.

behavioral standards Environmental, health, or safety standards that forbid conduct that is likely to have an adverse effect on human health or the environment or that significantly increases the risk of an accident; for example, a ban on outdoor burning of trash or leaves in an urban area or the prohibition of smoking around fuel-loading facilities.

bel A dimensionless unit to measure sound. It is defined in two ways: (1) as the common logarithm of a measured power divided by a reference power or (2) as the common logarithm of the ratio of a measured SOUND PRESSURE to a reference sound pressure. One bel equals 10 DECIBELs, the more commonly used unit. See also SOUND POWER LEVEL, SOUND PRESSURE.

BEN model A computer program that calculates the economic benefit a facility has gained by not complying with an environmental regulation. The calculation includes the cost of any required equipment, operating costs, discount rates, marginal tax rates, and the duration of the noncompliance. The USEPA may adjust statutory penalties for noncompliance by using the results of the model.

bench-scale tests Laboratory tests of potential cleanup technology conducted prior to field testing.

benchmark Concentrations of chemical pollutants that are used in calculating a HAZARD QUOTIENT, a measure of the likelihood that a chemical will cause an adverse effect. The selected value can be the NO-OBSERVED-ADVERSE-EFFECT LEVEL, water quality criteria, sediment quality criteria, or a similar standard.

benchmark station The location within a contaminated environment that will serve as the site from which samples will be collected to determine the likelihood that damage to individuals will occur if they contact the contaminants. See BENCHMARK.

beneficial use 1. In waste management, the application of a waste material in a constructive way, for example, using a former dredging material containment site as a public park. 2. Water used for a recognized societal purpose: public drinking supply, irrigation, wildlife habitat, recreation, and electric power generation. In law, a right to water can be conditioned on its beneficial use.

beneficiation In the recovery of minerals from mining geological deposits, the process of separating the desirable mineral ore from waste rock and other geological material. The first step in the purification of a mineral removed by mining a deposit of ore.

benefit-cost analysis See COST-BENEFIT ANALYSIS.

benign See BENIGN NEOPLASM.

benign neoplasm A tumor that does not differ greatly from the tissue from which it arose and that is not likely to invade other sites (metastasize).

benthic Referring to bottom-dwelling aquatic organisms. Compare PELAGIC, LITTORAL.

benthos The organisms living on the bottom of an ocean, river, lake, pond, or other body of water. Compare NEKTON, PLANKTON.

bentonite A porous, colloidal clay produced by the decomposition of volcanic ash. The clay can absorb considerable amounts of water and expands greatly in the process. Bentonite is commonly used to provide tight seals around well casing.

benzene The simplest aromatic hydrocarbon. Its chemical formula is C_6H_6. The compound is illustrative of the structural features of all aromatic compounds. Benzene has a long history of use in the chemical industry as a solvent and as a starting compound for the synthesis of a variety of other materials and is now also used

Benzene

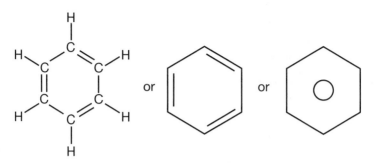

extensively in the rubber, paint, and plastic industries. The liquid is volatile, and emissions are regulated as toxic air pollutants. The major toxic effects of chronic exposure through inhalation, mostly an occupational hazard, concern the functioning of the bone marrow. A variety of blood disorders, such as aplastic anemia and leukemia, have been linked to excessive doses.

benzene hexachloride (BHC) Also known as hexachlorocyclohexane (HCH) and Lindane. An ISOMER of BHC is utilized as the insecticide Lindane. Various forms of this compound have been implicated in exposures resulting in neurological symptoms associated with depression of the central nervous system. BHC has also been shown to have carcinogenic properties in mice and to be bioaccumulated (increased in concentration) in the fatty tissue of humans.

benzine A mixture of ALIPHATIC hydrocarbons, such as gasoline or certain cleaning solvents. Not related to BENZENE.

benzo(*a*)pyrene (BaP) A carcinogenic POLYCYCLIC HYDROCARBON found in soot emitted by coal-burning facilities, in automobile particulate emissions, and in broiled or smoked meat and fish.

benzopyrene See BENZO(*a*)PYRENE.

Berkson's fallacy In epidemiology, the incorrect application of findings from studies using hospital or clinic populations as representative of the general population. The error arises from the possible self-selection of the hospital and clinic patients. Characteristics of those seeking medical treatment may differ significantly from those of the general population, and the patients at a particular hospital or clinic may be unrepresentative.

Berl saddles A type of packing used in a gas-cleaning device called a PACKED TOWER. The packing geometry is designed to maximize gas-liquid contact for pollutant removal efficiency. Other packing types include Pall rings, Lessing rings, Raschig rings, and Intalox saddles. See PACKING.

Berry, Wendell (1934–) American Poet and writer. Proponent of traditional rural life, opponent of modern industrial life. Two of his many works are *Home Economics* (1995), in which he indicts resource misuse, and *The Unsettling of America: Culture & Agriculture* (1996), which derides "industrial farming" and its damage to the environment and human culture.

beryllium (Be) An alkali earth metal that is hazardous when inhaled from airborne suspension. Causes a pathologic lung condition referred to as berylliosis after excessive, long-term occupational exposure in machine shops, ceramic and propellant plants, and foundries.

best available control measures (BACM) The most effective measures as determined by the USEPA for controlling small or dispersed particulates and other emissions from sources such as roadway dust, soot and ash from woodstoves, and open burning of brush, timber, grasslands, or trash.

best available control technology (BACT) Technology-based emission standards required by the USEPA for all new or modified major sources of air pollutants in geographical areas meeting the NATIONAL AMBIENT AIR QUALITY STANDARDS (also called PREVENTION OF SIGNIFICANT DETERIORATION areas). BACT is applied on a pollutant-specific basis; for example, a new source of sulfur dioxide emissions must apply BACT for sulfur dioxide if the area is clean with respect to sulfur dioxide levels. But if the facility is also a major source of carbon monoxide and the area does not meet the ambient standard for carbon monoxide, then a more stringent emission standard, LOWEST ACHIEVABLE EMISSION RATE (LAER), would be applied to the carbon monoxide emissions.

best available demonstrated technology (BADT) The level of EFFLUENT LIMITATION technology required by the 1972 CLEAN WATER ACT to be used in setting NEW SOURCE PERFORMANCE STANDARDS for new industrial direct discharges of water pollutants.

best available retrofit technology (BART) The level of air pollution control technology required on certain major existing sources (e.g., electric power plants) that may cause or contribute to visibility reduction in areas of the United States designated as Class I under the PREVENTION OF SIGNIFICANT DETERIORATION provisions of the CLEAN AIR ACT. Class I areas include major national parks and wilderness areas, mostly in the American West.

best available technology economically achievable (BAT) In water pollution control, the technological method required for all existing sources of toxic or NONCONVENTIONAL POLLUTANTS; BAT is set by industrial category or subcategory. The CLEAN WATER ACT specifies that the USEPA, in defining BAT, consider such factors as age of the facilities and equipment involved, process employed, engineering aspects of the control technique, process changes, cost of the reductions, and environmental impacts other than water quality, including energy requirements.

best conventional control technology (BCT) The level of water pollution control technology required of existing discharges for the treatment of CONVENTIONAL POLLUTANTS by the 1977 CLEAN WATER ACT.

best demonstrated available technology (BDAT) The treatment standard for hazardous wastes that, if met, allows the wastes to avoid the LAND DISPOSAL BAN.

best engineering judgment (BEJ) See BEST PROFESSIONAL JUDGMENT.

best management practices (BMP) 1. For facilities that manufacture, use, store, or discharge toxic or hazardous pollutants as defined by the 1977 CLEAN WATER ACT, a required program to control the potential spill or release of those materials to surface waters, such as dikes to contain tank overflows or heavy rainfall runoff. 2. Techniques to reduce NONPOINT SOURCE water pollution.

best practicable control technology (BPT) The level of EFFLUENT LIMITATION technology required by the 1972 CLEAN WATER ACT for existing industrial plants as of July 1, 1977; defined as the average of the best technology in each industrial category.

best professional judgment (BPJ) Used by environmental regulators to choose TECHNOLOGY-BASED treatment requirements for a water pollutant discharge permit when no national industry standards (such as BEST CONVENTIONAL CONTROL TECHNOLOGY) are available. See NATIONAL POLLUTANT DISCHARGE ELIMINATION SYSTEM.

beta counter A device used to measure IONIZING RADIATION. The counter is configured in such a way that the ionizing events caused by a BETA PARTICLE are selectively measured.

beta decay Decomposition of an unstable atom characterized by the loss of a particle with a mass approximately equal to that of an electron. The particle, termed a BETA PARTICLE, may carry a positive or a negative charge. Beta particles have a penetrating ability intermediate between that of ALPHA PARTICLES and GAMMA RAYS. Beta particles are a common cause of skin burns experienced by individuals exposed to mixed radiation sources.

beta particle Particle equivalent in mass to an electron emitted by certain substances undergoing RADIOACTIVE DECAY; a form of IONIZING RADIATION. Beta particles can carry a positive charge or a negative charge. Some emissions result when a neutron is converted to a proton in the atomic nucleus. This increase in the number of protons in the nucleus changes the ATOMIC NUMBER of the substance, and it therefore becomes a different element. For example, beta emission by a radioactive ISOTOPE of phosphorous (atomic number 15) converts the atom to sulfur (atomic number 16).

beta radiation The emission of a beta particle by an unstable atomic nucleus.

Beverage Industry Recycling Program (BIRP) Aluminum recycling initiative operated by the beverage industry to assist communities in organizing and managing recycling programs in an effort to increase the return of beverage cans for recycling. The program was initiated in response to the introduction of bottle bills requiring the customer to pay a refundable deposit when purchasing beverages. The bottle bills were intended to reduce litter as well as to promote recycling.

Bhopal, India The site of the pesticide manufacturing facility from which a large release of toxic gas, methyl isocyanate, occurred in December 1984. Approximately 3800 deaths have been attributed to the incident, which investigation has shown was caused by employee sabotage. Website: www.bhopal.com

bias An error in data gathering or analysis caused by faulty program design, mistakes by personnel, or limitations imposed by available instrumentation.

bicarbonate A compound containing the HCO_3^- group, for example, sodium bicarbonate ($NaHCO_3$), which ionizes in solution to produce HCO_3^-. See CARBONATE BUFFER SYSTEM.

bimetal Material composed of two different metals. For example, the commonly used food storage can consists of two layers: the exterior is a steel product and the inside lining is made of tin. Some beverage containers have steel bodies and aluminum tops.

bimodal distribution In statistics, a collection of observations with a large number of values found around each of two points. The distribution of the diameters of airborne particulate matter in urban areas, for example, is bimodal, with observations clumped around 0.2 and 10 micrometers.

binder 1. The chemical material that holds paint pigments together and attaches the paint film to a surface. It is nonvolatile—that is, not lost to the air as the paint dries. 2. The chemical material combined with the active ingredient in a pill for easier administration. Common binders in pills are starch, sugars, and gums.

binding energy In an atomic nucleus, the energy holding the neutrons and protons (collectively, NUCLEONS) together.

The atoms of heavier elements, those with larger numbers of protons and neutrons, have lower amounts of binding energy per nucleon because of the repelling force of the positively charged protons. Elements of ATOMIC NUMBER 83 and above are low enough in binding energy per nucleon to be UNSTABLE; in other words, they undergo RADIOACTIVE DECAY. In addition, FISSION, or splitting, of a heavy nucleus, such as uranium, creates lighter, more stable nuclei and releases the binding energy. Thus, the sum of the masses of the lighter fission nuclei is less than the mass of the heavier nucleus that splits; the difference in mass is represented by the binding energy released.

binding site In toxicology, the location in the tissues where a foreign chemical will attach and possibly accumulate, such as the liver, kidney, and plasma proteins. Also refers to the location on a protein molecule where the actual chemical bonding between a chemical and the protein occurs.

bioaccumulation The increase in concentration of a chemical in organisms that reside in environments contaminated with low concentrations of various organic compounds. For example, fish living in aquatic environments contaminated with compounds such as the CHLORINATED HYDROCARBONS absorb those compounds through the gills. Chemicals likely to be bioaccumulated are not readily decomposed in either the environment or an organism and are likely to be stored in the fatty tissue. Bioaccumulation is also the progressive increase in the amount of a chemical in an organism that results when the uptake, or absorption, of the substance exceeds its breakdown or excretion rate. In the complementary process of BIOLOGICAL MAGNIFICATION, an increase in chemical concentration in organisms is a result of the passage of the chemical through the food chain, not directly absorbed from the air, water, or soil, as in bioaccumulation. Also called bioconcentration, biological amplification, and biological concentration.

bioactivation A process that takes place in animal or human tissues during which a foreign substance is metabolized and transformed into a different INTERMEDIATE. In turn, the intermediate produced by the metabolic machinery of the body is more harmful than the original compound. Examples of compounds bioactivated in the body are benzene, chloroform, and parathion.

bioassay The use of living organisms to assess the adverse effects of an environmental sample. The organisms are placed in the sample (commonly, aquatic species in a water sample), and observed changes in the activity of the test organism are used as indicators of sample toxicity. See FATHEAD MINNOW; DAPHNID; GAMMARIDS.

bioavailability The magnitude and rate of availability of a DOSE of a chemical substance to body tissues. Some factors affecting bioavailability are the material's absorption, body distribution, metabolism, and excretion rates.

biocentric preservation A conservation philosophy that emphasizes as fundamental the right of all living organisms to exist and pursue their own interests. See ALTRUISTIC PRESERVATION.

biocentrism A life-centered philosophy holding that all living organisms have intrinsic value and rights regardless of whether or not they are useful to humans. For many holding this view, biodiversity is the highest ethical value in nature rather than being centered on human beings.

biochemical oxygen demand (BOD) The use of (or demand for) oxygen dissolved in water during the decomposition or metabolism of biodegradable organic compounds by microbes. The greater the amount of waste material (organic carbon) added to water, the greater will be the requirement for dissolved oxygen needed to convert the organic material to the mineral state (CO_2). Since dissolved oxygen is required by the organisms native to the body of water, BOD is a measure of the ability of a waste to cause damage in a receiving stream or lake. The customary units of the BOD measurement, which are not frequently expressed, are milligrams of oxygen utilized by one liter of wastewater incubated under the proper conditions for five days. The BOD of untreated municipal sewage is usually about 250, that of unpolluted water about 5. See CARBONACEOUS BIOCHEMICAL OXYGEN DEMAND, NITROGENOUS BIOCHEMICAL OXYGEN DEMAND.

biocide A chemical substance that kills living organisms. The designation is usually used to include materials that can kill desirable as well as undesirable organisms. PESTICIDE is formulated to kill undesirable organisms.

biocoenosis A community of animal and plant life.

bioconcentration The increase in concentration of a chemical in an organism that results when the uptake, or absorption, of the substance exceeds the rate of metabolism and excretion. See BIOACCUMULATION.

bioconcentration factor (BCF) Describes the accumulation of chemicals in aquatic organisms that live in contaminated environments. The factor is calculated by dividing the micrograms of the chemical per gram of aquatic organism by the micrograms of the same chemical per gram of water constituting the habitat.

bioconversion The extraction of energy from a BIOMASS directly, as in the combustion of wood or organic solid waste, or indirectly, as in the production of BIOFUEL.

biodegradable Describes a substance that can be metabolized into simpler components. The degradation or decomposition is usually performed by bacteria or microbes, referred to as DECOMPOSERS.

biodegradable plastic A type of plastic into which has been incorporated a nat-

ural polymer such as cornstarch that can be degraded by microorganisms. When items such as beverage yokes, fast-food packaging, and disposable diapers are made of such items, they will degrade to smaller pieces in the natural environment and present less of a litter problem when discarded inappropriately. The actual plastic polymers, however, do not decompose any more readily than plastic items that do not include the natural polymer.

biodegradation The metabolic breakdown of materials into simpler components by living organisms.

biodisc A large rotating cylinder containing surface features that allow for the growth of attached microorganisms. The cylinder revolves and contacts the wastewater along one side of the cylinder while the other side is exposed to the air. This promotes mixing of the wastewater and allows maximum oxygenation of the water, which stimulates decomposition of organic material dissolved or suspended in the water. The apparatus is used in some wastewater treatment plants as a type of SECONDARY TREATMENT.

biodiversity The variety inherent in natural biological systems. This variation is apparent on three levels. One, genetic diversity, is a measure of the degree of variation among the characteristics of a single species (e.g., the great variety among humans even though they all belong to the same species). The second, species diversity, is a measure of the number of different kinds of organisms that occupy a defined geographical area (e.g., stable rain forests are characterized by a great number of species in one acre). The third, ecological diversity, is represented by the great variety and complexity evident in undisturbed biological communities in terms of the number of different niches, trophic levels, and processes (e.g., a natural, undisturbed environment characteristically consists of many systems to capture radiant energy, sustain food webs, and recycle elements, whereas a disturbed system may have a restricted or diminished capability to carry out these natural functions).

biofuel Gaseous or liquid materials with a useful energy content produced from plant material or BIOMASS, for example, methane gas (BIOGAS), captured during the ANAEROBIC bacterial decomposition of organic material or ethyl alcohol (ethanol) produced from organic material by yeast. Biofuels are seen as possible substitutes for natural gas and gasoline.

biogas Methane gas produced during the ANAEROBIC decomposition of the remains of plants or animal wastes by bacteria.

biogenic Describing changes in the environment resulting from the activities of living organisms.

biogenic volatile organic compounds VOLATILE ORGANIC COMPOUNDS emitted to the air by vegetation.

biogeochemical cycling The flow of chemical substances to and from the major environmental reservoirs: ATMOSPHERE, HYDROSPHERE, LITHOSPHERE, and bodies of living organisms. As the materials move in the cycle, they often change chemical form, usually existing in a characteristic form in each reservoir. For example, carbon exists mainly as CARBON DIOXIDE in the atmosphere; mainly as CARBONIC ACID, BICARBONATE, or the CARBONATE ion when dissolved in water; and as more complex ORGANIC compounds in animals and plants. Essential materials that are recycled in the environment are carbon, nitrogen, phosphorus, and oxygen, among others. See CARBON CYCLE, NITROGEN CYCLE, HYDROLOGIC CYCLE, SULFUR CYCLE.

biogeochemistry The study of the transformation and movement of chemical materials to and from the LITHOSPHERE, the ATMOSPHERE, the HYDROSPHERE, and the bodies of living organisms.

biogeography The study of the geographic distribution of plants and animals.

biohazard The presence of microorganisms, such as bacteria, viruses, or fungi, that are capable of causing infectious diseases in humans.

bioindicator A living organism that denotes the presence of a specific environmental condition. For example, the presence of coliform bacteria identifies water that is contaminated with human fecal material.

biolistic gene transfer Physical method for introducing new genes into plant cells for the purpose of genetic engineering (i.e., adding genes from disparate sources). Target cells are bombarded with high-velocity microprojectiles coated with genetic material. The microprojectiles pass through the cell wall and cell membrane of the target cell and the new genes are injected into the target cell

biological additives Cultures of bacteria, enzymes, or nutrients that are introduced into an oil discharge or other waste to promote decomposition.

biological amplification See BIOACCUMULATION.

biological community See COMMUNITY.

biological concentration See BIOACCUMULATION.

biological contaminants Living organisms or derivatives of living organisms that can cause harmful effects when inhaled, swallowed, or otherwise taken into the body. Examples include bacteria, viruses, fungi, and animal or bird antigens. These kinds of materials are important components of indoor air pollution.

biological control Elimination or reduction of undesirable species, usually insects, by deliberate introduction of other living organisms that are PREDATORS, PARASITES, or PATHOGENS of the pest. The practice is usually considered an ecologically sound alternative to use of chemical PESTICIDE.

biological effects of ionizing radiation report (BEIR report) One of several reports issued by the Committee on the Biological Effects of Ionizing Radiation, established by the National Research Council of the United States to review the standards for IONIZING RADIATION exposure.

biological exposure index (BEI) Guidelines published by the AMERICAN CONFERENCE OF GOVERNMENTAL INDUSTRIAL HYGIENISTS (ACGIH) for assessing the hazard posed to healthy workers by chemical substances present in blood, urine, or breath. The guidelines are tied to the amount of the substance expected to be present in the body if inhalation exposure is eight hours per day, five days per week, at the THRESHOLD LIMIT VALUE/TIME-WEIGHTED AVERAGE concentration. See BIOLOGICAL MONITORING.

biological half-life The time required for one-half of an absorbed chemical substance to be biochemically degraded or excreted from the body of an animal.

biological integrity The ability of an environment to support and maintain a balanced, integrated, and functioning community of biological organisms.

biological magnification An increase in the concentration of heavy metals (such as mercury) or organic contaminants (such as chlorinated hydrocarbons) in organisms as a result of their consumption within the food chain of a particular habitat. Fish that consume large amounts of PLANKTON contaminated with mercury compounds retain the metal within their bodies as the greater portion of the food resource is metabolized; consequently, the amount of mercury in the fish gradually increases. Then, when a sea bird consumes a diet of contaminated fish, the metal in each fish is transferred to the tissue of the bird, and so on, up the food chain. Chemicals likely to undergo biological magnifi-

cation are not readily decomposed in the environment or metabolized by an organism and are usually stored in the fatty tissue of an organism. Compare to BIOACCUMULATION, which is the concentration of a chemical in an organism resulting from direct uptake from the environment (air, water, or soil) as opposed to through the food chain. Also called biomagnification.

biological medium The major compartments of an organism including blood, fatty tissue, lymphatic circulation, or gastrointestinal tract in which chemicals can be transported, stored, or transformed.

biological methylation The addition of a methyl group ($-CH_3$) to elemental or inorganic mercury (i.e., a mercury atom or ion) by bacteria, usually occurring in sediments within a water body. The methyl mercury produced is more toxic to humans than other forms of mercury and much more subject to BIOLOGICAL MAGNIFICATION.

biological monitoring In occupational health, the sampling and analysis of the blood, urine, or breath of workers to detect actual chemical exposure and absorption. The level of the chemical substance or a METABOLITE of the substance is compared with a BIOLOGICAL EXPOSURE INDEX, if one is available for the material, to determine its acceptability. Biological monitoring overcomes the problems of whether workplace air sampling is representative of actual human exposure and can warn of chemical overexposure before adverse health effects arise.

biological oxidation Any series of reactions in or by biological organisms that results in the metabolism, degradation, or decomposition of organic molecules. Biological processes that promote the decomposition of organic materials in wastewater, degradation of macromolecules discarded into the environment, detoxification of toxins in the liver, conversion of sugar into chemical energy in the body, and many other reactions in or promoted by living organisms. Processes that require the participation of living organisms. See OXIDATION.

biological stressor Nonnative species that is introduced into a particular environment that damages the natural biota of that system. Examples include zebra mussel, kudzu, nutria, and water hyacinth. These nonnative species can place unusual strain on native species and significantly alter the environment. The stresses are particularly acute in island ecosystems.

biological treatment The use of bacteria to degrade organic material. Treatment of municipal wastewater is a classic example of employing microorganisms to stabilize or degrade a waste material. See DECOMPOSITION and SECONDARY TREATMENT.

biological variability The observable differences among the individuals that constitute a given SPECIES. All humans belong to the same biological species, yet one can easily observe differences among individuals. Also, the differences that are observed when members of the same species are exposed to toxic or dangerous chemicals.

biological wastewater treatment The use of bacteria to degrade organic materials in wastewater. See SECONDARY TREATMENT.

biologically effective dose The internal dose of a toxic chemical required to cause a response or an adverse health effect. In order for a dangerous environmental pollutant released into the air, water, or soil to damage human health, the agent must gain entry into the body and accumulate at doses or concentrations that are sufficiently high to cause damage.

biologicals Living cultures of bacteria or fungi, active virus suspensions, vaccines, preparations consisting of products or parts of organisms, or related materials intended for use in diagnosing disease

states, immunization, treatment of disease, or similar functions.

biomagnification See BIOLOGICAL MAGNIFICATION.

biomanipulation The use of biological, chemical, or physical approaches to control the accumulation of plant nutrients in a lake. The process can include the introduction of new animal species to consume the algae overgrowth or the removal of phosphate-containing sediments, which promote algae growth.

biomarker A chemical present in the body used as a measurement of exposure to an environmental hazard. For example, blood continine levels are a biomarker for exposure to ENVIRONMENTAL TOBACCO SMOKE.

biomass Any biological material. In reference to alternative energy sources, mainly plants or parts of plants: harvested trees, leaves, limbs, and the like. In ecological studies, the DRY MASS of living organisms in a specified area, often expressed as grams of biomass per square meter.

biomass burning The use of biological material, such as trees, as a source of energy.

biome A broad regional area characterized by a distinctive climate, soil type, and biological community. Examples are desert, tundra, rain forest, and alpine biomes.

biometrics See BIOSTATISTICS.

biomonitoring The use of organisms to test an environmental MEDIUM or specific discharge for adverse effects. For example, selected fish species are placed in WASTEWATER effluent at various dilutions to determine aquatic toxicity. See BIOASSAY, WHOLE EFFLUENT TOXICITY TESTING.

bioregionalism A physiological outlook in which the organization of human activities is centered around natural geographic and ecological boundaries. This philosophy emphasizes a special sense of place and living within the natural resources of a specific locality. Living off the land without using outside resources.

bioremediation Use of biological processes to remove or detoxify pollutants from a contaminated environment. Biological organisms, particularly microorganisms, have a tremendous capacity to degrade, detoxify, and decompose a large variety of pollutants. The operation of a sewage disposal system offers an example of the potential of microorganisms to degrade various materials. The stimulating of natural microbial communities and the use of specific microorganisms to degrade crude oil resulting from oil spills illustrate the potential of this type of corrective action. Vascular plants have also been employed to remove or detoxify hazardous materials in abandoned waste disposal sites and in enhanced wastewater treatment systems.

biosensor An analytical device that couples biotechnology and electronic technology to measure some chemical in the environment. The biological component may consist of an enzyme, antigen, antibody, or microorganism; the electronic section may be an electrochemical device, optical transducer, or thermal detector. Together they permit the detection and/or quantification of a chemical or other agent.

biosolids The new, improved term for ACTIVATED SLUDGE.

biosphere The entire planetary ECOSYSTEM, including all living organisms and the parts of the Earth in which they live or that support them: the ATMOSPHERE, the HYDROSPHERE, and the LITHOSPHERE. Also called the ECOSPHERE. The term is also used to refer to only the living organisms on Earth, not to their physical and chemical environments.

biostabilizer A mechanical device that accelerates the formation of compost from a variety of starting materials including municipal solid waste and sewage sludge. Most systems operate by continually tumbling the solid material while maintaining optimal moisture and temperature conditions. See COMPOSTING.

biostatistics The methods for the mathematical analysis of data gathered relative to biological organisms. Also called biometrics.

biosynthesis The use of chemical energy by plants or animals to make carbohydrates, fats, or proteins. Synonymous with ANABOLISM.

biota The types of animal and plant life found in an area.

biotechnology The use of molecular biology techniques and tools to produce living organisms for beneficial use in novel ways. These procedures move beyond the traditional animal or plant breeding to produce biological strains endowed with traits that are of benefit to humans. Genes from different species frequently are introduced into target organisms. Applications include production of medicines and diagnostic aids, improvement of the nutritional or keeping quality of food crops, removal of toxins from the environment, and alteration of the natural resistance of plants to insecticides, insect pests, and plant pathogens. See also TRANSGENIC, GENETIC ENGINEERING.

biotic Of or relating to life, as in biotic components of an ECOSYSTEM. Compare ABIOTIC. See COMMUNITY.

biotic factors The influence or impact that other living organisms have on an individual.

biotic potential The upper limit of the ability of a species to increase in number; the maximal reproductive rate, assuming no limits on food supply or environmental conditions. A theoretical number that is not observed in nature for extended periods. See ENVIRONMENTAL RESISTANCE, LIMITING FACTOR.

biotransformation The metabolic conversion of an absorbed chemical substance, usually to a less toxic and more easily excreted form.

bioventing Bubbling air into contaminated groundwater to enhance aerobic microbial DECOMPOSITION. See IN SITU REMEDIATION.

birth rate The number of births per 1000 persons in a population in a given year as calculated by dividing the annual number of births by the midyear population, then multiplying by 1000.

bitumen The HEAVY OIL found in TAR SANDS, usually having a high sulfur content.

bituminous coal A soft coal with a relatively high energy content; the most abundant and widely used type of coal in the United State and chief fuel for electric power generation. See also ANTHRACITE, LIGNITE, SUBBITUMINOUS.

black box A part of a living or nonliving system that has a known or described function or role but for which the operational details or parts are unknown or omitted.

black liquor The alkaline liquid residue from cooking pulpwood to make paper.

black lung disease See SILICOSIS.

black oil A term for transported heavy oil or fuel oil, in contrast to transported refined products, such as gasoline, which are called clean oil.

blackbody A theoretical body with a surface that reflects no light and, thus, if not at a light-emitting temperature, would appear black in color. A blackbody is also a perfect emitter of ELECTROMAGNETIC RADIATION because it absorbs all incoming energy. INFRARED RADIATION emission cal-culations for the Earth or the human body make the simplifying assumption that they have the characteristics of a blackbody.

blackwater Sewage released from toilets. Contrast GRAYWATER.

blank A quality control sample that is processed to detect contamination introduced during laboratory tests designed to measure specific chemicals in samples collected from the environment. For example, if the system being examined is a lake or stream, a quantity of deionized water of the same volume as the sample drawn from the environment is processed as if it were an authentic sample. See TRIP BLANK.

blanketing plate Steel plates that are used to isolate or close off portions of ducts in exhaust systems.

blast gate A metal damper that can move back and forth in an air duct to achieve a balanced local exhaust ventilation system.

blood products Items recovered from human blood, including plasma, platelets, red cells, white cells, antibody, or a variety of serum proteins.

bloodworm A SLUDGE-eating worm typically found in water with a DISSOLVED OXYGEN content of less than one or two PARTS PER MILLION. A biological indicator of an environment that has been severely impacted by the addition of sewage or other organic wastes.

bloom In aquatic ECOSYSTEMS, the rapid growth of ALGAE. See ALGAL BLOOM.

blow-by In an automobile, the leakage of VOLATILE ORGANIC COMPOUNDS from the combustion chamber between the piston and the cylinder wall into the crankcase. To prevent the loss of these CRANKCASE BLOW-BY gases or vapors to the atmosphere, automobiles are now fitted with a POSITIVE CRANKCASE VENTILA-

TION system that routes the volatile organics back to the carburetor for combustion.

blowdown The water removed from COOLING TOWERS to prevent the accumulation of excessive amounts of dissolved salts that result from evaporation. The water that is lost is replaced by MAKEUP WATER.

blowout The uncontrolled release of oil, natural gas, or both from a well into the environment.

Blue Angel The world's first ECOLABELING program, begun in Germany in 1977. Its logo, the Blue Angel, is awarded to products that conserve resources, have a low impact on GLOBAL WARMING, and conserve energy. Website: www.ns.ec.gc.ca/g7/eco-ger.html

body burden The amount of a chemical material present in the body at a certain time.

body weight scaling The method of extrapolating DOSE-RESPONSE data from experimental animals to humans that adjusts the data on the basis of relative body weights. For example, if a rat is assumed to weigh 0.35 kilogram and a human 70 kilograms, then a dose of 1 gram for a rat is extrapolated to a human by using a scaling factor of 200, or a dose of 200 grams for the human. The dose for each is therefore about 2.857 grams per kilogram body weight. Compare SURFACE AREA SCALING FACTOR.

bog Area characterized by swampy conditions with waterlogged soils that tend to be acidic and to accumulate peat and other plant residues. Peat can be removed from these areas for use as combustion fuel. Cadavers that have been buried in bogs for hundreds of years and remained in relatively good condition have been found.

boiler water The water that is heated in an industrial device to produce steam.

boiling point The temperature at which the VAPOR PRESSURE of a liquid equals the ATMOSPHERIC PRESSURE. Commonly used as the temperature at which a liquid boils.

boiling-water reactor (BWR) A type of nuclear reactor power plant in which the water in contact with the core is allowed to boil. The water is thereby converted to steam, which is piped to drive a turbine, generating electricity. The other main type of nuclear reactor used for the production of electric power is the PRESSURIZED WATER REACTOR.

bole The trunk of a tree.

bomb calorimeter A laboratory device used to measure the energy value of a material (in calories) through combustion of the material.

bone seeker A radioactive material, such as strontium-90, that is metabolically incorporated into the bones after inhalation or ingestion and absorption.

boom A device deployed on a water surface to deflect or contain an oil spill. The booms are in sections, often 50 feet in length, that are connected as needed. One typically consists of an above-water freeboard (containment wall) and an underwater skirt attached to a float. A ballast weight is attached to the bottom of the skirt.

Components of a boom

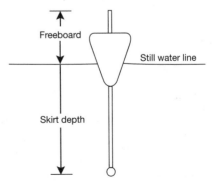

Freeboard

Still water line

Skirt depth

borehole Hole made with drilling equipment. A well.

boreal forest Northern forest, as in the boreal forest biome, characterized by evergreen conifers and long winters. The boreal forest is found in the northern parts of North America, Europe, and Asia.

botanical pesticide A plant-derived insecticide. A product consisting of a chemical produced naturally by plants and used to kill or injure insect pests. Such products are usually considered safer to use than synthetic pesticides.

bottle bill A law that mandates the use of returnable beverage containers, usually requiring that a deposit be collected when the beverage is purchased. When the containers are returned, the deposit is refunded or credited toward the next purchase. Advantages include litter reduction, reduction in solid waste volume, and conservation of natural resources.

bottled water Water meeting all applicable federal and state standards, sealed in a sanitary container, and sold for human consumption. The product cannot contain sweeteners or chemical additives other than flavors, extracts, and essences derived from fruit or spice added at levels less than 1% by weight of the final product. The product must be calorie-free and sugar-free and contain little or no sodium.

bottled water, carbonated A sparkling water containing carbon dioxide under pressure and characterized as bubbly or effervescent. The carbonated waters are commonly mineral waters containing specified amounts of dissolved solids, including sodium. Soda water, tonic water, and seltzer water are considered soft drinks since they may contain sugar and other ingredients.

bottled water, still Noncarbonated water. Same as BOTTLED WATER. Contrast BOTTLED WATER, CARBONATED.

bottom ash The solid material left on the bottom of an incinerator or a boiler after burning of solid waste or a boiler fuel. Compare FLY ASH.

bottom-land hardwoods Forested wetlands common along rivers in the southeastern United States. The areas are flooded part of the year and typically include stands of deciduous trees, those with broad leaves that are dropped in the winter. The regions provide valuable wildlife habitat. See HARDWOODS.

botulism A disease caused by the ingestion of food containing a TOXIN produced by the ANAEROBIC microorganism *Clostridium botulinum*. The toxin attacks the nervous system, with a fatality rate of about 60%. Primary risk arises from home-canned food, especially improperly prepared beef and pork products, corn, olives, beans, and spinach.

bound water Water molecules that are held tightly to soil or other solids. This water is not easily removed by normal drying and is not available for other purposes such as plant growth.

boundary layer The layer of the atmosphere from the Earth's surface upward in which frictional forces are significant. See EKMAN SPIRAL, GEOSTROPHIC WIND.

box model The representation of STOCKS and flows of material or energy in and through systems by boxes for the stocks and arrows from box to box for the flow directions and amounts. At equilibrium, the stock in a box divided by the rate of flow in or out is called the RESIDENCE TIME. The model can be applied to ECOSYSTEMS, lakes, the BIOSPHERE, a human or animal body, and other systems.

Boyle's law The law, by the English scientist Robert Boyle, that the volume of a perfect gas is inversely proportional to the pressure exerted on the gas, given a constant gas temperature. Expressed as

$pV = K$, where p is the pressure, V is the volume, and K is a constant.

brackish water Water with a salt content between 1000 and 4000 parts per million.

breakpoint chlorination A process used to guarantee the presence of sufficient chlorine concentrations in public supplies of drinking water for the protection of public health. The chlorine breakpoint is reached when a sufficient amount of chlorine is added to react with all OXIDIZABLE compounds in the water, such as ammonia. All additions after this breakpoint increase the levels of residual or unreacted chlorine available to eliminate dangerous microorganisms that may enter the water after treatment. Compare COMBINED RESIDUAL CHLORINATION.

breakthrough curve A graph of the volume of gas flowing through a fixed-bed adsorber (such as an activated charcoal filter) versus the concentration of the chemical substance to be adsorbed measured in the outlet gas—that is, the amount of chemical that escapes collection. As a carbon filter becomes saturated with the chemical collected from the gas stream, the concentration of the chemical in the outlet gas begins to increase; the point at which the outlet gas concentration exceeds the EMISSION STANDARD for the gas or vapor being collected is called the breakpoint. When the breakpoint is reached, the adsorbing material must be replaced or cleaned.

breathing zone Roughly from chest to head high, the preferred sampling height for occupational air contaminants.

breeder reactor A nuclear reactor that converts nonfissionable uranium-238 to the FISSIONABLE material plutonium-239 by arranging the CORE materials so that NEUTRONS produced in the reactor are captured, through a series of steps, by the uranium-238. Natural uranium ore contains only 0.7% fissionable material, the ISOTOPE uranium-235. The other 99.3% is the non-fissionable isotope uranium-238. Liquid sodium metal is normally employed as the fluid used to cool the reactor. The liquid sodium is heated because it is in direct contact with the reactor core. Subsequently, the hot sodium is pumped to other units designed to remove the heat and thereby maintain the temperature of the reactor at a safe level.

bremsstrahlung ELECTROMAGNETIC RADIATION emitted by charged particles as they are slowed upon impact or by passing through high-density material.

brine disposal Removal of water that contains high concentrations of salt.

British thermal unit (BTU) The quantity of heat required to raise the temperature of one pound of water one degree Fahrenheit. One BTU equals 252 CALORIES or 1054.4 JOULES.

bromate A chemical formed when OZONATION is used to remove PATHOGENS from drinking water; produced by a reaction of ozone with bromide in the treated water. One of the HALOACETIC ACIDS regulated as a DISINFECTION BY-PRODUCT under the SAFE DRINKING WATER ACT.

bronchioles The air-conducting tubes in the lungs; the two larger bronchi divide some 17 times into progressively smaller bronchioles, after which the gas exchange region begins. See ALVEOLAR REGION, MUCOCILIARY ESCALATOR.

bronchoconstriction Narrowing of the BRONCHIOLES, with a resulting increase in AIRWAY RESISTANCE. The condition is caused by exposure to excessive concentrations of certain air pollutants, such as sulfur dioxide.

Brower, David (1912–) American Activist in resource conservation. Held influential leadership positions in the SIERRA CLUB, FRIENDS OF THE EARTH, the Earth Island Institute, and League of Conservation Voters, among others.

brown coal   A term for LIGNITE.

brownfield   Abandoned or idle industrial site, usually in major cities of the North and Midwest, that cannot be used for development because of real or potential liability risks resulting from environmental contamination in the past, when regulations were less stringent. One of the major factors restricting economic development of the urban core of many large cities that once hosted heavy industry. Website: www.epa.gov/swerosps/bf/

Brownian motion   The random, nondirectional motion of small particles suspended in a gaseous or liquid medium, caused by the movement of the molecules constituting the medium. Viewed with a microscope, small particles of nonliving matter in a liquid suspension exhibit (random) Brownian motion, but bacteria or larger living cells move in a specific direction. Described by Robert Brown, a British botanist, 1773–1858.

brown lung disease   See BYSSINOSIS.

Brundtland Commission Report   A 1987 report issued by the United Nations World Commission on Environment and Development, headed by Dr. Gro Harlem Brundriand, then-premier of Norway. The study concluded that solutions to the global problems of environmental pollution and natural resource depletion require extreme changes in agricultural, banking, and industrial practices and reduced energy consumption and population growth. It urged movement toward SUSTAINABLE DEVELOPMENT, which emphasizes economic growth in an ecologically sound manner, especially in the less-developed countries.

BTEX   Acronym used to refer to the common hydrocarbon constituents of gasoline: benzene, toluene, ethylbenzene, and xylene. When gasoline migrates into groundwater, analytical tests for these chemicals can be used to detect its presence.

bubble   See BUBBLE POLICY.

bubble meter   A device for measuring GAS VOLUMETRIC FLOW RATE, which consists of an inverted titration burette attached to a pump, with a squeeze bulb filled with soapy water at the end. The bulb is squeezed, producing a soap bubble that travels up the buret with the flowing air. The timed travel of the bubble between marks on the burette gives the gas flow rate.

bubble policy   The USEPA air quality management program that allows air emissions of a particular pollutant from a plant or group of plants to be treated as if all individual emission points were under an imaginary bubble and exhausted at one emission point The total allowable emissions for a particular pollutant are then determined, and the facility manager can decide the most efficient mix of emission controls to use, as long as the overall limit is not exceeded. See also EMISSION REDUCTION CREDITS, EMISSIONS TRADING, OFFSET, NETTING OUT.

bubbler   An air-sampling device that collects a gaseous air contaminant by passing air through a chemical solution that reacts with and captures the material to be collected.

buffer   A chemical substance in solution that neutralizes an ACID OR BASE. Buffered solutions resist changes in pH. Surface waters and soils with chemical buffers are less susceptible to ACID DEPOSITION than those with poor buffering capacity.

buffer strips   Normally used in an agricultural context. Areas of grass or other vegetation planted between or below cultivated fields to prevent erosion.

buffer zone   Undeveloped, forested, or otherwise vegetated land between an industrial facility and the nearest residential area. Such zones are intended to lessen the impact of industrial operations on citizens and to reduce the risk of residing in an industrial zone.

buffered solution A solution containing a BUFFER.

buffering capacity The amount of acid or base that a given liquid or soil can neutralize.

build absolutely nothing anywhere near anyone (BANANA) The slogan sometimes attributed to environmental activists. See NOT IN MY BACKYARD; NOT IN MY TERM OF OFFICE; YES, IN MY BACKYARD, FOR A PRICE.

building cooling load The amount of heat (in BRITISH THERMAL UNITS) that must be removed from a building on an hourly basis to maintain indoor comfort. The greater the amount of heat produced by items such as lights, the greater the cooling load.

building envelope That part of a building in contact with the exterior environment, walls, roof, windows, and floor.

building-related illness See SICK BUILDING SYNDROME.

bulk liquids Liquid hazardous waste not in specific containers. Along with waste containing FREE LIQUIDS, these liquids are subject to the LAND DISPOSAL BAN.

bulk sample In asbestos control programs, a thumbnail-sized portion of a suspect ASBESTOS-CONTAINING MATERIAL that is collected for laboratory analysis. Other materials requiring bulk samples for laboratory analysis include SILICATES, POLYCHLORINATED BIPHENYLS, and fugitive grain dusts.

bulking sludge Suspended particulate material that does not settle to the bottom of a CLARIFIER, causing a rise in the level of SUSPENDED SOLIDS and BIOCHEMICAL OXYGEN DEMAND in water leaving a treatment facility. This condition is typically controlled by adjusting the MIXED LIQUOR SUSPENDED SOLIDS concentration and the FOOD-TO-MICROORGANISM RATIO. See also SLUDGE VOLUME INDEX.

bulky waste Large household discards such as furniture, tree stumps, major kitchen appliances, and junk automobiles. Household discards requiring equipment other than normal garbage or trash trucks for collection. See WHITE GOODS.

Bureau of Land Management (BLM) A section within the United States Department of the Interior responsible for the management of public lands termed national resource lands. These holdings (about 244 million acres) constitute lands that are not included within the national parks, national forests, or national wildlife refuges. The lands managed by the BLM are available for mining, grazing, and timber removal. Website: www.blm.gov

Bureau of National Affairs (BNA) A publishing company based in Washington, D.C., that issues timely reports on regulatory developments in several public policy areas, including environmental affairs. The reports are sometimes referred to as "banana" reports because of the acronym for the company. Website: www.bna.com

burning agents Additives that improve combustibility of the material to which they are added.

burnup 1. The percentage of uranium-235 or plutonium-239 used as fuel in a nuclear reactor that is lost as a result of fission. 2. The amount of heat obtained per unit mass of fuel.

butterfly damper A simple plate used to control the flow of gas through a duct by orienting it parallel to or perpendicular to the flow of the gas.

buy-back center A facility that pays for wastepaper, aluminum cans, plastics, and other recyclable items. The center delivers the material to a resource reprocessing operation or resells the material to a commodities broker for eventual reprocessing.

bypass To allow the release of exhaust gas or wastewater without passage through a pollution control device. Also, to allow drinking water through a treatment facility without adequate contaminant removal.

by-product A substance, other than the intended product, generated by an industrial process.

by-product material 1. A salable substance that is recovered from a pollution control device. 2. Any material made RADIOACTIVE as a result of exposure to external radiation sources. 3. ISOTOPE formed through the process of nuclear FISSION.

byssinosis A chronic lung disease resulting from long-term inhalation of high concentrations of cotton dust. Also called brown lung disease.

C

C3 plants A classification of plants based on the first intermediate produced as a result of the fixation of carbon dioxide. PHOTOSYNTHESIS in these plants produces an intermediate carbon compound with three carbons. Carbon dioxide is incorporated into a five-carbon compound, ribulose diphosphate, resulting in the production of two three-carbon compounds, phosphoglyceric acid. Plants with this type of metabolism are the most common and are more abundant in temperate climates of the middle latitudes. They include important food crops like beans, peas, wheat, oats, and rice. C3 plants exceed C4 PLANTS in photosynthesis under increasing concentrations of CARBON DIOXIDE. See GLOBAL WARMING.

C4 plants A classification of plants based on the first intermediate produced as a result of the fixation of carbon dioxide. The first product of PHOTOSYNTHESIS in these plants is an intermediate compound with four carbons. Carbon dioxide is added to a three-carbon compound, phosphoenolpyruvic acid, resulting in the formation of a four-carbon product, oxaloacetic acid. This type of metabolism is important in the adaptation of plants to hot and dry climates. The anatomy of the leaf in this type of plant is fundamentally different from that of C3 PLANTS. Sugarcane and corn are important representative examples. In general, C4 plants exhibit little increase in photosynthesis in response to increasing concentrations of CARBON DIOXIDE. See GLOBAL WARMING.

cadmium A toxic metal used in galvanizing, in nickel-cadmium batteries, and as a pigment. Nonoccupational human exposure results from cigarette smoking, air and water industrial discharges, and improper disposal of cadmium-containing waste. Body accumulation occurs mainly in the kidneys, where the protein metallothionein protects against low-level exposure by chemically binding to the cadmium.

calcination A process in which a material is CALCINEd.

calcine To heat a substance at a temperature below its fusion point but high enough to cause the loss of water or VOLATILE components, making the material crumbly. The process is often performed in a rotating cylindrical kiln.

caldera Large basin-shaped depression formed after explosive eruptions of volcanoes characterized by the presence of silica-rich magma. This type of magma is very thick and viscous, resulting in little lava flow and the consequent buildup of tremendous pressures under the dome that caps volcanic vents. The eruptions blow out large amounts of ash and magma, causing the collapse of the summit, forming a large basin-shaped depression, which may fill with water in later years. Crater Lake, Oregon, is a pronounced example of a caldera.

calibration The act of introducing known quantities to a measuring device to determine the response of the instrument.

calibration curve A set of responses given by a measuring instrument plotted against a set of known quantities introduced to the instrument The responses of the instrument to unknown quantities are then translated to measured quantities using the calibration curve, either by the instrument itself or by the analyst.

California Air Resources Board (CARB) The California state air pollution regulatory agency; sets the California motor vehicle emission standards, which are generally stricter than the federal limits. Website: www.arb.ca.gov

California Aqueduct A channel built to transport water from the region of California north and west of San Francisco to the south-central region of the state.

California list waste Certain liquid hazardous wastes banned from deep injection wells and, without prior treatment, from land disposal by regulations promulgated by the USEPA under the authority of the HAZARDOUS AND SOLID WASTE AMENDMENTS to the RESOURCE CONSERVATION AND RECOVERY ACT. See LAND DISPOSAL BAN.

calorie (cal) The amount of heat required to raise the temperature of one gram (one milliliter) of water one degree Celsius. When describing the energy content of food, spelled with a capital C and equal to 1000 calories (or 1 kilocalorie).

CAMEO® An integrated suite of computer programs developed by the USEPA and the NATIONAL OCEANIC AND ATMOSPHERIC ADMINISTRATION for use in planning and responding to chemical emergencies. CAMEO models are used by FIRST RESPONDERS to estimate possible evacuation zones. Website: www.epa.gov/ceppo/cameo/

Canadian deuterium-uranium reactor (CANDU) A NUCLEAR REACTOR of Canadian design that uses naturally occurring, unenriched URANIUM (about 0.7% of the fissionable isotope uranium-235) as the fuel and HEAVY WATER (containing deuterium, the heavy isotope of hydrogen) as the MODERATOR and coolant.

cancellation In pesticide regulation under the FEDERAL INSECTICIDE, FUNGICIDE, AND RODENTICIDE ACT, the process of gathering additional information to determine whether the REGISTRATION of a pesticide should be voided if a pesticide is suspected of posing "unreasonable adverse effects." A pesticide may be manufactured and used while a cancellation process is still open. See also SUSPENSION.

cancer A class of more than 100 diseases characterized by the uncontrolled growth of cells. Cancers are divided into groups based on the type of cells from which they arise: CARCINOMA, SARCOMA, LEUKEMIA, and LYMPHOMA.

cancer rate The number of people with cancers in a given area during a specified period, usually one year, divided by the number of persons in the area, multiplied by a constant, normally 100,000. For example, if 400 cancers occur in a community in one year and if the population of the community is 125,000, the cancer rate is 320 per 100,000. The rate is calculated to provide a common comparison of the INCIDENCE of cancers in communities of differing size and setting.

cancer potency factor (CPF) An estimate of the upper limit of the risk of contracting cancer per one unit of dose of a particular chemical; derived from the 95% upper-bound confidence limit of the slope of the DOSE-RESPONSE CURVE. This means that the true cancer risk is unlikely to be higher than the calculated risk and is likely to be lower; that is, a conservatively high estimate. Typical units are expressed as (milligrams per kilogram body weight per day)$^{-1}$ or as (micrograms per cubic meter of air)$^{-1}$, the latter often called the UNIT RISK ESTIMATE. The CPF times the actual or modeled dose, for example, in milligrams ingested per kilogram body weight per day, yields the lifetime cancer risk. Also called potency, slope factor, carcinogenic slope factor, potency slope factor, and $q1^*$.

cancer registry An organization, generally operated at a local, then state level, charged with the responsibility of maintaining records of the number and kinds of cancers that are diagnosed as well as the number of deaths resulting from cancers in a state. Provides a way to compare the INCIDENCE in a state with that in other regions of the country and to track the change in the numbers and types of cancers over time. See SURVEIL-LANCE, EPIDEMIOLOGY, AND END RESULTS (SEER).

cancer risk An estimation of the likelihood that an individual will develop a cancer over a lifetime.

cap In landfill management, the final, permanent layer of impermeable material placed on the surface as part of the CLO-SURE process. A cap is commonly a compacted clay or synthetic material designed to resist liquid infiltration.

capacity assurance plan A plan prepared by each state and submitted to the USEPA detailing the ability of the state to manage its hazardous waste for a 20-year period.

capacity factor A ratio used to describe the operation of a boiler. The actual or measured load on a boiler for a set period is compared with the theoretical rating of that boiler (that is, actual output per hour/theoretical output per hour × 100).

capillary action Movement of water or other liquid along a solid surface in response to the forces of adhesion, cohesion, and surface tension. The result is that when one end of a small tube is placed in water, the water will rise in the tube until the weight of the water in the tube is balanced by the adhesive forces. The smaller the diameter of the tube, the higher the water will rise. The process is important in various phenomena, such as the rise of water from the roots of a tree to the top branches, the migration of fluids upward through sediments, and the rise of water in a towel when one end is immersed in water.

captive breeding Maintaining animals in zoos, wildlife parks, and other facilities to produce stock for zoos, laboratories, wildlife parks, or similar facilities or for release into the wild. The programs are of particular importance to the maintenance of THREATENED SPECIES or ENDANGERED SPECIES.

capture 1. A process whereby a particle is added to a nucleus, with the possible release of energy; for example, the capture of a neutron by a nucleus and the subsequent emission of GAMMA RADIATION. 2. In groundwater HYDROLOGY, the difference between natural groundwater discharge from an AQUIFER and the artificial increase in the amount of water in the aquifer due to pumping of water from surface sources.

capture efficiency The fraction of organic vapors retained by a device intended to recover the vapors in a process or to prevent the release of an air pollutant.

capture velocity The air velocity at a point in front of a HOOD necessary to pull contaminated air from that point into the hood. Capture velocity varies with the release velocity of the airborne material to be captured and with the strength of competing airflows in the vicinity of the hood.

carbamates A group of NONPERSISTENT (degradable) synthetic insecticides that includes carbaryl (Sevin), Baygon, and Temik, among others. The carbamates are CHOLINESTERASE INHIBITORS and are generally selective in their toxicity.

carbaryl See SEVIN.

carbon (C) An element present in all materials of biological origin. The element contains six protons and usually six neutrons in the nucleus. Carbon atoms will bind with each other as well as with a host of other elements. Deposits derived from living sources, such as limestone, coal, oil, and natural gas, contain carbon as a principal element. Graphite and diamonds are composed almost exclusively of carbon.

carbon 14 (^{14}C) A RADIOACTIVE ISO-TOPE of carbon that emits beta particles when it undergoes RADIOACTIVE DECAY. The nucleus of a ^{14}C atom contains eight neutrons rather than the six neutrons found in ^{12}C, which is the most abundant isotope in nature. Carbon 14 is formed naturally in the atmosphere through the interaction of nitrogen gas in the air with cosmic rays. The atmospheric ^{14}C thus produced combines with atmospheric oxygen and is converted to CARBON DIOXIDE (^{14}CO$_2$). Because the radioactive form of carbon dioxide mixes with the normal, nonradioactive carbon dioxide in the atmosphere and because the radioactive and nonradioactive forms of carbon dioxide undergo the same chemical reactions, both forms can be found in living plant tissue as a result of PHOTOSYNTHESIS. While the plant is alive, the ratio of ^{14}C to ^{12}C remains about the same as time passes. However, when the plant dies, carbon dioxide from the air is no longer incorporated into plant mass. The radioactive form of carbon decays with a half-life of about 5760 years; consequently, the ratio between the two isotopes in the dead plant changes as time passes and the radioactive carbon decays. The changes in this ^{14}C:^{12}C ratio can be used to date materials. If a buried tree contains one-half of the radioactive carbon that one finds in a live tree (relative to the amount of normal carbon present), the tree is approximately 5760 years old. Materials older than 70,000 years cannot be aged by this technique since almost all of the radioactive carbon has decayed. Carbon 14 is also known as radiocarbon.

carbon adsorber A pollution control device that uses ACTIVATED CARBON to collect organic pollutants from a gas stream or from a wastewater stream. Organic compounds are collected on the surface of the activated carbon. See ADSORPTION.

carbon adsorption See CARBON ADSORBER.

carbon black A sooty material produced by incomplete combustion of hydrocarbons. Carbon black is made commercially under controlled conditions for use in pigments, tires, and inks.

carbon budget An analysis of any changes in the STOCKS and flows in, through, and out of the RESERVOIRS of the CARBON CYCLE.

carbon cycle The movement of carbon in various chemical forms from one planetary RESERVOIR to another.

carbon density The amount of carbon per unit area in an ecosystem, typically expressed as kilograms of carbon per square meter (kg [C]/m^2). Tropical forests have a carbon density of about 19 kg (C)/m^2 whereas deserts have a carbon density of about 0.3 kg (C)/m^2.

carbon dioxide (CO$_2$) A normal gaseous constituent of the atmosphere, the product of aerobic respiration, decomposition, and carbon fuel combustion. Because carbon dioxide absorbs INFRARED RADIATION, rising levels of carbon dioxide in the global atmosphere over the past century have prompted concerns about climatic change. See GREENHOUSE EFFECT.

Carbon Dioxide Information Analysis Center The primary repository for data pertaining to global climate change for the United States Department of Energy. The center is in the Environmental Sciences Division at Oak Ridge National Laboratory in Oak Ridge, Tennessee. The collection, storage, and analysis of data concerning the global carbon cycle, the effects of rising carbon dioxide concentrations on vegetation, and renewable energy sources constitute primary interests of the center. Website: http://cdiac.esd.ornl.gov

carbon filtration The passage of treated wastewater or domestic water supplies through ACTIVATED CHARCOAL to remove low concentrations of dissolved chemicals. See CARBON ADSORBER.

carbon flux Generally applied to calculations or estimations of the movement of carbon through or within an ecosystem. For example, the carbon flux of a forest is the balance between the amount of CARBON DIOXIDE fixed by photosynthesis in the forest and the amount of carbon dioxide produced by decomposition or respiratory metabolism in the same area.

carbon monoxide (CO) An odorless, colorless, and tasteless gas that is an air contaminant formed by the incomplete combustion of carbon fuels, such as wood, coal, or gasoline. Classed as an asphyxiant gas, absorbed carbon monoxide combines with hemoglobin in the blood to form CARBOXYHEMOGLOBIN, which impairs oxygen delivery to the tissues. The major source for human exposure is the automobile.

carbon oil A liquid fuel, derived from coal, used in lamps before the invention of electric lighting.

carbon polishing The removal of residual dissolved organic substances from wastewater by ADSORPTION on ACTIVATED CHARCOAL (GRANULAR ACTIVATED CARBON). A form of TERTIARY TREATMENT.

carbon sink Physical location in which carbon dioxide from the atmosphere accumulates as BIOMASS resulting from the fixation of carbon dioxide through photosynthesis, or deposits in the oceans as inorganic carbonates resulting from the building of shells of aquatic organisms like clams or from the precipitation of calcium carbonate. Such accumulations in trees or shells remove the carbon from the carbon cycle for long periods and reduce the amount of carbon dioxide in the atmosphere.

carbon source Term applied to the type of carbon compounds used by biological organisms to satisfy carbon requirements. For example, terrestrial plants commonly use carbon dioxide as a carbon source; a bacterium may use simple sugars to obtain needed carbon; and a lion obtains carbon from that available in the carcass of prey.

carbon tax A proposed tax on the carbon content of fuels, the purpose of which would be to decrease FOSSIL FUEL consumption and its attendant emissions of CARBON DIOXIDE, a GREENHOUSE GAS. A carbon tax has been suggested as a method for the United States to make the reductions of greenhouse gases required by the KYOTO PROTOCOL correspond to the FRAMEWORK CONVENTION ON CLIMATE CHANGE.

carbon tetrachloride A derivative of methane (CH_4) that is produced by substituting a chlorine atom for each hydrogen (CCl_4). Carbon tetrachloride is a common NONPOLAR SOLVENT that has significant industrial applications; it has also been used as a fumigant. Exposure to the HALOGENATED hydrocarbon has been linked to damage to the liver, kidney, and central nervous system. CARCINOGENIC properties have also been demonstrated.

carbon treatment In a drinking water purification process, the removal of COLLOIDS by ADSORPTION on ACTIVATED CHARCOAL. This step often improves color, taste, and odor.

carbonaceous biochemical oxygen demand The incubation of a sample of water or wastewater for a relatively short time in order to determine the biochemical oxygen demand. The short incubation, usually five days, is sufficient to detect only the microbial utilization of carbon compounds. A longer incubation (15 to 20 days) would also detect the oxidation of inorganic nitrogenous compounds (ammonia and nitrite) and the subsequent demand for molecular oxygen by chemoautotrophic bacteria. See BIOCHEMICAL OXYGEN DEMAND for a description of the basic concept associated with the measurement. See also NITROGENOUS BIOCHEMICAL OXYGEN DEMAND.

carbonate The CO_3^{-2} ION in the CARBONATE BUFFER SYSTEM. Combined with one proton, it becomes BICARBONATE, HCO_3^-; with two protons, CARBONIC ACID. The carbonate ion forms a solid

PRECIPITANT when combined with dissolved ions of calcium or magnesium.

carbonate aquifer An AQUIFER found in limestone and dolomite rocks. Carbonate aquifers typically produce HARD WATER, water containing relatively high levels of calcium and magnesium.

carbonate buffer system The most important BUFFER system in natural surface waters and wastewater treatment, consisting of a CARBON DIOXIDE-water-CARBONIC ACID-BICARBONATE-CARBONATE ion equilibrium that resists changes in pH. For example, if acid materials (hydrogen ions) are added to this buffer system, the equilibrium is shifted and carbonate ions combine with the hydrogen ions to form bicarbonate; bicarbonate then combines with hydrogen ions to form carbonic acid, which can dissociate into carbon dioxide and water. Thus the system pH is unaltered although acid has been added.

carbonate hardness See TEMPORARY HARDNESS.

carbonated water See BOTTLED WATER, CARBONATED.

carbonation, groundwater The dissolving of carbon dioxide in surface water as it percolates through the ground. The carbon dioxide reacts with water to form carbonic acid, a weak acid that causes the water to have a slightly acidic pH.

carbonic acid A mild acid formed by the dissolution of carbon dioxide in water. The carbonic acid content of natural, unpolluted rainfall lowers the pH of the rain to about 5.6.

carboxyhemoglobin (COHb) The chemical compound formed by the combination of absorbed CARBON MONOXIDE and the blood pigment hemoglobin. Hemoglobin normally combines with atmospheric oxygen drawn into the lungs to transport it from the lungs to body tissues. When carbon monoxide is chemically bonded to hemoglobin, the oxygen-carrying capacity of the blood is reduced. The COHb level is expressed as a percentage of all hemoglobin BINDING SITES occupied by carbon monoxide. Nonsmoking adults typically have an endogenous (formed within the body by the normal breakdown of HEME) COHb level of less than 1%. The USEPA NATIONAL AMBIENT AIR QUALITY STANDARD for carbon monoxide is set to prevent COHb levels from exceeding about 1.5%.

carboy A large glass jug or container, often with a wooden frame or case, used to store large amounts of liquid chemicals or to collect and transport large water samples.

carcinogen A chemical substance or type of radiation that can cause cancer in exposed animals or humans. There is no single definition of the evidence necessary to classify a substance as a carcinogen. Four sources of evidence are used: EPIDEMIOLOGY findings, long-term animal testing, short-term tests (such as the AMES TEST), and STRUCTURE-ACTIVITY RELATIONSHIPS. Separate guidelines and policies defining carcinogens have been issued by the USEPA, the Occupational Safety and Health Administration, the Consumer Product Safety Commission, and the Food and Drug Administration. These agencies and the INTERNATIONAL AGENCY FOR RESEARCH ON CANCER classify chemicals by the degree of evidence available as to their carcinogenicity. Some 150–200 chemicals and several physical agents (ionizing radiation, ultraviolet radiation) appear on various lists as having sufficient or limited evidence of carcinogenicity; about 40 of these are classified as human carcinogens. Note that of the roughly 9 million known chemical substances, only about 7000 have been tested for carcinogenicity. Carcinogens vary in their estimated ability to induce cancer. See CANCER POTENCY FACTOR, NATIONAL TOXICOLOGY PROGRAM.

Carcinogen Assessment Group (CAG) A section of the USEPA that uses risk extrapolation models to produce quantitative estimates of the risk of exposure to environmental CARCINOGENS. Risk extrapolation models use data from animal exposure to high doses of a chemical to derive the human risk of exposure to low doses of the chemical. The CAG is responsible for many UNIT RISK ESTIMATES, which result from risk extrapolation.

carcinogenesis The generation of cancer. Most cancers are derived from changes in the deoxyribonucleic acid (DNA) or gene structure of a single cell (mutations). These changes are then passed on by inheritance mechanisms to the cells that descend from this single cell. The types of changes can be categorized into three primary areas: (1) those caused by chemical agents that bring about changes in the chemical structure of DNA, (2) those caused by ionizing radiation that result in changes in cellular chromosomes, and (3) those caused by viruses that introduce new DNA into cells. A single change produced by any of the three general mechanisms is not sufficient to produce a cancer cell. Several random changes are needed to convert cells from the normal to the cancerous state. As a consequence, cancer develops slowly from a population of mildly irregular cells. The agent or chemical that causes the first heritable change in the gene structure of the cell is termed the initiator (it is usually a mutagen). After the action of the initiator, several exposures to other agents, termed promoters (which are not necessarily mutagens), are required to change one of the descendants of the first irregular cell to a tumor cell. The initiators and promoters must act in the proper sequence and with proper timing to bring about the complete transformation of a normal cell to a tumor cell. The identification of initiators and promoters is not always possible with any degree of certainty. For example, agents such as tobacco smoke appear to act as mutagens and initiate various tumors, and reproductive hormones seem to function as promoters and stimulate the continued development of other tumors. The

environment and lifestyle of an individual also play a significant role in the development of most cancers; however, the identification of the specific factors within the environment of an individual that are responsible for tumor development is not always possible. See MUTATION, MUTAGEN, CARCINOGEN, ONCOGENIC, and DEOXYRIBONUCLEIC ACID.

carcinogenic Describing a chemical substance or type of radiation that can cause cancer in exposed animals or humans. See CARCINOGENESIS.

carcinogenic activity indicator (CAI) A numerical cancer potency indicator for chemical substances, calculated by dividing the additional percentage of test animals found to have tumors by the lifetime does of the tested chemical, in molecules per kilogram body weight. This calculation is done for each point on a DOSE-RESPONSE CURVE. CAIs are useful for comparisons of chemicals, such as pesticides, that have possible substitutes. The CAIs reveal the lowest-risk material appropriate for the application.

carcinogenic potency factor (CPF) See CANCER POTENCY FACTOR.

carcinoma A tumor belonging to the most common group of cancers, those arising in epithelial tissues, such as the lining of the lung, gastrointestinal tract, or other organs, such as the skin.

carnivore An animal that eats only meat, such as the wildcat or wolf. Compare HERBIVORE, OMNIVORE.

Carnot engine A theoretical device for converting heat into work (force) in which a piston moves in a cylinder without friction. The movement of the piston is performed by gas compression and expansion, which is caused by a heat source and a HEAT SINK, which absorbs the excess heat produced. The model illustrates that the efficiency of the engine depends on the absolute temperatures of the heat source and the heat sink.

carrier An inert liquid or solid in which an active ingredient or toxic substance is dissolved or suspended. The inert material is not toxic; however, the carrier is used to facilitate application of the agent in the case of a pesticide or to facilitate the entry of the agent into the body in the case of a medication.

carrier gas An inert gas, such as helium, that is employed as a medium for the transport of low concentrations of some active ingredient. The analysis of some chemicals requires that a carrier gas be used to transport the agent in question through the analytical instrument.

carrying capacity (K) In ecology, the maximum number of organisms that an ECOSYSTEM can support over an extended period without significant degradation of the ecosystem.

carryover Liquid particles introduced to a gas stream by the turbulent air-gas contact in a SCRUBBER. Carryover is removed by a MIST ELIMINATOR placed downstream from the scrubbing device. Also called ENTRAINMENT.

Carson, Rachel (1907–64) American Marine biologist and author of *Silent Spring,* the 1962 book on pesticide (especially dichlorodiphenyltrichloroethane [DDT]) threats to the environment that marked the beginning of the political ascendancy of the environmental movement in the United States. Other works include *Under the Sea Wind* (1941), *The Sea around Us* (1951), and *The Edge of the Sea* (1954).

cascade impactor A device used to determine the size distribution of airborne particles. Air is drawn through a series of plates, each plate removing particles of a smaller diameter. The mass of the particles caught on each plate is measured to give a mass size distribution.

case-control study A type of epidemiological (disease-related) study that compares the past exposure of two groups to a particular environmental factor. The cases are a group of persons identified as having a certain disease, and the controls are members of a group similar to the cases in terms of age, sex, race, and other factors who do not have the disease. Any significantly greater exposure to the factor in cases compared with the controls may indicate a causative link between the agent and the disease. Also called a retrospective study.

cash crop Plants or plant products from a farm produced for the purpose of generating income. For example, a farm may be involved in soybean cultivation to provide a monetary income (the cash crop) and hay to use on site to feed the farm animals (not a cash crop).

cash-out A type of MIXED FUNDING AGREEMENT in which the USEPA manages the cleanup of the waste site and a POTENTIALLY RESPONSIBLE PARTY is not involved other than contributing its share of the cost.

casing In water or oil and gas wells, a solid steel or polyvinyl chloride (PVC) pipe defining the outer diameter of the well. Other pipe may be run inside the casing.

cast-iron pipe (CI pipe) The type of pipe used in household water systems before copper and polyvinyl chloride (PVC) pipes were introduced.

catabolism The biological breakdown of materials into their simpler components; decomposition. Performed by DECOMPOSER organisms, mainly bacteria and fungi. Compare ANABOLISM.

catalysis The operation of a CATALYST to speed up a chemical reaction.

catalyst A chemical substance that allows a chemical reaction to take place more readily, for example, at a lower temperature or pressure, without itself being consumed or altered by the reaction. An ENZYME is a biological catalyst, which enhances reactions within living organisms.

catalytic converter An air pollution control device, installed in an automobile exhaust system, that reduces the levels of HYDROCARBONS and CARBON MONOXIDE exiting the tailpipe. The CATALYST enhances the OXIDATION of hydrocarbons to carbon dioxide and water vapor and the oxidation of carbon monoxide to carbon dioxide. THREE-WAY CATALYSTS also include in the device a reducing catalyst that converts NITRIC OXIDE and NITROGEN DIOXIDE to nitrogen gas.

catalytic cracking A process for converting high-molecular-weight hydrocarbons found in crude oil to smaller hydrocarbon molecules by heating in the presence of a metal compound that serves as a catalyst (a substance that speeds up the rate of a chemical reaction without entering into the reaction itself). Through the use of a catalyst, the cracking process can be done at lower temperatures and pressures and with greater control than would be possible otherwise. The process increases the amount of gasoline and other light hydrocarbons that can be produced from heavy crude oil.

catalytic incinerator An air pollution control device that oxidizes or degrades VOLATILE ORGANIC COMPOUNDS by using a CATALYST to promote the combustion process. This type of incinerator can operate at lower temperatures than the standard combustion incinerators, thereby lowering fuel and operating costs.

catastrophic system Used to describe a biological system that unexpectedly loses stability suddenly. The sudden demise of a mature forest, the crash of a fishery resource, and a rapid explosion in the numbers of an insect pest are common examples of such population shifts.

catchment area The area that draws surface runoff from precipitation into a stream or urban storm drain system; synonymous with WATERSHED.

categorical exclusion Types of actions determined by federal agencies to have no significant environmental impacts and therefore requiring no ENVIRONMENTAL ASSESSMENT or ENVIRONMENTAL IMPACT STATEMENT under the NATIONAL ENVIRONMENTAL POLICY ACT.

categorical pretreatment standard A technology-based standard of treatment of the processing of wastewater at an industrial facility before the water is released into a public sewer system. The purpose is to require the adoption of the best technology available for such activity. See PRETREATMENT.

categorical variable A qualitative variable created by classifying observations into categories. For example, a series of temperature measurements could be classified into the categorical variables low, normal, and high, with low defined as less than 10° C, normal between 10° C and 30° C, and high greater than 30° C. Many statistical techniques are inappropriate for categorical variables: an average, for example. Compare QUANTITATIVE VARIABLE.

cathode The positive pole of an ELECTROLYTIC cell or a battery. When the battery is connected in a circuit, electrons flow from the ANODE to the cathode.

cathodic protection A method to protect iron or steel tanks, pipes, or other structures from corrosion, to prevent leaks and spills. Iron or steel corrosion is caused by OXIDATION, which is a loss of electrons. A metal like zinc or magnesium is connected to serve as a source of replacement electrons for those lost by the iron or steel of the tank or pipe. The zinc or magnesium ANODE is oxidized instead of the iron or steel CATHODE.

cation A positively charged ION.

cation exchange The displacement of one CATION for another, often on the surface of a soil or clay particle. Cations of sodium or potassium may be ADSORBED to clay; if groundwater containing other metal cations (like lead, cadmium, or zinc) flows over the clay, the metal cations displace the sodium and potassium ions, releasing them to the groundwater but removing the other metals from it. This natural cleansing process is important in assessing the hazard posed by a leaking waste site. See CATION EXCHANGE CAPACITY.

cation exchange capacity (CEC) The amount of the positively charged ions (cations) of calcium, magnesium, potassium, sodium, hydrogen, or aluminum that soil or sediment can hold. The value is expressed in milliequivalents per 100 grams of soil. The larger the number, the more cations the soil or sediment can hold. Generally, the higher the content of clay and organic material, the greater will be the CEC and the greater the potential to hold plant nutrients and metal pollutants. See CATION EXCHANGE.

caustic Alkaline, BASIC. Also used to refer to the heavily used industrial chemical sodium hydroxide, which is a strong BASE or source of OH^- ions.

cavitation The generation and subsequent collapse of bubbles on the surface of an impeller or the gate of a valve. The processes can cause pitting or damage to the surface of these structures.

cell The unit of biological structure and function that can stand alone and carry out all the fundamental life processes. A cell is an independent unit that can obtain nutrients from the environment, derive energy from organic materials, reproduce exact copies of itself, and release waste products into the surrounding environment. No structure derived from a cell, such as the nucleus, mitochondria, or chloroplast, can carry out all of the life functions in an independent fashion. Viruses are life-forms that are simpler than cells in structure and function; however, they are not capable of independent metabolism and cannot reproduce unless a cell is involved. All cells are bound by a plasma membrane, composed of lipids and proteins, and nucleic acids (deoxyribonu-

cleic acid [DNA]) necessary for inheritance functions. Bacteria and simple plants and animals consist of only one cell, which is capable of independent existence. In higher plants and animals, similar cell types form tissues, which in turn form organs.

cell, landfill A completed waste storage area in a landfill, separated from other cells by COVER material.

cell culture The growing of animal or plant tissue in an artificial medium contained within a test tube or bottle.

cellular respiration 1. As a general term, used to indicate the sum of the processes that a biological cell employs to obtain chemical energy from organic nutrients such as sugar and the utilization of chemical energy to carry out the life functions of the cell. 2. As a specific term, the transfer of electrons from a substrate that is oxidized during cellular metabolism to the ultimate acceptor of those electrons employing a series of oxidation-reduction reactions carried out by the cytochrome system. High-energy intermediates are produced during the process. AEROBIC organisms use molecular oxygen (O_2) as the final acceptor of electrons producing water (H_2O). Some ANAEROBIC bacteria that use a respiration system employ a substitute for molecular oxygen such as nitrate (NO_3^-), sulfate (SO_4^{2-}), or carbonate (CO_3^{2-}), producing nitrous oxide (N_2O), hydrogen sulfide (H_2S), or methane (CH_4), respectively.

cementing Pumping of a cement SLURRY into a well or behind the casing of a well.

centerband frequency See OCTAVE BANDS.

centerline concentration In estimating the off-site impact of an air pollution source using a GAUSSIAN PLUME MODEL, the model's estimated concentrations along a line directly downwind from the source; estimated levels decrease with increasing distance on both sides of the centerline concentration.

Centers for Disease Control (CDC) An agency of the Public Health Service, United States Department of Health and Human Services. The lead federal agency in developing disease prevention and control, environmental health, health promotion, and health education as well as surveillance of disease patterns through epidemiological (disease-related) data collection, analysis, and distribution. CDC publishes and distributes *Morbidity and Mortality Weekly Report,* which is a major source of public health information. The agency is headquartered in Atlanta, Georgia, and was formerly known as the Communicable Disease Center. Website: www.cdc.gov

centistoke A unit of KINEMATIC VIS-COSITY equal to 1×10^{-6} square meter per second.

central collection point Site used to collect waste generated at sundry locations. The collected waste is then transported for the prescribed treatment prior to disposal. Examples of this type of system include medical waste collection from locations throughout a city, community hazardous waste collection, and a central drop-off point for bulky waste.

central nervous system The brain, spinal cord, and associated nerves. The functioning of this system is disrupted by many pesticides, chlorinated organic compounds, and other toxic contaminants added to the environment

centrifugal collector An air pollution control device that collects particles in an airstream by using CENTRIFUGAL FORCE. See CYCLONE.

centrifugal fan A fan commonly used in industrial exhaust ventilation systems. The fan blades turn inside a housing, as in vacuum cleaners and hair driers.

centrifugal force An apparent force on a rotating object, directed outward from the center of rotation.

centrifugal pump A device that moves water with a rotating impeller (a series of metal vanes) surrounded by a casing. This compact, simple pump is the most common type used in water distribution and treatment.

centrifugation The separation of materials of different densities in a CENTRIFUGE.

centrifuge A device that employs CEN-TRIFUGAL FORCE to separate a mixture into components by their relative densities, especially to separate SUSPENDED SOLIDS from liquids.

CERES principles See COALITION FOR ENVIRONMENTALLY RESPONSIBLE ECONOMIES.

cesium-137 (^{137}Cs) One of the important RADIOACTIVE alkali metals. A common product of nuclear weapon explosions and nuclear reactors. In humans, the isotope is absorbed rapidly and is distributed throughout the body, where it enters into reactions that normally involve potassium. Both cesium-137 and its decay products release energetic BETA and GAMMA RADIATION, which can cause significant whole-body radiation damage. Deaths resulting from acute exposures are usually due to dysfunction of bone marrow.

cesspool An underground cistern used for sewage disposal in areas not served by a community sewage collection system. This disposal method is generally prohibited in the United States.

chain of custody The documented transfer(s) of an environmental sample from the person doing the collecting to any transporters until it reaches the person performing any required analysis, to ensure that no contamination or substitution occurs.

chain reaction In nuclear processes, the sequence of atomic FISSION created by a neutron impacting an unstable nucleus, causing fission, which produces additional neutrons, which can impact other unstable nuclei, causing more fission, and so on. The energy that is released through a controlled chain reaction is captured and converted to electricity by a nuclear power station. See NUCLEAR REACTOR.

change of state The change of a chemical substance from one physical state (solid, liquid, or gas) to another as a result of temperature or pressure changes; in the environment, the change is usually caused by a temperature change. The changes of state of water and the associated heat transfers are especially important in environmental heat regulation, including the moderation of ecosystem weather extremes and, on a larger scale, the movement of heat poleward at the surface of the Earth. A change of state in water is also vital for human body heat regulation and removal of heat from industrial processes.

channelization Flood control or navigation projects that straighten, widen, or deepen surface water channels such as streams, rivers, or canals. Adverse environmental consequences can include increased SEDIMENTATION, bank erosion, increased flooding, decrease in BIOMASS, and (surface) SALTWATER INTRUSION.

chaotic system An environment that exhibits variability that precludes predictions on a short time scale with any degree of certainty. Environments in which the populations of some organisms fluctuate in ways that are not explainable through routine mathematical predictions.

chaparral A BIOME with hot, dry summers and rainfall mainly in the winter months. Vegetation consists of shrubs and small evergreen trees. Chaparral communities are found around the Mediterranean Sea, in central and southern California, in coastal Chile, in southern Australia, and in southern Africa.

characteristic hazardous waste Materials defined as hazardous waste by their possession of one or more of the following characteristics, as defined in regulations issued by the USEPA: IGNITABILITY, CORROSIVITY, REACTIVITY, or TOXICITY. Included in the *code of federal regulations*, Title 40, Part 261, Subpart C. See LISTED HAZARDOUS WASTE.

characterization of ecological effects The systematic evaluation of the ability of a chemical or other hazard to cause harm in or to disrupt the natural environment. A formal framework for the identification of the precise nature of the hazard, a quantification of the amount of material or extent of exposure, and the characterization of the risks associated with the agent are included in the analysis. See ECOLOGICAL RISK ASSESSMENT.

characterization of exposure Within the systematic evaluation of the ability of a chemical or other hazard to cause harm in or to disrupt the natural environment, the section that evaluates the interaction of the foreign agent with one or more parts of a specified environment. See ECOLOGICAL RISK ASSESSMENT.

charcoal filter See CARBON FILTRATION.

charge rate The rate at which material is introduced to an incinerator; typically expressed in pounds per hour.

Charles's law The gas law stating that the volume of a given amount of gas is directly proportional to the absolute temperature of the gas, given a constant pressure. Expressed as $V = TK$, where V is the volume, T is the absolute temperature, and K is a constant. Charles's experimental law can be expressed as $V = V_0 (1 + \alpha v \Theta)$, where V is the volume at temperature Θ in °C, αv is a constant equal to $1/273.15$, and V_0 equals the gas volume at 0° C. Also called the Gay-Lussac law.

check dam A small dam across a stream that decreases stream velocity and thereby reduces erosion of the banks and bottom while increasing sediment deposition.

chelating agent Chemical compound that has the ability to bind strongly with metal ions. Such agents are used to enhance the excretion of various toxic metals. For example, ethylenediaminetetraacetic acid (EDTA) is administered to treat lead poisoning. The EDTA chemically binds to the lead and the lead-EDTA complex is excreted.

Chemical Abstracts Service Registry Number (CAS number) A unique number assigned by the Chemical Abstracts Service, a division of the American Chemical Society, to each chemical substance and used for positive identification. A chemical may be known by several different names, but it has only one CAS number. The USEPA often uses CAS numbers to describe clearly the substances referred to in its regulations.

chemical asphyxiant A gas that deprives the body of proper oxygen absorption, such as carbon monoxide, or a chemical material that prevents oxygen use at the cellular level, such as cyanide. See CYANOSIS and SIMPLE ASPHYXIANT.

chemical bond The forces that hold atoms together. The two chief types of chemical bonds are the ionic bond and the covalent bond. The ionic bond results form the transfer of electrons from one atom to another as occurs in many inorganic salts such as sodium chloride, common table salt. In some cases, ionic bonds are easily disrupted by dissolving the salt in water. The covalent bond results from the sharing of electrons between atoms as occurs in the water molecule or in most organic molecules. The covalent bond is much more difficult to disrupt.

chemical builders Inorganic phosphate compounds (such as tripolyphosphate) added to detergents to facilitate the formation of complexes of calcium or magnesium ions that may be present in water supplies. The binding of the ions with the inorganic phosphate compound reduces the precipitation (solidification) of the detergent molecules in HARD WATER.

chemical case Classification used for the purposes of review and regulation of pesticides. Those commercial preparations that have the same active ingredient, or the same active ingredient with a slight modification, are grouped for examination in the licensing process.

chemical emergency The actual or threatened release of a toxic chemical in amounts that represent an immediate acute threat to human health or to the environment.

chemical energy The potential energy residing in the bonds of chemical compounds. For example, solar energy is converted by photosynthesis to chemical energy in the form of carbohydrates. The chemical energy originating in the Sun is stored in the bonds holding the carbohydrate molecule together.

chemical equilibrium The condition in which chemical reactions take place equally in the forward and reverse directions. For $A + B \rightleftarrows C + D$, the concentrations of A, B, C, and D are constant.

Chemical Hazard Information Profiles **(CHIPs)** A publication of the USEPA, Office of Toxic Substances, containing, for a particular chemical, estimates of occupational, consumer, and environmental exposure; human health and environmental effects; and pertinent standards and regulations. CHIPs are available for more than 200 chemicals.

Chemical Hazard Response Information System (CHRIS) A set of manuals used for assessing the health, safety, and environmental hazards posed by chemical releases, emphasizing spills to surface water. The manuals were developed for the United States Coast Guard. The computer model version of CHRIS is the Hazard Assessment Computer System (HACS).

Chemical Manufacturers Association (CMA) A 190-member group of chemical producers that supports research, workshops, and technical symposia on the environment, health, and safety of chemical manufacturing and distribution. Operates the CHEMICAL TRANSPORTATION EMERGENCY CENTER (CHEMTREC) to support responses to chemical transportation accidents. Adoption and implementation of RESPONSIBLE CARE, a management initiative intended to improve the health, safety, environmental performance, and community outreach of member companies, are required for membership in the organization. The name of the organization has been changed to the American Chemistry Council. Website: www.cmahq.com

chemical oxygen demand (COD) A chemical measure of the amount of ORGANIC substances in water or wastewater. A strong oxidizing agent, acid, and heat are used to OXIDIZE all carbon compounds in a sample. Nonbiodegradable and recalcitrant (slowly degrading) compounds, which are not detected by the test for BIOCHEMICAL OXYGEN DEMAND, are included in the analysis. The actual laboratory measurement involves a determination of the amount of oxidizing agent (typically, potassium dichromate) that is reduced during the reaction.

chemical stressors Those chemicals, released into the environment through the actions of humans, capable of causing adverse reactions among plants and animals.

chemical toilet A toilet facility in which waste is retained and chemicals, such as lime, are added to disinfect the sewage and/or to control odors.

Chemical Transportation Emergency Center (CHEMTREC) A service operated by the CHEMICAL MANUFACTURERS ASSOCIATION that provides timely information about chemicals involved in a hazardous material transportation incident. CHEMTREC personnel can send detailed information to FIRST RESPONDERS via the HAZARD INFORMATION TRANSMISSION (HIT) system and, if necessary, will contact the producer of the

material(s) for additional guidance. The 24-hour telephone number of the center is 800-424-9300.

chemical treatment General term used for the addition of a chemical agent to a waste, commonly wastewater, to remove, inactivate, precipitate, or otherwise render less harmful some dangerous substance or organism, for example, the addition of lime to wastewater to remove phosphates by precipitation as calcium phosphate.

chemical weathering The gradual decomposition of rock by exposure to water, atmospheric oxygen, and carbon dioxide.

chemigation The application of agricultural chemicals (fertilizer, herbicide, pesticide) via irrigation water.

chemiluminescent detector An analytical instrument that measures the level of a particular wavelength of light emitted by the chemical reaction occurring when a gaseous compound is introduced to a sample of ambient air. The intensity of the light emission is directly proportional to the concentration of a particular gas in the air sample and can be calibrated to determine the gas concentration. This method is used for the determination of OZONE and NITROGEN DIOXIDE levels in the atmosphere.

ChemNet An Internet-based organization providing a vehicle for communication among the buyers and sellers of the chemical and pharmaceutical industries. Information is available concerning markets and supplies of frequently traded chemicals and pharmaceuticals, trading contacts and partners, trade shows and conferences, visibility for small companies, and a vehicle for advertisement. Website: www.chemnet.com

chemoautotrophs Microorganisms that derive biologically useful energy from the oxidation of inorganic chemical substances, usually ammonia, sulfur, nitrite, and ferrous iron. The bacteria that oxidize iron sulfide to iron sulfate and sulfuric acid, the biological activity responsible for ACID MINE DRAINAGE.

chemodynamics The study of the transport, conversion, and fate of chemical substances in air, water, or soil, including their movement from one medium to another.

chemosphere The region of the upper atmosphere including the MESOSPHERE and upper STRATOSPHERE in which various sunlight-driven chemical reactions occur. See ATMOSPHERE.

chemosynthesis A type of metabolism characteristic of some bacteria. The organism can satisfy carbon requirements for the utilization of carbon dioxide and obtain energy from the oxidation of inorganic compounds. Examples include those bacteria that subsist through the oxidation of hydrogen, sulfur, ammonia, or iron to obtain the energy needed for metabolism. See AUTOTROPH. Contrast PHOTOSYNTHESIS.

chemotrophs See CHEMOAUTOTROPHS.

Chernobyl Ukraine site of the April 1986 nuclear power plant disaster. Gross operator errors led to an explosion and fire and continentwide radioactive contamination. From 50 to 100 million CURIES are estimated to have been released, with yet-unknown adverse human health consequences.

child-resistant packaging (CPR) Containers that protect children (or adults) from injury or harm resulting from the inadvertent contact with or consumption of some dangerous material. The term is most frequently applied to the packaging of medicines and pesticides for distribution to the public.

chimera An organism that contains genetic information from two different species. Such organisms are produced through GENETIC ENGINEERING technol-ogy. Most commonly used to refer to mammal species that have been produced by the fusion of cells from two different species prior to fetal development.

China syndrome A popular but misleading term for the potential catastrophic result of a nuclear core overheating to the extent that the floor of the containment vessel melts and the now-molten core sinks toward the other side of the Earth (toward China, for a reactor in the United States). Actually, the molten core would stop sinking a few meters below the surface. See MELTDOWN.

chisel plowing A type of reduced-tillage farming requiring minimal disturbance of the ground before planting of crops. A chisel plow, a curved blade resembling a chisel, is used to gouge a trench in the earth into which seeds are planted. This technique leaves the great majority of plant residues from previous crops on the surface to help prevent both erosion and the growth of grass. Chisel plowing differs from traditional methods using the moldboard plow, which turns all of the soil, burying most if not all of the crop residues from previous crops.

chi-square test A method used to determine whether STATISTICALLY SIGNIFICANT differences exist in frequency data from separate groups, such as whether the number of persons in an exposed group who exhibit a certain adverse health effect is statistically different from the number of persons in an unexposed (control) group with the adverse effect.

chloracne An acnelike eruption of the skin caused by exposure to certain chlorinated aromatic organic compounds.

chlor-alkali Describing an industrial facility that produces CHLORINE gas and sodium hydroxide (caustic) by passing an electric current through salt water. Some chlor-alkali plants have used mercury cells, from which mercury can be emitted to the environment, in this process.

chloramines Compounds containing nitrogen, hydrogen, and chlorine, formed by the reaction between hypochlorous acid (HOCl) and ammonia (NH₃) and/or organic amines in water. The formation of chloramines in drinking water treatment extends the disinfecting power of chlorine. Also called COMBINED AVAILABLE CHLORINE.

chloramine-T A chlorine-containing material used to add chlorine to drinking water.

Chlordane A CHLORINATED HYDROCARBON insecticide chemically related to ALDRIN and ENDRIN. The agent was widely used to treat soil around homes for termite control because it is persistent (long-lasting) in the environment. The agent is moderately toxic to mammals and has been shown to cause cancer and to alter fertility in some. Currently, the use of this pesticide is severely limited in the United States.

chlorides Negative chlorine ions, Cl⁻, found naturally in some surface waters and groundwaters and in high concentrations in seawater. Higher-than-normal chloride concentrations in freshwater, due to sodium chloride (table salt) that is used on foods and present in body wastes, can indicate sewage pollution. The use of highway deicing salts can also introduce chlorides to surface water or groundwater. Elevated groundwater chloride levels in drinking water wells near coastlines may indicate SALTWATER INTRUSION.

chlorinated 1. Describing an organic compound to which atoms of chlorine have been added. 2. Describing water or wastewater that has been treated with either chlorine gas or a chlorine-containing compound.

chlorinated dibenzofurans (CDBF) A class of highly toxic aromatic halogenated hydrocarbons resembling the chlorophenols and dioxins.

chlorinated hydrocarbons Synthetic chemical substances containing chlorine, hydrogen, and carbon. The addition of chlorine to ORGANIC compounds causes these materials to break down slowly in the environment. The chlorinated hydrocarbons are also a class of insecticides, which includes dichlorodiphenyltrichloroethane (DDT), mirex, ALDRIN, KEPONE, LINDANE, heptachlor, toxaphene, and many others.

chlorinated organics Synonym for CHLORINATED HYDROCARBONS.

chlorinated solvent A liquid material usually consisting of an organic hydrocarbon to which chlorine has been added. Examples include carbon tetrachloride, methylene chloride, and trichloromethane. This class of solvent is especially useful for removal of oil and grease from fabrics and metals. Likewise, they pose a significant health risk to those individuals who are exposed to these agents because of the ability of solvents to dissolve fatty material, causing damage to cells. Excessive exposure to high concentrations can cause central nervous system depression, and some chlorinated solvents are listed as actual or possible human carcinogens. See SOLVENT.

chlorination The process of adding chlorine to water or wastewater in order to kill or inactivate dangerous microorganisms or viruses. Chlorine in various forms, such as chlorine gas, bleach, or solid chlorine-containing compounds, can be used.

chlorination, breakpoint See BREAKPOINT CHLORINATION.

chlorination, combined residual See COMBINED RESIDUAL CHLORINATION.

chlorine (Cl₂) One of a group of elements classified as the halogens. Chlorine, the most common halogen, is a greenish yellow gas with an irritating odor. Chlorine is very reactive: it forms salts with metals, forms acids when dissolved in water, and chemically combines readily with hydrocarbons. Various forms of chlorine are used to disinfect water. CHLORINATED HYDROCARBONS are used widely as PESTICIDES and industrial chemicals. Freon, a CHLOROFLUOROCARBON, is a synthetic material containing chlorine. Chlorine is produced by the electrolysis of brine (concentrated salt solution).

chlorine, combined available See COMBINED AVAILABLE CHLORINE.

chlorine, free available See FREE AVAILABLE CHLORINE.

chlorine, free residual See FREE RESIDUAL CHLORINE.

chlorine-contact chamber Apparatus used in the disinfection of domestic wastewater after treatment and just before release. The chamber assures vigorous mixing of the agent used to add chlorine (commonly chlorine gas) with the treated wastewater to ensure mixing of the chlorine with the water so that dangerous microorganisms of fecal origin are killed or inactivated. The process is intended to protect public health. See CHLORINATION.

chlorine demand The amount of chlorine that must be added to purify drinking water; the amount required to react with all dissolved and particulate materials and inorganic ammonia in the water.

chlorine residual Chlorine added to drinking water in excess of the amount needed to react with organic and inorganic materials suspended or dissolved in the water. This chlorine is available to eliminate microorganisms that enter the water distribution system after treatment.

chlorite A chemical formed when chlorine dioxide is used to remove PATHOGENS from drinking water. One of the HALOACETIC ACIDS regulated as a DISINFECTION BY-PRODUCT under the SAFE DRINKING WATER ACT.

chlorofluorocarbons (CFCs) A class of simple hydrocarbon derivatives in which chlorine and fluorine are substituted for some or all of the hydrogens (e.g., CCl_2F_2). They are commonly called freons. As a group, these compounds are VOLATILE, nonreactive, noncorrosive, and nonflammable. They have been used widely in consumer products (propellants in aerosol sprays and coolants in refrigerators and air conditioners) and in industrial applications (electronics and blown Styrofoam manufacture). However, the chlorofluorocarbons have been implicated in the reduction of the ozone content of the stratosphere (OZONE LAYER) and are GREENHOUSE GASES: that is, they absorb outgoing infrared radiation from the Earth. For these reasons their use is being phased out around the world. See OZONE LAYER DEPLETION, GREENHOUSE EFFECT.

chloroform A simple halogenated hydrocarbon ($CHCl_3$) obtained from the chlorination of methane (CH_4). Once used in human anesthesia, chloroform remains an important industrial chemical. It is one of the more common halomethanes produced during the chlorination of water. Low concentrations in drinking water promote kidney and liver damage in animals.

chlorophenols See PENTACHLOROPHENOL.

chlorophyll The green pigment in plants that absorbs a portion of incoming sunlight for use in PHOTOSYNTHESIS.

chloroplasts The structures within a plant cell that contain CHLOROPHYLL.

chlorosis The yellowing of plant leaves, indicating loss of CHLOROPHYLL or a reduction in the number of CHLOROPLASTS. The condition can be caused by a deficiency of iron, magnesium, sulfur, or nitrogen; disease; insufficient sunlight; or certain air pollutants. Air pollutants that can cause chlorosis to susceptible vegetation include sulfur dioxide, fluorides, and ethylene.

cholera An infectious waterborne disease that is characterized by severe diarrhea and the resultant dehydration and electrolyte imbalance. The disease is caused by bacteria of the genus *Vibrio*. Outbreaks are associated with contamination of surface waters with human FECAL material.

cholinesterase See ACETYLCHOLINE-STERASE, CHOLINESTERASE INHIBITORS.

cholinesterase inhibitors Chemical substances that inactivate the enzyme cholinesterase, resulting in nerve dysfunction. Normal transmittal of nerve impulses across synapses, the junctions connecting nerve fibers to each other and connecting nerves to muscles, is accomplished by the chemical acetylcholine. After acetylcholine moves across the synapse to relay the nerve impulse, cholinesterase breaks down the acetylcholine, which is later re-formed to carry another nerve signal. If acetylcholine is not broken down—that is, if cholinesterase has been inhibited—then the nerve stimulation is excessive and can lead to twitching, convulsions, and death. The nerve gases developed for chemical warfare and their relatives, the ORGANOPHOS-PHATE insecticides, are cholinesterase inhibitors.

chromatogram The pattern formed on or in an ADSORBENT material or on a printed output when closely related chemicals are separated by CHROMATOGRAPHY.

chromatography A process used to separate and/or identify similar compounds by allowing a solution of the compounds to migrate through or along a substance that selectively adsorbs the compounds in such a way that materials are separated into zones. The simplest example of the process is separating the colors in ink by placing a drop of the ink on a napkin and dampening the cloth. As water migrates up the napkin, it moves and separates the different dyes.

chromosomal nondisjunction The failure of CHROMOSOMES to separate during cell division. As a result, daughter cells do not have the necessary number of chromosomes for normal functioning.

chromosome A threadlike structure in the cell nucleus, composed of deoxyribonucleic acid (DNA), proteins and the linearly arranged GENES. Each chromosome contains coiled DNA molecules. The number of chromosomes in the nucleus is characteristic of the species. Humans have 46.

chronic effects In environmental health, long-lasting, usually sublethal adverse impacts an human health associated with the exposure to environmental toxins. Analogous to chronic back pain caused by an injury sustained in childhood.

chronic exposure In toxicology, doses, usually at low concentrations, that extend for long periods, from six months to a lifetime.

chronic toxicity Adverse health effect that either is the result of CHRONIC EXPOSURE or is permanent or long-lasting, as in scarring of lung tissue.

chrysotile In the past, the most widely used form of ASBESTOS in the United States. Also called white asbestos.

cilia Tiny hairlike projections from the surface of some single-celled organisms. Used for movement. See CILIATED MUCOSA, CILIATES.

ciliate The common name of a type of protozoan or single-celled animal that moves with the aid of short, hairlike projections termed cilia. These organisms are important members of the community of organisms that carry out mineralization, the conversion of organic macromolecules to simple inorganic chemicals.

ciliated mucosa The lining of the respiratory tract in which tiny moving hairlike projections called cilia move mucus upward. Inhaled particles are swept out of the lungs if caught by the ciliated mucosa.

cinder cone A distinctive type of small volcano constructed almost entirely from ash and dust blown from a central vent. The internal structure is composed of compacted layers of ash sloping away from the summit.

circle of influence The outer edge delineating the impact of the pumping of water from an aquifer. The boundary around a well at which the water level in the aquifer falls if the pumping rate exceeds the rate at which water can migrate through the geological deposit containing the water. See CONE OF DEPRESSION.

circle of poisons Exposure to a pesticide through the importation of foods contaminated with a pesticide that is manufactured in the United States but cannot legally be used on U.S. crops. The pesticide is exported, used on crops in some foreign country, and transported back into the country on imported foods.

cistern A large tank used to store water, usually rainwater, for later use. The use of open cisterns contributed to the spread of yellow fever because they make ideal systems for the reproduction of mosquitoes. Cisterns have also been linked to the incidence of disorders of the GASTROINTESTINAL TRACT when the stored water is used for drinking.

citizen suit provision A feature of many federal environmental statutes that allows private citizens or organizations to file suits involving enforcement of pollution control regulations, to challenge regulations of the USEPA, or to force the administrator of that agency to perform a nondiscretionary duty, that is, a specific action required of him or her by a federal statute. See also STANDING, LEGAL.

claims-made insurance policy See ENVIRONMENTAL IMPAIRMENT LIABILITY POLICY.

Clapeyron-Clausius equation See CLAUSIUS-CLAPEYRON EQUATION.

clarification The process of removing PARTICULATE MATTER from wastewater. Normally, the water is allowed to stand, thereby facilitating the settling of the particles.

clarifier In a drinking water facility or wastewater treatment plant, the unit that removes suspended PARTICULATE MATTER. See PRIMARY CLARIFIER, SECONDARY CLARIFIER.

Class 1 area A designation under the PREVENTION OF SIGNIFICANT DETERIORATION (PSD) provisions of the CLEAN AIR ACT. Class I areas include national parks, wilderness areas, national monuments, or similar areas of special national significance where existing air quality is protected, even if the air is far cleaner than the levels allowed by the NATIONAL AMBIENT AIR QUALITY STANDARDS (NAAQS).

Class I substance Classification of CHLOROFLUOROCARBONS on the basis of the ozone-depletion potential as authorized under the 1990 amendments to the CLEAN AIR ACT. The chlorofluorocarbons with an ozone-depletion potential of 0.2 (1.0 represents a maximum) or greater were placed on the list of Class I substances. Included were the common freons, CFC-11 and CFC-12, with depletion potentials equal to 1.0; 13 other CFCs; 3 halons; carbon tetrachloride and methyl chloride. Class I substances are being phased out under provisions of the MONTREAL PROTOCOL.

Class I, II, III, IV, and V injection wells Classifications of the USEPA that determine the permit requirements of an INJECTION WELL: Class I, well into which liquid hazardous wastes or other fluids are pumped down, with the fluids injected into an underground formation below the lowest underground source of drinking water that is within a one-quarter-mile radius of the well; Class II, a well used to dispose of fluids produced by oil and gas wells, to introduce fluids for ENHANCED OIL RECOVERY or for liquid hydrocarbon storage; Class III, a well used to pump fluids underground for mineral extraction; Class IV, a well used to reinject treated fluid from a SUPERFUND cleanup site into or above an underground formation within a one-quarter-mile radius of the well; Class V, wells not included in Classes I–IV, mainly shallow industrial disposal wells or RECHARGE wells.

Class II substance   HYDROCHLOROFLUOROCARBONS listed by the 1990 amendments to the CLEAN AIR ACT; scheduled for phaseout under provisions of the MONTREAL PROTOCOL.

classifier, air   See AIR CLASSIFIER.

Clausius-Clapeyron equation   The relationship used to calculate the change in the vapor pressure of a liquid as the temperature changes. The equation, also known as the Clapeyron-Clausius equation, can take the form

$$\ln pv = n\frac{\Delta H_v}{RT} + B$$

where $\ln pv$ is the natural logarithm of the vapor pressure of the liquid, ΔHv is the latent heat of vaporization, R is the universal gas constant, T is the absolute temperature, and B is a material-specific constant.

clay liner   A layer of clay soil that is added to the bottom and sides of a pit designed for use as a disposal site for potentially dangerous wastes. The clay prevents or reduces the migration of liquids from the disposal site.

clay pan   A tightly compacted layer of natural soil that restricts the migration of liquids into the underlying strata.

Clean Air Act (CAA)   The basic federal air pollution control statute. First passed in 1963 after a 1955 federal statute authorizing research and technical assistance. The 1965 and 1967 amendments initiated automobile and stationary source standards. Major amendments in 1970 and 1977 provide for the NATIONAL AMBIENT AIR QUALITY STANDARDS, the STATE IMPLEMENTATION PLAN process, the PREVENTION OF SIGNIFICANT DETERIORATION program, emission standards for automobiles, NATIONAL EMISSION STANDARDS FOR HAZARDOUS AIR POLLUTANTS, and minimal technology standards for new or modified sources (NEW SOURCE PERFORMANCE STANDARDS). The 1990 amendments include provisions for operating permits for stationary sources (TITLE V PERMIT), a phaseout of ozone-layer-depleting chemicals, ACID RAIN controls, tradable emission credits, a system of ranking NONATTAINMENT AREAS by severity of air pollution, stricter auto emission standards, and new AIR TOXICS controls (MAXIMUM ACHIEVABLE CONTROL TECHNOLOGY).

Clean Air Scientific Advisory Committee (CASAC)   A seven-member independent panel established by the CLEAN AIR ACT to review the basis (CRITERIA DOCUMENT) for each NATIONAL AMBIENT AIR QUALITY STANDARD every five years.

clean fuels   Fuels that can be substituted for gasoline, including natural gas, methanol, ethanol, or liquefied petroleum products such as propane and butane. These fuels produce less air pollution than gasoline. See ALTERNATIVE FUELS.

clean oil   See BLACK OIL.

clean room   A room maintained in a dust-free condition. A clean room is used to manufacture electronic components that would be damaged by contamination. Access usually requires special clothing and decontamination.

Clean Water Act (CWA)   The basic federal water pollution control statute. The Water Quality Act of 1965 began setting water quality standards, and the 1966 amendments increased federal funding for sewage treatment plants. The 1972 amendments established a goal of zero toxic discharges and "fishable" and "swimmable" surface waters. Additional amendments were passed in 1977 and

1987. The enforceable provisions included TECHNOLOGY-BASED effluent standards, administered through the NATIONAL POLLUTANT DISCHARGE ELIMINATION SYSTEM, for POINT SOURCES; a CONSTRUCTION GRANTS PROGRAM to build or upgrade municipal sewage treatment plants (now a revolving fund); a regulatory system for spills of oil or hazardous waste; a PRETREATMENT control program; STORM WATER RUNOFF controls; and a wetlands preservation program.

cleaner technologies substitutes assessment A method developed by the USEPA for evaluating the comparative risk, performance cost, and resource conservation aspects of technologies developed as alternatives to chemicals currently used by specific industry segments. The assessment supports cooperative joint efforts among trade associations, businesses, public interest groups, and academia to help certain businesses select environmentally sound products, processes, and technologies.

cleanup In hazardous waste management, the decontamination of water, soil, or an aquifer that is determined to contain concentrations of a leaded or spilled substance that threatens the public health or the environment.

clear cutting The removal of all trees in an area without regard to size or species. The process leaves large tracts of land without substantial vegetation, with a resulting increase in erosion. The practice also allows the planting of economically valuable tree species throughout the cleared tract.

clearwell An underground tank holding treated drinking water before distribution to customers.

climate The weather patterns in a particular region, generalized over a long period.

climate change An alteration of area or worldwide long-term weather patterns, as in the case of GLOBAL WARMING.

Climate Change Detection Project Operated by the United Nations, a project instituted in 1989 by the Commission for Climatology Working Group on Climate Change Detection under the auspices of the WORLD METEOROLOGICAL ORGANIZATION. The objective of the project involves the use of climate information and other data to provide an assessment of the interpretation and applicability of databases for the detection of climate change on global and regional scales. Website: www.wmo.ch/web/wcp/wcdmp/ccd.html

climax The last stage in ecological SUCCESSION. Relatively more stable, with a greater SPECIES DIVERSITY than earlier, nonstable stages.

climax community In a specific habitat, used to describe the biota that remains stable over a long period. An agricultural field that is abandoned becomes overgrown with plants that change over time as the field ages. When those changes cease and the populations of plants and animals remain about the same over a long period, the community is described as having reached the climax stage. See COMMUNITY.

clinker Solid residue formed in an incinerator from various noncombustible materials such as glass or metal.

clone 1. To make many exact copies of something. 2. A single-celled organism that divides to produce a colony of genetically identical offspring (e.g., the growth of bacteria). 3. An individual plant produced from a single somatic (nonsex) cell. 4. An individual animal produced from the transfer of the nucleus from a somatic (nonsex) cell to an egg from which the nucleus has been removed and the subsequent implantation of the modified cell into another animal for gestation. 5. Many copies of a fragment of DNA. See ORGANISMAL CLONING.

cloning See ORGANISMAL CLONING.

closed canopy A forest in which the foliage at the crown or tops of the trees covers at least 20% of the ground area. Such forests often consist of mature trees that have overlapping foliage, reducing the amount of sunlight reaching the forest floor. Compare OPEN CANOPY.

closed-cycle cooling A process in which cooling water used in an industrial process or in the generation of electrical energy is not discharged into receiving streams, where direct discharge can have adverse effects, but is circulated through COOLING TOWERS, ponds, or canals to allow the dissipation of the heat and reuse of the water. Compare OPEN-CYCLE COOLING.

closed loop In wastewater treatment, an enclosed process of recycling, reclaiming, or reusing wastewater for purposes other than drinking.

closed-loop recycling Remanufacture of a discarded item back into the original material. For example, discarded aluminum beverage cans are collected, shredded, melted, and re-formed into new beverage cans. Compare OPEN-LOOP RECYCLING.

closed system In physics, a system that does not exchange matter or energy with the surroundings. In ecology, a system exchanging energy, but not matter, with the surroundings. The planet Earth is a closed system in the ecological sense, absorbing and radiating solar energy, while recycling matter within the biosphere. Compare OPEN SYSTEM.

closed water loop A process in which decontaminated or cooled wastewater is not discharged into a receiving stream but is reused. Any water lost during the process by evaporation or binding with some material is replaced by makeup water. See CLOSED-CYCLE COOLING. Compare with OPEN WATER LOOP.

closure Any action prescribed by regulations implementing the RESOURCE CONSERVATION AND RECOVERY ACT that must be performed at facilities operated to dispose of hazardous waste, industrial nonhazardous waste, or municipal solid waste if waste will no longer be received for treatment or disposal. The actions include, among many other things, the placement of a final cover on the buried waste, the establishment of a long-term groundwater monitoring program, and the filing of a notice in state property records that a waste facility has been closed at the location. The monitoring and property record notice are also termed postclosure actions.

closure plan The written document, for a specific hazardous waste facility, outlining CLOSURE. See POSTCLOSURE PLAN.

cloud feedback The opposing effects of clouds on the Earth's climate. More cloud cover means an increased ALBEDO, which lowers incoming solar energy. Lower incoming energy means lower air temperatures, which reduce evaporation and cloud formation. But more cloud cover also means more absorption of outgoing INFRARED energy and a greater retention of heat in the atmosphere, which increase evaporation and cloud formation. These changes are imperfectly captured by computer simulations of climate change. See GENERAL CIRCULATION MODELS. Also called cloud-climate feedback.

Club of Rome An informal international organization begun in 1968 by Aurelio Peccei, an Italian industrialist. The purpose of the group is to foster understanding of the finite and interdependent nature of the natural resources available in the world. The organization sponsored the Project on the Predicament of Mankind, which depended on computer models to predict future availability of critical resources and which led to the publication of *The Limits to Growth* in 1972 and *Mankind at the Turning Point* in 1974. Subsequently, many other reports on global problems and the future, including energy, waste management, education, and microelectronics, were issued. These books and reports revived and expanded the MALTHUSIAN philosophy beyond agricultural production. Humanity was scheduled to run out of gold by 1981, mercury by 1985, petroleum by 1992, and natural gas by 1993, to mention a few of the dire events predicted, none of which happened. Website: www.clubofrome.org

clustering In EPIDEMIOLOGY, the (actual or perceived) grouping in time or space of cases of a disease. Common causes of disease clusters are often difficult to identify conclusively, especially if the alleged cause is a low level of environmental contamination.

coagulation The grouping together of solids suspended in air or water, resulting in their PRECIPITATION. Coagulation is encouraged in wastewater treatment plants by the addition of ALUM, ferrous sulfate, and other materials. See COLLOID, FLOCCULATION, PRIMARY TREATMENT.

coal A solid fossil fuel found in layers beneath the surface of the Earth. The resource is mined and used primarily as a fuel to generate steam for the production of electricity. Coal is graded on the basis of heat content and classified as ANTHRACITE, BITUMINOUS, SUBBITUMINOUS, or LIGNITE. See SURFACE MINING and RECLAMATION.

coal-cleaning technology The application of physical, biological, or chemical processes to remove from mined coal agents that contribute to air pollution or ash accumulation when coal is use as a fuel. The processes associated with cleaning include milling, dewatering, drying, chemical extraction, floatation, sonic enhancement, and agglomeration. The removal of organic sulfur residues is an important consideration because of the potential for the release of sulfur dioxide when coal is burned as a fuel.

coal gasification The conversion of solid coal to a low-energy gas mixture containing mainly methane, hydrogen, and carbon monoxide. The process is not currently economically competitive with production of NATURAL GAS; however, it provides a potential alternative fuel in the event of a shortage in crude oil supplies.

coal liquefaction The conversion of solid coal to a liquid fuel. The process is not yet economically competitive with PETROLEUM production; however, it provides a potential alternative fuel in the event of a shortage in crude oil supplies in the future.

coal tar A crude mixture of aromatic hydrocarbons produced from the destructive distillation of coal. The mixture is usually a viscous material that can be used as a fuel and as a tar for roads and roofing. It can also be further refined to produce a large array of chemicals, including creosotes, phenols, naphthalenes, and similar aromatic compounds. Products such as dyes, resins, perfumes, and flavoring agents are also prepared from coal tar.

coal-tar creosote The black liquid residue that remains after the processing of the crude mixture produced from the destructive distillation of coal. The processing removes economically useful materials such as phenols, naphthalenes, and other aromatic compounds from coal tar. The creosote is a crude mixture of various heavy oils.

coal washing The use of water to classify coal by size and density, with the removal of rock and some sulfur contaminants.

coal workers' pneumoconiosis (CWP) A chronic lung disease characterized by the accumulation of fibrous connective (scar) tissue in varying degrees of severity (FIBROSIS); the condition is caused by long-term overexposure to coal dust containing significant amounts of FREE SILICA and is also associated with chronic inhalation of coal dust that has very low free-silica content. See SILICOSIS.

Coalition for Environmentally Responsible Economies (CERES) A wide array of religious, environmental, and labor organizations who together promote environmental practices to be implemented by industry. The 10 CERES principles are often set forward as shareholder resolutions in corporation proxy statements.

coarse screen   See BAR RACK.

Coastal Zone Management Act (CZMA)   A 1972 federal law, amended in 1980, that provides guidance and financial assistance to voluntary state and local coastal management programs. Goals of the programs include the protection of natural resources and the management of land development in coastal areas, along shorelines, and on shorelands (extending inland as far as a strong influence on the shore is expected). The state programs established under the act vary widely in their approach and application.

cocarcinogen   In the two-stage model of carcinogenesis, a chemical substance that enhances the INIATION stage by increasing the BIOAVAILABILITY of a CARCINOGEN, decreasing the metabolic detoxification of the carcinogen, or inhibiting deoxyribonucleic acid (DNA) repair, among other actions. Cocarcinogens differ from PROMOTERS, which act after the genotoxic initiation stage is complete to enhance the growth of a converted cell into a tumor.

cochlea   The spiral-shaped section of the inner ear connected to the auditory nerve. The nerve endings within this organ are damaged by excessive noise, with a resulting impairment of hearing.

co-composting   COMPOSTING operation that processes widely different discards into soil conditioner, for example, sewage sludge with municipal solid waste or animal manure with wood chips.

Code of Federal Regulations (CFR)   The annual compilation of all current regulations that have been issued in final form by any federal regulatory agency; the publication is organized by subject titles. Most environmental regulations are found under Title 40, Protection of the Environment. Occupational health and safety regulations are under Title 29, Labor.

codistillation   Evaporation of a chemical that accompanies evaporation of water.

coefficient   1. In mathematics, a number that multiplies another quantity or variable, such as the 7 in 7*c*. 2. In physics, a number or ratio that expresses the relationship between two quantities, given certain conditions.

coefficient of entry   In industrial ventilation, the ratio of the actual airflow rate into a HOOD to the ideal flow rate. The ideal flow rate would result if all hood STATIC PRESSURE were converted to VELOCITY PRESSURE, without losses. The coefficient is commonly calculated by dividing the measured hood velocity pressure by the hood static pressure and taking the square root of the result.

coefficient of haze (Coh)   A measurement of air visibility derived from the darkness of the stain on a white paper tape through which the air has been filtered. The value is usually expressed as the number of Coh units per 1000 linear feet of air, with 1 per 1000 corresponding to clean air on a bright day and 5 per 1000 corresponding to significant visibility reduction due to smoke.

coefficient of permeability   See HYDRAULIC CONDUCTIVITY.

coevolution   Simultaneous EVOLUTION of two or more species of organisms that interact in significant ways, for example, a certain species of flowering plant and a specific species of hummingbird that feeds primarily on the flower of that plant.

cofactor   In biochemical reactions, a small molecular weight substance needed to assist in a reaction promoted by an ENZYME. Many of the vitamins and minerals that are taken as food supplements serve as cofactors in the metabolic reactions catalyzed by protein enzymes. See CATALYST.

coffin   A strong, shielded container used for transporting radioactive materials.

cofire   To burn more than one type of fuel in a boiler, for example, burning coal and scrap tires in the same boiler to generate electrical energy.

cogeneration   The use of steam or heat to process materials and generate electricity. For example, high-pressure steam may be routed through an electricity-generating turbine before application in industrial processes, reducing the electricity demand on a central power station.

cohort   In EPIDEMIOLOGY, a group of people sharing one or more characteristics. A birth cohort consists of all persons born within a certain period, usually a year. A group of persons exposed to similar levels of a toxic substance during a similar period is a cohort.

cohort study   An epidemiological study that follows two groups, one exposed to a suspected disease risk factor, the other not exposed, and compares disease rates.

coke   A solid carbon residue resulting from distillation of coal or petroleum. The product is used as a fuel and as a reducing agent in steel making. The volatile materials emitted to the air by coke ovens used in steel mills include known human carcinogens and are regulated by the USEPA.

cold side   That part of the process in a petrochemical facility requiring the cooling of a gas to facilitate separation and purification of products on the basis of the condensation temperature of the chemical. Generally requires the use of rotating equipment to facilitate the compression and cooling of hot gases produced in other parts of the operation. Contrast HOT SIDE.

cold-side ESP   An ELECTROSTATIC PRECIPITATOR (ESP) designed for flue gases less than 400° F (204° C) and located

downstream from the AIR PREHEATER. The lower flue gas temperature means a smaller air volume and, thus, a smaller, less expensive unit. This type of precipitator is not as effective as a HOT-SIDE ESP on fly ash from low-sulfur coal combustion.

cold temperature CO   A standard for automobile emissions of carbon monoxide (CO) to be met when the exhaust system is at temperatures of 20° and below. At these temperatures, the pollution control system is not as effective in removing CO from the exhaust and emissions are higher than those released when the engine is at operating temperature. See CATALYTIC CONVERTER.

coliform bacteria   Gram-negative, rod-shaped bacteria, including primarily FECAL COLIFORM, found in the digestive tract; other forms are found in soil and water. The presence of coliforms in water, in other liquids, or on surfaces is used to indicate the possible presence of fecal material.

collecting surfaces   The collection electrodes, either tubes or flat plates, that provide the attraction surface area in an ELECTROSTATIC PRECIPITATOR.

collection efficiency   An expression of the performance of air pollution control equipment, or the percentage reduction in pollution concentration entering the device compared with the pollutant concentration in the exhaust air. It is calculated by subtracting the pollutant concentration in the exhaust from the pollutant concentration in the incoming air and dividing by the incoming air concentration. This result is multiplied by 100 for expression as percentage efficiency (by weight).

collection system   1. The underground pipe network that channels domestic sewage to a sewage treatment plant. 2. The underground pipes that capture and transport LEACHATE to the surface for treatment or disposal (LEACHATE COLLECTION SYSTEM).

colloids Particles with diameters of 1–1000 nanometers (10^{-9} meter) dispersed into a gaseous, liquid, or solid medium. Colloidal particles suspended in water or wastewater cannot be removed by filtration or sedimentation unless the process is preceded by COAGULATION to increase the particle size.

colluvial Describing eroded material found at the bottom or on the lower slopes of a hill.

colony count A method for the quantification of bacteria in an environmental sample. A portion of a liquid sample or a dilution thereof is spread across the surface of a suitable solid nutrient medium and allowed to incubate. The number of bacterial colonies that develop on the surface is counted and the necessary mathematical calculations made to compute the number of bacteria per unit volume of the sample. The technique is based on the assumption that one bacterial cell will grow and divide to produce one colony on the surface of the nutrient medium.

colony-forming unit Usually applied to the quantification of fungi in a sample of air, water, soil, or other material obtained from the environment. A specified amount of sample is placed on the surface of a solid medium that will support the growth and development of fungi. The subsequent development of filamentous growth characteristic of fungi is taken to mean that a spore or mycelium fragment was in the environmental sample. Also used when bacteria contained in an environmental sample may be in the form of a many-celled chain or other type of clump. See COLONY COUNT.

colorimetry Methods of chemical analysis in which a change in color and/or color intensity is the indication of the presence, concentration, or both of a particular material. The color change is normally measured with an electronic device.

combined available chlorine Chlorine present in water as CHLORAMINES; produced by COMBINED RESIDUAL CHLORINATION. Compare FREE RESIDUAL CHLORINE.

combined cycle generation A system designed to increase the efficiency of a gas TURBINE. The otherwise-wasted heat energy from the hot gases used to drive the turbine is extracted with heat exchangers and used to produce steam for a conventional steam turbine that can generate electricity.

combined residual chlorination The drinking water treatment method that involves the addition of chlorine to water at levels sufficient to produce, in combination with ammonia and/or organic amines, a COMBINED AVAILABLE CHLORINE residual. This chlorine residual maintains the disinfecting power of a treatment throughout the water distribution system. Another approach to water chlorination is BREAKPOINT CHLORINATION.

combined sewer A water drainage pipeline that receives surface runoff as well as sanitary or industrial wastewater. Combined sewer systems are common in the older cities in the northeastern United States. See COMBINED SEWER OVERFLOW.

combined sewer overflow The release of wastewater from a sewer system that collects and transports both sanitary waste from homes and businesses and storm waters resulting from rain runoff. When the system is not capable of holding and treating large volumes characteristic of heavy rainfall, the water overflows, untreated or only partially treated, into the receiving body of water.

combustible gas indicator (CGI) A safety device used to measure the concentration of a flammable vapor or gas in air. The output is usually presented as the percentage of the concentration of the gas required for burning or exploding in the presence of an ignition source. See LOWER EXPLOSIVE LIMIT.

combustible liquid A liquid with a FLASH POINT above 100° F and below 200° F.

combustible material Any substance that will burn under ordinary circumstances.

combustion A rapid chemical reaction of a fuel with oxygen that produces heat and light. The combustion of carbon fuels (wood, coal, natural gas, petroleum products) produces a mixture of exhaust gases that includes water vapor, carbon dioxide, nitrogen, and oxides of nitrogen.

combustion air Air blown into a furnace to provide oxygen for the combustion of fuel.

combustion products Gases, solids, or other material produced during the burning of some substance.

comfort chart A graph of different combinations of air temperature, RELATIVE HUMIDITY, and air motion showing the percentage of test subjects feeling comfortable under various conditions. The chart includes different comfort zones for summer and winter and is useful for the design and operation of air heating and cooling devices. The chart is applicable only to the culture and climate in which the test subjects' comfort zones were recorded.

commingled recyclables All recyclable items that have been collected in a household and placed in a single container for collection. The items must be separated by the recycling agency before delivery to a facility that converts the used material into a new product.

command and control regulation The use of detailed standards, regulations, permit provisions, penalties, and so forth to meet a legislative mandate. Under this regulatory model, the legislature passes a law delegating to an administrative agency (such as the USEPA) the authority to write and enforce exacting rules for the regulated community to follow. This is the primary approach taken in the United States to solve environmental problems.

command post See INCIDENT COMMAND POST.

commensalism A form of species interaction in which one species is benefited but the other is unaffected; for example, shellfish may provide shelter to other, more mobile organisms and are neither harmed nor benefited in the process.

comment period The time allowed for public comment on regulations proposed by a federal administrative agency in accordance with the provisions of the ADMINISTRATIVE PROCEDURE ACT; usually 60 days after publication in the *federal register*.

commercial solid waste A category of municipal discards produced at wholesale businesses, retail establishments, service industries, office buildings, and multifamily residential structures. Usually collected by private solid waste management companies and disposed of at facilities rated to receive household solid waste.

comminutor A mechanical device that cuts and shreds solids as wastewater enters a treatment plant.

Commission for Environmental Cooperation (CEC) An agreement among the governments of Canada, Mexico, and the United States begun under the North American Agreement for Environmental Cooperation (NAAEC) to establish common ground for environmental problems and to prevent environmental/trade disputes. Website: www.cec.org

common ion effect The decreased solubility of an ionized salt caused by the addition of a chemical that ionizes to form an ion that is the same as one formed by the salt. For example, if chemical AB forms A^+ and B^- ions and chemical BC, which forms C^+ and B^- ions, is added, then B^- is the common ion. The additional

B⁻ ions from chemical CB increase the ion product [A⁺][B⁻] such that it reaches the solubility product (the maximal amount of the two ions that can be present in a solution), after which the chemical AB will precipitate. The effect is important in the removal of ions from solution in water treatment.

Common Sense Initiative A USEPA approach that examines and manages pollution problems by industry sector, not by environmental medium (air, water, land). Environmental protection legislation (CLEAN AIR ACT, CLEAN WATER ACT, RESOURCE CONSERVATION AND RECOVERY ACT) has, in the past, encouraged the medium-specific approach.

communal resource management system Cooperative system or custom (maintained either informally through tradition or formally through a legal system) by which some finite resource of value to a group of people (e.g., water, land, fishery, or forest) is maintained over the long term in a way that ensures the continued availability of the resource. Contrast TRAGEDY OF THE COMMONS.

communicable disease A disease for which the causative pathogenic organism is readily transmitted by person-to-person contact, FOMITES, water, food, or air.

community In ecology, the populations of all plant and animal species present in an ecosystem. Also called the biotic community.

Community Awareness and Emergency Response (CAER) The first Code of Management Practices of the RESPONSIBLE CARE program of the CHEMICAL MANUFACTURERS ASSOCIATION established to encourage, at the local level, planning for emergencies resulting from releases by industrial facilities of hazardous materials that impact nearby communities.

community ecology A specialized branch of the study of the relationships among organisms and between organisms and their surroundings in which all types of organisms, regardless of the kind, type, or species, in a defined geographical region are the subject of investigation. An alternate approach to the study of ecology would be the examination of the relationships governing the activities and properties of a single species in some defined geographical region.

community water system A public water system with 15 or more connections and serving 25 or more year-round residents and thus subject to USEPA regulations enforcing the SAFE DRINKING WATER ACT.

compacted solid waste Solid waste after COMPACTION. The waste may be shredded first, compacted, and formed into bales. See SHREDDING, BALER.

compaction 1. The mechanical volume reduction (increase in density) of solid waste by the application of pressure. 2. The application of pressure to soil or clay, reducing the permeability to liquids.

compaction ratio The ratio of the volume of solid waste before COMPACTION to the reduced volume after compaction.

comparative risk assessment Methodology employing science, policy, and economic analysis as well as stakeholder participation to identify and address the areas of greatest environmental risks and to prioritize environmental problems. The risk ranking produced by this process can be used to achieve the greatest risk reduction (using the criteria used for the ranking) and is touted as a way to synthesize the otherwise medium-specific environmental protection programs driven by the enabling statutes such as the CLEAN AIR ACT and the CLEAN WATER ACT.

compartment A conceptual unit of an ecosystem (e.g., BIOMASS) or the body or an organism (e.g., the liver) in and through which a chemical moves; used in mathematical models. See BOX MODEL, RESIDENCE TIME, STOCKS.

compartment model   See BOX MODEL.

compatibility   A condition describing materials that can be mixed without adverse environmental effects or risks to human health. For example, soft drinks and water are compatible because they can be mixed without adverse consequences, and metallic sodium and water are not compatible because mixing the two promotes the release of hydrogen, which frequently leads to fires.

compensation depth   See COMPENSATION POINT.

compensation point   The point under water at which plant PHOTOSYNTHESIS just equals RESPIRATION. The water depth defines the lower boundary, where the rate of oxygen production by photosynthesis equals the rate of oxygen consumption by respiration. Also called compensation depth.

competing risks   Potential adverse outcomes that become more probable as one reduces another, indirectly related risk. For example, a successful program to encourage bicycle commuting would reduce the risk from air pollution–related human health effects and the environmental risks attendant on GLOBAL WARMING but would increase the risk of injuries and death from bicycle accidents.

competition   In ecology, the interaction among species or individuals of the same species in which they struggle to obtain the same food, space, or other essentials.

competitive exclusion   The hypothesis stating that when organisms of different species compete for the same resources in the same habitat, one species will commonly be more successful in this competition and exclude the second from the habitat.

complete carcinogen   In the two-stage model of carcinogenesis, an exposure containing materials that act as INITIATORS and PROMOTERS. Cigarette smoke is usually considered to be a complete carcinogen.

completed test   The third, and last, part of the examination of water for the presence of bacteria of fecal origin. Cultures that are scored as positive in the earlier steps of the analysis (CONFIRMED TEST) are subjected to a verification by inoculating appropriate media (eosin methylene blue agar plates) and performing GRAM-POSITIVE/GRAM-NEGATIVE staining of isolated colonies. See also PRESUMPTIVE TEST.

complex   The incorporation into or combination of CATIONS with other molecules in such a way that the cations are no longer available to enter into reactions with other charged molecules.

complex terrain   Land in the vicinity of an air pollution source with an elevation greater than the source's stack height. Complex terrain must be taken into account when using AIR QUALITY DISPERSION MODELS to estimate the impact of a source on nearby RECEPTORS.

compliance assurance monitoring (CAM)   Monitoring required by provisions of the CLEAN AIR ACT. This monitoring is intended to demonstrate that a major source of air pollution is operating within the emission limits of its TITLE V PERMIT.

compliance coal   Coal that is sufficiently low in sulfur content that less than 1.2 pounds of SULFUR DIOXIDE is produced per 1 million BTU of coal heat input. Also known as low-sulfur coal or clean coal.

compliance coating   Coating, such as paint, that conforms to regulations that address the content of VOLATILE ORGANIC COMPOUNDS. These volatile constituents are often used as solvents and contribute to air pollution as they enter the atmosphere after application to a surface.

compliance cycle   A system organized and directed by the USEPA for stabilizing the monitoring of water supply systems for adherence to drinking water standards. The frequency with which each potential pollutant must be monitored by drinking water systems varies from once every three months to yearly to once per three-year period to once every nine years. To establish order in the schedule, January 1, 1993, was selected as a starting date and each facility was to proceed through three three-year cycles to complete all monitoring requirements. Then the cycle was to be repeated after nine years.

compliance monitoring Collecting, verifying, and evaluating data on air or water quality to determine whether the concentrations of pollutants are below those levels specified by operating permits of a facility. Facilities must participate in a self-monitoring program and supply the results to regulatory agencies.

compliance monitoring program The extensive follow-up groundwater monitoring required at a TREATMENT, STORAGE, OR DISPOSAL facility if the DETECTION MONITORING PROGRAM indicates a possible leak from the hazardous waste at the site. Data from UPGRADIENT WELLS are compared with data from DOWNGRADIENT WELLS for specific chemicals to help determine the source and extent of any groundwater combination.

compliance order/action An official communication from an administrative agency that requires a pollution source to conform to environmental protection regulations by following a COMPLIANCE SCHEDULE. Usually issued after a NOTICE OF VIOLATION.

compliance point The physical location with respect to a hazardous waste secure landfill from which groundwater samples will be taken to determine compliance with the USEPA groundwater protection standard.

compliance schedule An agreed date or series of milestone dates for a facility to reduce emissions or otherwise be in compliance with environmental regulations; negotiated with the facility by a state environmental protection agency or the USEPA. See COMPLIANCE ORDER.

component separation The separation of municipal solid waste into categories including newsprint, white paper, cardboard, plastic, food waste, glass, ferrous metals, aluminum, yard waste, leather, and rubber products.

composite liquid waste sampler (COLIWASA) A device used to collect samples from containers such as 55-gallon drums, which may have several layers of liquids. The device obtains a representative sample of the entire column of liquid, ensuring that each layer is analyzed.

composite sample A representative water or wastewater sample made up of individual smaller samples taken at intervals.

composite volcano The most common type of continental volcano. Characterized by high, steep cones composed of layers of thick silica-rich lava alternating with layers of ash and dust. A depression at the summit usually indicates the position of the vent. Eruptions of composite volcanoes can be extremely violent because of the thick, gas-rich nature of the underlying magma, which allows for the buildup of high pressures.

compost The material produced by COMPOSTING, useful as a soil conditioner.

composting The controlled degradation by AEROBIC microorganisms of organic materials in solid waste to produce COMPOST, a soil conditioner and fertilizer. Wetted solid waste is stacked in piles or rows, which are periodically turned to ensure that sufficient oxygen is present for the DECOMPOSERS. The process is conducted on scales ranging from backyard heaps to tractor-using operations at municipal solid waste processing facilities.

composting facilities A plant location where suitable organic wastes are ground or shredded, mixed with a bulking agent, and allowed to decompose in the presence of air by using static piles or mechanical tumbling until the residue degrades into a humuslike soil conditioner. See COMPOSTING.

compound A substance made up of two or more ELEMENTS in a fixed proportion by weight. The various elements can be separated only by chemical reactions, not by physical means. The physical and chemical properties of a compound are a result of the chemical combination of the elements and are not those of the individual elements. Compare MIXTURE.

comprehensive assessment information rule (CAIR) A chemical substance regulatory program implementing the TOXIC SUBSTANCES CONTROL ACT. The act requires manufacturers to submit detailed information on a chemical, including data about its potential environmental fate (i.e., persistence, transport, and distribution) and release, to the USEPA for use in determining the risk to human health and the environment. The CAIR is replacing the information submission requirements under the PRELIMINARY ASSESSMENT INFORMATION RULE (PAIR).

Comprehensive Environmental Response, Compensation, and Liability Act (CERCLA) The statute, also known as the SUPERFUND law, establishes federal authority for emergency response and cleanup of hazardous substances that have been spilled, improperly disposed, or released into the environment. The primary responsibility for response and cleanup lies with the generators or disposers of the hazardous substances (see POTENTIALLY RESPONSIBLE PARTIES), with a backup federal response using a trust fund (see HAZARDOUS SUBSTANCES SUPERFUND). The legislation was enacted in 1980 and significantly amended in 1986 (SUPERFUND AMENDMENTS AND REAUTHORIZATION ACT). See NATIONAL CONTINGENCY PLAN.

Comprehensive Environmental Response, Compensation, and Liability Information System (CERCLIS) A computerized system containing the basic information about and current status of a site being cleaned up under the NATIONAL CONTINGENCY PLAN, such as a Superfund hazardous waste site.

comprehensive general liability policy (CGL policy) An insurance policy covering a broad range of potential liabilities arising in a policy year, including claims resulting from sudden or accidental releases of pollutants, but not gradual leaks or normal pollutant emissions, which fall under the POLLUTION EXCLUSION CLAUSE. See also ENVIRONMENTAL IMPAIRMENT LIABILITY POLICY.

compressed natural gas (CNG) An alternative fuel that is a possible replacement for gasoline. The use of CNG reduces the impact of automobiles on air quality because hydrocarbons in the exhaust are minimal and because vapors released directly to the atmosphere from the fuel do not participate significantly in the generation of ground-level ozone. However, use of the fuel does result in the release of nitrogen oxides into the atmosphere.

concentration The amount of a chemical substance in a given amount of air, water, soil, food, or other medium. The value can be expressed as mass of the chemical in a given mass of the medium, the volume of the chemical in a given volume of the medium, or the mass of the chemical in a given volume of the medium. For gaseous air contaminant concentrations, two expressions are appropriate: a volume/volume ratio and a mass/volume ratio. The volume/volume ratio units are typically PARTS PER MILLION (volume), equivalent to one liter of pollutant per one million liters of air, or parts per billion (volume), one liter of pollutant in one billion liters of air. The mass/volume units are typically micrograms of pollutant per cubic meter of air. For airborne PARTICULATE MATTER, only mass/volume units are used, typically micrograms of particulate per cubic meter of air. In water, mass/volume and mass/mass ratios are used; the volume and mass of the aqueous medium are easily interchanged because one liter of water has a mass of one kilogram. Typical units are milligrams of pollutant per liter of water, which is the same as parts per million (mass), or micrograms of pollutant per liter of water, which equals parts per billion (mass). Soil and food concentrations are mass/mass ratios, in milligrams of a chemical per kilogram of medium, which is the same as parts per million (mass), or micrograms of a chemical per kilogram of medium, equal to parts per billion (mass). See the Appendixes for additional information.

concentration factor See BIOCONCEN-TRATION FACTOR.

concentration gradient An expression of the change in the concentration of a material over a certain distance. Chemicals diffuse from areas of higher concentration to areas of lower concentration; the diffusion rate increases with an increase in the concentration gradient. This principle is used in the removal of pollutants from exhaust gases (SCRUBBERS) and water effluents (PACKED TOWER AERATION). The gaseous or liquid material into which the pollutant is diffusing is replenished rapidly to maintain its low concentration and, thus, a high collection efficiency. See FICK'S FIRST LAW OF DIFFUSION.

concurrent flow The arrangement of material flow in systems designed to remove specific chemicals (such as sulfur oxides) from stack gases. The liquid containing the substance that absorbs or reacts with the undesirable gas enters the gas stream flowing in the same direction as the stack gas. Opposite of COUNTERCURRENT FLOW.

condensate Liquid condensed from the vapor or gaseous state. In natural gas production, the liquid components sometimes present in the gas stream exiting a well or gases such as propane or butane that are readily condensed to form LIQUEFIED PETROLEUM GAS (LPG) fuel.

condensation The change of a gas or vapor to a liquid. At atmospheric pressure, the process is caused by the removal of heat from the gas or vapor. The amount of heat removed from (or released by) a unit of gas or vapor to cause it to become a liquid is called the heat of condensation, which is numerically equal to the HEAT OF VAPORIZATION of the liquid.

condensation nuclei SUBMICROMETER particles naturally present in the atmosphere on which water vapor condenses to form droplets.

condenser A heat-extraction device used in a steam engine or turbine to condense the steam to a liquid.

conditional registration Refers to a special circumstance allowing the use of a new pesticide product under the provisions of the Federal Insecticide, Fungicide, and Rodenticide Act. This type of registration is a temporary certification that can be applied before final registration is granted. A product containing a previously unregistered active ingredient may be conditionally registered for use only if the administrator of the USEPA determines that such a classification is in the public interest, that a reasonable time for conducting the testing required for formal registration has not elapsed, and that use of the product for the time allotted for conditional use does not present an unreasonable risk to the public or to the environment.

conditionally exempt small-quantity generator A classification of a hazardous waste generator applied by the USEPA to facilities that generate less than 100 kilograms of hazardous waste per month. This classification exempts facilities from some regulations concerning the handling and disposal of hazardous waste on site. See SMALL-QUANTITY GENERATOR.

conductance The ability of a material to transmit electricity; the opposite of resistance.

conduction The transfer of heat by direct contact.

conductivity See HYDRAULIC CONDUCTIVITY.

cone of depression A drop in the WATER TABLE caused by a groundwater withdrawal rate exceeding the recharge rate. The resultant shape of the water table resembles an inverted cone.

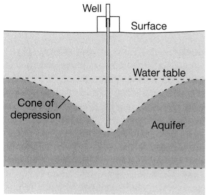

Cone of depression

cone of influence See CIRCLE OF INFLUENCE, CONE OF DEPRESSION.

cone penetrometer technology (CPT) Technology used to measure soil characteristics on the basis of penetration resistance. Sensors in a device with a cone-shaped end measure the tip resistance and the friction on the side walls, producing real-time data as the probe is pushed into the ground.

confidence interval A specified range of probability for a particular population statistic. For example, the statement "the 95% confidence interval for the mean dissolved oxygen (DO) level is 10.4–12.4 milligrams per liter" indicates that, in terms of information from a population sample, the true value for the mean DO level lies between 10.4 and 12.4 milligrams per liter and that the chance that this range does not contain the true mean is 5%, or 5 in 100.

confidence limits The end points of a CONFIDENCE INTERVAL.

confidential business information (CBI) Material that contains trade secrets, commercial information, or financial data that have been claimed as confidential by the manufacturer of a chemical or pesticide under review by the USEPA. Such material is needed during the evaluation of the product but is not disclosed to the public by the agency.

confidential statement of formula (CSF) A list of ingredients for a new formulation of a pesticide. The list is required by the USEPA during its consideration of REGISTRATION of the product but is not made available to the general public.

confined aquifer An AQUIFER located between two relatively impermeable layers of material with the groundwater confined under pressure significantly greater than atmospheric pressure. Also called an artesian aquifer.

confining layer A layer of underground rock or clay with low HYDRAULIC CONDUCTIVITY (permeability) that is located above and/or below an AQUIFER. Also called a confining bed.

confirmed test The second stage in the examination of water for the presence of bacteria of fecal origin. Culture findings that are positive on the first portion of the testing procedure (PRESUMPTIVE TEST) are inoculated into tubes of brilliant green lactose bile broth and examined for fermentation when incubated at 35° C for 48 hours. If fermentation is present, a third stage, the COMPLETED TEST, is performed.

confluence The location at which two streams or bodies of water merge to become one.

confluent growth Describes a circumstance encountered in attempts to quantify bacteria in some environmental sample in which the number of bacterial colonies on the laboratory medium is so great that they merge into one large mass. Such results are of no utility in attempting to determine bacterial numbers. See COLONY COUNT.

confounding variable A characteristic, habit, exposure, or the like that is linked to a disease risk and is found unequally in two groups in an epidemiological study, therefore confusing the comparison of the groups. For example, in a study of workplace lung cancer risk in coal miners compared with gold miners, if 75% of the sample population of coal miners were smokers but only 25% of the gold miners smoked, a researcher finding excess lung cancer risk in coal miners could not conclude that coal mining poses greater lung cancer risk than gold mining. If the study design failed to account for smoking, then it is possible that the researcher could conclude incorrectly that coal mining was a risk factor, failing to consider the confounding variable.

congener A chemical compound, person, or organism that resembles another in appearance or properties.

congenital In environmental health, describing a disease or risk factor that is wholly or primarily related to an organism's genetic makeup.

conifer One of the group of (mostly evergreen) cone-bearing trees or shrubs, such as pines, spruces, and firs. They are softwoods. Compare HARDWOODS.

coning Describing a PLUME from a smokestack that is being dispersed in the horizontal and vertical directions; the behavior of a plume that is best simulated by the GAUSSIAN PLUME MODEL. Compare FANNING, FUMIGATION, LOFTING, LOOPING, TRAPPING.

connate water Inactive fossil water that was trapped within the rock structure at the time of rock formation. This class of water has not been surface water. This type of fossil water is of interest as a potential source of water in missions to the Moon and planets.

consent decree A binding agreement by both parties in a lawsuit that settles the questions raised by the case; no additional judicial action is required.

conservation Careful and organized management and use of some natural resource or land area, emphasizing applied scientific principles. See also PRESERVATION.

conservation easement A legal restriction placed on a property owner limiting the use of the property to activities that are compatible with the long-term protection of environmental interests.

constituent concentrations in waste extract table (CCWE table) A listing of hazardous waste treatment standards expressed as the concentration of certain hazardous waste substances in an extract of the treated waste assembled by the TOXICITY CHARACTERISTIC LEACHING PROCEDURE. If the extract concentration is less than the specified CCWE concentration, the treated waste may be disposed of in a landfill. See LAND DISPOSAL BAN. See also CONSTITUENT CONCENTRATIONS IN WASTES TABLE.

constituent concentrations in wastes table (CCW table) A listing of treatment standards expressed as substances present in a hazardous waste after treatment. If the results of an analysis of the treated waste (not a liquid extract, as in the CONSTITUENT CONCENTRATIONS IN WASTE EXTRACT TABLE) show the concentrations to be below the values in the CCW table, the waste may be disposed of in a landfill. See LAND DISPOSAL BAN.

constituent(s) of concern Specific chemicals that are selected for evaluation in the assessment of the pollution status of an area or location.

constructed wetlands Artificial WET-LANDS built in basins or channels that use the natural cleansing processes performed by vegetation, soil, and microorganisms to treat wastewater.

construction and demolition waste Category of solid waste produced during repair and remodeling of structures, demolition of buildings, construction projects, and highway repair. Disposal of such debris is subject to a very limited regulatory burden.

construction ban 1. In air quality management, the prohibition of new construction or major modifications to existing facilities within a pollutant-specific NONATTAINMENT area, that is, a geographic zone defined as having air quality that does not meet one of the NATIONAL AMBIENT AIR QUALITY STANDARDS. The USEPA can implement the ban if, in its judgment, the state has submitted an inadequate STATE IMPLEMENTATION PLAN to demonstrate attainment of the standard for that pollutant. Only facilities that will emit the pollutant for which the area is designated as NONATTAINMENT are affected. 2. In environmental management, rules against development (roads, housing, shopping centers, dams) that might exacerbate noise problems, harm threatened or endangered species, disrupt wilderness areas, or have other damaging effects.

construction grants program A program administered by the USEPA under the CLEAN WATER ACT that provided federal matching funds to build or upgrade PUBLICLY OWNED TREATMENT WORKS (municipal sewage treatment plants) to a level of SECONDARY TREATMENT, thus reducing the discharge of water pollutants derived from sanitary wastewater. Now phased out and replaced with a REVOLVING FUND.

constructive metabolism See ANABOLISM.

consumer confidence report (CCR) An annual summary provided by COMMUNITY WATER SYSTEMS to their customers describing the quality of their drinking water. Required by the 1996 amendments to the SAFE DRINKING WATER ACT.

Consumer Product Safety Commission (CPSC) An agency of the federal government, established in 1973, responsible for issuing and ensuring compliance with regulations designed to protect the public from potentially hazardous household items, such as toys, lawn mowers, and flammable fabrics. The CPSC has banned the following known or suspected carcinogens from certain consumer products: asbestos, benzene, vinyl chloride, and fire retardant in children's sleepwear (Tris). Webstite: www.cpsc.gov

consumers In ECOLOGY, organisms that gain energy by eating other organisms. The place of a consumer in the FOOD CHAIN is defined by what it eats. HERBIVORES eat plants and are PRIMARY CONSUMERS; a human can be a primary consumer by eating plants or a SECONDARY CONSUMER by eating an animal that feeds on plants.

consumption of water After water is withdrawn for industrial, commercial, or agricultural use, the degradation of the quality of the water or the loss of a significant portion of it through evaporation, EVAPOTRANSPIRATION, or contamination by a substance such that the water cannot be reused. See WITHDRAWAL OF WATER.

consumptive use of water See CONSUMPTION OF WATER.

contact chamber An enclosed vessel in which a gas is mixed with (and commonly absorbed by) a liquid, usually water.

contact hazard A chemical material with irritant or corrosive properties that, in case of a spilling accident, can be harmful to persons if it comes into contact with their skin or eyes.

contact inhibition A property of normal animal cells that results in the stopping of cellular growth when one cell touches or contacts another cell. A type of control that limits the growth and reproduction of normal cells. The inhibition does not occur in tumor cells, which continue to grow even when crowded.

contact pesticide Usually an insecticide that kills insect pests when the insect physically touches the agent. Ingestion of the agent is not necessary for the destruction of the insect pest.

contact process A basic process for the industrial manufacture of sulfuric acid. Oxidized sulfur, as sulfur dioxide, is converted to sulfur trioxide by using a catalyst; the sulfur trioxide is then absorbed into a sulfuric acid liquid.

contact stabilization A version of the ACTIVATED SLUDGE PROCESS in which raw sewage is mixed with activated sludge in a small contact tank for about one-half hour, time enough for the microorganisms to absorb the wastewater organic material but not to decompose it. The wastewater and sludge mixture is then separated in a CLARIFIER, and the wastewater is discharged. The sludge settling in the clarifier is pumped to a stabilization tank, where about three hours of microbial decomposition of the absorbed organics takes place. The sludge is then returned to the contact tank for mixing with incoming raw sewage.

contact urticaria An allergic reaction of the skin caused by direct contact with some substance; the reaction is characterized by intense itching of red, swollen, fluid-filled patches. See also ALLERGEN.

containment building A term employed in the nuclear industry to identify the building housing the nuclear reactor and associated equipment The building is constructed to confine any RADIOACTIVE substances that may be released from the reactor in the event of an accident.

containment vessel A pressurized steel vessel that houses the CORE of a nuclear reactor. This vessel contains the liquid used to cool the core and is designed to confine any RADIOACTIVE substances that may be released from the reactor core in the event of an accident.

contaminant A substance, chemical, or microorganism that makes a medium (air, water, soil, food) impure, infected, radioactive, or lower in quality. If the contaminant exists at an excessive CONCENTRATION, then the medium might be polluted. An agent that is out of place.

continental crust See CRUST.

continental drift According to the theory of plate tectonics, the movement of the Earth's continents as the crustal plates on which they rest move across the surface of the semiliquid mantle of the Earth.

continental shelf The seafloor sloping away from the continents at an angle of about one degree. Commonly defined as the shoreline area under water less than 200 meters deep and usually extending about 70 kilometers. At the outer edge of the shelf, the seafloor drops sharply.

contingency plan In environmental management, a detailed document describing an organized, planned, and coordinated course of action to be followed in the event of a fire, explosion, or release of a hazardous substance that could threaten human health or safety or the integrity of the environment. See NATIONAL CONTINGENCY PLAN, NATIONAL RESPONSE TEAM, RISK MANAGEMENT PLAN, LOCAL EMERGENCY PLANNING COMMITTEE.

continuity equation The relation, based on the conservation of mass, that equates the VOLUMETRIC FLOW RATE Q of an incompressible fluid in a duct or pipe to the product of the fluid velocity V and the cross-sectional area A of the duct or pipe: $Q = VA$. If the area increases, then the velocity must decrease, and conversely. The equation is also applied to liquid flow through a system: the flow in, Q_{in}, flow out, Q_{out}, and change in the storage volume, Δ Vol, for a given time must be in balance: $Q_{in} - Q_{out} = \Delta$ Vol storage volume. For example, if water flow into a reservoir is 3000 cubic meters per day and flow out is 1000 cubic meters per day, then during one day the reservoir volume must increase by 2000 cubic meters, excluding evaporation and seepage losses.

continuous analyzer 1. An instrument that samples and analyzes ambient air on a continuous basis, producing short-term (typically five-minute) averages recorded on a computer disk or a paper tape. **2.** Laboratory instrumentation that allows the operator to load a number of samples at one time. The instrument automatically draws the proper amount of material from each sample, adds the necessary reagents, promotes the required reaction between the material of interest and the reagents, monitors the result of the test, and provides the operator with a data printout of the results.

continuous discharge A routine release of a pollutant into the water or air during normal operations at a facility. Such discharges are usually made under a permit authorizing the release.

continuous emission monitoring (CEM) The continuous sampling and analysis of gases, particulate matter, or OPACITY by monitors placed inside smokestacks. This type of arrangement provides more useful information than STACK SAMPLING, which may be conducted only every several years. Also called in situ monitoring.

continuous-feed reactor (composting) A method of converting yard litter, sludge, or other solid refuse into humic, soil-like material. The solid waste is moved through the apparatus in such a way that the conversions are done rapidly, with the solid refuse added to the front end of the reactor and humic substances removed from the back in a continuous fashion. This contrasts to the BATCH METHOD, in which the proper composting mixture is prepared, placed in a vessel or static pile, allowed to compost for a specific period, then removed.

continuous-flow microbiological system An operation in which liquid wastes or media are added to a decomposition or growth vessel to allow for the growth and metabolism of bacteria on a continuous basis. The reactants are added to the vessel and the products are removed on a continuous basis. This system contrasts to the BATCH METHOD, in which the reactants are added at the start, the reaction is allowed to proceed, and the entire contents of the reaction vessel are discharged at the end of the growth period.

continuous-flow system A system having an uninterrupted, time-varying input and output of material in one of two ways. Also described as a plug-flow model. The model assumes the effluent from the system is discharged in the same order as the material entering the system, with little mixing. It is appropriate for systems that have an extended length and small cross-sectional area, such as small streams and pipes. The completely mixed-flow model assumes that material flowing through the system is uniformly mixed while in the system and thus the discharge is physically and chemically the same as the system contents. This model is appropriate for most lakes, reactors, tanks, and other mixing vessels.

continuous noise A constant sound that varies less than five DECIBELs during a noise measurement or evaluation.

continuous process An industrial production method in which raw materials or reactants move continually into a reaction vessel or other area and continually combine or react, and from which the product is continually withdrawn. The release of pollutants from a continuous process is likely to be continuous as well but may be less than that from a BATCH PROCESS because the continuous process is most often enclosed.

continuous variable A quantitative variable that can take an infinite number of values in a set (for example, the mass of a soil sample), but practically the values are limited by the accuracy of the measurement method. See also DISCRETE VARIABLE.

contour ditch An irrigation ditch that follows the elevation of the land.

contour line See ISOPLETH.

contour mining A type of surface mining of coal in which coal is removed from relatively shallow deposits by removing the earth and rock covering the deposit. The mining proceeds around the natural topographical features of a hill or mountain. See also SURFACE MINING.

contract lab Private company or laboratory operating under an agreement signed with the USEPA authorizing the unit to conduct sampling of any medium, engage in laboratory analyses, or conduct research or monitoring tasks on behalf of the agency. See CONTRACT LABORATORY PROGRAM.

contract laboratory program (CLP) An operation of the USEPA and participating private laboratories that provides analysis of environmental samples as part of the cleanup of hazardous waste disposal sites. The contract laboratories must meet strict USEPA quality standards. See COMPREHENSIVE ENVIRONMENTAL RESPONSE, COMPENSATION, AND LIABILITY ACT; SUPERFUND AMENDMENTS AND REAUTHORIZATION ACT.

contribution rights Authorized under the COMPREHENSIVE ENVIRONMENTAL RESPONSE, COMPENSATION, AND LIABILITY ACT (CERCLA), the ability of a POTENTIALLY RESPONSIBLE PARTY (PRP) or other entity that spends money to clean up a hazardous waste site to sue other PRPs for their share of the waste cleanup costs. Contributions can be allocated by a court on the basis of the relative masses of the waste, sometimes weighted by its toxicity.

Because CERCLA allows the government to hold one PRP liable for the *entire* cleanup cost, contribution rights define the method for that PRP to be reimbursed from its fellow PRPs. See JOINT AND SEVERAL LIABILITY, COST RECOVERY.

control group A group of humans, animals, or plants that is not treated or exposed in an epidemiological study or in an experiment conducted to determine the effect of some agent or condition. For example, in an experiment to determine the effect of high temperature on the activity of flies, the experimental group would be maintained at some elevated temperature while the control group would be held at ambient temperatures.

control rods Long cylinders of neutron-absorbing material, such as cadmium or boron, that are part of a nuclear reactor CORE. The rods are lowered or raised as required to control the rate of NUCLEAR FISSION.

Control Techniques Guidelines (CTGs) A series of USEPA documents providing technical and economic information on the control of VOLATILE ORGANIC COMPOUNDS; the guidelines are used by state regulatory agencies to set limits for volatile organic compound emissions for existing sources in ozone NONATTAINMENT AREAS.

controlled area In a facility using radioactive materials, a defined area within which worker exposure to IONIZING RADIATION is monitored and controlled by an employee specifically responsible for radiation protection.

controlled reaction A chemical reaction taking place at desirable rates and under temperature and pressure constraints maintained to ensure safety and produce the desired product.

controlling variable The material or factor that determines the response of a process or experiment, given certain conditions. For example, the most important input controlling the yield of a crop in an area may be identified as the level of soil nitrogen. Note that if adequate nitrogen is added, then the controlling variable for crop yield will shift to some other factor, such as soil mixture. Related to LIEBIG'S LAW OF THE MINIMUM, LIMITING FACTOR.

convection The transfer of heat by a moving fluid, such as air or water.

convection currents Rising or sinking of water or air due to temperature differences. Hot air at the surface of the Earth rises because it has a lower density than the surrounding air. Cold water flowing into warmer surface water sinks because of the lower density of underlying waters that are warmer. The circulations associated with such movements can be considered as processes that equalize temperature in an environment and promote mixing.

Convention on Biodiversity A 1992 international agreement to conserve the natural HABITAT of species, thus protecting biological DIVERSITY; promotes SUSTAINABLE DEVELOPMENT. Supports monitoring of biological resources and cooperation among member states. More at www.biodiv.org.

Convention on International Trade in Endangered Species of Wild Fauna and Flora (CITES) A 1973 treaty that binds nearly 100 signing countries to establish a system of permits for the exporting and importing of endangered species or products of organisms that are endangered. The CITES protected species list contains about 1200 animals and plants. The list can be found at Title 50, Part 23, Section 23, of the *code of federal regulations*. Website: www.wcmc.org.uk:80/CITES/eng/

conventional pollutants Under the CLEAN WATER ACT, the following pollutants: materials exerting a BIOCHEMICAL OXYGEN DEMAND, FECAL COLIFORM, SUSPENDED SOLIDS, oil and grease, and extremes of pH. The act specifies that the definition of a conventional pollutant is not limited to these five, but no other categories have been added.

conventional systems In the management of municipal wastewater, gravity sewers that have been used to collect municipal wastewater that is conveyed to a facility that uses PRIMARY TREATMENT and SECONDARY TREATMENT to process the water prior to discharge into a waterway.

convergence 1. In meteorology, the flowing together of air masses. Air converges toward the center of low-pressure systems, where it rises. See DIVERGENCE. 2. Another name for CONVERGENT EVOLUTION.

convergent evolution The development of plant or animal species that are similar in appearance and ecosystem function in separate, but environmentally similar, geographic regions. The species similarity is explained by independent adaptation to the same environmental conditions. Also called convergence. See ECOLOGICAL EQUIVALENTS.

conveyance loss Water or other liquid lost in pipes, channels, conduits, or other systems used for transport. The loss can result from leakage, chemical reactions, or evaporation.

cool desert Land area at high latitudes or elevation characterized by low moisture levels and infrequent precipitation. Rather than being the hot sandy deserts that are common nearer the equator, these regions remain relatively cool and are characterized by a rocky terrain.

coolant Any gas (air) or liquid (water, antifreeze) used to conduct heat away from some source, such as a nuclear reactor or engine.

cooling pond A holding pond into which hot water is discharged to allow cooling prior to reuse or discharge.

cooling tower A large structure designed to transfer heat from industrial or electric power plant cooling water to the atmosphere. Wet cooling towers operate by introducing the water directly to the air and forcing evaporative cooling. Dry cooling towers retain the warm water within a piping system and allow heat transfer by CONDUCTION and CONVECTION only.

cooling tower blowdown   See BLOWDOWN.

cooling water   Water used to remove waste heat from industrial processes or turbines used to produce electricity.

cooperative agreement   In environmental research and planning, a contract whereby the USEPA transfers funds, equipment, property, or anything of value that is needed by a state agency, university, or nonprofit organization to engage in studies, cleanup, or other authorized tasks.

coppicing   The cutting back of trees or bushes to cause shoots or sprouts to grow. The new growth is then harvested. Coppicing is an efficient method for the controlled harvest of firewood.

coprophagous   Describing organisms other than bacteria that feed on fecal matter, such as certain insects.

coprostanol   A LIPID-soluble material, the presence of which indicates contamination with fecal material. Used as a measure of the extent of contamination of river or lake sediments with sewage.

coral reef   An underwater feature frequently associated with oceanic islands in clean, warm tropical seas. Composed of the calcareous (calcium carbonate) structures produced by innumerable sessile animals. These structures are living formations built as successive generations accumulate on the top of earlier community members. The system depends on the productivity of tiny photosynthetic algae that have a symbiotic relationship with the animals forming the coral and are thus limited to clear waters. Corals are very productive and provide habitat for many marine life-forms; however, they are sensitive to changes in water clarity and other problems related to the actions of humans. See ALGAE, PHOTOSYNTHESIS, SYMBIOSIS.

core   The part of a nuclear reactor where FISSION takes place; the location of the FUEL RODS.

core region   The primary industrial, population, or economic section of a country. That geographical region dominating the social, political, and financial life of a country. Also, that region where the natural environment is potentially subjected to the greatest impact.

core sample   A sample of soil or sediment taken with the aid of a pipe or tube that is pushed or driven into the soil or sediment. The soil that is removed is taken from the sampling device in a cylindrical section in which the vertical positioning of layers of sediment is maintained.

Coriolis force   The apparent force, resulting from the Earth's rotation, that deflects air or water movement. Winds seen from a moving frame of reference (the Earth) are deflected to the right in the Northern Hemisphere and to the left in the Southern Hemisphere. The force influences the circulation of air in the atmosphere and the movements of water in the oceans.

cornucopian   Describing a core belief grounded in optimism, taking as a symbol the horn of plenty from Greek mythology. This outlook maintains that human ingenuity and enterprise will continue to provide the means to improve the human condition and solve problems associated with environmental issues such as pollution, population growth, and depletion of finite resources. Contrast MALTHUSIAN.

cornucopian fallacy   Description applied by critics to an optimistic outlook regarding the future of the human race.

Those who hold a core belief grounded in pessimism contend that the optimists are engaging in either wishful thinking or deliberate denial concerning the problems related to such matters as pollution, population growth, and resource depletion. See MALTHUSIAN.

corporate average fuel economy (CAFE) standard Implemented by the Energy Policy and Conservation Act of 1975 in response to the petroleum supply disruptions of 1973–74, a federal requirement that the sales-weighted average fuel consumption of new passenger cars and light trucks sold by each automobile manufacturer greatly increase; failure to comply results in monetary penalties. The current CAFE standard is 27.5 miles per gallon for passenger cars and 20.7 miles per gallon for light trucks, including sport utility vehicles.

Corps of Engineers (COE) The U.S. Army agency under the Department of Defense responsible for navigation and flood control projects, including extensive dam construction. The COE has jurisdiction over WETLANDS conservation and dredge-and-fill (SECTION 404 PERMITS) activities under the CLEAN WATER ACT. Website: www.usace.army.mil

corpse drift tests Used after a marine oil spill to estimate the number of birds that died but were not recovered. Bird carcasses are placed in the water and the number that wash ashore are counted.

corrective action management unit (CAMU) A designated area of a facility within which wastes are allowed to be consolidated without triggering a violation of an otherwise-applicable LAND DISPOSAL BAN.

corrective action/order Under the RESOURCE CONSERVATION AND RECOVERY ACT (RCRA) permit program for TREATMENT, STORAGE, OR DISPOSAL facilities, the required cleanup of any releases of hazardous substances from the facility relating to activities occurring before RCRA became effective.

corridor When applied to wildlife management, a strip of natural habitat between two nature preserves that allows for the migration of wildlife from one area to another. Such access is important in preventing the genetic isolation and INBREEDING of wildlife confined to a relatively small area.

corrosivity A characteristic used to classify a hazardous waste. A waste is corrosive (and therefore hazardous) if the pH is less than or equal to 2.0 or greater than or equal to 12.5.

cosmic radiation See COSMIC RAYS.

cosmic rays IONIZING RADIATION, both electromagnetic energy and particles, originating in outer space. Secondary cosmic rays, produced in the atmosphere by incoming cosmic rays, are the immediate source of human cosmic radiation exposure, which makes up about 30%–40% of a typical BACKGROUND RADIATION dose and about 10%–15% of the average American's annual dose of ionizing radiation from all sources.

cost-benefit analysis (CBA) A calculation of the present value of the costs and benefits associated with a project or action. Nonmonetary costs or benefits, such as illnesses or deaths prevented, are converted to monetary units for the analysis. Such an analysis is required of all major federal regulations by EXECUTIVE ORDER 12866.

cost-benefit ratio The results of a COST-BENEFIT ANALYSIS expressed as costs divided by benefits or as benefits divided by costs (benefit-cost ratio).

cost-effectiveness analysis (CEA) A study to determine the lowest-cost method of achieving a certain defined goal.

cost recovery A lawsuit brought by a responder to the excessive discharge or spill of a HAZARDOUS SUBSTANCE or oil against one or more of the POTENTIALLY RESPONSIBLE PARTIES (PRPS), that is, the entities that caused the spill or discharge. As authorized under the COMPREHENSIVE ENVIRONMENTAL RESPONSE, COMPENSATION, AND LIABILITY ACT, the responder-plaintiff is entitled to reimbursement for all necessary expenses of the cleanup or response action in accordance with the NATIONAL CONTINGENCY PLAN.

cost sharing A concept allowing joint financing of research, site cleanup, sewer installation, or other projects in which the federal government contributes part of the financing needed to complete the project and the local entity contributes part of the cost. For example, the operating costs related to equipment and supplies for a research project undertaken at a university may be funded by the federal government, and the university provides the buildings, infrastructure, and personnel required to complete the project.

coulomb (C) The SI unit of electric charge corresponding to a current of one ampere in one second (1 C = 1 A s).

Council on Environmental Quality (CEQ) A group within the Executive Office of the president established by the NATIONAL ENVIRONMENTAL POLICY ACT (NEPA). The CEQ administers the ENVIRONMENTAL IMPACT STATEMENT process begun by the NEPA. The CEQ also issues an annual report on environmental quality. Website: www.whitehouse.gov/CEQ/

countercurrent flow The arrangement of material flow in a system designed to remove specific chemicals (such as sulfur oxides) from STACK GASES. The liquid containing the substance that absorbs or reacts with the undesirable gas and the stack gas move through the system in opposite directions, and collection of the contaminant occurs when the streams collide. See also CONCURRENT FLOW.

counting chamber A special device used to facilitate the counting of cells by microscopic techniques.

coupled In GLOBAL WARMING studies, describing the models that attempt to simulate the energy exchanges between the atmosphere and the oceans.

covalent bond A strong attraction between atoms that results from the sharing of electrons between two atoms. A single covalent bond results from the sharing of a pair of electrons by two atoms, each atom contributing one electron to the pair. The bonds between carbon atoms in organic molecules serve as examples. Covalent bonds are much more difficult to break than the weaker attractions between atoms illustrated by hydrogen bonds and ionic bonds.

cover See COVER MATERIAL.

cover crop Plants cultivated on an agricultural field to protect the area from erosion between crop seasons, to provide shelter to immature seedlings, or to add nutrients (e.g., nitrogen) to depleted soil.

cover material In the management of solid waste at a sanitary landfill, soil or other agents used to cover the deposits made each day. The fresh solid waste deposited daily must be covered with soil (six inches) or some substitute at the end of the working day to prevent the problems associated with trash migration, fires, odors, birds, flies, and rodents. Also called cover or daily cover.

cracking The utilization of heat and/or a CATALYST to reduce high-molecular-weight hydrocarbons in crude oil to smaller molecules, which are the constituents of products such as gasoline, ethylene, or heating oil.

cradle-to-grave A phrase used to describe the comprehensive management of hazardous waste under the RESOURCE CONSERVATION AND RECOVERY ACT. Substances meeting the definition of hazardous waste are regulated from their point of generation to their final treatment and/or disposal.

crankcase blow-by   See BLOW-BY.

criteria   Descriptive factors taken into account by the USEPA in establishing standards regulating the release of pollutants into the air, water, and ground. These factors can be used as guides by state agencies in establishing their own standards.

criteria document   A compilation of human health and environmental effects used by the USEPA to set NATIONAL AMBIENT AIR QUALITY STANDARDS for the CRITERIA POLLUTANTS.

criteria pollutants   The six air pollutants constituting the NATIONAL AMBIENT AIR QUALITY STANDARDS set by the USEPA using CRITERIA DOCUMENTS. The criteria pollutants are carbon monoxide, nitrogen dioxide, particulate matter, lead, sulfur dioxide, and ozone.

critical   Describing a nuclear reactor or weapon in which a self-sustained fission chain reaction takes place. In the case of a nuclear reactor, the term describes normal operation and does not imply any emergency or unusually dangerous condition.

critical aquifer protection area (CAPA)   A RECHARGE ZONE for certain SOLE SOURCE AQUIFERS designated for additional protection under the SAFE DRINKING WATER ACT.

critical factor   In the environment, the single factor that is closest to a tolerance limit of a specific species. For example, primary food supply may be such a factor if the amount available is barely sufficient to provide for the nutrition of the species, or the temperature of the environment may be if conditions are at the maximal tolerance limit of a species. Any decrease in the primary food supply in the former case or any elevation of the temperature in the latter case will result in the demise of the species.

critical mass   The smallest amount of FISSIONABLE material that will allow a self-sustaining nuclear CHAIN REACTION without the aid of other MODERATORS.

critical organ   An organ that an animal cannot live without, such as heart or brain, in contrast to organs that can be lost without loss of life, such as the spleen.

critical point   The location downstream from a waste discharge at which the DISSOLVED OXYGEN of the water is lowest; the lowest point on an OXYGEN SAG CURVE.

critical reactor   A nuclear reactor that is sustaining a spontaneous nuclear chain reaction. Such a reactor is engaged in normal operations, and the designation does not imply any emergency or unusually dangerous condition.

critical thinking   In the environmental arena, the opposite of blind, unquestioning acceptance of the opinions and assertions of others as they relate to issues concerning the protection of the environment and the promotion of human health. An informed, deliberate, reason-based, and reflective contemplation and evaluation of opinions and information related to ecological matters.

critical wind speed   In air pollution dispersion calculations, the wind speed corresponding to the highest estimated ground-level air concentration downwind from a source, such as a smokestack. The critical wind speed depends on atmospheric stability and the height of release of the air contaminant.

crocidolite   A form of asbestos-containing mineral, referred to as blue asbestos.

cross-connection   Any actual or potential point where a drinking water supply may be contaminated with an unapproved water system or other contamination source. The integrity of a drinking water supply depends on the elimination of real or potential sources of contamination after the water is placed into a distribution system.

cross contamination 1. A problem associated with the integrity of samples collected from environmental media. Integrity is threatened by sample contact with chemicals from other samples, other locations, or other reagents. **2.** The movement of chemicals or constituents from a sewer system, pipeline, surface drainage, or other source into a distribution system used to transport drinking water to the customers of a municipal system. **3.** The movement of underground contaminants from one aquifer or groundwater level to another as a result of the drilling of a well.

cross-media pollution The transfer of chemical contaminants from one environmental medium (air, water, soil) to another.

cross-sectional method In EPIDEMIOLOGY, a study designed to determine the prevalence of disease in a population along with existing population characteristics, such as the finding that people who watch more than six hours of television daily have a higher-than-average rate of heart disease. This snapshot approach is not useful for answering cause-and-effect questions; nor does it indicate which came first, the disease or the characteristic.

crosswind In an AIR QUALITY DISPERSION MODEL, indicating the direction perpendicular to a line directly downwind from the source. The greater the crosswind distance from CENTERLINE CONCENTRATION, the lower the estimated pollutant concentration. Also termed the y-direction; the x-direction is directly downwind.

crown fire A forest fire among the tops of trees.

crucible furnace A furnace that raises the temperature of a small porcelain container (crucible), reducing the contents to ash. The device is used to process particulate samples for determination of the ash content (for example, the amount of metals) that will remain after burning at high temperatures.

crude ecological density The number of organisms per unit area or per unit volume of total space in an ecosystem. Compare with ECOLOGICAL DENSITY, which includes only the total HABITAT area.

crude oil A complex mixture of liquid hydrocarbons in the unrefined state as produced from underground formations.

crude oil fraction The hydrocarbon compounds in a CRUDE OIL that have boiling points within a certain range. PETROLEUM is separated into its components in an oil refinery by fractional distillation: the oil is heated, and the compounds that will vaporize within a certain temperature range are removed and condensed, yielding various products such as gasoline and kerosene.

crude rate In public health statistics, a rate that includes the entire population in the population at risk. Specific rates narrow the population at risk to a certain age group, sex, race, or other category.

crumb rubber 1. Product from the recycling of used tires. The tires are shredded and the wire removed to produce a crumbly material consisting of rubber fragments. The product can then be used to supplement asphalt for road paving, construction of ground cover at playgrounds, or other uses. **2.** The final product in the creation of synthetic rubber for the manufacture of tires and other applications. The raw synthetic rubber product is in the form of small colorless flakes that can be reformulated with the input of carbon black and other additives, then cross-linked to yield the final commercial product.

crust The cool, outermost layer of the surface of the Earth, floating on a hot, more dense, semiliquid mantle and inner core. The outer layer consists of two basic types of material and is fragmented or broken unevenly into seven major and several minor pieces termed plates. The two basic types of crustal material are the continental crust composed of granitic substances and the oceanic crust composed of basaltic substances. The continental crust is the thicker of the two and floats higher on the mantle because of a lower density than that of the thinner oceanic crust. As the plates composing the crust migrate across the surface of the mantle, earthquakes and volcanic activity are generated.

Cryptosporidium Genus name of a protozoan organism responsible for some waterborne diseases. The protozoan is commonly associated with the fecal material of livestock and some pets, and infection of humans is usually the result of transfer from the animal to humans. The disease state that results from presence of the organism, cryptosporidiosis, is characterized by an unusually watery, greenish, and offensive diarrhea accompanied by severe abdominal pain, cramps, fever, and vomiting. The disease is self-limiting in most cases but potentially lethal to malnourished children and acquired immunodeficiency syndrome (AIDS) patients. Prevention involves filtration of water and limiting of contact with domestic livestock.

crystalline silica The mineralogical form of silica, or silicon dioxide, which causes SILICOSIS after prolonged exposure. Exposure to noncrystalline silica, or amorphous silica, does not cause silicosis. The three types of crystalline silica are quartz, tridymite, and cristobalite. Also called free silica.

cubic meter (m^3) A metric volume equal to 1000 LITERS, 264.2 gallons, or 35.31 cubic feet.

cullet Mixed scrap glass.

cultural eutrophication The excessive addition of plant nutrients into aquatic ecosystems by human activities. A typical result is the growth of aquatic weeds and algal slime to such an extent that use of the water by humans is prohibited.

culture A term employed in MICROBIOLOGY to designate the growth and multiplication of bacteria or fungi in a closed artificial container.

culture dish A shallow device for the cultivation of microorganisms on a solid nutrient medium (an agar) prepared in the laboratory. The most common device is a petri dish measuring 100 millimeters by 15 millimeters.

culture media Solid or liquid substances prepared and sterilized in the laboratory to provide the nutrients needed for the growth of bacteria or fungi.

cumulative probability distribution A diagram of a cumulative distribution function $F(x)$, which is the probability that a random variable x will be less than or equal to a certain value of x. For example, the diagram could depict the annual probability that an hourly ozone concentration will be less than or equal to x, which could be created for a particular location on the basis of past experience.

cumulative working-level months (CWLMs) See WORKING-LEVEL MONTH.

Cunningham correction factor A modification of STOKES'S LAW that increases the terminal SETTLING VELOCITY for particles smaller than about five micrometers in diameter, accounting for the ability of the particle to slip between gas molecules in the air.

curbside collection A community recycling program requiring residents to separate recyclable items from the remainder of household solid waste and to place the separated items along the street for pickup by the municipal service. See SOURCE SEPARATION.

curie (Ci) A unit of RADIOACTIVITY corresponding to 3.7×10^{10} decays per second. RADON levels, however, are expressed as the radon decay rate per volume of air, usually picocuries per liter of air. A picocurie is 1×10^{-12} curie.

cut diameter The diameter of particles collected with 50% efficiency by a particulate control or air-sampling device, or so that about half of the particles are captured.

cutting oil An oil or oil-water emulsion used to reduce heat and friction when operating metal-working machines. Prolonged skin contact can cause occupational acne.

Cuyahoga River "The River That Caught Fire." An oil slick on the stretch of this river running through Cleveland, Ohio, ignited in June 1969, burning two railroad trestles. This and other rivers had burned before, but the incident was often cited as a dramatic example of the need for water pollution control, which took the form of the federal CLEAN WATER ACT of 1972 and a successful regional plan to clean up the river and Lake Erie.

cyanosis A condition in which the skin appears blue caused by insufficient oxygen in the blood.

cycles per second The number of wave cycles passing a given point in one second; a FREQUENCY unit. The frequency of sound waves is sometimes still given in cycles per second. For electromagnetic radiation, one cycle per second is termed one HERTZ.

cycling load facility A power station that generates electrical energy to meet a fluctuating daily demand. See also BASE LOAD FACILITY.

cyclone In meteorology, a counterclockwise air circulation with a low-pressure center; tornadoes and hurricanes are intense cyclones. See also ANTICYCLONE.

cyclone collector A cylindrical or cone-shaped air-cleaning or air-sampling device designed to remove particles from an airstream by CENTRIFUGAL FORCE. The particulate-containing air enters the top of the cyclone and spins downward, throwing the particulate outward against the cyclone wall. The particles fall into a collection hopper, and the cleaned air exits the top of the cyclone.

cytotoxin Any toxic material that kills cells.

D

daily cover   See COVER MATERIAL.

Dalton's law   The law of partial pressures, discovered by the English scientist John Dalton, stating that a mixture of nonreacting gases or vapors exerts a pressure equal to the sum of the pressures that would be exerted by each gas or vapor alone. For example, the mixture of gases in the atmosphere exerts an ATMOSPHERIC PRESSURE, which is equal to the total individual pressures of nitrogen, oxygen, argon, carbon dioxide, water vapor, and other components.

damage risk criterion   That noise level above which permanent hearing loss is likely to occur.

damp   In coal mining, a poisonous or explosive gas in a mine. From the German *Damf,* for "vapor." Carbon monoxide is known as WHITE DAMP, and methane is known as FIREDAMP.

damper   In-duct movable plates used to adjust air VELOCITY PRESSURE and thereby balance the system airflow in the local exhaust ventilation system of a workplace.

dangerously reactive material   A substance that will react with itself or with air or water to produce a condition hazardous to individuals or the environment, for example, metallic sodium.

daphnid   Small freshwater crustacean of the species *Daphnia magna* or *D. pulex.* The organisms are used in the laboratory testing of pollutants for toxicity to aquatic biota.

Darcy's law   An equation stated by the French engineer Henri Darcy in 1856 that relates groundwater velocity to the product of the HYDRAULIC CONDUCTIVITY (PERMEABILITY) of an AQUIFER and the slope of the WATER TABLE (the HYDRAULIC GRADIENT). The velocity V is given by

$$V = -K \frac{\Delta H}{L}$$

where K is the hydraulic conductivity and ΔH is the loss of HYDRAULIC HEAD over the distance L.

dark green technology   An application or innovation that has direct and intentional environmental benefit, for example, a new SCRUBBER design to control air pollutants. Compare LIGHT GREEN TECHNOLOGY.

daughter   The material resulting from the RADIOACTIVE DECAY of a parent material. For example, radon-222 is a daughter of radium-226; the decay products of radon-222 are its daughters. Also called progeny.

day-night sound level (L_{dn})   The weighted equivalent sound level in DECIBELS for any 24-hour period. Ten decibels are added to the sound levels recorded from 10 P.M. to 7 A.M. to account for the greater disturbance typically caused by nighttime noise. A measure of community noise; not used for workplace evaluations.

de micromis   Under hazardous waste cleanup regulations, describing a POTENTIALLY RESPONSIBLE PARTY (PRP) with a minimal waste contribution to a HAZARDOUS WASTE site, much less than a DE MINIMIS contributor, usually less than

0.001% by volume. These parties are given an opportunity to pay a small monetary settlement to be released from SUPERFUND liability and from the CONTRIBUTION RIGHTS of other PRPS.

de minimis 1. Quantities of pollutants or other chemical substances that are small enough to be exempt from environmental regulations. From *de minimis non CURAT lex,* "the law does not concern itself with trifles." 2. Under hazardous waste cleanup regulations, describing a POTENTIALLY RESPONSIBLE PARTY (PRP) with a minimal waste contribution to a HAZARDOUS WASTE site, usually 1% or 2% of the whole by volume, although toxicity can be a factor. De minimis parties may negotiate settlement agreements that include a small monetary payment and release from SUPERFUND liability and from the CONTRIBUTION RIGHTS of other PRPs. See INNOCENT LANDOWNER, DE MICROMIS.

dead end The end of a water main used to deliver water to residents served by a municipal water company. Because the end of the system is typically remote from the primary source of water and because water tends to remain in these isolated section for a long time, problems related to a reduced amount of chlorine residual are common. The water may have a disagreeable odor and taste, be cloudy, and have a higher than normal bacterial load.

debarment The disqualification of a business from receiving government contracts; can be applied for many reasons, including excessive violations of the CLEAN WATER ACT or the CLEAN AIR ACT.

death phase The terminal stage of growth of bacteria in laboratory culture. After the bacteria in the culture media have exhausted the supply of available nutrients, growth (cell division) stops. When the cells can no longer maintain viability, cell death results and the number of viable cells decreases.

death rate The number of deaths per 1000 persons in a population in a given year. The rate is calculated by dividing the annual number of deaths by the midyear population, then multiplying by 1000.

debt for nature swap The exchange of a country's international debts for its agreement to take specific actions to conserve and protect its natural resources, such as establishing nature reserves. Conservation groups have served as mediaries, buying the (often overdue) debt at a deep discount, then offering to cancel the debt upon receiving the conservation commitment from the debtor country. The swaps have been applied in a number of Latin American and African countries.

decay product An element or isotope resulting from radioactive decay; it may be stable or radioactive (undergo further decay). See RADIOACTIVE SERIES.

decay series An illustration of the series of elements (DAUGHTERS) produced during the radioactive decay or decomposition of an unstable element until it reaches the point where it is no longer unstable or radioactive.

decibel (dB) A unit for expressing SOUND PRESSURE LEVEL or SOUND POWER LEVEL. The logarithmic decibel scale for sound pressure levels extends from 0 (the hearing threshold) to over 130 (causes pain); normal speech is at about 60 dB.

decibels, A-weighting network (dBA) The frequency-weighted SOUND PRESSURE LEVEL that best matches the noise sensitivity of the human ear. The overall measured sound pressure level is a sum of the sound pressures measured at the various frequencies composing a sound. Noise meters can be set to the A-weighting network, which is derived from the FLETCHER-MUNSON CONTOURS and reflects the relative insensitivity of the human ear to lower-frequency sounds. The meter adjusts downward the sound measurements at the lower frequencies to compute a total dBA. Noise standards are almost always in dBA units.

deciduous Describing vegetation that loses leaves seasonally. Compare EVERGREEN.

deciduous forest biome See TEMPERATE DECIDUOUS FOREST.

decline spiral A deterioration of a species, community, or ecosystem in response to environmental damage. The rate of collapse accelerates beyond routine expectations after the disruption of normal ecological functions.

decommissioning Taking a power plant, industrial facility, machinery, or vehicle out of use. The monetary costs and environmental hazards of decommissioning a nuclear electric power plant are considerations in the debate over the expansion of the nuclear power industry.

decomposers A community of organisms, including bacteria and fungi, that metabolically break down complex organic matter into simpler materials. Used extensively to treat waste from domestic sewage.

decomposition The breakdown of a complex material into simpler materials. The complex material can be organic or inorganic, and the decomposition can be caused by heat, sunlight, water, chemicals, or metabolism. Metabolic decomposition is performed by DECOMPOSERS.

decontamination The physical or chemical removal of potentially toxic materials (chemical or radioactive substances) to produce an acceptable level.

decontamination factor The ratio of the amount of a given radioactive substance entering a cleaning process to the amount exiting the process.

deep ecology A perspective on environmental problems that emphasizes the interrelatedness of the Earth and its biota, the equal importance of all species, and the need for radical social, economic, and political changes to improve and maintain environmental quality.

deep-well injection Pumping of a waste liquid into geological strata below ground. See CLASS I, II, III IV, AND V INJECTION WELLS.

defect action level (DAL) The allowable amount of insects or insect body parts in food, set by the United States Food and Drug Administration (for example, 50 insect fragments and 2 rodent hairs per 100 grams of peanut butter).

Defenders of Wildlife American conservation organization dedicated to species protection and habitat preservation. Active enforcer of the ENDANGERED SPECIES ACT; international focus through the CONVENTION ON INTERNATIONAL TRADE IN ENDANGERED SPECIES OF WILD FAUNA AND FLORA (CITES). With headquarters in Washington, D.C. Membership in 1999, 180,000. Website: www.defenders.org

deficiency disease A disorder caused by the lack of an essential nutrient in the diet (such as scurvy, caused by a vitamin C deficiency).

definitive test A test of the acute or chronic toxicity of a chemical to an aquatic test organism, usually expressed in concentration-response units, such as a median lethal concentration (LETHAL CONCENTRATION: 50%) or the percentage of the test organisms that exhibited a certain response (EFFECTIVE CONCENTRATION: $x\%$).

defoliant A herbicide that removes leaves from trees and shrubs.

deforestation The removal of trees and other vegetation on a large scale, usually to expand agricultural or grazing lands. Global deforestation has contributed to the rise in atmospheric CARBON DIOXIDE levels over the past century. See GREENHOUSE EFFECT.

degasification A step in the processing of drinking water before distribution to the public. The removal of dissolved gases.

degenerative disease A disorder characterized by the gradual loss of some vital function, such as nerve function. For example, chronic exposure to low levels of pesticides may cause the gradual loss of motor nerve function.

degradation 1. A decrease in the quality of the environment 2. The chemical, physical, or biological breakdown of a complex material into simpler components.

degreasing The removal of grease and oil from wastewater, machinery, or similar items.

degree-day A unit expressing the extent to which a daily mean air temperature is above (or below) a certain standard. For example, three daily temperatures of 78° F would be recorded as 18 degree-days over a standard of 72° F. Seasonal or annual degree-day totals for an area are used to estimate heating (or, at times, cooling) requirements for a building or other facility.

de-inking A necessary step in the processing of newsprint and other paper for recycling. The used paper is converted to a pulp and treated with alkali, causing the fibers to swell. As the fiber swells, the ink moves off the pulp and can be removed, and the cleaned fibers are processed for the manufacture of new paper products.

deionization The removal of all charged atoms or molecules from a material, such as water. For example, the removal of salt from water involves the removal of sodium ions (Na^+) and chloride ions (Cl^-). The process commonly employs one resin that attracts all positive ION and another resin to capture all negative ions.

deionized water Water that has been passed through resins that remove all IONS.

Delaney clause A 1958 amendment to the federal FOOD, DRUG, AND COSMETIC ACT that prohibits the inclusion in food products of any additive that is known to cause cancer in animals or humans. Amended by the FOOD QUALITY PROTECTION ACT OF 1996. In response to a 1992 federal court ruling that banned PESTICIDE RESIDUES from food if the pesticide showed evidence of carcinogenicity, Congress changed the FOOD, DRUG, AND COSMETIC ACT definition of FOOD ADDITIVE to exclude pesticide residues. The USEPA tolerances for pesticides are still in effect. See TOLERANCES, PESTICIDE.

delegated state See DELEGATION.

delegation In environmental regulation, the transfer of authority to administer a federal program to a state agency. Oversight and final approval of a state's actions are retained at the federal level. See PRIMACY.

delisting The formal removal of a chemical substance from a USEPA list of regulated materials, such as hazardous wastes or hazardous substances. This would be done, for example, if recent experimental or clinical evidence indicated that a substance were less hazardous than originally thought.

demand In an economic sense, the amount of a product or service that customers are willing and able to pay for, given the current price. If the price of the product or service does not reflect the environmental costs created by its production, the demand will often be too high, leading to environmental problems. See EXTERNALITY.

demand-side waste management The change in the characteristics of generated waste as customers use purchasing decisions to make known to manufacturers a preference for environmentally sound products and packaging made from recycled material, products that result in the least possible amount of waste, and items containing no hazardous substances.

demineralization The treatment of drinking water to remove dissolved minerals. The process is useful in reducing water hardness.

demister See MIST ELIMINATOR.

demographic transition The pattern of population growth exhibited by the now-developed countries during the 19th and early 20th centuries. During the first of three stages, society experiences historically high birth rates and high death rates, especially in the young. Population growth is low. The second stage reflects the economic development of a country and the resulting higher standard of living; birth rates stay high, but death rates for infants and children decline sharply. This stage is a period of rapid population growth. The third stage arrives with a fall in birth rates in the developed country to match lowered death rates. Population growth returns to a low level.

demography The statistical study of human population size, growth, density, and age distribution.

demolition waste See CONSTRUCTION AND DEMOLITION WASTE.

denaturing In nuclear fuel management, the addition of NONFISSIONABLE material to nuclear reactor BY-PRODUCTS to make the material unfit for use in a nuclear bomb.

dendritic Treelike. See also ARBORESCENT.

denitrification The removal of nitrate ions (NO_3^-) from soil or water. The process reduces desirable fertility of an agricultural field or the extent of undesirable aquatic weed production in aquatic environments. See DENITRIFYING BACTERIA.

denitrifying bacteria Bacteria in soil or water that are capable of anaerobic respiration, using the nitrate ion as a substitute for molecular oxygen during their metabolism. The nitrate is reduced to nitrogen gas (N_2), which is lost to the atmosphere during the process.

dense nonaqueous phase liquid (DNAPL) Liquid hydrocarbons and chlorinated hydrocarbons that do not dissolve in water. Mixtures of water and these liquids separate into two phases with the hydrocarbons in the bottom layer. Since these agents have a specific gravity greater than 1 (the specific gravity of water), they migrate to the bottom of an aquifer when involved in the contamination of groundwater.

density (ρ) **1.** Mass per unit volume. Common units are kilograms per cubic meter ($kg\ m^{-3}$), grams per cubic centimeter, pounds per cubic yard, and pounds per cubic foot. **2.** Number of individuals of a particular species per unit area of land.

density-dependent factor An influence on population growth that increases with size (density) of the population. Examples include PARASITISM, PREDATION, and COMPETITION. See also DENSITY-INDEPENDENT FACTOR.

density-independent factor An influence on population growth whose strength is not affected by the size (density) of the population. The impact of seasonal cold weather on plant populations is an example. See also DENSITY-DEPENDENT FACTOR.

deoxyribonucleic acid (DNA) The macromolecule containing the genetic information governing the properties of individual cells and organisms; the genes of a cell. Exact copies of the genes of each cell are transferred to each daughter cell during cell division. Changes in the structure of DNA (mutations) can occur naturally or result from exposure to certain chemicals or radiation. Some changes result in the death of the cell, and others are passed on to future generations in the form of an altered gene. Protection provided to individuals undergoing medical X rays is designed to prevent alteration in the DNA structure of gametes. See MUTAGEN and MUTATION.

Department of Energy (DOE) A federal executive department, created in 1977, with responsibility for energy policy of the United States. Website: www.doe.gov

depauperate Describing a nutrient-poor ecosystem or an area with low SPECIES DIVERSITY.

dependency load See DEPENDENCY RATIO.

dependency ratio The ratio of non-workers (children and retirees) to workers in a human population: the higher the ratio, the greater the dependency load. The ratio is most useful when applied to industrial economies in which children are not employed and large numbers of older people depend on public pensions financed by taxes on the younger (working) generation. Also called dependency load.

dependent variable A characteristic or condition of an object, person, population, system, and so on, that changes its value or degree with changes in another, independent variable. In epidemiology, the disease risk associated with an exposure, habit, or condition. For example, the increase in lung cancer risk with the number of cigarettes smoked daily. Compare INDEPENDENT VARIABLE.

depleted uranium Uranium that contains less than 0.7% of the fissionable isotope uranium-235.

deposition 1. The washout or settling of material from the atmosphere to the ground or to surface waters. 2. The absorption or ADSORPTION in the respiratory tract of inhaled gases, vapors, or particles.

depressurization (of structures) The lowering of atmospheric pressure inside a building relative to the atmospheric pressure outside a building by the operation of fireplaces or other appliances that exhaust indoor air but do not supply adequate makeup air. Outdoor air pollutants migrate to the interior more readily under these circumstances.

depuration A process during which an organism, such as an oyster, eliminates dangerous chemicals or microorganisms when placed in uncontaminated water.

derived-from rule The USEPA regulatory provision that any waste derived from the treatment, storage, or disposal of a HAZARDOUS WASTE is itself a hazardous waste.

dermal absorption/penetration The process by which an agent enters the circulatory system of an individual through movement of material through the skin. Substances that are FAT-SOLUBLE can penetrate the skin with relative ease. The technology associated with medicines administered by the skin patch rely on this type of transport. Dimethylsulfoxide, phenol, and some pesticides serve as examples of organic substances that can enter the body through this mechanism. See DERMAL ROUTE, DERMAL TOXICITY.

dermal route Direct contact of a chemical with the skin. If it is absorbed to the blood, potential adverse effects are DERMATITIS, PHOTOSENSITIZATION, and SYSTEMIC toxicity. See PRIMARY IRRITANT, SECONDARY IRRITANT, PERCUTANEOUS.

dermal toxicity 1. The adverse effects of agents that gain entry into the body by transport across the skin. The agents enter the circulatory system after contact with the skin and exert an effect on a target organ within the body. 2. An agent that damages the skin upon contact, for example, a strong acid or alkali. See DERMAL ABSORPTION/PENETRATION, FAT SOLUBLE.

dermatitis Inflammation or irritation of the skin. The condition can be caused by allergic reactions associated with contact with some plants or excessive contact with irritant chemicals, such as strong acids.

dermatosis Any abnormality or disease of the skin, including DERMATITIS. The condition can be caused by overexposure to toxic chemical agents, bacterial or viral infection, or excessive exposure to sunlight.

desalination The removal of salts from water to allow it to be used for drinking, irrigation, or industrial processing. The two main desalination techniques are DISTILLATION and OSMOSIS.

descriptive epidemiology Studies of the distribution of human disease that describe disease rates or prevalence in various subpopulations in terms of age groups, sex, race, occupation, and so forth without attempting to identify specific causes or environmental exposures that may increase disease risk in these groups. See also COHORT STUDY, CASE-CONTROL STUDY.

desert A BIOME characterized by low moisture levels and infrequent precipitation, usually defined as less than 10 inches of rain per year. Air temperatures are extreme and fluctuate widely. Deserts are very sensitive to environmental disturbance and recover very slowly, if at all, after severe damage.

desertification A process whereby land that is covered with vegetation is converted to desert. The term is generally applied to the production of artificial deserts where people have intensified the problems caused by droughts through overgrazing of marginal land, repeated burning of natural vegetation, intensive farming of arid land, aggressive removal of trees, and prolonged irrigation of arid land for agricultural use.

design capacity In waste management, the number of tons that a solid waste burning facility is engineered to process in 24 hours of continuous operation.

design flow In wastewater treatment, the average flow of wastewater that a treatment facility is built to process efficiently, commonly expressed in millions of gallons per day (MGD).

desorption The removal of a substance that has been absorbed or adsorbed by another material.

desiccant A chemical that absorbs moisture. These agents are added to packaging to prevent moisture damage to contents, and some can also be used as insecticides since they upset the water balance in insects after consumption.

design for the environment (DfE) A voluntary partnership between the USEPA and industry, research institutions, universities, and others to encourage risk reduction and prevention in products and processes. Part of the POLLUTION PREVENTION philosophy inherent in recent legislation, such as the Pollution Prevention Act of 1990, that places emphasis on the reduction of risk and environmental impacts in the earliest stages of design work. Website: www.epa.gov/opptintr/dfe/

design value In air quality management, the fourth-highest OZONE reading in a three-year period. Used to draft the ozone CONTROL STRATEGY in the STATE IMPLEMENTATION PLAN. Higher design values imply that more stringent control will be necessary to achieve the NATIONAL AMBIENT AIR QUALITY STANDARD for ozone. See PHOTOCHEMICAL OXIDANTS.

designated facility 1. A location for the treatment, storage, and disposal of hazardous waste operating under a state and USEPA permit. 2. The location given on a manifest by a generator of hazardous waste for the ultimate destination of a waste shipment. See MANIFEST SYSTEM.

Designated National Authority (DNA) In countries participating in the United Nations PRIOR INFORMED CONSENT PROCEDURE program, the government contact(s) that reviews chemicals to be imported and used in that country.

designated pollutant An air pollutant not regulated as part of the CLEAN AIR ACT AIR TOXICS control program, not listed by the NATIONAL EMISSION STANDARD FOR HAZARDOUS AIR POLLUTANTS program, and not among the six pollutants assigned a NATIONAL AIR QUALITY STANDARD but subject to NEW SOURCE PERFORMANCE

STANDARDS. Examples of pollutants falling under this category are acid mists, total reduced sulfur compounds, and fluorides. See NEW SOURCE REVIEW.

designated uses Those purposes for which surface water systems are used, which determine the state water quality standards that apply. Under the provisions of the CLEAN WATER ACT, states are required to classify streams, lakes, rivers, and other bodies of water according to a primary use then to achieve and maintain environmental conditions so that the intended use can be made of those systems. Examples of designated use include fishing, swimming, public water source, and irrigation.

destratification Vertical mixing of a pond or lake resulting from an equalization of the temperature profile in the water column from the top to bottom. See FALL TURNOVER, SPRING TURNOVER, STRATIFICATION.

destroyed medical waste Regulated MEDICAL WASTE that has been rendered unrecognizable through thermal treatment, shredding, or grinding prior to final treatment and disposal.

destruction and removal efficiency (DRE) An expression of incinerator performance in terms of the percentage of a particular incoming chemical material that is destroyed. Calculated as $(I - O/I) \times 100$, where I is the input rate of the material to be incinerated and O is the output rate of the material. See FOUR NINES, SIX NINES.

destructive distillation The heating of organic matter, such as coal, oil, or wood, in the absence of air or oxygen. The process produces volatile substances that are removed and recovered. The solid residue that remains is a mixture of carbon resembling COKE and ash.

desulfuration A metabolic oxidation of a sulfur-containing organic compound within a biological system, resulting in the incorporation of molecular oxygen into the molecule and the concurrent elimination of sulfur.

desulfurization The removal of sulfur-containing compounds, such as sulfur dioxide, from a gas exhaust.

detectable leak rate From a storage tank, the smallest leak that can be discerned with a reasonable probability of detection during a test. Expressed in terms of gallons per hour or liters per hour.

detection level See DETECTION LIMIT.

detection limit The smallest amount of a particular chemical that can be detected by a specific analytical instrument or method. See INSTRUMENT DETECTION LIMIT, METHOD DETECTION LIMIT, QUANTITATION LIMIT.

detection monitoring program Groundwater monitoring at the boundary of a TREATMENT, STORAGE, OR DISPOSAL FACILITY (POINT OF COMPLIANCE) to detect any contamination caused by leaks from the hazardous waste at the facility. The materials for which the samples must be analyzed (the INDICATOR PARAMETERS/CONSTITUENTS) are specified in the facility permit.

detention basin A relatively small storage lagoon for slowing storm water runoff; it is filled with water for only a short time after a heavy rainfall. See also RETENTION BASIN.

detention time The interval in which storm water is retained in a detention basin.

detergent 1. A synthetic water-soluble compound capable of holding dirt in SUSPENSION, emulsifying oils, or acting as a wetting agent; these compounds are commonly used in household cleaning agents. 2. An oil-soluble substance added to motor oils for the purpose of holding foreign substances in suspension.

detritivore Animal (such as an insect), that feed on particulate material derived from the remains of plants or animals. See DETRITUS FOOD CHAIN.

detritus Dead organic matter derived from plant or animal body parts and excretions. Material that accumulates on the surface of the ground or the bottom of bodies of water.

detritus food chain A feeding pattern in which an animal community survives through the consumption of decaying organic plant matter. Compare GRAZING FOOD CHAIN.

deuterium An isotope of hydrogen that contains one proton and one neutron in the nucleus of the atom. Also called heavy hydrogen.

Deutsch-Anderson equation The basic equation used in the design of ELECTROSTATIC PRECIPITATORS; the collection efficiency of the precipitator η is $\eta = 1 - e^{-w(A/Q)}$, where A is the total collection plate surface area, Q is the exhaust gas VOLUMETRIC FLOW RATE, w is the particle DRIFT VELOCITY, and e is the base of the natural logarithm. Named for W. Deutsch and E. Anderson.

dew point The temperature at which, for a given water vapor content and constant pressure, condensation of water from the atmosphere commences. The temperature at which any gas begins to condense.

dewatering Physical removal of water from sludge by pressing, CENTRIFUGATION, or air drying. The resulting product can be composted, used as landfill, or burned.

diatomaceous earth A geological deposit consisting of the remains of microscopic, unicellular plants called DIATOMS. The material is easily powdered and is often employed as a filter for treatment of public water supplies and for many other applications.

diatoms A distinctive group of unicellular plants (algae) that produce cell walls composed of silicate minerals. The plates of the cell wall form overlapping halves termed frustules. Diatoms are distributed in almost every environment on Earth and constitute some of the oldest known fossil records. See DIATOMACEOUS EARTH.

dibenzofurans A family of toxic chlorinated organic compounds consisting of two benzene rings connected by two bridges, one a carbon-to-carbon link, the other a carbon-oxygen-carbon link. Along with DIOXINS, these compounds are present in minute amounts in the air emissions from solid and/or hazardous waste incinerators. Often called simply furans.

dibenzo-*para*-dioxin See TETRACHLORODIBENZO-*para*-DIOXIN.

dichlorophenoxyacetic acid (2,4-D) A chlorophenoxy herbicide that is one of the most familiar chemicals, used for control of broadleaf weeds and woody plants; an active ingredient, along with TRICHLOROPHENOXYACETIC ACID (2,4,5-T), in Agent Orange, used as a defoliant during the Vietnam War. The chemical kills plants by acting as a plant hormone. Toxicity for animals requires relatively large doses.

dichlorodiphenyldichloroethane (DDD) A major METABOLITE produced during the biological degradation of dichlorodiphenyltrichloroethane (DDT). The material has some insecticide properties and is accumulated in organisms along with the parent compound. Less toxic to mammals than DDT.

dichlorodiphenyldichloroethene (DDE) One of the major products created during the metabolism of DDT within biological systems. The material has high lipid solubility and accumulates along with DDT in fatty tissue of animals. Like the parent compound, DDE is recalcitrant and demonstrates a persistence in the environment Many of the adverse effects on birds ascribed to DDT are actually the result of DDE. DDE degrades the metabolism of birds, resulting in reproductive failure.

dichlorodiphenyltrichloroethane (DDT)
One of the best known of the chemical insecticides. First synthesized in 1874, the insecticidal property was discovered by P. Mueller, a Swiss chemist, in 1939. Used widely during World War II and since to control lice that spread typhus and to reduce other VECTOR-BORNE diseases, such as mosquito-carried malaria and yellow fever. DDT and its degradation products have been shown to be persistent in the environment and to be accumulated by organisms within adipose tissue. Wild birds are adversely affected by DDT, producing eggs with very thin shells, thereby becoming unable to reproduce. Use of the agent has been severely restricted in the United States since 1972, but DDT is still applied in many other countries. It is less toxic to mammals than ENDRIN or the CARBAMATE insecticides.

Dichlorodiphenyltrichloroethane

dieback A dramatic decline in the number of individuals within a population of organisms.

dieldrin An insecticide belonging to the chlorinated cyclodiene class, which also includes aldrin and chlordane. Dieldrin is absorbed directly through skin contact and rapidly combines with blood serum proteins. The chemical accumulates in the fatty tissues of the body. Dieldrin is a neurotoxin that causes convulsive responses in high doses. Damage to both liver tissue and liver function has also been observed.

diethylstilbestrol (DES) A hormone used as a medication from the 1940s until 1970 (when it was banned) to prevent miscarriages in high-risk pregnancies. DES has been found to cause unusual vaginal cancers in young women and abnormal development of the genitalia of both men and women born to mothers who took the hormone during early stages of pregnancy.

diffraction A change in the AMPLITUDE or phase of waves that strike an object.

diffusion The movement of a material within a supporting medium, such as the movement of molecules of a pollutant in a wastewater outfall into the surrounding water or the movement of sugar molecules as a sugar cube dissolves in undisturbed water. Diffusion varies in response to changes in the concentration of the material. See GRAHAM'S LAW.

diffusion coefficient See FICK'S FIRST LAW OF DIFFUSION.

diffusivity The proportionality constant used to determine the mass movement of a chemical in a particular medium in response to changes in concentration.

digester A water pollution control device used to enhance the microbial decomposition or MINERALIZATION of PARTICULATE ORGANIC MATTER. The device operates by way of ANAEROBIC DECOMPOSITION of the waste.

digester gas The gas produced as a result of the microbial decomposition of PARTICULATE ORGANIC MATTER under anaerobic conditions. Methane and hydrogen are major components.

digestion The process of MINERALIZING or decomposing PARTICULATE ORGANIC MATERIAL to lessen the impact of the addition of domestic waste to streams.

dike 1. In risk management, a short wall of concrete, soil, or other material erected around a tank to contain a spill. 2. When used to describe a geological feature, a formation that extends vertically and cuts across geological strata in areas adjacent to old volcanoes. The dike is formed by the forcing of magma from below into a fracture in the overlying strata, forming a semivertical structure that cuts across geological strata from earlier deposits. The dike normally radiates away from ancient volcanic necks, may range in width from a few centimeters to hundreds of meters, and may radiate hundreds of meters from the former volcano. After erosion, a dike is usually a long narrow ridge. Compare SILL.

diluent A substance used to dilute a solution or suspension.

dilution factor The extent to which the concentration of some solution or suspension has been lowered through the addition of a DILUENT. The term is frequently used to describe the extent to which a sample must be diluted prior to the quantification of bacteria within the sample. Usually expressed as a negative exponent of 10 (10^{-3}) or a fraction (1/1000).

dimictic lake A stratified lake that undergoes two OVERTURNS each year. The water in lakes stratifies in response to differences in the temperatures of surface and deep waters. The surface water is warmer because of radiant heating by the Sun, and the bottom water is cooler and therefore denser. The waters in these two layers (termed the EPILIMNION on the surface and HYPOLIMNION on the bottom) are separated by a boundary referred to as the THERMOCLINE. This layering is disrupted in response to variation in air temperature (and surface water temperature) with the changing seasons. As the epilimnion cools, it sinks, mixing the water within the lake. Compare MEROMICTIC LAKE.

diminishing returns See THEORY OF DIMINISHING RETURNS.

dinitrophenols A family of chemical compounds that includes dinitrophenol (DNP), dinitroorthocresol (DNOC), and 2-*sec*-butyl-4,6-dinitrophenol (DINOSEB). The compounds have been used for a variety of purposes ranging from weed control to weight control (DNP). The acute toxic effects on humans generally are related to the ability of the compounds to uncouple the metabolic conversion of carbohydrates to the chemical energy used in cellular metabolism within the body. The result is an increase in the rate of metabolism that leads to restlessness, sweating, flushing of the skin, and fever. Chronic overexposure may cause fatigue, anxiety, and weight loss, among other problems.

dioxin See TETRACHLORODIBENZO-*para*-DIOXIN.

direct combustion reheat The injection of hot combustion gases into a flue gas to increase the buoyancy of the exhaust exiting the smokestack.

direct discharger A facility that releases an effluent into a stream or other waterway, as contrasted to a facility that sends an effluent to a publicly owned treatment works for treatment prior to discharge. Those facilities that release effluents directly into waterways operate with permits granted under the National Pollution Discharge Elimination System. See POINT SOURCE. Compare INDIRECT DISCHARGE.

direct filtration Technology employed for processing groundwater or surface water for use in a public water supply. Coagulants are added, the water mixed vigorously, and the mixture is then filtered to remove suspended particulate materials. The technology eliminates the time-consuming sedimentation process. See RAW WATER, WATER TREATMENT.

direct photolysis The transformation or decomposition of a material caused by the absorption of light energy.

direct plating A method used to identify fungi present in soil or litter. A known weight of soil or litter is distributed over the surface of an appropriate medium and incubated to facilitate the growth of fungi.

After development of the fungal colonies, the organisms are counted and identified. The method is used in a similar manner to quantify specific bacteria present in water or wastewater.

direct push Technology used to examine soil characteristics by driving, pushing, and/or vibrating small-diameter hollow steel rods into the ground. See CONE PENETROMETER TECHNOLOGY (CPT).

directivity A characteristic of a sound emission that is higher in some directions from the source than in other directions because the sound is emitted from one side of a machine, for example, or because the machine is placed against a wall. See DIRECTIVITY FACTOR.

directivity factor (*Q*) A numerical expression of the DIRECTIVITY of a noise. Expressed as $Q = P_0/P_1$, the ratio of the sound power (P_0) of a small spherical source emitting sound equally in all directions to the sound power (P_1) of the actual source, where each sound power produces the same measured SOUND PRESSURE LEVEL at a certain distance and angle.

directivity index Ten times the common LOGARITHM of the DIRECTIVITY FACTOR ($DI = 10 \log Q$); the index is used in the calculation of expected noise levels from a source, such as in the vicinity of a machine.

discharge The volume of water flowing past a point in a stream or pipe for a specific time interval.

discharge monitoring report (DMR) The form filed by holders of NATIONAL POLLUTANT DISCHARGE ELIMINATION SYSTEM (NPDES) permits that contains the results of the required periodic pollutant analyses of their water discharges.

discount rate A factor applied to future monetary payments or receipts (and any environmental costs or benefits, if put in monetary units) to account for the delay in the payment or receipt. The discount rate applied is a function of the value of immediate receipt and use of the money, the interest rate that could be earned during the year, and the possible inflationary decline in the money's value. The discount rate formula is a rearrangement of the equation to compute compound interest, $Vn = V_0(1 + r)n$, where V_0 is the value at time period zero, r is the interest rate, n is the number of compounding periods, and Vn is the value at the end of n periods. Thus, $V_0 = Vn/(1 + r)n$ is applied to discount back to a present value V_0, where r is called the discount rate, Vn is the future value to discount, and n is the number of discount periods. For example, if \$1.00 (*Vn*) is to be received one year from now ($n = 1$), and the sum of the considerations specified implies a discount rate of 10% ($r = 0.10$), then the present value (V_0) of \$1.00 to be received in one year is \$1.00/(1 + 0.10)[1], or 90.9 cents. The choice of discount rate can be the deciding factor when comparing current and future benefits with current and future costs of a proposed project.

discrete variable A quantitative variable that is limited to a finite or countable set of values, regardless of the accuracy of the measurement method. There can be no intermediate values between the numbers representing a discrete variable. For example, the number of persons residing in the state of Nevada must be represented by a whole number and exclude decimal fractions. See also CONTINUOUS VARIABLE.

disease A change from a state of health. A negative alteration in the condition of a person or other organism resulting from destabilizing effects of poor nutrition, infectious organisms, chemical contamination, or mental abnormality.

disinfection The killing of dangerous bacteria or other microbes in water, in wastewater, in the air, or on solid surfaces.

disinfection by-products (DBPs) Chemicals formed when chlorine, chlorine dioxide, or ozone is used to remove

PATHOGENS from drinking water. Despite the enormous public health benefits of chemical disinfection, DBPs may pose some health risk themselves. The DBPs are TRIHALOMETHANES, HALOACETIC ACIDS, CHLORITE, and BROMATE.

disintegration A spontaneous change in the nucleus of a RADIOACTIVE element that results in the emission of some type of radiation and the conversion of the element into a different (DAUGHTER) element.

dispersal The breaking up, spreading out, or distribution of some material released from a concentrated source to a more diffuse distribution within the environment.

dispersants Chemicals added to a material, such as crude oil, to promote the formation of smaller aggregates (break up a concentrated discharge in the environment).

dispersion The act of breaking up concentrations of some agent into a more diffuse distribution, for example, the spreading and mixing of a smokestack plume as it travels downwind. See AIR QUALITY DISPERSION MODEL.

dispersion coefficients Variables in the GAUSSIAN PLUME MODEL of horizontal and vertical dispersions of an air pollutant. Their value increases with increasing downwind distance from the air pollutant source and with increasing atmospheric turbulence. Larger values for the dispersion coefficients decrease the estimate of the average downwind pollutant concentration generated by the model. See AIR QUALITY DISPERSION MODEL.

dispersion model See AIR QUALITY DISPERSION MODEL, GAUSSIAN PLUME MODEL.

dispersion parameters See DISPERSION COEFFICIENTS.

displacement savings In waste management, financial savings realized by replacing natural gas or electricity from

utility companies with METHANE gas produced and recovered from a solid waste landfill.

disposal All activities associated with the long-term handling and disposition of solid waste of all types. Ultimate disposition may involve placement in a sanitary landfill or incineration.

disposal facility A location dedicated to the handling, treatment, or ultimate interment of solid or hazardous waste. The operation may be a landfill or incinerator.

disposal pond A small, usually diked, enclosure that is open to the atmosphere and into which a liquid waste is discharged. See LAGOON.

disposal site See DISPOSAL FACILITY.

dissociation The partial or total separation of certain molecules into IONs when dissolved in water; for example, carbonic acid in water produces a positively charged hydrogen ion and a negatively charged bicarbonate ion. See DISSOCIATION CONSTANT.

dissociation constant At equilibrium, the ratio of the molar concentrations of the ions produced by the DISSOCIATION (separation) of a molecule to the molar concentration of the undissociated molecule. For acids, the ratio is abbreviated Kq and is often expressed as pKa, the negative logarithm of Ka For bases the terms are Kb and pKb. For example, carbonic acid dissociates into hydrogen ions and bicarbonate ions. At 25° C, Ka, the ratio of the molar concentrations of hydrogen ions and bicarbonate ions to the molar concentration of carbonic acid, is 4.47×10^{-7} and pKa is 6.35.

dissolved air flotation A technique used to separate oil and SUSPENDED SOLIDS from water. Air is bubbled upward in a tank, carrying the oil and solids to the surface. This frothy top layer is removed by a skimmer.

dissolved gases Gases that are in solution homogeneously mixed in water.

dissolved organic carbon (DOC) A measure of the organic compounds that are dissolved in water. In the analytical test for DOC, a water sample is first filtered to remove particulate material, and the organic compounds that pass through the filter are chemically converted to carbon dioxide, which is then measured to compute the amount of organic material dissolved in the water. See TOTAL ORGANIC CARBON.

dissolved organic matter (DOM) See DISSOLVED ORGANIC CARBON.

dissolved oxygen (DO) The amount of molecular oxygen (O_2) dissolved in water. The units, often not expressed, are milligrams of oxygen per liter of water. DO is an important measure of the suitability of water for aquatic organisms. A level of 8 or 9 represents the concentration that one would expect to encounter in streams that have not been polluted with the organic waste common in domestic sewage. Waters with a dissolved oxygen value of 4 and below are not suitable for habitation by many forms of animal life.

dissolved oxygen sag curve See OXYGEN SAG CURVE.

dissolved solids Primarily, the inorganic salts in solution (homogeneously mixed) in water. These chemicals cannot be removed by filtration and must be recovered by evaporation of the water.

distillation A process of separation and/or purification of the components of some substance based on differences in their boiling points. The components within a mixture of two liquids can be separated if the two liquids have different boiling points. The mixture is heated, and the material with the lower boiling point is converted to a vapor first. That vapor can then be condensed to produce a liquid that consists primarily of the single liquid.

distillation tower An apparatus used to separate the components of crude oil into different fractions, depending on their relative boiling points. See DISTILLATION.

distilled water Water that has been purified by the DISTILLATION process. Water that contains various chemicals or ions in solution is heated to boiling and the water vapor is condensed. The process leaves behind various inorganic ions and produces water that is free of dissolved salts.

distribution In toxicology, the transport and diffusion of an absorbed material within the body. The body distribution of a compound is a function of the size, water or fat solubility, and ionization, among other factors, associated with the material.

disturbance In environmental management, any activity or event that disrupts natural ecosystems, biological communities, or specific populations of organisms, or an activity or event that significantly alters the physical environment.

dithiocarbamates A group of compounds of low toxicity used as fungicides to protect seeds and vegetable products. The two most common are ferric dimethyldithiocarbamate (Ferbam) and zinc dimethyldithiocarbamate (Ziram).

diurnal Daily; exhibiting a daily cyclical pattern.

divergence In METEOROLOGY, the flow of air in different directions. In high-pressure systems, air diverges from the center, where other air masses sink to replace it

divergent plate boundary In the theory of plate tectonics, a boundary between two plates that make up the crust of the Earth. The boundary is characterized by a chasm between the two plates, filled with molten rock from within the Earth.

diversion 1. The use of part of the water in a stream for irrigation or as a drinking water supply. **2.** A structure used to control the flow of water across a surface for the purpose of preventing erosion. **3.** The movement of part of the municipal solid waste produced in a community into recycling, reuse, or composting programs to lessen the amount of material that must be disposed of by incineration or use as landfill.

diversity In ecology, a measure of the number of different species, along with the number of individuals in each representative species, in a given area. Undisturbed environments tend to be characterized by high diversity, whereas polluted environments tend to exhibit low diversity. See RICHNESS, EVENNESS.

diversity index A mathematical expression that depicts species DIVERSITY in quantitative terms. The Shannon-Weaver index is a widely used measure.

DNA See DEOXYRIBONUCLEIC ACID.

DNA hybridization Use of a segment of deoxyribonucleic acid (DNA) to detect specific genes by forming complexes of DNA derived from more than one source. The DNA of a cell consists of two long chains of nucleotides that are held together as a result of specific binding between bases (adenine to thymine and guanine to cytosine) and that are intertwined to form a double helix. The ordered sequence of bases is different in every organism, and the two interwoven strands of a single organism are termed *complementary* because they will bind to each other. This property can be used to look for specific DNA sequences or genes. A preparation of a single strand of the DNA from the gene in question, sometimes termed a probe, can be combined with a DNA mixture. If a complementary piece of DNA is present, a complex consisting of pieces of DNA from the two sources (a hybrid) is formed.

Dobson unit (DU) A measure of the amount of ozone in a column of air starting from the ground and extending up through the atmosphere. One Dobson unit equals 2.7×10^{16} ozone molecules in the air column above one square centimeter of the Earth's surface. Dobson units are determined by measuring the degree of ULTRAVIOLET light absorption above a point on the Earth. Used to describe the extent of OZONE LAYER DEPLETION.

dolomite A natural mineral consisting of calcium magnesium carbonate, $CaMg(CO_3)_2$. Also referred to as limestone, a form of marble, and dolomitic lime. Upon heating, carbon dioxide is released from the mineral.

domestic sewage Wastewater and solid waste that are characteristic of the flow from toilets, sinks, showers, and tubs in a household.

domestic waste See DOMESTIC SEWAGE.

domestic water Water used within a household.

dominant gene A gene that is expressed even though it may be paired with a matching recessive gene. When male and female gametes fuse in the reproductive process, individual gametes furnish one member of each pair of chromosomes that will determine the genetic makeup of the offspring. If the gametes are from plants and if plant height is governed by one gene, the possible genes may be designated *h* for the short variety and *H* for the tall variety. In the case where *H* is dominant and *h* is recessive, a plant with gene pair *Hh* would be tall, since *H* is dominant and would be expressed despite the presence of the gene *(h)* that codes for the short variety of plant.

Donora episode The week-long STAGNATION period of high air pollution levels in Donora, Pennsylvania, during October 1948. See EPISODE.

donor-controlled flow In BOX MODELS, a flow of a substance from one compartment to one or more others. The movement depends on the STOCK of the substance in the originating compartment.

dose The amount of chemical agent or radiation to which an organism is exposed, or the amount the organism absorbs over a specific period.

dose commitment The amount of radiation to which the body is exposed through ingestion and retention of radioactive substances.

dose-distribution factor A consideration of the effects of some dangerous substance in the body when the substance is concentrated in specific areas rather than randomly or evenly distributed.

dose equivalent (DE) A measure of EFFECTIVE RADIATION DOSE, computed as $DE = D(QF)(DF)$, where D is the absorbed dose, QF is a QUALITY FACTOR, and DF is a distribution factor. Typically expressed in REMS or SIEVERTS.

dose rate The amount of a chemical or radiation to which a person is exposed, or that is absorbed, per unit of time.

dose response An adverse effect in an organism attributable to a particular physical or chemical agent.

dose-response assessment A description, using the DOSE-RESPONSE RELATIONSHIP between certain levels of exposure to a chemical or physical agent and the anticipated adverse effects, of the extent of disease, injury, or death resulting from an estimated exposure to a population.

dose-response curve A plot showing the changing response in a group of organisms at various levels of exposure to a chemical or physical agent.

dose-response relationship The quantitative relationship between an exposure to a physical or chemical agent and a subsequent biological effect. The exposure to a chemical substance can be described as the mass ingested or inhaled or as the

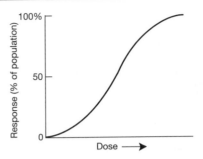

Dose-response relationship

mass actually absorbed into the body tissues; the dose units can be in mass of the chemical per unit body weight on a one-time basis or mass per unit body weight per day for an extended period or for a lifetime. The response can be assessed in terms of the severity of biological effect observed or the portion of a test group exhibiting a particular adverse effect, usually compared with an unexposed (control) group.

dosimeter A device to measure the amount of radiation to which a person is exposed. It can be a simple FILM BADGE or a complicated ionization meter.

dosimetry The measurement of the amount of radiation to which a person is exposed or that a person absorbs.

double bottoms Watertight compartments along the bottom of a vessel that provide extra protection against cargo spills. See OIL POLLUTION ACT OF 1990.

double recessive The circumstance in which recessive genes in an organism are expressed because they are not paired with a dominant gene that masks the recessive character. When male and female gametes fuse in the reproductive process, individual gametes furnish one member of each pair of chromosomes that will determine the genetic makeup of the offspring. If the gametes are from plants and if plant height is governed by one gene, the possible genes may be designated h for the short variety and *H* for the tall variety. In the case where *H* is the DOMINANT GENE and *h* is the RECESSIVE GENE, a plant with gene pair *hh* (double recessive) would be short since the dominant *H* gene is not present to force the development of the tall variety.

doubling dose The amount of a specific radiation exposure needed to produce a doubling of the natural rate of appearance of genetic or somatic (cellular) abnormalities.

doubling time The time required for a population to double in size.

downgradient well One or more monitoring wells placed to sample groundwater that has passed beneath a facility with the potential to release chemical contaminants into the ground; one of a network of groundwater-monitoring wells required by RESOURCE CONSERVATION AND RECOVERY ACT regulations for landfills or surface impoundments that treat, store, and/or dispose of hazardous waste. Results of *testing* downgradient well water are compared with data from an UPGRADIENT WELL to determine whether the facility may be contaminating the groundwater.

downwash See STACK DOWNWASH.

draft In air quality control, a flow of gas or vapor created by a pressure difference. See FORCED DRAFT, INDUCED DRAFT, NATURAL DRAFT.

draft environmental impact statement (DEIS) An unfinished version of an ENVIRONMENTAL IMPACT STATEMENT that is sometimes circulated for review.

drain field See LEACHING FIELD.

drainage Water that is in excess of the quantity that will infiltrate the ground and consequently flows off the ground into the nearest collection area.

drainage basin That extent of land that drains water into some lake, river,

stream, or other body of water. The quality of the water entering the lake or stream depends on the nature of the soils and/or pollutants in the area draining into the lake or stream. See WATERSHED.

drainage well A borehole designed to transport water from the surface into a subterranean geological deposit. Such systems provide for rapid migration of surface pollutants into groundwater.

Draize test The application of a chemical into an albino rabbit's eye to test its potential irritative ability to the human eye. In vitro tests are being considered as a replacement.

drawdown The vertical difference in the water table of an undisturbed aquifer and the water table in the same aquifer after the removal of water by pumping from wells.

dredge and fill permit See SECTION 404 PERMIT.

dredging The removal of mud and sediment from the bottom of a stream, river, channel, lake, or other body of water. The activity may be done to improve water flow, facilitate boat or ship traffic, remove contaminated sediment, or achieve some other purpose related to deepening of a waterway. Dredging may contribute to water contamination by disturbing and resuspending pollutants buried in sediment, destroy BENTHIC communities, or decrease water quality through increases in the TURBIDITY of the water. Dredging in coastal areas may allow SALTWATER INTRUSION, damaging freshwater habitats. The disposal of the sediment (termed dredge spoils) removed from the waterway also presents complicated environmental choices, especially when dredging involves the removal of contaminated materials.

drift Mist that escapes from cooling towers used in CLOSED-CYCLE cooling systems.

drift nets In the last several decades, large-scale nylon monofilament nets that were strung for miles in the ocean by fishing boats. Large numbers of sea birds, albacore, blue sharks, turtles, and dolphins were shown to be caught and killed in the nets, leading to multiplying bans on their use.

drift velocity For particles collected by an ELECTROSTATIC PRECIPITATOR, the rate at which the particles move toward the collection plates. The measurement is used in electrostatic precipitator design and in estimation of particle collection efficiency. See DEUTSCH-ANDERSON EQUATION.

drill cuttings Rock or other materials forced out of the borehole as a well is drilled.

drilling fluid A dense fluid material, often containing bentonite clay and barite, used to cool and lubricate a well-drilling bit, seal openings in the wall of the borehole, transport DRILL CUTTINGS to the surface, reduce drill pipe friction, and control well pressure.

drinking water, bottled Water sold to the public in sanitary containers for consumption. The water may be from any approved source, including a public water supply, provided that disinfection and filtration processes have been applied and the system meets the contaminant standards established under the Clean Water Act. The water must by calorie-free and sugar-free and contain no sweetening or chemical additives other than flavoring extracts and essences. The product may be sodium-free or contain small amounts of sodium. Bottled water products must comply with standards established by the Food and Drug Administration. See BOTTLED WATER; BOTTLED WATER, CARBONATED; BOTTLED WATER, STILL.

drinking water supply Water provided for use in households, from surface supplies (rivers, lakes, and reservoirs) or subsurface supplies (aquifers). The distribution of water to households is regulated under the SAFE DRINKING WATER ACT of 1974 as amended.

drip irrigation A conservative use of water to irrigate a crop by using a pipe or hose that delivers a small amount of water to each plant over a long period rather than spraying a large amount over a complete field or flooding furrows between rows of crop plants. The technology uses less water and reduces the possibility of salt accumulation.

droplet A small airborne liquid particle that is larger than liquid AEROSOL and therefore settles out of the atmosphere relatively quickly.

drop-off recycling program A community recycling program requiring citizens to take items for recycling to a designated location for collection. Compare CURBSIDE COLLECTION.

dry adiabatic lapse rate The ADIABATIC LAPSE RATE for air not saturated with water vapor, or 0.98° C per 100-meter rise (0.54° F per 100 feet). Expressed as

$$-\frac{dT}{dz} = \gamma d$$

where T is air temperature, z is altitude, and γd the dry adiabatic lapse rate. Compare WET ADIABATIC LAPSE RATE.

dry alkali injection See DRY SCRUBBER.

dry-bulb temperature The temperature reading from an ordinary thermometer; the reading is used with the WET-BULB TEMPERATURE to compute RELATIVE HUMIDITY.

dry deposition The introduction of acidic material to the ground or to surface waters by the settling of particles containing sulfate or nitrate salts. Compare WET DEPOSITION.

dry gas meter A device used in the field to measure gas flow rates in air sampling systems and in the laboratory to calibrate other flow meters. The gas being monitored should be free of moisture and corrosive material to prevent damage to internal parts of the meter.

dry mass The mass of a sample after all water has been removed.

dry scrubber 1. An air pollution control device that removes SULFUR DIOXIDE from stack gases by injecting a finely divided dry chemical reagent such as limestone, trona (Na_2CO_3), or nahcolite ($NaHCO_3$) into the flue gas. The sulfur dioxide reacts with the injected chemical to form a solid particulate, which is captured in a BAGHOUSE. 2. A gravel bed filter used to collect particulate matter from gas streams. See also FLUE GAS DESULFURIZATION.

dry weather flow The lowest volumetric flow of water in a stream; occurs during the driest months of the year. See 7Q10.

Dubos, Rene (1901–82) American (French-born) Bacteriologist. Author of classic works in environmental health: *So Human an Animal* (1968), *Man Adapting* (1965), and *Only One Earth: The Care & Maintenance of a Small Planet* (1972), with BARBARA WARD.

Ducks Unlimited North American wetlands and waterfowl conservation organization; founded in 1937; 1999 membership 553,000. Headquarters in Memphis, Tennessee. Website: www.ducksunlimited.org

duff Partially decomposed organic material found on a forest floor.

dump See OPEN DUMP.

dust Suspended solid particles in the atmosphere. Natural dust sources include wind erosion and volcanic eruptions; human sources are activities involving the crushing, grinding, or abrasion of materials.

dust bowl A semiarid geographical area in which cultivation has led to extensive wind erosion of topsoil. It is often used to refer to the area of the United States that in the mid-1930s experienced soil loss and dust storms, including parts of Texas, New Mexico, Oklahoma, Colorado, Utah, and Kansas.

dustfall The amount of PARTICULATE MATTER that settles out of the atmosphere at a particular location within a certain time. Measured with a DUSTFALL BUCKET, it is a crude measure of air quality and is no longer widely used.

dustfall bucket An open bucket or jar placed to collect larger-diameter (greater than 30-MICROMETER) particulate matter settling out of the air. The mass of the dust collected during a certain period is translated to dustfall units of mass per area per time, such as tons per square mile per month.

Dutch elm disease Fatal infectious disease of elm trees (genus *Ulmus*). The disease is caused by a fungus (genus *Ceratocyslis*) and is spread from tree to tree by the elm bark beetle (genus *Scolytus*).

dynamic equilibrium state The condition of a system in which inflow of materials or energy equals outflow.

dynamic viscosity See VISCOSITY.

dysentery A disorder of the gastrointestinal tract characterized by severe diarrhea with blood and pus in the feces. The disease frequently results from an infection by bacteria of the genus *Shigella*.

dystrophic Describes the water in a lake that is high in humic substances and plant degradation products, resulting in a brown color. Plant life and animal life are typically sparse, and the water has a high OXYGEN DEMAND.

E

e The base of the NATURAL LOGARITHM, approximately equal to 2.718282.

early reductions program A feature of the 1990 amendments to the CLEAN AIR ACT; it gives sources of AIR TOXICS a six-year extension to comply with MAXIMUM ACHIEVABLE CONTROL TECHNOLOGY (MACT) if they voluntarily reduce their emissions of a listed hazardous air pollutant by 90% before the MACT standard is announced.

Earth Day When first held April 22, 1970, the day was observed on American college campuses with demonstrations and "teach-ins" on environmental issues and was an early indication of the extent of political support for the environmental movement in the United States. Commemorated on a lesser scale in subsequent years until the widely observed 20th anniversary in 1990, which coincided with a renewed worldwide concern for environmental issues. For Earth Day activities see Earth Day Network (www.earthday.net).

Earth Island Institute An environmental quality and wildlife protection organization founded in 1982 by DAVID BROWER. Membership of 20,000 in 1999. Headquarters in San Francisco. Website: www.earthisland.org

Earth Summit A meeting organized by the United Nations Conference on Environment and Development, held in Rio de Janeiro, Brazil, in June 1992. Attended by representatives of over 170 countries, who adopted agreements on forestry practices, biological DIVERSITY, GLOBAL WARMING, and SUSTAINABLE DEVELOPMENT. See AGENDA 21.

Earthwatch A global environmental data system begun by the UNITED NATIONS ENVIRONMENT PROGRAM in 1972. See GLOBAL ENVIRONMENT MONITORING SYSTEM, INTERNATIONAL ENVIRONMENTAL INFORMATION SYSTEM, INTERNATIONAL REGISTER OF POTENTIALLY TOXIC CHEMICALS.

ecesis The successful settlement of a plant or animal in a new habitat through colonization, migration, or introduction by humans.

Eckenfelder's equation One of several equations predicting performance of a TRICKLING FILTER wastewater treatment system. Given the filter area and depth, the volumetric flow of wastewater, and the BIOCHEMICAL OXYGEN DEMAND of the wastewater entering the treatment plant, the equation estimates the reduced biochemical oxygen demand of treated wastewater. One form of the equation is

$$St = S_0 \exp\left[- KD(A/Q)x\right]$$

where St is the biochemical oxygen demand of the treated wastewater, S_0 is the influent biochemical oxygen demand, exp is the base of the natural logarithm, K is a rate constant, D is the trickling filter depth, A is the filter area, Q is the wastewater flow rate, and x is an empirical constant determined by the type of filter medium. Named for W. W. Eckenfelder.

ecocentric A type of ENVIRONMENTALISM taking the view that large-scale ecological processes such as evolution, adaptation, and biogeochemical cycling are the most important aspects of nature. Moral values and rights are ascribed to ecological processes, and the whole is considered more important than the individual parts.

Consequently, efforts to conserve and protect entire natural systems should be given priority. Those holding this view challenge the industrial culture's view of humans as the center of nature.

ecofeminism A type of ENVIRONMENTALISM that views environmental problems as manifestations of a worldview riddled by hierarchical (male) domination systems based on race, colonialism, class, gender, and sexuality. Those holding this view seek to integrate various protest groups into a movement arrayed against a culture that is based on the domination of nature by humankind. The degradation of the natural environment is seen to spring from the same forces responsible for the exploitation and mistreatment of women, children, racial minorities, and homosexuals. When considering the relationship between humankind and nature, ecofeminists seek to build a culture based on pluralistic, nonhierarchical, and caring relationships that emphasize kinship and appropriate reciprocity.

ecojustice A concept holding that injustice in the social order in terms of restrictions of the rights of fellow human beings and lack of integrity in the natural order spring from the same source. In addition to considering the rights of human beings, we should consider the rights of other species to exist.

ecolabeling Information or logos on product labels that notify consumers that the product is "environmentally friendly": it is made of recycled material, is recyclable, is produced from renewable resources, produces lower emissions, for example. See GREEN SEAL, BLUE ANGEL, RECYCLING SYMBOL.

ecologic fallacy In EPIDEMIOLOGY, an erroneous association between a characteristic or an exposure common to a group and a health effect or a condition present in that group. For example, although most residents of a county might drink bottled water, an elevated rate of gastrointestinal illness in the county may not necessarily be associated with bottled water exposure. The ecologic fallacy is prevented by developing disease rates for subsets of a group, thereby preventing the potentially false assumption that average or majority group characteristics apply equally to individuals making up the group.

ecological density The number of organisms per unit area or per unit volume of a segment or compartment of a larger, complete ecosystem. This segment is referred to as a specific HABITAT within the larger system. An example would be the mud habitat on the floor of a lake ecosystem. The lake would include not only the bottom mud but also the water, shoreline, and other smaller habitats within the complete lake ecosystem. The density is also reportable as the weight of BIOMASS within the specific habitat. Contrast with CRUDE ECOLOGICAL DENSITY, which includes the biota in all compartments of a complete lake ecosystem.

ecological economics A transdisciplinary field of thought that applies the insights of ecology to economic analysis. A major concern of those who work in this area is the equitable distribution of resources and rights within the present human generation and among the present and future generations.

ecological efficiency The effective transfer of useful energy from a food to the animal consuming the food. The term is normally applied to the feeding structure of animal communities in the natural environment, and the transfer is usually expressed as a percentage; 10% is the usual energy transfer. See TROPHIC LEVEL, FOOD WEB, FOOD CHAIN.

ecological equivalents Unrelated species of plants or animals, living in different geographical locations, that are similar in appearance and perform similar ecosystem functions. For example, the zebra of Africa and the buffalo of North America both occupy grasslands and are grazers. A product of CONVERGENT EVOLUTION.

ecological impact The effect that some activity has on living organisms of a defined area.

ecological indicator An individual species or a defined assemblage of organisms that serves as a gauge of the condition of the environment. For example the bacterium *Escherichia coli* indicates the presence of sewage in water, and the mussel *Mylilus edulis* lives in polluted water.

ecological integrity See RESILIENCE STABILITY, RESISTANCE STABILITY.

ecological niche The activities and relationships of an organism within its environment. See NICHE.

ecological pyramid A construct depicting the relationships among organisms of various producing and feeding groups within a specified area or location. The diagram attempts to show the transfer of useful energy from one feeding group to the next in a semiquantitative fashion. The amount of plant BIOMASS in an area is normally placed at the bottom, and the mass of each succeeding feeding group is placed on top, with CARNIVORES occupying the highest level. The result is a figure resembling a pyramid. The diagram is also referred to as an energy pyramid when the caloric content of the organisms in each trophic (feeding) level is used to construct it. See PYRAMID OF BIOMASS.

ecological risk assessment The application of a formal framework, analytical process, or model to estimate the effects of human actions on the natural environment and to interpret the significance of those effects. The analysis includes hazard identification, exposure and dose-response assessments, and an evaluation of the level of risks involved.

ecological succession See SUCCESSION.

ecological sustainability An environmental management approach that stresses the mutual importance and inter-dependence of ecological, social, and economic components of a region.

ecology The study of the relationships among organisms and between organisms and their surroundings.

economic poison Chemicals used as pesticides as well as those used to remove the leaves from crops to facilitate harvesting.

economic threshold In the management of insect pests in an agricultural setting, the point at which the cost of damage done by insect pests exceeds the cost of pest control. Beyond this point, the money spent on the application of pesticides and other measures is justified on the basis of potential economic losses that will result if control measures are not implemented. Spending $100 to prevent $10 worth of damage to crops makes little economic sense, whereas spending $100 to prevent $1000 worth of damage seems more wise. The economic threshold in this example would be projected crop damage at the $100 level.

economizer A device in a boiler that warms FEEDWATER by using the hot exhaust gases from the furnace.

ecosphere Those environments that harbor living organisms. See BIOSPHERE.

ecosystem A level of organization within the living world that includes both the total array of biological organisms present in a defined area and the chemical-physical factors that influence the plants and animals in it. The study of an ecosystem demands that the investigator consider all biological and nonbiological variables operative within a defined area.

ecosystem management An integration of ecological, economic, and social goals in a comprehensive approach to management of natural resources. Implicit in the approach is a recognition that sustained progress toward social goals cannot be made in a deteriorating environment.

The general principles of this approach are managing whole landscapes, watersheds, or regions over ecological timescales; considering human needs and promoting sustainable economic development; maintaining biological DIVERSITY and natural biogeochemical processes; utilizing cooperative institutional agreements; integrating science and management; assuring stakeholder and public involvement; encouraging collective decision making; and adapting management decisions to experience. The goals of the approach revolve around accommodating human use to the viability of the natural environment.

ecosystem restoration The active manipulation of nature to recreate species composition and biological community structure as close as possible to the state that existed before environmental degradation attributable to humans. For example, the acreage of a farm originally carved out of the prairie would be managed in such a way that the area would revert to an environment resembling a natural prairie to the extent possible with the technology and practices available. The terminology is also applied to the recovery of natural systems damaged by natural events, for example, the rebuilding of coastal marsh lost to erosion.

ecotone A zone of transition between two well-defined vegetated areas.

ecotourism The use of natural wilderness resources to attract visitors by providing travel, cultural exploration, and experiences in a wild setting. Visitors viewing wildlife, a forest, or another natural setting produce income for the native inhabitants and the natural resources are maintained for future use. This differs from the model of economic development, which involves resource extraction without replacement and its attendant environmental degradation. Achieving this ideal of SUSTAINABLE DEVELOPMENT of a resource is difficult, especially if the local area becomes a popular stop for tourists.

ectoparasite An organism (a parasite) that damages a host by attaching itself to the outside of the host organism, for example, a leech attached to a person.

ectotherm An organism that lacks the ability to regulate internal body temperature. The internal body temperature of this type of organism reflects the temperature of the environment. Fish are ectotherms. Also referred to as poikilothermal or cold-blooded. Compare ENDOTHERM.

edaphic Describing soil characteristics, such as water content, pH, texture, and nutrient availability, that influence the type and quantity of vegetation in an area.

edaphic climax The climax stage of an ecological SUCCESSION in which the COMMUNITY is in equilibrium with localized soil or climatic conditions.

eddy diffusion The mixing of clean air and contaminated air by parcels (eddies) of air moving in a random, irregular manner. The eddies that cause this dilution are produced by MECHANICAL TURBULENCE and/or THERMAL TURBULENCE. Turbulent eddies that are roughly the same size as the pollutant mass are most effective at dilution. If the eddies are small relative to a plume of contaminated air, they will only dilute the outside edges of the plume; if they are much larger than the plume, the entire plume will be moved without significant dilution.

edema The accumulation of an excessive amount of fluid in body tissues, characteristic of damaged tissue.

edge effect The observed increase in the number of different species along the margins of two separate communities of vegetation, the ECOTONE.

effective concentration dose The amount of a chemical or physical agent that, after exposure, absorption, and metabolism, actually reaches a target organ or tissue and is able to cause an adverse effect.

effective half-life The time required for the amount of a RADIOACTIVE element or compound absorbed by a living organism to be reduced by a factor of 50%. The decreased radioactivity in the body is a combination of the natural decay of the radioactive element and the excretion of the element or compound by biological processes.

effective radiation dose See DOSE EQUIVALENT.

effective stack height In air pollution dispersion calculations, the height at which the pollutants begin their descent to the ground; the sum of the physical stack height and the length of the PLUME RISE. Gas exiting a stack rises above the stack exit by VERTICAL INERTIA due to the buoyancy of the stack gas. Higher exit velocities and higher exhaust gas temperatures cause a higher plume rise. The greater the effective stack height, the greater air mass for the pollutants to mix in and disperse, with resulting lower ground-level concentrations downwind of the stack.

effluent Wastewater that flows into a receiving stream by way of a domestic or industrial POINT SOURCE.

effluent limitation An amount or concentration of a water pollutant that can be legally discharged into a water body by a POINT SOURCE, expressed as the maximum daily discharge, the maximum discharge per amount of product, and/or the concentration limit in the wastewater stream, as a 24-hour or 30-day average. The applicable TECHNOLOGY-BASED standard is set by the USEPA by INDUSTRIAL CATEGORY or subcategory but differs between new and existing sources and by broad types of water pollutants: CONVENTIONAL POLLUTANTS; toxic pollutants; nontoxic NONCONVENTIONAL POLLUTANTS; dredge and fill wastes; and heat discharges.

effluent standard The maximal amounts of specific pollutants allowable in wastewater discharged by an industrial facility or wastewater treatment plant. The standards are set for individual pollutants and apply across industrial categories. The term can be contrasted with EFFLUENT LIMITATIONS, which are set for individual pollutants by INDUSTRIAL CATEGORY.

E_h An expression of REDOX POTENTIAL.

Ehrlich, Paul R. (1932–) American Entomologist, ecologist. Author of *The Population Bomb* (1968), predicting Malthusian disaster; also, with his wife, Anne, and John Holdren, *Ecoscience: Population, Resources, Environment* (1970); and with Anne: *Extinction: The Causes & Consequences of the Disappearance of Species* (1981). See SIMON, JULIAN.

Ekman dredge A device for sampling bottom sediments from lakes or streams with a slow current. A metal box, typically six by six by six inches, the bottom of which can be opened and closed by the movement of spring-loaded jaws. The box, mounted on a pole or line and with jaws locked in the open position, is driven into soft sediments. The jaws are then closed to trap sediments within the box sampler for removal to the surface. The Ekman dredge is most useful in the sampling of soft mud sediments for the examination of biota entrained within or in the sampling of soft sediments for the analysis of both the physical properties of the sediment and the chemical contaminants that may be in the mud.

Ekman spiral The change in wind direction with altitude caused by the varying effect of surface friction. At high altitudes (above 700 meters), wind direction is unaffected by surface friction and is determined by a balance between PRESSURE GRADIENT and CORIOLIS forces. At lower altitudes, surface friction slows wind velocity, thereby reducing the Coriolis force, and the wind turns toward the lower pressure. The change in wind direction from the surface to the top of the friction layer can be represented by a spiral.

Ekman water bottle A tubular device used to sample water at selected depths. The tube is open at both ends as it is lowered by a line but flaps at the ends snap shut when a weight is dropped down the line, thereby enclosing the water sample derived from the specified depth.

El Niño Spanish for "the boy," specifically the Christ child. Named by fishermen who noted that in some years surface water temperatures were higher beginning in December. The affected area is off the coast of South America in the equatorial latitudes. The warmer water can last for several months or, in some cases, several years. The degree of warming is also variable. El Niño has been linked to large-scale changes in weather patterns.

electricity The form of energy arising from the movement or accumulation of electrons. The movement of electrons (called electric current) produces a magnetic field, a phenomenon used to convert electrical energy to mechanical energy in electric motors. Conversely, generators use mechanical energy to move a magnetic field, producing an electric current in a conductor.

Electric Power Research Institute (EPRI) An industry organization, based in Palo Alto, California, that sponsors research and development in technologies for the production, distribution, and use of electricity, including the control of environmental impacts. Website: www.epri.com

electrolysis The passage of an electric current through an electrolyte, causing the migration of the positively charged ions to the negative electrode (the cathode) and of the negatively charged ions to the positive electrode (the anode).

electrolyte Any compound that dissociates into IONS when dissolved in water. The solution that results will conduct an electric current. For example, table salt (NaCl) is an electrolyte.

electrolytic Pertaining to ELECTROLYSIS or an ELECTROLYTE.

electrolytic recovery An electrochemical technology designed to decrease pollution and enhance recycling of metal ions from process baths and solutions. Metal ions are removed from waste streams by passing the waste through an electrolytic cell consisting of closely spaced anodes and cathodes. The technology is employed primarily in the metal-finishing industry for process waste containing cadmium, copper, nickel, silver, and gold.

electromagnetic radiation Wave functions that are propagated by simultaneous periodic variations of electric and magnetic field intensity. Types of electromagnetic radiation range from those of short wavelengths, such as X rays and gamma rays, to those of long wavelengths, such as radio waves. Also included are ultraviolet radiation, visible light, infrared radiation, and microwaves. The shorter the wavelength, the more dangerous is the radiation to humans. All types of electromagnetic radiation travel through a vacuum at the speed of light.

electromagnetic spectrum The range of types of ELECTROMAGNETIC RADIATION in wavelength (or frequency) order.

electromagnetic terrain conductivity A noninvasive survey of soil for the detection of buried metal objects and chemical contaminants that conduct an electrical current. The method is especially useful in the detection of electrically conductive LEACHATES and contaminant plumes.

electron One of the elementary particles of an atom. Negatively charged electrons surround the positively charged nucleus of an atom and occur in numbers equal to the number of protons in the nucleus. An electron has a mass 1/1837 that of a proton.

ELECTROMAGNETIC SPECTRUM (APPROXIMATE)		
Radiation	*Wavelength (m)*	*Frequency (Hz)*
Gamma radiation	10^{-13}–10^{-10}	10^{19}–10^{21}
X rays	10^{-12}–10^{-9}	10^{17}–10^{20}
Ultraviolet radiation	10^{-9}–10^{-7}	10^{15}–10^{18}
Visible light	10^{-7}–10^{-6}	10^{14}–10^{15}
Infrared radiation	10^{-6}–10^{-4}	10^{12}–10^{14}
Microwaves	10^{-4}–1	10^{9}–10^{12}
Radio waves	1–	–10^{9}

electron capture detector (ECD) A sensitive detector used in conjunction with a gas chromatograph, an instrument used to measure small amounts of compounds in samples collected from sources in the environment. The chromatograph employs a gas stream to carry and separate the organic compounds of interest. As the carrier gas (normally nitrogen) flows through the detector, the gas is ionized by a radioactive substance. The electrons resulting from this ionization migrate to an anode, creating a voltage. When the compounds in question are encountered in the carrier gas, the electric current is interrupted. The decrease in the current is proportional to the concentration of solute in the gas. The device can detect as little as 10^{-14} gram of organochlorine pesticide.

electron volt (eV) A unit equivalent to the amount of energy gained by an electron passing through a potential difference (the difference in the strength of an electric field between two points) of one volt in a vacuum. One electron volt is equal to 1.6×10^{-12} erg or 1.6×10^{-19} joule.

electrophoresis A technique used to separate, identify, and quantify proteins and similar macromolecules as these molecules migrate within gel or cellulosic substrates under the influence of an electric current.

electrostatic precipitator (ESP) A device to control the release of particles from an exhaust into the atmosphere. An electric charge is first applied to the particles, and the charged particles are then collected on the surface of oppositely charged plates.

element A chemical substance that cannot be broken down to simpler units without changing its chemical properties. The atoms of an element all have equal numbers of protons and electrons. The list of 103 known elements includes oxygen, carbon, nitrogen, iron, silver, and gold. See the Appendixes.

elementary neutralization unit A device for treating wastes that are defined as hazardous because of their CORROSIVITY.

elevation head The potential energy in a hydraulic system, represented by the vertical distance between the hydraulic system (pipe, channel, and so on) and a reference level, expressed in length units. The sum of the elevation head and PRESSURE HEAD is the HYDRAULIC HEAD. See also HEAD, TOTAL.

elution The removal of an adsorbed material from a GAS CHROMATOGRAPH column by a gas stream.

elutriation The removal of particulate matter from a fluid stream by gravitational settling as the fluid moves upward.

EMAP data See ENVIRONMENTAL MONITORING AND ASSESSMENT PROGRAM.

emergency core cooling system (ECCS) A complete system of pumps, piping, water reservoirs, and heat exchangers that serves as a safety device in nuclear reactors. The system is designed to remove excessive heat from a nuclear reactor should the normal core cooling mechanism fail.

emergency exemption An action permitted by the provisions of the Federal Insecticide, Fungicide, and Rodenticide Act by which the USEPA grants a temporary exemption to a state or other federal agency allowing the use of a pesticide product that is not registered (officially approved for use). The circumstances under which the exemption can be granted usually involve serious pest problems against which registered pesticides are not effective. The USEPA performs a risk assessment as part of the exemption process. See REGISTRATION.

Emergency Planning and Community Right-to-Know Act (EPCRA) See TITLE III.

emergency removal action Actions taken outside regulatory restraints, but following the NATIONAL CONTINGENCY PLAN, to remove materials like leaking drums or explosive chemicals that pose an imminent threat to human health or the environment.

Emergency Response Notification System (ERNS) A computer database of all oil and HAZARDOUS SUBSTANCE spills reported to the NATIONAL RESPONSE CENTER. The database has been maintained by the USEPA and the Department of Transportation since 1987. For data descriptions and searches see www.epa.gov/ERNS.

emergent plant A plant that grows in shallow water with the root system in the sediment under the water and the upper vegetation rising above the water.

emigration The departure of individuals from a POPULATION in a specific area or country. Compare IMMIGRATION.

emission allowance See EMISSION STANDARD, TRADABLE EMISSION ALLOWANCE.

emission factor A numerical estimate of the mass of one or more air contaminants produced for a given amount of material processed by an industrial facility or, in the case of transportation sources, per mile driven. It is important to note whether the emission factor is for an uncontrolled source or one with properly functioning air pollution control equipment. This factor is used to arrive at a rough estimate of the total air emissions for a facility or geographical area. See EMISSION INVENTORY.

emission inventory A compilation of estimated air pollutant emissions, by pollutant, from smokestacks, automobiles, and other sources in a given area. The inventory is usually conducted by an environmental regulatory agency.

emission inventory questionnaire (EIQ) The form sent by an environmental regulatory agency to all facilities in a given jurisdiction for the production of an EMISSION INVENTORY. The questionnaire usually contains detailed questions about types, quantities, and locations of air emissions.

emission reduction credits (ERCs) Quantifiable permanent reductions in air pollutant emissions, beyond those legally required, placed in USEPA-approved accounts (banks). The reductions can be used as part of a BUBBLE POLICY emission limitation or as an internal OFFSET and, in some cases, traded or sold to other facilities. The use of the term *banks* does not imply monetary transfers.

emission standard The maximal legal amount of a particular pollutant that may be released into the air from a pollutant source. The standard applicable to a particular source depends on the type of pollutant, the time when the source began operation, the location of the source, and the quality of the air in the surrounding area.

emissions Pollutant gases, particles, or liquids released into the atmosphere.

emissions trading The internal exchange or sale to other companies of surplus air pollutant EMISSION REDUCTION CREDITS in accordance with USEPA regulations for the BUBBLE POLICY, the OFFSET program, the NETTING OUT for modifications, or the EMISSION REDUCTION CREDITS program.

emissivity The ratio of the amount of energy actually radiated by an object to the maximum amount of energy the body would radiate if it were a BLACKBODY. For blackbodies, emissivity and ABSORPTIVITY are equal. See KIRCHHOFF'S LAW.

emphysema A progressive, debilitating disorder of the lungs characterized by damage to and stretching of the air spaces of the ALVEOLI or tiny air sacs of the lungs. The air sacs enlarge, the walls are damaged, and the individual alveoli fuse into a larger structure with a lower surface area to volume ratio, which limits the exchange of gases with the blood. The damaged area loses elasticity and ability to recoil normally when a person attempts to expel air from the lungs. Consequently, the residual volume of air remaining in the lungs on exhaling increases, and the person is not able to draw in a sufficient volume of fresh air (and oxygen) on the next breath. The condition is commonly associated with the use of cigarettes.

empirical Based on experience or observation, as opposed to theory.

emulsifier Additive that facilitates the mixing of two or more liquids that normally do not mix, for example, an agent that promotes the mixing of oil and water.

encapsulation A hazardous waste treatment process that permanently encloses the waste with materials such as special adhesives or concrete to prevent escape of dangerous substances into the environment. See SOLIDIFICATION; STABILIZATION, WASTE.

endangered species A species of plant or animal that is presently in such small numbers that the species is in danger of disappearing from either all or a significant part of its natural range. See ENDANGERED SPECIES ACT.

Endangered Species Act (ESA) The 1973 federal law establishing procedures for the listing of species of plants or animals as ENDANGERED SPECIES or THREATENED SPECIES. The law also prohibits federal agencies from engaging in projects that place in jeopardy the continued existence of species that are threatened or endangered. The prohibition extends to projects that involve either federal funds or the federal permitting process. The United States Fish and Wildlife Service, Department of the Interior, is responsible for placing and removing species on the endangered and threatened lists, which include species in habitats worldwide. In the United States, about 930 animals and plants are listed as endangered and around 270 are threatened. Outside the United States, 518 animal and plant species are listed as endangered and 39 as threatened. The lists are found at http://endangered.fws.gov. See CONVENTION ON INTERNATIONAL TRADE IN ENDANGERED SPECIES OF WILD FAUNA AND FLORA.

endangerment assessment Under the COMPREHENSIVE ENVIRONMENTAL RESPONSE, COMPENSATION, AND LIABILITY ACT, a study to determine both the chemical nature and the extent of contamination at a site on the NATIONAL PRIORITIES LIST and the risks to human health and the environment presented by that contamination. The analysis is done as part of the REMEDIAL INVESTIGATION/FEASIBILITY STUDY (RI/FS) of a hazardous waste site.

endemic Describing a disease or characteristic commonly found in a particular region or group of people; a disease constantly present at low levels in an area.

endocrine disruptors See ENVIRONMENTAL HORMONES.

endoergic Describing a reaction or process that absorbs energy (ENDOTHERMIC). Compare EXOERGIC.

end-of-pipe technology/treatment (EOP) **1.** A wastewater management system designed to remove or reduce pollutants added to water during some industrial process. A facility's discharge PERMIT dictates the required reduction in pollutant levels before an effluent can be released into the environment; end-of-the-pipe means that it is the last chance to remove a pollutant from wastewater before release into a receiving stream. **2.** An air-pollution control device on an exhaust system or a catalytic converter on an automobile; in both cases treating exhaust gases rather than modifying the process so that the pollutant is not generated during operations. Contrast POLLUTION PREVENTION.

endogenous Originating within an organism. Compare EXOGENOUS.

endoparasite A parasite that lives within the host and damages the host as a result of its presence, for example, a tapeworm living within a person's gastrointestinal tract. Compare ECTOPARASITE.

endotherm An organism that has the ability to maintain a constant body temperature through physiological mechanisms. Mammals are endotherms. Also referred to as a homeotherm or warm-blooded. Compare ECTOTHERM.

endothermic Describing a reaction or process that absorbs energy or heat. Compare EXOTHERMIC.

endpoint A biological or chemical change used as an indication of the effect of a chemical or physical agent on an organism. If the anticipated change occurs, the measured or estimated dose or exposure associated with the change is recorded. Endpoints range from the more objective and easily determined, such as the death or malformation of the organism, to the more subtle or subjective, such as altered enzyme levels or behavioral abnormalities.

endrin An insecticide belonging to the class of compounds referred to as chlorinated polycyclic hydrocarbons. Structurally similar to aldrin and chlordane, endrin is a neurotoxin that has also been demonstrated to cause birth defects and to affect the immune system adversely. The pesticide is persistent in the environment, and use in the United States is banned.

energy The capacity to do work or produce a change, for example, to move an object from one place to another or to change the temperature of an object. Energy cannot be destroyed or created; however, the form of the energy can be changed. For example, the potential energy available in coal as a result of its chemical composition can be converted to heat energy through burning, and the heat energy released through burning can be subsequently converted to electrical energy by generating steam to spin a turbine, producing electric power. Sources of energy used to generate electricity include the potential energy associated with fossil fuels (coal, oil, and natural gas), water held behind dams, and heat released in nuclear reactors.

energy cost The amount of energy required to exploit a resource, for example, the quantity of energy required by an animal to obtain food or the energy used to remove coal from a deposit.

energy flow The path of energy as it moves through the various components of a COMMUNITY of organisms. The path includes the input of solar energy, the nergy captured by photosynthesis, the utilization of energy by various animal groups, and the loss of heat from the community.

energy management system A control system capable of monitoring internal environmental conditions and system loads and making adjustments in the heating, ventilation, and air conditioning in a structure. The automatic control system conserves energy while maintaining comfort in the structure.

energy pyramid See ECOLOGICAL PYRAMID.

energy recovery Incineration of waste material, commonly MUNICIPAL SOLID WASTE, that would otherwise be discarded. The heat from the incineration is used to generate steam to operate a generator producing electrical energy. Thus, energy that would have been lost is recovered and converted to a usable form.

Energy Research and Development Administration (ERDA) The former federal agency responsible for nuclear power research and encouraging new energy technologies. The agency was created in 1974 when the responsibilities of the ATOMIC ENERGY COMMISSION were divided into separate agencies, one (ERDA) dedicated to energy development, and the other, the NUCLEAR REGULATORY COMMISSION, dedicated to safety and regulatory activities. ERDA's responsibilities were transferred to the DEPARTMENT OF ENERGY in 1977.

Energy Star 1. A labeling program created by the USEPA and the Department of Energy to promote energy conservation that, by lessening electricity demand, will improve air quality. 2. The program also extends to voluntary partnerships between the USEPA and businesses to promote energy conservation in buildings (this includes, but is not limited to, the GREEN LIGHTS program). Energy Star also works with home builders in the design and construction of more energy-efficient houses.

energy subsidy The energy input required to produce one unit (calorie) of food energy in agriculture. Modern high-yield agriculture requires fossil-fuel energy inputs for the production of pesticides and fertilizers and for the manufacture and operation of farm machinery. About 0.5 calorie of energy input is required to yield 1.0 calorie of corn in the United States.

However, *delivery* of 1.0 calorie of corn to the consumer may use another 8 to 12 calories in processing and distribution.

enforceable requirements Provisions included in a PERMIT granted to a facility under an environmental control statute such as the CLEAN WATER ACT or the CLEAN AIR ACT by the USEPA. Failure to meet the limitations imposed by the permit could result in a NOTICE OF VIOLATION or a COMPLIANCE ORDER.

enforcement (environmental) Legal actions taken by the USEPA, a state agency, or a local entity to obtain compliance of a facility with environmental laws, rules, and regulations. A guide to federal environmental enforcement, known as the *yellow book*, is found at http://es.epa.gov/oeca/fedfac/yellowbk/.

enforcement decision document (EDD) The former term for a RECORD OF DECISION.

engineering controls In occupational health and safety, the use of control devices or techniques that reduce the level of the chemical or physical hazard at the source. Emphasized by the OCCUPATIONAL SAFETY AND HEALTH ADMINISTRATION. Compare ADMINISTRATIVE CONTROLS, PERSONAL PROTECTIVE EQUIPMENT (PPE).

enhanced inspection and maintenance (I&M) Called for by the 1990 amendments to the CLEAN AIR ACT, an INSPECTION AND MAINTENANCE (I&M) program that includes additional vehicle testing to determine the efficiency of emission controls. The enhanced inspections are targeted for those communities that have had difficulty meeting ambient air quality standards for automotive pollutants, especially OZONE.

enhanced oil recovery Techniques for the removal of the remaining thick, heavy oil from reservoirs after PRIMARY RECOVERY and SECONDARY RECOVERY. A typical operation may involve steam injection into the reservoir to reduce the viscosity and provide pressure to force the oil into collection wells.

enrichment A process by which the abundance of fissionable isotopes of uranium or plutonium is increased to create a nuclear fuel that will support a sustained nuclear reaction. Uranium recovered by the refining of uranium ore does not contain a sufficient abundance of the fissionable isotope (uranium-235) to be used in most nuclear power plants or in nuclear weapons. The relative abundance of uranium-235 must be increased before the material is suitable for use in reactors or weapons.

enteric organism In the broadest sense, any microorganism whose normal habitat is the human gastrointestinal tract. The term is frequently used in a more narrow sense to signify the GRAM-NEGATIVE bacteria that inhabit the gastrointestinal tract. It is also used to describe enteric viruses that are excreted in human feces and that cause disorders associated with the consumption of food contaminated with fecal material (hepatitis virus, for example).

enterococcus Any GRAM-POSITIVE oval or round bacterium (coccus) that normally resides in the human gastrointestinal tract, frequently belonging to the genus *Streptococcus*.

enteropathogenic organism Any GRAM-NEGATIVE bacteria capable of causing illnesses or dysfunction of the human gastrointestinal tract. The disorders range from mild upsets to diarrhea and dysentery to typhoid and cholera.

enterotoxin Specific toxins or enteropathogenic organisms secreted by some bacteria cause dysfunction of the human gastrointestinal tract. The agents are a type of EXOTOXIN.

entrainment 1. The capture of solid particles, liquid droplets, or mist in a gas stream. 2. In SCRUBBERS, the liquid aerosol in the gas stream exiting the control device. 3. The incidental trapping of fish

and other aquatic organisms in the water used for cooling electrical power plants.

entrainment separator See MIST ELIMINATOR.

entrainment velocity The gas velocity that keeps particles of a given size and density suspended (dispersed) and causes deposited particles of that type to become resuspended; typically applied to airflow in ducts carrying airborne particles.

entropy (S) The availability of energy to do work. When one form of energy is converted to another (for example, when coal is burned to produce heat), entropy increases. Also used as a thermodynamic measure of the randomness or disorder within a system. The higher the entropy, the more disordered is the system.

environment An aggregate of the conditions that make up the surroundings of an individual or community. The components of an environment include climate; physical, chemical, and biological factors; nutrients; and social and cultural conditions. These influences affect the form and survival of individuals and communities.

environmental assessment A preliminary study to determine the need for an ENVIRONMENTAL IMPACT STATEMENT.

environmental audit 1. An internal investigation of company compliance with environmental regulations. 2. Study of a site prior to a real estate transaction to uncover potential environmental liability associated with the property, such as the prior improper disposal of hazardous waste in the ground. See PHASE I ENVIRONMENTAL ASSESSMENT.

Environmental Council of the States (ECOS) A cooperative organization of state government environmental managers established in 1993 that promotes innovative implementation of environmental regulations and state involvement as partners with the USEPA in the shaping of national environmental policy and serves as a clearinghouse for solutions and approaches to environmental issues. Website: www.sso.org/ecos

Environmental Defense Fund (EDF)

A U.S. environmental organization active in legal, economic, and scientific aspects of environmental issues; the fund has been responsible for a number of important environmental law cases in the United States, such as *Environmental Defense Fund, Inc. v. Ruckelshaus,* leading to the banning of the pesticide dichlorodiphenyltrichloroethane (DDT). In 1999, membership numbered about 300,000. Website: www.edf.org

environmental dispute resolution

See ALTERNATE DISPUTE RESOLUTION, REG-NEG.

environmental ethic

A value system that judges human actions in terms of whether they harm, sustain, or improve environmental quality.

environmental exposure

See DOSE, EXPOSURE.

environmental fate

The result of the physical, chemical, and/or biological changes in a chemical after release into the environment.

environmental hormones

Chemical pollutants that have the potential to substitute for, or interfere with, natural hormones. Excessive exposures to these substances can influence reproductive success or cause tumor growth. For example, a degradation product of dichlorodiphenyltrichloroethane (DDT) limits the reproductive success of birds by promoting a reduction of estrogen levels in the blood of female birds.

environmental impact statement (EIS)

A report required by the NATIONAL ENVIRONMENTAL POLICY ACT detailing the consequences associated with a proposed major federal action significantly affecting the environment.

environmental impairment liability policy (EIL policy)

An insurance policy offering bodily injury and property damage coverage for gradual releases of pollutants for any claims made during a policy year, no matter when the release occurred. The EIL policy approach fills the gap left by POLLUTION EXCLUSION CLAUSES. Also called a claims made insurance policy.

environmental incident

The release or threatened release of a chemical, physical, or biological agent that may cause harm to human health or the environment. Government teams or contractors respond to prevent or minimize harm. See REPORTABLE QUANTITY.

environmental indicators

An organism, usually a plant species, that has an unusual sensitivity to alterations in the environment. Consequently, the presence or absence of the species provides a sign of an environmental imbalance. For example, the disappearance of some lichens indicates air pollution caused by sulfur dioxide, and the absence of normally present trout is an indication of a reduction of water quality in a stream.

environmental justice

A merger of social forces that combines the civil rights movement with environmental protection to demand a safe, healthy life-giving environment for all, regardless of economic status or ethnicity. The inclusion of race as an environmental issue grows out of the observations that poor minorities live in close proximity to industrial sites and that people of color are subjected to a disproportionately high level of health risk because of where they live and work. On a deeper level, the merger is an effort to use racial issues to subvert industrialism. Industrialism is rearticulated as class discrimination and institutional racism rather than as an agent of progress. The effort seeks to link industrialism not only to the domination of nature but also to the domination of people. The environmental justice movement differs from the nature-first groups in that the former is local in scope;

deals with human spaces rather than the natural environment; concerns people and the places where they live, work, and play; and generally adds people to the mix of environmental protection.

environmental lapse rate The negative or positive change in air temperature with change in altitude. The rate varies with changes in solar insulation, high- or low-pressure systems, wind motion, and time of day. See also ADIABATIC LAPSE RATE, MIXING HEIGHT.

Environmental Leadership Program (ELP) A voluntary pilot program instituted in 1994 by the USEPA to provide recognition to industrial facilities for utilizing effective environmental management systems, taking innovative approaches to assuring compliance with environmental regulations, and emphasizing POLLUTION PREVENTION practices. Part of the USEPA efforts to build working relationships with STAKEHOLDERS. Facilities apply to participate. Website: http://es.epa.gov/elp/

environmental lien A legal attachment to property used to secure payment of costs arising from response actions, cleanup, or other remediation activities after the release of hazardous substances or petroleum products. The objective is to ensure that the facility or company responsible for a release shares some of the response costs.

environmental literacy The capacity to understand the concepts, terminology, and issues associated with destruction of and protection of the natural environment and the place of humans in the process. The capacity to understand the issues, writings, images, terminology, and options associated with the protection of the natural environment and human health. The motivation for compiling this dictionary.

environmental management hierarchy A ranking of the actions an industry or administrative agency should take to reduce adverse impacts of pollution on

human health and the environment, in descending priority: prevention, RECYCLING, treatment, and release/disposal. See POLLUTION PREVENTION.

environmental medium Air, water, or ground into which pollutants are released.

Environmental Monitoring and Assessment Program (EMAP) A USEPA program that attempts to use environmental monitoring data to assess the health and sustainability of natural ecosystems. See ENVIRONMENTAL RISK ASSESSMENT. Website: www.epa.gov/emap/

Environmental Monitoring for Public Access and Community Tracking (EMPACT) A USEPA data-sharing program begun in 1996 to ensure that citizens have access to environmental quality information about their area that is timely (including real-time monitoring data, where available), accurate, and in a form that is understandable. The U.S. Geological Survey and the National Oceanic and Atmospheric Administration also provide data. Communities choose their preferred method(s) to receive EMPACT information: the Internet, television, radio, newspapers, community meetings, and others. Website: www.epa.gov/empact/

environmental monitoring methods index (EMMI) The official USEPA database of analytical methods for environmental pollutants. The index includes approved methods for sample collection, storage, and preservation; extraction of the pollutant from the environmental MEDIUM; and analysis (chemical detection). Available on CD-ROM.

environmental near miss See NEAR MISS.

Environmental Protection Agency (EPA) An independent executive agency of the federal government, established in 1970, responsible for the formulation and enforcement of regulations governing the release of pollutants and other activities that may adversely affect the public health or environment. The agency also approves and monitors programs established by state and local agencies for environmental protection. Some states have used the acronym in the title of their environmental management agency; for example, CAL/EPA is the California Environmental Protection Agency. See REORGANIZATION PLAN NUMBER 3.

environmental racism 1. The purposeful restriction of individuals in minority racial groups to polluted or degraded environments, commonly close to industrial facilities that add significant amounts of pollutants to the environment. 2. Ascribing of racial motives to operators of industrial facilities because of the observation that poor minorities live in close proximity to industrial sites and that people of color are subjected to a disproportionately high level of health risk because of where they live and work. This circumstance is seen as the result of deliberate actions taken by members of the majority race. Also see ENVIRONMENTAL JUSTICE.

environmental resistance The forces of nature (predators, drought, and so on) that tend to maintain populations of organisms at stable levels.

environmental resource Any tangible item available in nature and needed by an organism or process.

environmental response team (ERT) A group of individuals with special training and equipment who provide assistance in the event of spills or releases that threaten human health or the environment. They provide decontamination equipment, hazard evaluation, sampling and analysis, cleanup technologies, water supply decontamination, and removal of contaminated material, among other services and resources.

environmental science The application of the laws, theories, and concepts of science to issues related to degradation of the natural environment and to the human condition.

environmental tobacco smoke (ETS) Indoor emission of a combination of gases and particles from smoking of cigarettes, pipes, or cigars. These emissions can occur at nuisance levels or unhealthy concentrations. See also PASSIVE SMOKING.

environmentalism Active participation in attempts to publicize, define, and solve problems associated with human population growth, pollution, technology, and resource utilization. Aspects of the natural environment are taken as normative, and the application of technology (chemical manufacturing, nuclear power, fossil fuel use) and human intervention in natural environments (pesticide application, hydroelectric power) are seen as unacceptable threats to human health and the functioning of the natural environment. Also, the use of environmental issues to advance a specific social, economic, or political agenda.

enzyme A protein compound that acts as a CATALYST for biochemical reactions. Enzymes are sensitive to changes of temperature, pH, and other substances in the environment

EP toxicity See EXTRACTION PROCEDURE TOXICITY TEST.

EPA identification number The number assigned to an off-site disposal facility under the RESOURCE CONSERVATION AND RECOVERY ACT (RCRA). Also called the RCRA number.

epidemic A condition or disease that affects many individuals within a community at about the same time. An epidemic is usually characterized by a sudden onset and rapid spread of disease throughout the community; requires a sufficiently large number of individuals susceptible to the condition.

epidemiology The science that deals with the incidence and distribution of human disease or disorders.

epidermal cells The cells forming the outer layer of the skin or the thin outer covering of plant leaves, stems, and roots.

epifluorescence A type of microscopy in which the specimen under examination is illuminated by the projection of ultraviolet light through the microscope from above the microscope stage. The technology differs from standard microscopy, in which the specimen is illuminated from below the slide with white light. The observer sees the object as light emitted from the specimen by fluorescence mechanisms (glowing) stimulated by the ultraviolet light. The use of special dyes can be required.

epigenetic 1. Describing a cancer mechanism that does not involve the direct interaction of a chemical with cellular genetic material (deoxyribonucleic acid [DNA]). 2. Describing a geological deposit that has been changed from its original state by outside influences, for example, the conversion of accumulated shell deposits to limestone.

epilimnion The upper layer in a lake in which the water is stratified by temperature. Large standing bodies of water tend to form layers as a result of nonuniform heating by the Sun. The water at the surface, the epilimnion, is warmer, whereas the water at the lower depths, the HYPOLIMNION, is colder. The boundary between these two layers is the THERMOCLINE. The water in these distinct layers is of different densities, and the layers remain separate until the water temperatures change or are forced to mix by physical forces such as wind. See also FALL TURNOVER, SPRING TURNOVER, TURNOVER.

epiphyte A plant that grows on the surface of another plant without damaging the plant that serves as the host for the epiphyte. For example, single-celled diatoms grow on the surface of marsh grasses without apparent damage to the marsh grass.

episode A period of extremely high air pollutant concentrations, lasting for several days to a week or more, associated with an abnormally high incidence of res-piratory disease and, in the worst cases, an increased death rate in the affected area. The condition is caused by air STAGNATIONS, or persistent INVERSION conditions, that allow pollutant concentrations to increase. Sites of historic air pollution episodes include Meuse Valley, Belgium, in December 1930; Donora, Pennsylvania, in October 1948; and London, England, in December 1952. Major episodes have not occurred in the United States or Western Europe since the 1960s thanks to the installation of air pollution controls on major stationary sources (smokestacks) and the widespread switch in the energy source for domestic space heating from coal to natural gas or electricity generated at a central power station. See DONORA EPISODE.

epithelium A membranous, protective tissue layer that covers the surface of body organs and that lines various body cavities.

equalization See FLOW EQUALIZATION.

equalization basin See FLOW EQUALIZATION.

equal-loudness contours See FLETCHER-MUNSON CONTOURS.

equation of state An equation derived from a combination of Boyle's law and Charles's law that describes the relationships among the pressure, density, and temperature of a gas: pressure = gas constant × density × absolute temperature, $P = RrT$, where R is the UNIVERSAL GAS CONSTANT, r is density, and T is the ABSOLUTE TEMPERATURE.

equilibrium A STEADY-STATE condition in which flow in equals flow out. In a BOX MODEL, the condition in which there is neither net gain nor loss of the amount of matter or energy in each box or compartment.

equilibrium ethic An outlook that attempts to establish a middle ground between the total development of a potential resource by economic interests and the absolute preservation of that resource in the natural condition.

equivalent method An analytical procedure deemed by the USEPA to be equivalent to the standard, or official, method for the determination of pollutant concentrations, as approved by the agency. An individual or organization must demonstrate the accuracy and reliability of the new method before it can be certified as an equivalent method. See REFERENCE METHOD.

equivalent weight (EW) For dissolved compounds, the MOLECULAR WEIGHT divided by the number of hydrogen or hydroxyl ions in the undissolved compound. For elements, the ATOMIC WEIGHT divided by the VALENCE.

erosion The process of wearing away of a surface by physical means. The term is usually applied to the loss of soil through the agency of wind or water.

erythema An abnormal redness of the skin due to irritation or tissue damage.

Escherichia The genus name of a type of bacteria whose normal habitat is the colon in humans and other warm-blooded animals. The organism is GRAM-NEGATIVE, ferments lactose at 37° C, and can grow with or without molecular oxygen. If members of this genus, referred to as fecal coliforms, are found in water, the water is considered to be contaminated with fecal material. Some strains of this genus cause severe diseases of the gastrointestinal tract.

estimated environmental concentration (EEC) In pesticide regulation, the concentration of the pesticide in aquatic or terrestrial ecosystems where NONTARGET ORGANISMS may be exposed.

estuary Coastal waters where seawater is measurably diluted with freshwater; a marine ecosystem where freshwater enters the ocean. The term usually describes regions near the mouths of rivers and includes bays, lagoons, sounds, and marshes.

ethanol A two-carbon primary alcohol, with the chemical formula CH_3CH_2OH. The pharmacologically active ingredient in alcoholic beverages and spirits. Used extensively as a solvent and as an OXYGENATE in gasoline. Also known as ethyl alcohol, grain alcohol, and alcohol.

ethylene dibromide (EDB) A substance used for fumigating foodstuffs for protection against insects and nematodes. EDB is found as a residual in some treated grains and has been implicated as a carcinogen in test animals.

ethylene dichloride (EDC) $(C_2H_4Cl_2)$ A colorless, oily liquid that has an odor similar to that of chloroform and that is slightly soluble in water. Produced by reacting hydrochloric acid or chlorine gas with ethylene. The most common use is in the manufacture of VINYL CHLORIDE, the monomer used in the manufacture of POLYVINYL CHLORIDE (PVC). Minor uses include application as a dispersant in the manufacture of rubber and plastic. Formerly used as a metal degreaser. Acute inhalation exposures in humans cause nerve and kidney damage as well as respiratory distress, cardiac arrhythmia, and nausea. Chronic exposure to this chemical over long periods causes damage to the liver and kidneys. EDC probably causes cancer in humans; however, the potency of the chemical in this respect is low.

etiology The cause of a disease or abnormal condition.

eucaryotic Describing a cell that has a well-defined nucleus confined by a nuclear membrane or a multicellular organism that comprises such cells. The cell type for all organisms except the bacteria and blue-green algae. Compare PROCARYOTIC.

euphotic zone The upper layer of a body of water that is penetrated by sunlight. Photosynthesis or primary production takes place in this layer. Compare BATHYL ZONE, ABYSSAL ZONE.

euploid A cell that has the correct number of chromosomes. If a normal cell contains 22 chromosomes, a euploid cell would have either that number or some exact multiple of that number, like 44. See ANEUPLOID.

eury- A prefix that means "wide" or "broad." A euryhaline organism, for example, is able to live in an environment with large changes in salinity. *Eurytherm* is used to describe an animal that can live in an environment with large changes in temperature.

eutectic chemical Describes a chemical, commonly an inorganic salt, that undergoes a phase change from solid to liquid when heated (as does water) and reconverts to a solid when cooled. Unlike water, the inorganic salts useful in heat exchangers designed to capture and transfer solar energy do not undergo large volume changes as they change phase from a solid to liquid and back. Using these chemicals, a large amount of heat energy can be stored in a relatively small space.

eutrophic Describing a river, lake, stream, or other body of water enriched with excessive amounts of plant nutrients, such as nitrates and phosphates. Such environments are characterized by the excessive growth of aquatic plants. Compare MESOTROPHIC, OLIGOTROPHIC.

eutrophication The addition of excessive plant nutrients to a river, lake, stream, or other body of water. The nutrients in excess are usually nitrates or phosphates, and the process leads to prolific growth of aquatic plants.

evaporation A change of state from liquid to gas. Some molecules in a liquid have enough energy to escape to the gas phase. The rate increases with temperature.

evaporation pond Surface enclosure exposed to the atmosphere used to allow for the volatilization of water to thicken or dry some substance, for example, sewage sludge.

evaporative cooling Cooling of a liquid, such as water, by allowing a portion to evaporate. The process is important in the operation of COOLING TOWERS used to cool heated effluents from power plants as well as in the cooling of the human body through the evaporation of perspiration. The process is more effective than convection cooling. See HEAT OF VAPORIZATION.

evapotranspiration The combined action of evaporation (a physical process that converts liquid water to a gas) and transpiration (the loss of water vapor from plants).

evenness A mathematical expression that describes the distribution of individuals in a COMMUNITY among the species represented. A community has high evenness if there are equal numbers of individuals in the species present. Conversely, a community is described as having a low evenness if most of the organisms are members of one species and other species are represented by one or a few individuals. One calculation method for evenness is

$$\frac{1}{\Sigma (Ni/N)^2}$$

where Ni is the number of individuals in each species and N is the total number of individuals.

event tree A graph of all possible outcomes and their probabilities after an initiating event. The probability of an outcome, such as an accidental death, can be estimated by combining all of the event probabilities that lead to a fatality. The representation is used in accident and risk analysis. Compare FAULT TREE.

evergreen Plant that retains its leaves throughout the year. The term is most commonly applied to trees such as spruces and firs. Compare DECIDUOUS.

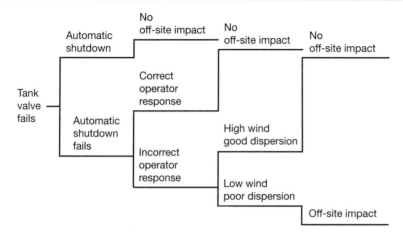

Event tree for off-site impact from a tank valve failure

evolution   A scientific theory proposing that higher forms of life have descended from lower forms by way of natural mechanisms. The process by which the characteristics of a POPULATION of organisms change over time in response to natural genetic variation within the population and to selection forces of nature acting on that population. The selection forces include any of a multitude of natural factors that have potential negative impact on the survival of an organism, for example, predators or parasites, nutrient supply, climatic changes, or physical isolation from a suitable habitat. See NATURAL SELECTION.

ex situ conservation   Protection or enhancement of BIOLOGICAL DIVERSITY outside species' natural HABITATS. Compare IN SITU CONSERVATION.

exceedance   1. A measured environmental concentration for an appropriate AVERAGING TIME that violates a published AMBIENT STANDARD. 2. The release of a pollutant from an industrial or waste treatment facility at levels constituting a violation of permit restrictions.

exception report   Under the MANIFEST SYSTEM, a notification that must be sent to the regional administrator of the USEPA by a hazardous waste generator if the manifest accompanying waste shipped off-site by the generator for treatment and/or disposal has not been returned to the generator within 45 days of the waste shipment. The manifest system documents hazardous waste in transit to its ultimate treatment and/or disposal site.

excess air   The air supplied for a combustion process over the amount theoretically required for complete burning. Usually expressed as a percentage, for example, an incinerator operating at 80% excess air has an air/fuel molar ratio of 1.8:1. Lowering excess air levels helps control the formation of THERMAL NOx. See STOICHIOMETRIC RATIO.

exchange reaction   A process to soften water by altering the ionic composition through the trading of ions. A resin is saturated with sodium ions; then, as water containing divalent ions, such as calcium or magnesium, flows over the resin, the sodium ions are released into the water and the calcium or magnesium ions are adsorbed by the resin.

exchangeable cation   A positively charged ion loosely bonded to soil particles. The ion is not removable by pure water; however, it is readily exchanged for the sodium ions in a neutral salt SOLUTION. The soil particle adsorbs the sodium ion and releases the exchangeable cation.

excursion 1. Used to indicate the release of some pollutant at levels beyond that allowed by regulations or an operating permit. 2. Ambient pollutant concentrations that exceed the allowable standard for the required AVERAGING TIME.

Executive Order 12291 An order issued by President Ronald Reagan in 1981 requiring administrative agencies to prepare a REGULATORY IMPACT ANALYSIS for all major regulatory actions. Repealed by President Clinton in September 1993. See EXECUTIVE ORDER 12866.

Executive Order 12498 An order issued by President Ronald Reagan in 1985 requiring all federal regulatory agencies to submit to the OFFICE OF MANAGEMENT AND BUDGET an annual report summarizing agency objectives for the coming year, including the issuance of any major regulations. Together with EXECUTIVE ORDER 12291, the action was an attempt to reduce uncertainty and costs associated with federal regulations and standards. Repealed by President Clinton in September 1993. See EXECUTIVE ORDER 12866.

Executive Order 12866 An order issued by President Clinton in September 1993 that required administrative agencies to "assess all costs and benefits of available regulatory alternatives," including quantitative costs and benefits and those "that are difficult to quantify," which in the environmental realm would potentially include certain intangible adverse health effects and environmental degradation. Further, "in choosing among alternative regulatory approaches, agencies should select those approaches that maximize net benefits (including potential economic, environmental, public health and safety)." Therefore, although EXECUTIVE ORDER 12291 AND EXECUTIVE ORDER 12498 were repealed, COST-BENEFIT ANALYSIS

retains an importance in environmental regulation. The Clean Air and Clean Water Acts contain provisions that explicitly exempt certain standards from cost-benefit analysis.

exempted aquifer Part or all of an underground geological deposit holding an amount of water suitable for use as a drinking water supply; however, the groundwater is not used as a source of drinking water. Such deposits are exempt from certain regulations that apply to aquifers that serve as sources of drinking water. The classification of a system as an exempted aquifer signifies that the aquifer is not currently used as a source for drinking water and the system cannot now or in the future serve as a source of drinking water.

exempted solvent An organic compound that is not subject to VOLATILE ORGANIC COMPOUND controls under the CLEAN AIR ACT; the chemical does not contribute significantly to the photochemical reactions leading to the production of ground-level ozone.

exfiltration The movement of air out of a building as a result of wind or temperature differences. Compare INFILTRATION.

exhaust gas recirculation (EGR) An automobile emission control system that mixes a small amount of the exhaust gases with the fuel-air mixture entering the engine via the intake manifold. This lowers the combustion temperature and reduces the formation of OXIDES OF NITROGEN (NOx).

exhaustible resources Those items society removes from the natural environment in support of human existence and that are present in finite amounts and not replaced on a time scale sufficient to replenish the amount removed. The mineral resources and fossil fuels are prime examples. Contrast RENEWABLE RESOURCE.

existence value Nonmonetary value assigned to wildlife. A valuation that goes beyond the value assigned to a species in terms of economic, recreational, or aesthetic considerations. The importance placed on the continued survival of a species beyond what it can do for humans.

exoergic Describing a reaction that gives off energy or heat (same as EXOTHERMIC). Compare ENDOERGIC.

exogenous Originating outside an organism. Compare ENDOGENOUS.

exothermic Describing a reaction that gives off energy or heat. Compare ENDOTHERMIC.

exotoxin A protein produced and released by the bacteria that cause certain diseases in humans, for example, botulism and tetanus. These proteins are extremely toxic in microgram quantities.

exp When followed by a number or expression that represents an exponent, the base of the NATURAL LOGARITHM; the same as *e*.

expected deaths In EPIDEMIOLOGY, the calculated number of deaths over a given time in a study group if members of that group died at the national average rate for persons of the same age, sex, and race. The number of actual deaths in the study group is compared with the number of deaths expected to determine whether the two groups differ by a statistically significant amount.

expedited removal action (ERA) The cleanup of a hazardous waste disposal site without preparation of a REMEDIAL INVESTIGATION/FEASIBILITY STUDY; performed when a release or a threatened release of a hazardous substance requires immediate efforts to protect public health or the environment.

experimental concentration—percent (EC_x) In a test of the toxicity of a chemical to aquatic life, the experimentally determined concentration of the chemical that is calculated to affect X percent of the test organisms.

expert testimony See EXPERT WITNESS.

expert witness In a judicial proceeding, a person testifying on a subject about which he or she has extensive knowledge and training. Federal Rules of Evidence govern the use of expert witnesses. See RULE 702, TOXIC TORT.

explosive limits See LOWER EXPLOSIVE LIMIT, UPPER EXPLOSIVE LIMIT.

exponential decay The decline in the number of a population, amount of a pollutant, level of radioactivity, and so forth, according to the exponential function $N = N_0 e^{-kx}$, where N is the amount left after decay, N_0 is the initial amount, k is a constant, and x is a variable such as time, altitude, or water depth.

exponential growth 1. Growth in the size of a population in which the number of individuals increases by a constant percentage of the total population each period. When the number of individuals in the population is plotted against time, the increase appears as a J-shaped curve; when the logarithms of the numbers of individuals are plotted against time, the increase appears as a straight line. 2. The growth of bacteria in culture where division is by binary fission. Represented by the equation $N(t) = N_0 e^{kt}$, where $N(t)$ is the population size at time t, e is the base of the natural logarithm, and k is a constant. This stage is followed by a STATIONARY GROWTH PHASE. See LOG PHASE.

exposure Contact between a chemical, physical, or biological agent and the outer surfaces of an organism. Exposure to an agent does not imply that the agent will be absorbed or will produce an effect.

exposure assessment An estimate of the actual or anticipated contact of a chemical, physical, or biological agent with a population, including the route(s), frequency, concentration, and duration of the exposure and the number of individuals exposed at various levels.

exposure pathway How a pollutant moves from the source of release through air, water, food, or soil to natural organisms or humans.

exposure route How a pollutant in the environment gains entry into an organism. The possibilities include inhalation, ingestion, or absorption through the skin.

external cost See EXTERNALITY.

external radiation The exposure of the body to IONIZING RADIATION where the source of the radiation is located outside the body.

externality The cost or benefit of some activity that affects persons not involved directly with the activity. Called a negative externality if costs are imposed, for example, a decrease in value of residential property near a new industrial facility, or a positive externality, as in the aesthetic and economic benefits accruing to the neighbors of a person repainting his or her home. The goal of environmental quality management is to internalize externalities, that is, push the cost of pollution back to the polluter.

extinction The elimination of every individual within a particular species.

extinction coefficient A variable used in KOSCHMIEDER'S RELATIONSHIP to determine visual range in the atmosphere. The extinction coefficient $bext$ is expressed as $bext = brg + bag + bscatp + bap$, where brg is scattering by gas molecules (RAYLEIGH SCATTERING), bag is absorption by gas molecules, $bscatp$ is scattering caused by air particulate matter, and bap is absorption by particles.

extirpate To destroy or remove completely, as in the extinction of a species as a result of overhunting.

extractable organic halogens (EOX) Organic compounds combined with any members of the chemical halogens (mainly chlorine, bromine, and fluorine) that can be removed from a soil or sludge sample with the solvent ethyl acetate. The organic halogens represent a class of unusually toxic materials.

extractable organics Organic chemical compounds that can be removed from a water sample by the solvent methylene chloride under conditions of pH greater than 11 or less than 2. Organic compounds in water represent a class of pollutants that are potentially toxic materials.

extraction procedure toxicity test (EP toxicity test) A laboratory test in which a solid waste material is treated to leach out certain toxic metals and/or pesticides. Formerly the official toxicity characteristic test method for a hazardous waste, the EP toxicity test has been replaced by the TOXICITY CHARACTERISTIC LEACHING PROCEDURE (TCLP).

extrapolation The prediction of outcomes, in particular circumstances, beyond known experience or experimental observations but based on existing empirical data.

extremely hazardous substance (EHS) One of about 400 chemicals listed by the USEPA under the Emergency Planning and Community Right-to-Know Act of 1986. The act called for, among other things, community planning for the accidental release to the atmosphere of toxic materials from industrial facilities or along transportation routes and their possible adverse effects in the nearby community. Chemicals were placed on the list on the basis of acute toxicity and annual production volume. See also TITLE III.

Exxon Valdez The vessel that ran aground in March 1989, spilling 11 million gallons of CRUDE OIL into Prince William Sound, Alaska. Prompted the passage of the OIL POLLUTION ACT OF 1990.

F

F factor The ratio of the volume of gas produced by the combustion of a fuel to the energy content of a fuel. The gas volume used can include all combustion gases (wet F factor), all gases excluding water vapor (dry F factor), or only carbon dioxide (carbon F factor). The wet F factor is in wet standard cubic meters per joule, the dry F factor is in dry standard cubic meters per joule, and the carbon F factor is in standard cubic meters per joule. The appropriate factor is used in calculations of air pollutant emission rates for particulate matter, sulfur dioxide, nitrogen dioxide, and others.

F waste Material from nonspecific sources defined as LISTED HAZARDOUS WASTE by the USEPA in the *Code of Federal Regulations*, Title 40, Part 261.31. See also K WASTE, P WASTE, U WASTE, HAZARDOUS WASTE.

fabric filters Filter bags made of Teflon, nylon, cotton, or glass fibers used to remove particulates from industrial exhaust gases. The material collected on the filters is periodically shaken or blown off and falls to a hopper for disposal or recycling. See BAGHOUSE.

facepiece The part of a respirator covering all or part of the wearer's face. The facepiece, which contains connectors for inflowing and outgoing air, must make an airtight seal to operate properly.

facilitator A person trained in the art of drawing diverse groups to a consensus on an issue; mediator. Facilitators may be used in ALTERNATE DISPUTE RESOLUTION or NEGOTIATED RULE MAKING.

facilities plans In wastewater management, documentation required by the USEPA in support of the construction of a municipal wastewater treatment facility. The documents include plans and studies related to community needs, cost-effectiveness of various alternatives, environmental impact statements, and descriptions of the facility, costs, and construction schedule.

facility emergency coordinator The individual at an industrial facility designated as the INCIDENT COMMANDER in the event of a release of dangerous material, a fire, or an explosion at the installation. This person is responsible for the formulation and execution of contingency plans for emergencies.

facultative bacteria Microorganisms able to grow either with or without molecular oxygen (O_2).

fall turnover The exchange of top and bottom waters of a stratified lake promoted by the cooling of the surface water during the fall of the year. The surface water cools in response to the falling air temperature and becomes denser as a result. When the surface water cools to a temperature lower than that of the bottom water, the upper layer of water sinks, forcing the bottom water to the surface. Compare SPRING TURNOVER.

fallout 1. Radioactive particles introduced to the atmosphere by a nuclear accident or explosion. The particles can be transported thousands of miles before settling to the ground. 2. Any solid matter emitted to the atmosphere by human activities (smokestack or tailpipe particulates) or natural processes (dust storms, volcanoes, forest fires), then returning to the surface by gravitational settling.

fallow Describing arable land that is left uncropped for a growing season; a part of the rotation practiced under nonirrigated agriculture. The land stores water during the fallow period, and the next crop has available a two-year water supply and more nutrients.

false negative An erroneous test result that labels a chemical or individual as not having a certain property or condition when in fact the property or condition is present. For example, the false determination that a chemical is not a CARCINOGEN when it actually is a carcinogen. Compare FALSE POSITIVE.

false positive An erroneous test result that labels a chemical or individual as having a certain property or condition when in fact the property or condition is not present. For example, a false test result stating that a person is infected with the human immunodeficiency virus (HIV) when in fact the virus is not present. Compare FALSE NEGATIVE.

fan curve A graphical depiction of the volumetric airflow of a fan for different STATIC PRESSURES at a given fan turning rate in revolutions per minute (rpm). A family of curves is typically generated with a separate curve for each of a series of rpm values.

fanning Describing a PLUME from a smokestack that is being emitted into a TEMPERATURE INVERSION, allowing little dispersion of the plume in the vertical plane. Compare CONING, FUMIGATION, LOFTING, LOOPING, TRAPPING.

far field The area away from a sound source in which the SOUND PRESSURE LEVEL is inversely proportional to the square of the distance from the source, according to the INVERSE SQUARE LAW. Compare NEAR FIELD.

far infrared Electromagnetic radiation with wavelengths from 20 micrometers to 1000 micrometers. See INFRARED RADIATION.

fast breeder reactor A nuclear reactor that produces a significant quantity of fissionable material (plutonium-239) when uranium-238 absorbs FAST NEUTRONS produced during the fission process. See BREEDER REACTOR.

fast neutron A neutron with energy exceeding 1×10^5 electron volts as it is ejected from a nucleus that undergoes fission. Compare SLOW NEUTRON.

fathead minnow A fish species (*Pimephales promelas*) often used in aquatic BIOASSAYS.

fat soluble Describing a material that dissolves (is stored) in fat. Fat solubility is a characteristic of a compound that is more likely to undergo BIOACCUMULATION or BIOLOGICAL MAGNIFICATION.

fault tree A graph of the events or failures necessary for an accident or other adverse effect to occur. The representation is developed by starting with the undesired outcome, such as loss of radiation in a nuclear power plant accident, and working backward to describe what events or sequences of events can lead to the outcome. The technique is used in accident risk analysis and in development of strategies for risk reduction. Compare EVENT TREE.

fauna A general term for the animal life of an area or region. Compare FLORA.

feasibility study A detailed technical, economic, and/or legal review of a specific proposed project at a particular location to outline all potential costs, benefits, and likely problems. If the project is a hazardous waste site cleanup, the process is called a REMEDIAL INVESTIGATION/FEASIBILITY STUDY (RI/FS).

fecal bacteria Any type of bacteria whose normal habitat is the colon of

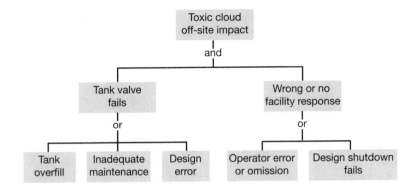

Fault tree for a toxic material release with off-site impact

warm-blooded mammals, such as humans. These organisms are usually divided into groups, such as FECAL COLIFORMS or FECAL STREPTOCOCCI.

fecal coliform A type of bacteria whose natural habitat is the colon of warm-blooded mammals, such as humans. Specifically, the group includes all of the rod-shaped bacteria that are nonspore forming, gram-negative, and lactose fermenting in 24 hours at 44.5° C and that can grow either with or without oxygen. The presence of this type of bacteria in water, beverages, or food is usually taken to mean that the material is contaminated with solid human waste. Bacteria included in this classification represent a subgroup of the larger group termed coliforms.

fecal material Solid waste produced by animals and discharged from the gastrointestinal tract. Also referred to as feces or solid excrement.

Fecal streptococci A type of bacterium whose natural habitat is the colon of warm-blooded mammals, such as humans. The group includes those bacteria that are gram-positive, are spherical, and occur in chains. This type of bacterium is catalase negative and capable of growth in media containing 6.5% sodium chloride. Most of these organisms are in the species *Streptococcus faecalis*.

fecundity A measure of the capacity of an organism to produce offspring. Organisms classified as fecund have the potential to produce many offspring in a short time frame (a roach, for example).

Federal Energy Regulatory Commission (FERC) An independent federal administrative agency, governed by five commissioners, responsible for regulating the sale of electricity or natural gas in interstate commerce, construction and operation of oil and gas pipelines, and licensing of nonfederal hydroelectric power plants. Website: www.ferc.fed.us/

Federal Environmental Pesticide Control Act (FEPCA) The 1972 amendments to the FEDERAL INSECTICIDE, FUNGICIDE, AND RODENTICIDE ACT that gave the USEPA authority to register pesticides and to suspend or cancel registration of any pesticide found to threaten human health or the environment unduly.

Federal Food, Drug, and Cosmetic Act See FOOD, DRUG, AND COSMETIC ACT.

Federal Hazardous Substances Act (FHSA) The federal statute providing for health and safety review of the design and content of consumer products. Implemented and enforced by the CONSUMER PRODUCTS SAFETY COMMISSION.

Federal Hazmat Law See HAZARDOUS MATERIALS TRANSPORTATION LAW.

Federal Implementation Plan (FIP) Under the CLEAN AIR ACT, a document prepared by the USEPA that describes the actions to be performed by a state to comply with provisions of the CLEAN AIR ACT. Prepared only if the STATE IMPLEMENTATION PLAN does not meet the minimal statutory requirements.

Federal Insecticide, Fungicide, and Rodenticide Act (FIFRA) The federal law that regulates the manufacture and use of pesticides. The law was first passed in 1947 and amended in 1972, 1975, and 1978. Under FIFRA, pesticides must be registered with and approved for use by the USEPA before they are sold. Marketing approval may require the pesticide container label to include directions for proper use and disposal, and the purchase and/or use of some chemicals is restricted to certified (trained) applicators. See REGISTRATION, SUSPENSION, CANCELLATION.

Federal Land Policy and Management Act (FLPMA) A 1976 federal law that, together with the TAYLOR GRAZING ACT, outlines policy concerning the use and preservation of public lands in the United States. The FLPMA gives the Bureau of Land Management, in the Department of the Interior, the responsibility to manage all public rangelands not within national forests or national parks, applying MULTIPLE USE policies. The act also grants the federal government power to control the environmental consequences of mining on public lands.

Federal Motor Vehicle Control Program (FMVCP) The complete collection of federal initiatives designed to control pollution emitted from motor vehicles. Included are tailpipe emission standards, regulatory control of evaporative fuel losses, development of testing methods, and guidance to state INSPECTION AND MAINTENANCE programs.

Federal Power Commission (FPC) The predecessor federal agency to the FEDERAL ENERGY REGULATORY COMMISSION (FERC), established in 1920. The authority was transferred to the new agency in 1977.

Federal Register As required by the ADMINISTRATIVE PROCEDURE ACT, a daily publication of the U.S. government that contains federal administrative agency proposed RULES, final rules, and other executive branch documents. Final regulations and standards are annually codified by subject in the CODE OF FEDERAL REGULATIONS. Website: www.gpo.gov/nara/cfr

Federal Test Procedure (FTP) The methods prescribed by the USEPA for the stationary testing of the efficiency of emission control devices or the fuel economy (miles per gallon) of automobiles.

Federal Water Pollution Control Act (FWPCA) See CLEAN WATER ACT.

feedback Corrective information or a signal generated within a self-regulating system or process that is intended to induce a change in that system or process. Feedback relationships are important aspects of computer models intended to simulate natural processes.

feedlot A confined area in which cattle are held and fed to promote maximal weight gain prior to marketing. The large quantities of animal waste produced can lead to pollution problems in nearby water bodies. See BIOCHEMICAL OXYGEN DEMAND.

feedstock The bulk chemical resources needed in the petrochemical industry for the creation of product. Feedstocks may be natural materials (for example, salt or natural gas) or manufactured chemicals used to make other manufactured chemicals (ethylene, propylene).

feedwater Boiler water that is converted to steam during the operation of a fossil-fuel or nuclear power plant. The water is boiled by heat transferred from the furnace or reactor core, and the steam is employed to drive turbines for the generation of electricity.

feldspar Rock formed by the slow cooling of molten magma, constituting about half of the Earth's crust. Crystalline rock composed primarily of aluminum silicates.

fen A flat, marshy area that tends to have waterlogged soil fed by groundwater and runoff. This type of area often accumulates peat.

fenceline The outer perimeter of the property of an industrial facility, the closest possible site of exposure to the public. In the management of a CHEMICAL EMERGENCY, as long as hazardous concentrations of a spill or release do not migrate past the outer property line, the responsibility for managing the incident rests primarily with the personnel at the facility. If the impact moves beyond the outer property line, a HAZARDOUS MATERIAL RESPONSE TEAM from other industries or local government units may be called in by the FIRST RESPONDERS. For routine emissions, the point of compliance with the NATIONAL AMBIENT AIR QUALITY STANDARDS or the estimated levels using an AIR QUALITY DISPERSION MODEL is the fenceline.

fenceline concentration The concentration of an air contaminant measured just outside the perimeter of the property of an industrial facility, the closest possible site of exposure to the public. See FENCELINE.

Fenton's reagent A mixture of iron (Fe^{+2}) and hydrogen peroxide (H_2O_2) that can be used to oxidize chemically or to degrade toxic organic chemicals in soil. Soil contaminated with a variety of chlorinated aromatic compounds can be mixed with the two ingredients to effect the conversion of the organic material to carbon dioxide and chloride ions. The reagent is useful in the cleanup of contaminated soil.

feral animal A domesticated animal living in the wild.

fermentation A type of bacterial or yeast metabolism (chemical reaction) characterized by the conversion of carbohydrates to acids and alcohols, usually occurring in the absence of molecular oxygen.

fermentation tube method A technique for the examination of water or wastewater for the presence of fecal bacteria. Portions of a water sample are inoculated into culture tubes containing a growth medium that has lactose. Coliform bacteria produce acid and gas by the fermentation of lactose when incubated at certain temperatures.

ferrous metal Magnetic metal derived from iron and steel. A classification important in the recycling of municipal and industrial waste. These materials range from bimetal (tin) food cans, appliances, barrels and drums, piping, strapping, and junk cars to decommissioned ships. Ferrous metal represents the most frequently recycled commodity in terms of tons recycled, and old steel is an important ingredient in the manufacture of new steel.

fertile atoms Nonfissionable isotopes that absorb neutrons then decay to a fissionable material. The most common fertile atom is uranium-238, which decays to fissionable plutonium-239.

fertile isotope See FERTILE MATERIAL.

fertile material A material that is not capable of undergoing nuclear fission but can be converted to a fissionable material by irradiation in a nuclear reactor. The most common fertile isotopes are uranium-238 and thorium-232, which are converted to the fissionable materials plutonium-239 and uranium-233, respectively.

fertility rate In an animal population, the average number of offspring born to females. Within human populations, only those females of childbearing age are considered. It is calculated as the number of live births in a calendar year divided by the number of women of ages 15–44 in the population, then multiplied by 1000.

fetal alcohol syndrome A variety of birth defects associated with the consumption of alcoholic beverages by pregnant women. The symptoms range from mild to severe and include both mental abnormalities and physical malformations. Common facial disorders include broadened nasal bridge, upturned nose, thinned upper lip, eye and ear abnormalities, and excessive body and facial hair. The majority of modern mental retardation is attributable to this disorder. Outcomes include reduction in intelligence quotient (IQ), learning disabilities, speech and language disorders, an inability to relate cause and effect, behavior disorders, impulsiveness, short attention span, and psychomotor dysfunction. The syndrome is 100% preventable and carries significant personal and societal cost.

fetotoxic Describing a chemical substance or other agent that has adverse effects on a developing fetus.

fibers per cubic centimeter (f/cc) Common units for the expression of airborne asbestos concentration. The asbestos in air is present in the form of countable fibers. A cubic centimeter is equal to one-millionth of a cubic meter, or a cube about 0.4 inch on a side.

fibrosis The formation of an excessive amount of fibrous tissue in the lung resulting from overexposure to certain insoluble particulate materials. Fibrotic particulates include rock and coal dust, fine sands, and asbestos. The accumulation of the fibrous tissue reduces the efficiency of oxygen transfer in the lungs. See SILICOSIS, ASBESTOSIS, CRYSTALLINE SILICA.

Fick's first law of diffusion A law stating that the rate of diffusion of one material through another is proportional to the cross-sectional area of diffusion, the concentration gradient, and a diffusion coefficient. The value of the diffusion coefficient depends on the size and electric charge of the diffusing substance, the type of material the diffusing substance is moving through, and the absolute temperature. The law is expressed as

$$M/A = -D(dC/dX)$$

where M is the mass transfer rate, A is the cross-sectional area, D is a diffusion coefficient, and dC/dX is the CONCENTRATION GRADIENT.

field blank A sample container carried to and from the sample collection site but not used for taking environmental samples. This filter, collection tube, or other container is analyzed along with the actual samples to detect contamination that may occur during sample collection and transport.

fill material Inert material such as soil, stone, concrete, or gravel used to pack a low area for the future development of the property.

film badge A personal monitoring device used to determine an individual's radiation exposure level. The small badge is worn on the clothing and may contain one or more layers of photographic film, which is examined after an appropriate interval to determine exposure.

filter Any medium through which liquids or gases are passed to remove particulate material or other undesirable constituents.

filter cake 1. The solids or semisolids deposited on a filter as a fluid is moved through it. 2. The remaining solids or semisolids on a filter after the fluid in a material is extracted by negative pressure.

filter strip Term applied to a vegetated area of land placed between a source of potentially polluted water and a receiving stream. As the water flows through the vegetated area, sediments, inorganic plant nutrients, organic materials, or other pollutants are captured or taken up by the plants, preventing pollution of the receiving stream.

filterable Description of particles that are sufficiently small to allow their passage through filters capable of retaining most particles. For example, a filterable virus is one that passes through a filter that normally retains bacteria.

filtering velocity The speed at which air or fluid moves through a paper, synthetic, or sand filter for the purpose of removing particulate material from the air or liquid. The units used are volume measure (such as cubic meters, gallons, or liters) per unit time (days, seconds, or hours).

filtrate The liquid that passes through some filter medium. For example, if a sample of polluted water is passed through a paper filter that retains the particulate material, the dissolved pollutants are recovered in the filtrate. When coffee is made by using a paper filter to hold the ground beans, the coffee that results is actually a filtrate.

filtration treatment Most commonly used to describe the processing of raw water prior to drinking as in slow sand filtration or the treatment of wastewater as in a trickling filter.

final clarifier A gravitational settling tank installed as part of some wastewater treatment plants to be used after the biological treatment step. The tank functions to remove SUSPENDED SOLIDS. Also called a secondary clarifier.

financial assurance The RESOURCE CONSERVATION AND RECOVERY ACT requirement that owners and operators of hazardous waste treatment, storage, or disposal facilities demonstrate to the permitting authority that funds will be available to meet the estimated expenses of CLOSURE and POSTCLOSURE activities at the site.

finding of no significant impact (FONSI) A document prepared by a federal agency presenting reasons why an activity or development project will not have an appreciable effect on the human environment and for which an ENVIRONMENTAL IMPACT STATEMENT will therefore not be prepared.

fines Very small airborne particles, usually less than two MICROMETERS in diameter.

finished water Water that has completed a purification or treatment process. Compare RAW WATER.

fire-climax community A habitat in which the population of plants (and consequently the animals present) is maintained by periodic fires. Examples include managed pine forests where ground fires are set intentionally to prevent broadleaf hardwoods from developing and the prairie systems in which the dominance of grasses is maintained by periodic fires.

fire ecology The active participation by humans in the management of forest and prairie resources by using fire as a tool. Fire has been and is an important environmental factor in the development and maintenance of natural ecosystems. As an understanding of ecology has advanced, the importance of using fire in a proactive way to manage forests and prairie systems has been realized. Controlled fires reduce the accumulation of decaying organic matter and litter that serve as fuel for future catastrophic fires, release plant nutrients, control invasion by nonnative exotic species, release and/or activate seeds, and clear competitive overgrowth.

fire point The lowest temperature at which a liquid vaporizes at a sufficient rate to support continuous combustion. Compare FLASH POINT.

firedamp Methane gas found in underground coal mines. See also DAMP.

firestorm An intense fire characterized by rapid burning and strong winds generated by convection currents produced by the fire.

firm capacity For public drinking water supplies, the system delivery capacity with the largest single water well or production unit out of service.

first draw The water that flows when a tap is first opened, especially at a residence first thing in the morning. The water that flows out first has been in the pipes all night and is likely to have the highest concentration of lead and other substances derived from the plumbing.

first flush The first runoff draining from the watershed (fields, streets, parking lots, ditches, and so on) when it rains. The concentration of plant nutrients, sediments, sand, oil, trash, and other pollutants is much higher than the runoff produced at later times. Controlling the first flush is critical to limiting pollution from runoff. See NONPOINT SOURCE.

first law of thermodynamics A law stating that during any chemical or physical change in a closed system, energy is not destroyed or created but is changed from one form to another. Its expression for a closed system is $Q = \Delta U + W$, where Q is the net heat absorbed by the system, W is the work performed, and ΔU is the change in internal energy. For example, chemical energy may be changed to heat energy during the burning of fossil fuels. The law is also referred to as the law of conservation of energy.

first-order reaction A chemical reaction in which the rate of reaction is directly proportional to the concentration of one of the reactants, not to any other chemical within the reaction mixture. Compare ZERO-ORDER REACTION.

first responder(s) As defined in the OCCUPATIONAL SAFETY AND HEALTH ADMINISTRATION regulations protecting workers who respond to a release or threatened release of a HAZARDOUS SUBSTANCE: (1) an individual or group trained to issue the initial warnings and response calls (awareness level) or (2) an individual or group trained in the actions necessary to protect human health or the environment from the threatened or actual release (operations level). First responders do not prevent, contain, or clean up the release of hazardous materials, which are duties of a HAZARDOUS MATERIAL RESPONSE TEAM. See HAZWOPER training.

first third See LAND DISPOSAL BAN.

First World The industrialized or economically developed countries of the world, including the United States, Canada, Japan, Australia, United Kingdom, France, and Germany. These countries consume a significant portion of the natural resources of the Earth.

Fish and Wildlife Service An agency, created with the United States Department of the Interior in 1940, responsible for fish and wildlife management on federal lands and for protection of these resources from possible harmful activities of other government agencies. The Fish and Wildlife Service compiles the list of endangered and threatened species authorized by the ENDANGERED SPECIES ACT. Website: www.fws.gov/

fish kill Large numbers of dead fish in a relatively small segment of a water body, almost always caused by a lack of required DISSOLVED OXYGEN, not by toxic chemicals.

fish ladder A series of small pools arranged in an ascending fashion to allow the migration of fish upstream past constructed obstacles, such as dams.

fish protein concentrate (FPC) A dry flour or paste derived from processing fish, often fish considered undesirable for direct consumption. Rich in protein, the product has been proposed as a dietary supplement for populations in less-developed countries, but cost and residual fishy taste have prevented its acceptance.

fission The splitting of the nucleus of an atom into two or more nuclei with the concurrent release of neutrons and a large amount of energy. The process is induced in a nuclear reactor to produce energy and in the detonation of nuclear weapons.

fission products The nuclei produced when elements, such as uranium-235, undergo fission. The fragments are ISO-TOPES of various elements and frequently are RADIOACTIVE. In a nuclear reactor, as these fragments accumulate, the efficiency of the reactor declines. Fission products constitute the nuclear waste in the spent fuel rods that are removed from the reactor.

fissionable material A HEAVY ATOM that can be split by the absorption of SLOW NEUTRONS. The three most common fissionable materials are uranium-233, uranium-235, and plutonium-239. Fissionable materials are used as fuel in a NUCLEAR REACTOR and in nuclear weapons.

fissure A long, narrow crack or split in a solid material, as in the cracks left in the Earth as the result of an earthquake.

fixation 1. In ecology, the incorporation of CARBON DIOXIDE into organic compounds by PHOTOSYNTHESIS. 2. In waste management, increasing of the stability of a waste material by involving it in the formation of a stable solid derivative. Stabilization lessens the potential harmful effect of a pollutant.

fixed carbon 1. The solid nonvolatile portion of organic waste material left after combustion, excluding ash and moisture. 2. The amount of carbon dioxide converted to plant biomass by the process of photosynthesis.

fixed-location monitoring Sampling of air or water from one location over a relatively long period.

fixed solids See SUSPENDED PARTICULATE.

flame ionization detector (FID) An analytical device used in GAS CHROMATOGRAPHY. The detector burns the gas containing the chemicals extracted from an environmental sample, which produces a current of ions or electrons proportional to the amount of specific organic materials present.

flame retardant A chemical added to cloth to prevent or retard rapid burning.

flameless furnace See GRAPHITE FURNACE.

flammable limits See LOWER EXPLOSIVE LIMIT, UPPER EXPLOSIVE LIMIT, FLAMMABLE RANGE.

flammable liquid Under United States Department of Transportation regulations, any liquid with a FLASH POINT less than 100° F (38° C).

flammable material Any solid, liquid, or gas that burns rapidly when ignited.

flammable range The temperature range between a material's LOWER EXPLOSIVE LIMIT and its UPPER EXPLOSIVE LIMIT.

Flannery decree The 1976 consent decree subsequently codified by the 1977 amendments to the CLEAN WATER ACT that required the USEPA to develop best available technology standards for over 20 industrial categories involving 65 classes of toxic water pollutants, called the PRIORITY POLLUTANTS. Named for the judge in the case, Thomas Flannery, U.S. district court judge for the District of Columbia.

flare A tall stack used for the routine combustion of waste gases or for the burning of materials that must be routed from a chemical reaction vessel or refining process during UPSET conditions.

flash point The lowest temperature at which a flammable liquid produces a sufficient amount of vapor to ignite with a spark.

Fletcher-Munson contours The results of sound loudness observations on young male subjects by Fletcher and Munson in 1933, expressed as plots of the SOUND PRESSURE LEVELS (in decibels) at particular frequencies that are perceived as equal in loudness. The subjects were given a 1000-hertz reference tone (for example, at 40 decibels) and were asked to adjust the decibel level of another tone of different frequency until that frequency sounded as loud as the reference tone. The test group consistently adjusted lower-frequency sounds (less than 1000 hertz) upward to make them equal to the reference tone in loudness; for example, a 60-decibel 200-hertz sound was equated to a 40-decibel 1000-hertz sound. At 1000 to 5000 hertz little adjustment occurred; there was some upward adjustment of tones greater than 5000 hertz. The contours illustrate the relative insensitivity of the human ear to lower frequencies. The DECIBELS, A-WEIGHTING NETWORK (DBA) scale is derived from the 40-decibel Fletcher-Munson contours. See PHONS.

floating roof (lid) A roof that floats on the hydrocarbon liquid in an oil refinery storage tank, moving up and down with the level of crude oil or fuel in the vessel. The absence of vapor space above the liquid and the maintenance of a tight seal between the tank wall and the roof greatly reduce hydrocarbon evaporation losses to the atmosphere.

floc In wastewater treatment, the particles formed by the COAGULATION of even smaller particles, or COLLOIDS.

flocculation In wastewater treatment, the rapid mixing of chemicals into the wastewater to enhance the formation of FLOC. Particles must be of sufficient size for removal, filtration, or sedimentation.

floodplain The land area bordering a river that floods when the river overflows.

flora A general term for the plant life in an area or region. Compare FAUNA.

flotation Any act taken to facilitate some material's floating on water or other liquid. For example, some mineral ores are separated by flotation technology in which finely powdered ores are separated from rock by using air to force the ore to float while the rock sinks. SUSPENDED SOLIDS are separated during sewage treatment processes by using tiny air bubbles to move fine particulate material to the water's surface, from which they can be skimmed off and removed.

flow control In waste management, a state or local government requirement that municipal solid waste collected in their jurisdiction must be taken to a certain waste disposal site, especially a WASTE-TO-ENERGY facility. The dedicated waste flow is to ensure that the facility operates at its planned level and remains financially sound. Flow control laws have been invalidated by the U.S. courts.

flow equalization The process of smoothing the amount of water or other fluid entering a process over time. For example, the amount of rainwater falling on an industrial facility varies considerably over time, ranging from none to millions of gallons in a short period. In order to treat the water properly before release, the water falling during storm events must be captured and held so that the amount entering the water-treatment process can be equalized over time to allow for sufficient treatment prior to release.

flow rate The amount of water or other fluid flowing through a pipe, canal, stream, or river; expressed as the amount of material moving over a specified period, such as cubic meters per second, gallons per minute, or millions of gallons per day.

flow regime The pattern of movement of runoff across an area of land that drains into a lake or stream, or the movement of water across an irrigated agricultural field.

flow system See CONTINUOUS-FLOW SYSTEM.

flow-through test A test of the toxicity of a chemical to aquatic organisms. The test solution containing the chemical passes through the test chamber and is not recycled. See STATIC TEST, SEMISTATIC TEST.

flue A conduit for the passage of smoke or exhaust gases.

flue ash See FLUE DUST.

flue dust Very small particles carried along in the gases that exit a furnace during combustion or from a metal-smelting operation.

flue gas Hot gases produced during combustion within a furnace. The gas is exhausted through the FLUE, carrying FLUE DUST and gaseous combustion products such as CARBON MONOXIDE and SULFUR DIOXIDE.

flue gas conditioning The addition of chemicals, such as ammonium salts, to an exhaust gas to raise the RESISTIVITY of the FLY ASH in the airstream before the mixture enters an ELECTROSTATIC PRECIPITATOR. Greater resistivity allows the particles to maintain an electrostatic charge and promotes collection efficiency in the electrostatic precipitator. Used particularly with low-sulfur coals.

flue gas desulfurization (FGD) The removal of sulfur-containing compounds, such as sulfur dioxide, from exhaust gas. The most widely used of many sulfur dioxide removal processes is limestone scrubbing, in which powdered limestone (calcium carbonate) is injected into an exhaust gas and sulfur dioxide is absorbed and neutralized, creating calcium sulfite and calcium sulfate as by-products. See DESULFURIZATION, DRY SCRUBBER.

flue-gas scrubbing See FLUE GAS DESULFURIZATION; SCRUBBER, IMPINGEMENT; SCRUBBER, SPRAY; SCRUBBER, VENTURI.

fluidized bed combustion A burning process designed to promote the efficient combustion of coal. The coal is first powdered, then made to flow as a liquid by the injection of a rapidly moving stream of gas. Burning is carried out while the solid powder is in this fluidlike state. Limestone is added to absorb sulfur dioxide created from sulfur contaminants in the coal.

fluidizing air A rapidly moving stream of air that is injected into a bed of powdered material to impart a fluidlike state to the powder. See also FLUIDIZED BED COMBUSTION.

flume A narrow trough or chute for carrying water.

fluorescence The emission of light or other electromagnetic energy caused by the excitation of an atom.

fluorescence detector A detector used with LIQUID CHROMATOGRAPHY. A sample stream is subjected to shortwave radiation, wavelength-specific for a certain substance, which causes any of that substance in the sample to reemit (fluoresce) longer-wavelength energy that can be measured by the detector to indicate the amount of substance in the sample.

fluoridation The addition of FLUORIDE to drinking water in an effort to reduce tooth decay among the human population.

fluoride A negative ion formed from the element fluorine or a compound containing fluorine. Fluoride combines with tooth enamel to render the enamel less soluble in acid environments. Fluoride compounds are added to public water supplies to prevent tooth decay. Fluorine is a halogen; chemical symbol F.

fluorocarbons See CHLOROFLUOROCARBONS.

fluorosis A disorder of the teeth or bones associated with excessive consumption of FLUORIDE; characterized by mottling and brittleness.

flux 1. The flow rate of mass, volume, or energy per unit of cross-sectional area that is perpendicular to the flow. 2. The movement of dissolved and suspended matter into and out of a marsh as the tides flood and ebb.

flux density In a nuclear reactor, the number of neutrons passing through a given unit of area per unit of time, usually through one square centimeter per second. For a plane that rotates inside a sphere to meet the neutrons at right angles, $\varphi = \Delta N/\Delta a\, \Delta t$, where φ is the flux density, Δa is the area of the circular plane, ΔN is the number of neutrons, and Δt is the time of measurement.

fly ash Small solid particles of noncombustible residue produced by the burning of a fuel such as coal. The particles are carried from the combustion process by the flue gas. Compare BOTTOM ASH.

fogging Applying insecticide by the release of fine droplets in a product that resembles smoke. The process can be used to reduce mosquito or fly populations in a community. A single-family residence can also be treated by the fogging process. Often the home is enclosed in a plastic material and the insecticide fog blown into the enclosure.

fomite Any object contaminated by pathogens from a diseased person.

food additive Any chemical added to a food product during processing to enhance shelf life, appearance, flavor, or nutritional content. The list of common additives includes agents that prevent caking of the food (aluminum calcium silicate), act as chemical preservatives (ascorbic acid, propionic acid, and sodium sulfite), emulsify (desoxycholic acid and propylene glycol), provide dietary supplement (biotin, leucine, and zinc sulfate), bind other materials (calcium acetate, potassium citrate, and tartaric acid), and

stabilize (agar-agar and guar gum). Food additives are regulated by the FOOD AND DRUG ADMINISTRATION under authority of the FOOD, DRUG, AND COSMETIC ACT. See DELANEY CLAUSE, GENERALLY RECOGNIZED AS SAFE.

Food and Agricultural Organization (FAO) An office of the United Nations founded in 1945 with a mandate to improve nutrition and standards of living and to increase agricultural productivity, with a special focus on rural populations. The FAO is the largest autonomous agency within the United Nations, with 180 member nations and a staff of over 4000. The FAO provides information and advice on food production to governments, offers direct development assistance, and serves as an international forum for debate on food and agriculture issues. Website: www.fao.org.

Food and Drug Administration (FDA) A federal agency within the United States Department of Health and Human Services that regulates the quality and safety of foods, food colors and additives, drugs, and cosmetics. Established by the FOOD, DRUG, AND COSMETIC ACT of 1938, the role of the agency has been further defined in subsequent amendments. Website: www.fda.gov

food chain The flow of CARBON and ENERGY within a specified area as a result of the feeding sequence of organisms within a COMMUNITY. The organisms are divided into TROPHIC LEVELS, which depend on how an individual organism obtains its food. The first level in the chain is occupied by the green plants, termed PRIMARY PRODUCERS; those animals that consume the plants are termed HERBIVORES or CONSUMERS and are placed in the second trophic level; the animals that eat other animals are termed CARNIVORES or SECONDARY CONSUMERS and are placed in the highest feeding level. The transfer of materials or mass from one trophic level to the next is approximately 10% efficient. In a simple food chain consisting of grass (primary producer) to rabbit (herbivore) to owl (carnivore), 1000 pounds of grass would be needed to support 100 pounds of rabbit, which, in turn, would support 10 pounds of owl. The food chain thus described is referred to as a grazing food chain since it is based on the consumption of live, standing biomass (the grass). Other food chains are based on the consumption of the remains of dead organisms. The primary consumers in such food chains, referred to as DETRITUS-based, consist of bacteria, fungi, and various worms and insects (termed the DECOMPOSER community). The next highest trophic level consists of those carnivores that consume these decomposers. See ECOLOGICAL PYRAMID.

food chain crops Plants that, if contaminated by environmental pollutants, directly or indirectly cause human exposure to the contaminants. These include plants grown for human consumption, those ingested by animals that are ultimately part of the human diet, and tobacco.

food chain efficiency An expression of the net transfer of useful energy from a food to the animal consuming it. See ECOLOGICAL EFFICIENCY.

food color Any dye or pigment that imparts color to food. There are nine synthetic color additives certified (approved) for American use by the FOOD AND DRUG ADMINISTRATION (FDA). Pigments from natural sources do not require FDA approval. Color additives are also used in drugs and cosmetics. Labels indicate the certified color and its use: FD&C means food, drug, and cosmetics; D&C and "external D&C" also are used.

Food, Drug, and Cosmetic Act (FDCA) First passed as the 1906 Food and Drug Act, with major amendments in 1938, 1958, 1962, and 1976, the basic federal law concerning the sanitary condition and safety of food, including FOOD ADDITIVES, and the efficacy and safety of drugs and cosmetics. The act is administered by the FOOD AND DRUG ADMINIS-

TRATION (FDA). See DELANEY CLAUSE, GENERALLY RECOGNIZED AS SAFE.

food intoxication A pathological condition in humans or animals that is caused by the consumption of food that contains a toxin. For example, botulism and common food poisoning are caused by the production of toxic compounds during the growth of specific types of bacteria in food prior to consumption.

Food Quality Protection Act (FQPA) A 1996 federal statute that amended the FEDERAL INSECTICIDE, FUNGICIDE, AND RODENTICIDE ACT and the FOOD, DRUG, AND COSMETICS ACT, providing new regulation of PESTICIDE RESIDUES in food, with an emphasis on protection of infants and children.

food waste A class of municipal solid waste consisting of the remains of uneaten food or of the process of food preparation at residences or commercial establishments such as cafeterias, restaurants, and industrial kitchens.

food web The interrelationship among the biological organisms in a COMMUNITY according to the transfer of useful energy from food resources to organisms eating those resources. A FOOD CHAIN depicts a simple linear transfer from one organism to another; however, most animals eat more than one type of food and the food web model illustrates the complexity of feeding patterns within the natural environment. For example, a single resource such as grass may serve as food for insects, mice, rabbits, and deer, and the mice may in turn be eaten by snakes, owls, and foxes.

food-processing waste Solid waste that remains after the conversion of a raw agricultural item into a commercial product. The discards left after processing, for example, corn shucks and cobs left after the corn is prepared for canning at a commercial facility.

food-to-microorganism ratio (F/M ratio) The ratio of organic material load to the microorganism mass in the aeration tank of a wastewater treatment facility. The ratio is calculated as

$$F/M = (Q)(BOD)/(MLSS)(V)$$

where Q is the flow rate of RAW SEWAGE, BOD is the BIOCHEMICAL OXYGEN DEMAND of that sewage, $MLSS$ is the MIXED LIQUOR SUSPENDED SOLIDS concentration, and V is the aeration tank volume. Commonly expressed as kilograms of BOD per kilogram MLSS per day.

foot-pound   A unit of WORK equivalent to 1.356 JOULES.

force mains   Pipes in which wastewater is pumped under pressure; the system is used in some areas that have small elevation changes with distance and therefore need to augment gravity flow. Contrast GRAVITY FLOW. See PRESSURE SEWER.

forced draft (FD)   The pushing or forcing of gases through an enclosed area (such as a combustion chamber) by the use of a fan or blower. See INDUCED DRAFT, NATURAL DRAFT.

forced expiratory volume (FEV$_1$)   The volume of air a person can exhale in one second; a lung function test.

forced oxidation   A chemical process in which pollutants in an exhaust or discharge are forced into contact with air or pure oxygen to convert them to a stable form.

forced vital capacity (FVC)   The maximum volume of air a person can exhale after a maximum inhalation; a lung function test.

forcing functions   In ecology, important factors determining the composition of natural ecosystems, such as temperature, rainfall, sedimentation, timing and amount of watershed runoff, nutrient levels, and solar radiation.

Forest and Rangeland Renewable Resources Planning Act (FRRRPA)   See NATIONAL FOREST MANAGEMENT ACT.

forest management   Administration and planning of forest resources employing rational, interdisciplinary scientific principles. Depending on the intended use of the resource, the effort may include harvesting of trees, restoration of degraded land, multiple use scenarios, pest control, and the maintenance of a diverse biological community.

Form R   The annual report of routine and accidental chemical releases from certain facilities required by Section 313 of the Emergency Planning and Community Right-to-Know Act (TITLE III).

formaldehyde   A simple one-carbon aldehyde having the formula CH_2O. The chemical is a colorless gas, produces a pungent odor, and acts as an irritant. A solution of 40% formaldehyde in water, termed formalin, was commonly used as a biological preservative. The gas is a primary irritant that contributes to the odor and eye irritation of PHOTOCHEMICAL AIR POLLUTION and indoor air pollution at retail establishments handling particleboard or fabric. The irritation associated with formaldehyde results from the solubility of the gas in water and the resulting reactions in the mucous membranes of the eyes and nose. Formaldehyde is used as an industrial chemical in the manufacture of various polymers, including urea formaldehyde insulation foam.

formalin   See FORMALDEHYDE.

fossil fuels   Crude oil, natural gas, peat, coal, or other hydrocarbons that are derived from the remains of plants and/or animals that were converted to other forms by biological, chemical, and physical forces of nature.

four nines   The destruction requirement for selected PRINCIPAL ORGANIC HAZARD CONSTITUENTS in waste incinerators: 99.99%. See DESTRUCTION AND REMOVAL EFFICIENCY, SIX NINES, TRIAL BURN.

Fourth World 1.   A classification applied to the poorest countries. Characterized by a poorly functioning market economy, ineffective government management, and little prospect of economic development. 2. Ethnic, linguistic, and/or religious groups with a common culture residing within a nation state with dissimilar characteristics. The Fourth World group may seek political autonomy, including its own territory.

fracture In geology, a break or crack in a rock formation promoted by stresses within the rock.

Framework Convention on Climate Change A treaty signed by nations at the EARTH SUMMIT although negotiated prior to that meeting. Signing nations agreed to stabilize and reduce GREENHOUSE GAS emissions that may contribute to GLOBAL WARMING. Specific greenhouse gas reductions are specified in the KYOTO PROTOCOL.

free available chlorine The sum of the hypochlorous acid (HOCL) and hypochlorite ion (OCL⁻) concentrations in water. These forms of chlorine are actively involved in the destruction of bacteria, viruses, and other organisms that cause disease and participate in other chemical reactions that take place in water.

free field Conditions in which sound can be measured without interference by echoes from barriers in the test area. Actual free-field conditions rarely exist outside special sound chambers, but nearly free-field conditions are present if the distance to barriers is great enough to have no significant influence on the sound measurements.

free liquids Liquids capable of migrating from waste and contaminating groundwater. Hazardous waste containing free liquids may not be disposed of in landfills. See PAINT FILTER LIQUIDS TEST, LAND DISPOSAL BAN.

free-living Describing a species of plant or animal that is capable of living, reproducing, or carrying out a specific function without the direct assistance of a plant or animal of a different species. Compare SYMBIOTIC.

free moisture Liquid that drains freely from solid waste by the action of gravity only.

free radicals Unstable atoms or molecules with at least one unpaired electron; these materials are highly reactive and thus are short-lived. They act as important intermediates in PHOTOCHEMICAL AIR POLLUTION, PHOTOLYSIS, CHLOROFLUOROCARBON depletion of the OZONE LAYER, combustion processes, and polymerization reactions.

free residual chlorine The FREE AVAILABLE CHLORINE level present after the destruction of ammonia and the reduction in CHLORAMINE residuals by the progressive addition of chlorine (BREAKPOINT CHLORINATION). The residual available for disinfection in a water distribution system practicing free residual chlorination. Compare COMBINED AVAILABLE CHLORINE.

free silica See CRYSTALLINE SILICA.

freeboard The vertical distance between the waste contained in a tank or surface IMPOUNDMENT and the top of the enclosure or container.

freons A family of compounds containing carbon, chlorine, and fluorine with a typical chemical formula of CF_2Cl_2 (freon-12). The compounds are gases at room temperature and are used as the carrier gas in aerosol cans (in some countries outside the United States), as foaming agents and solvents, and as the heat transfer gas in refrigerators and air conditioners. Freon production is being phased out around the world under the provisions of the MONTREAL PROTOCOL and subsequent international agreements. See CHLOROFLUOROCARBONS.

frequency For electromagnetic radiation or sound waves, the number of wave cycles passing a point in one second. Formerly expressed as cycles per second; now expressed as HERTZ. One hertz equals one wave cycle per second.

freshwater Water without significant amounts of dissolved sodium chloride (salt). The opposite of seawater. Water characteristic of rain, rivers, ponds, and most lakes.

freshwater ecosystem Describing the plants, animals, and physical properties of rivers, lakes, and ponds. These differ markedly from the biota in a seawater environment. An aquatic environment characterized by a lack of salt.

friability The degree to which a solid can be crushed and powdered.

friable Describing a solid that is easily crushed and powdered. ASBESTOS-CONTAINING MATERIAL that can be crushed by hand pressure (friable) poses an inhalation hazard and is the target of federal and state regulatory programs.

Friends of the Earth (FOE) A conservation and environmental organization, founded in 1969, dedicated to the preservation, restoration, and wise use of natural resources. United States headquarters in Washington, D.C., affiliates in 37 countries. Through the Friends of the Earth Foundation, the organization promotes public education and monitors enforcement of environmental policies. In 1999 membership in the United States numbered about 35,000. Website: www.foe.org

front end recovery A centralized solid waste treatment process in which mechanical or manual separation of paper, glass, metals, and/or organic matter suitable for composting is performed on the collected waste before further processing.

frontier mentality An outlook reflecting the frontier past of America when land, forests, and wild animal populations seemed to be infinite. The view that assumes that the world has unlimited natural resources for human use regardless of the consequences to the natural environment or to the biosphere.

frost heave The uneven rise in a ground surface caused by the accumulation of ice in the subsurface soil.

fuel Coal, oil, natural gas, or products thereof that are used to power equipment or provide heat. ENRICHED URANIUM used to provide energy in nuclear reactors.

fuel assembly A bundle of about 200 fuel rods, each of which contains pellets of ENRICHED URANIUM. These clusters are placed in the CORE of a nuclear reactor to provide the fissionable material needed to power the reactor. Also called fuel elements.

fuel cell A device in which hydrogen and oxygen combine to produce an electric current. Used to power crewed spacecraft and considered as a power source for low-emission vehicles.

fuel cycle The steps in the production of ENRICHED URANIUM for use in a nuclear reactor and the handling of fuel elements after they are removed from the reactor. The complete cycle includes the mining of uranium ore, purification of the uranium, enrichment of purified uranium in the fissionable isotope, manufacture of pellets of enriched uranium, fabrication of fuel rods and fuel assemblies, use of the fuel in reactors, recovery of reusable uranium and other elements from used rods, and disposal of waste material generated in the process. See FUEL ROD, FUEL ASSEMBLY, FUEL ENRICHMENT, FUEL REPROCESSING, FISSIONABLE MATERIAL, ISOTOPE.

fuel economy standard See CORPORATE AVERAGE FUEL ECONOMY (CAFE) STANDARD.

fuel elements See FUEL ASSEMBLY.

fuel enrichment Increasing the abundance of FISSIONABLE MATERIAL. When uranium is purified from geological deposits, the product consists of a mixture of about 0.7% uranium-235 and 99.3% uranium-238. In such mixtures, the fissionable ISOTOPE (uranium-235) is not sufficiently concentrated to support a sustained fission reaction in common reactors. The amount of fissionable material can be increased by purifying the uranium-235 or by adding plutonium-239.

fuel NO$_x$ Nitrogen oxides formed by the oxidation (combustion) of organic nitrogen present in coal or oil. Compare THERMAL NO$_x$.

fuel reprocessing The recovery of usable uranium and other elements from FUEL RODS that have been removed from a nuclear reactor. The amount of fissionable material decreases and the amount of fission products increases when fuel rods have been employed in a nuclear reactor for about three years. As a consequence, the efficiency of the fission reaction diminishes to the point that the used rods must be removed. These used rods can be processed so that the substantial amount of usable uranium that remains can be reclaimed and other useful and waste elements removed.

fuel rod A long tube that contains pellets of enriched uranium dioxide used to fuel a nuclear reactor. These rods are commonly constructed of a zirconium alloy or stainless steel and are bundled into a FUEL ASSEMBLY of about 200 rods.

fugitive emission Any gas, liquid, solid, mist, dust, or other material that escapes from a product or process and is not routed to a pollution control device.

fume Finely divided airborne solids formed by the condensation and solidification of material emitted as a vapor or gas; usually irritating and offensive at high concentrations. Contrary to popular use, gasoline "fumes" are not fumes.

fumigant Any substance that is used as a gas, particulate, vapor, or smoke to kill pests (insects or rodents) in foodstuffs or structures.

fumigation 1. In pest management, the application of a FUMIGANT to a material or area in order to kill pests or dangerous organisms. 2. In air quality management, high ground level pollutant concentrations downwind from a smokestack that result when daytime THERMAL TURBULENCE (hot air rising from the ground) breaks up a TEMPERATURE INVERSION that held the PLUME aloft. Compare CONING, FANNING, LOFTING, LOOPING, TRAPPING.

fundamentally different factors (FDFs) A type of variance from a water pollutant EFFLUENT LIMITATION that may be granted under provisions of the Clean Water Act if a permit applicant can show that the plant or facility is fundamentally different from the facilities, equipment, or operating conditions used by the USEPA when setting the TECHNOLOGY-BASED standard for the industry category or subcategory as a whole.

fungi A diverse group of plants that are not capable of photosynthesis but obtain carbon nutrition from the absorption of carbon compounds from the environment. The category includes organisms ranging from yeast to molds to mushrooms and organisms that have broad ability to degrade complex organic molecules in the environment. Fungi are responsible for the spoiling of food (moldy bread) and rotting of wood and may be useful in the biological oxidation of toxic and hazardous substances such as certain chlorinated organic compounds. Most fungi require the oxygen in air for growth. Fungi are important in the release of spores into the atmosphere, thereby contributing to INDOOR AIR POLLUTION, SICK BUILDING SYNDROME, and allergic responses termed hay fever. A few fungi can cause infectious diseases in humans.

fungicide Any substance that kills FUNGI or molds.

furans See DIBENZOFURANS.

furrow irrigation The addition of water to support the growth of agricultural crops by flooding the fields and allowing water to flow along the furrows between rows. Contrast DRIP IRRIGATION.

fusion A reaction that results from combining nuclei of small atoms to form larger atoms. Joining the nuclei of two atoms requires atomic collisions at very high temperatures and pressures, with a significant release of heat energy. The process is the underlying force for the release of energy from stars and from the detonation of a hydrogen bomb. Also referred to as a thermonuclear reaction.

G

gabion A wire mesh container filled with rocks used as a barrier to retard soil erosion.

gage pressure See GAUGE PRESSURE.

Gaia hypothesis The proposition that the composition and temperature of the atmosphere are products of interrelated activities in the BIOSPHERE, especially those of microorganisms, and that the biosphere behaves as a single self-regulating organism. Gaia was the ancient Greek goddess of the Earth. The hypothesis was developed by the British scientist James Lovelock and the American biologist Lynn Margulis.

gallons per capita per day (GPCD) An expression of the average rate of domestic and commercial water demand per person, usually computed for public water supply systems. Depending on the size of the system, the climate, whether the system is metered, the cost of water, and other factors, public systems in the United States experience a demand rate of 60 to 150 GPCD.

game fish Usually refers to finfish like trout or bass sought by sport fishermen. Generally considered a valuable resource because of the economic activity associated with sport fishery.

gamete A reproductive cell containing one copy of each gene (haploid) needed to provide the genetic information required for the development of an offspring. When gametes from the female (egg) are combined with or fertilized by those of the male (sperm), a zygote containing two copies of each gene (diploid) is produced. The zygote can develop into an offspring.

gamma decay Spontaneous decomposition of the nucleus of an unstable (RADIOACTIVE) element characterized by the release of GAMMA RAYS, an energetic form of radiation that presents a significant health risk at excessive exposures.

gamma radiation See GAMMA RAY.

gamma ray A type of ELECTROMAGNETIC RADIATION, produced by some RADIOACTIVE substances, with a very short wavelength and high energy level. This type of radiation is strongly penetrating, potentially harmful to living things, and more energetic than X rays. Also referred to as GAMMA RADIATION. See GAMMA DECAY.

gammarids Any of three species of the amphipod *Gammarus (G. fasciatus, G. pseudolimnaeus, G. lacustris)* used in the laboratory analysis of the toxicity of pollutants to aquatic animals. The organism is used to determine the toxicity of liquid effluents without chemical analysis to determine the precise chemical composition of the toxicant. See WHOLE EFFLUENT TESTING, BIOASSAY.

gap analysis A technique employed in the conservation of biological DIVERSITY, the protection of regional species RICHNESS, and wildlife management by examining areas (gaps) between protected HABITATS. In some instances, the areas outside the parks, reserves, refuges, or game management regions have a richer collection of species of interest than those inside the protected regions. Consequently, the outside regions may be fruitful habitat for investigations.

garbage Waste material, typically from domestic and commercial sources.

gas One of the three states of matter. In a gaseous state, there is little attraction between the particles, which have continual, random motion. The gas has no fixed shape or volume, can expand indefinitely, and assumes the shape of the space in which it is held. Gas is also easily compressed, with the random collisions between particles exerting pressure on the walls of the container.

gas barrier A layer of material placed on the top and/or sides of a landfill to prevent the off-site migration of gas (and odors) produced by microbial degradation of the buried waste. Typical barrier materials are compacted clay and synthetic membranes.

gas constant See UNIVERSAL GAS CONSTANT.

gas chromatogram A graphic representation produced by the detector used in GAS CHROMATOGRAPHY. The identity and amount of contaminants can be determined from such an output.

gas chromatograph The analytical instrument used to perform GAS CHROMATOGRAPHY.

Graphic chart of instrument output

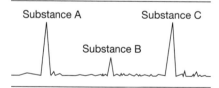

Gas chromatograph

gas chromatography (GC) An analytical technique that can yield both qualitative and quantitative evaluations of sample mixtures of volatile substances. The compounds of interest are separated by using an inert gas to flush a sample preparation through a column packed with a substance that selectively absorbs and releases the volatile constituents. A device is used to detect the level of each compound as it exits the column.

gas chromatography/mass spectrometry (GC/MS) A sensitive and accurate analytical technique, used mainly for organic compounds, in which the gas effluent from a GAS CHROMATOGRAPH is piped to a MASS SPECTROMETER for additional analysis, which typically includes precise identification of the organic constituents in the sample.

gas contacting device Equipment designed to encourage substantial mixing of liquid and gas streams.

gas-cooled reactor (GCR) A nuclear reactor in which the heat produced by fission reactions is removed from the reactor CORE by a gas, as opposed to air, water, or liquid sodium. The gases most commonly used are helium and argon. See also HIGH-TEMPERATURE GAS REACTOR, WATER-COOLED REACTOR.

gas stripping Technology used to remove volatile liquids or gaseous pollutants from water or to recover products or reactants from liquid streams in an industrial setting. A carrier gas such as nitrogen or air is bubbled through wastewater or other liquid to force the removal of volatile chemicals that are then collected or released to the atmosphere.

gaseous diffusion A method used to separate isotopes of uranium. Uranium oxide that is prepared from uranium ore is first converted to uranium hexafluoride, a gaseous mixture containing both of the naturally occurring isotopes of uranium, uranium-235 and uranium-238. The isotopes are then separated on the basis of differences in their diffusion properties.

gasification See COAL GASIFICATION.

gas laws The relationships among gas temperature, pressure, and volume. See BOYLE'S LAW, CHARLES'S LAW, IDEAL GAS LAW.

gas-liquid chromatography See GAS CHROMATOGRAPHY.

gasohol An automobile fuel that is a blend of alcohol (usually ethyl alcohol) and gasoline. The typical mixture contains 10% alcohol. The blend was developed to conserve supplies of gasoline in the United States; however, the mixture did not gain wide public acceptance in the 1980s, and use was largely discontinued. Brazil remains the world's leading consumer of gasohol fuel.

gas-phase absorption spectrum The pattern of radiation absorbance by chemical compounds in a gaseous mixture. Gases absorb radiation in certain frequency ranges. The spectrum is used to monitor some air pollutants.

gas-to-cloth ratio (G/C) See AIR-TO-CLOTH RATIO.

gas volumetric flow rate See VOLUMETRIC FLOW RATE.

gastroenteritis A disorder of the stomach, small intestine, or colon that is characterized by nausea, vomiting, and/or diarrhea. The condition is caused by a variety of bacteria and viruses.

gastrointestinal tract The organs of the alimentary system through which food passes that extends from the mouth to the anus. The term is also applied specifically to the stomach, small intestine, and colon.

gauge See GAUGE PRESSURE.

gauge pressure The pressure exerted by gases relative to ATMOSPHERIC PRESSURE, commonly expressed in millimeters of mercury (mm Hg), pounds per square inch (gauge), or inches of water. If atmospheric pressure is 760 mm Hg and gauge pressure is 20 mm Hg, then the total (or absolute) pressure is 780 mm Hg.

Gaussian plume model A basic AIR QUALITY DISPERSION MODEL based on an assumed normal distribution of vertical and horizontal downwind concentrations from a pollutant source. The official USEPA computer programs used to estimate air concentrations are Gaussian plume models. See DISPERSION COEFFICIENT.

Gay-Lussac's law See CHARLES'S LAW.

gavage In toxicological testing, the introduction of the test chemical to an animal via a stomach tube.

gene A specific segment of DEOXYRIBONUCLIC ACID (DNA) located on a CHROMOSOME. The individual segments can code for a specific protein, a specific type of RIBONUCLEIC ACID (RNA), or a recognizable trait of an organism. Hundreds of genes can be located on a single chromosome. A GAMETE has one copy of each gene (haploid); all other cells of an organism have two copies of each gene (diploid). Mutagens to which an organism is exposed cause permanent changes in the gene structure.

gene addition The application of the technology associated with molecular biology to correct a genetic mutation in an organism. Under some circumstances, the genetic constitution of an organism can be corrected by the introduction of a "good" copy of the damaged gene. The introduced gene integrates into the chromosomes of the recipient organism at some place other than the location of the damaged gene. Regardless of the site of integration of the introduced gene into the damaged organism (see GENE REPLACEMENT), the genetic defect of the recipient organism is corrected. See GENETIC ENGINEERING.

gene bank See SEED BANK.

gene pool The total genetic information within a population of plants or animals that is capable of interbreeding.

gene replacement The application of the technology associated with molecular biology to the correction of a genetic defect in an organism. The genetic defect can be considered to be caused by a mutation that damaged a gene, preventing the normal functioning of that gene. Under some circumstances, the damaged gene can be replaced by the introduction of a cloned gene produced in the laboratory into the chromosome of the recipient organism. The introduced gene takes the place of the damaged unit, thereby "curing" the defect. Compare GENE ADDITION. See GENETIC ENGINEERING.

General Accounting Office (GAO) An agency operating as an arm of the United States Congress with responsibility for auditing and reviewing government expenditures and programs; the office determines whether federal agencies are spending public funds as intended by the legislation that authorizes the expenditures. Website: www.gao.gov

General Agreement on Tariffs and Trade (GATT) An international agreement intended to encourage free trade among member states by lowering tariff barriers and to provide a forum within which to resolve trade disputes. GATT has been the subject of environmental controversy because its free trade rules have been interpreted to prevent nations from using import restrictions to solve global or transnational environmental problems, such as the U.S. ban on imports of tuna that were caught in nets that killed dolphins. The concern is that international differences in waste management, pesticide controls, air and water pollution standards, and species protection policies will be harmonized at the expense of environmental protection. See also NORTH AMERICAN FREE TRADE AGREEMENT (NAFTA), NORTH AMERICAN AGREEMENT FOR ENVIRONMENTAL COOPERATION (NAAEC), *MAQUILADORAS*, AGENDA 21, EARTH SUMMIT.

general circulation model (GCM) Computer programs that simulate oceanic or atmospheric circulation and temperature. GCMs are used for weather (short-term) forecasts and for predictions of climate (long-term) change, such as GLOBAL WARMING.

General Duty Clause A section of the OCCUPATIONAL SAFETY AND HEALTH ACT providing that "each employer shall furnish . . . a place of employment . . . free from recognized hazards that are causing or are likely to cause death or serious physical harm to his employees." The clause is used by the Occupational Safety and Health Administration to force correction of workplace conditions not specifically covered by codified health and safety regulations.

generally recognized as safe (GRAS) A classification of FOOD ADDITIVES. The FOOD, DRUG, AND COSMETIC ACT of 1958 provided "grandfather" approval for most additives in common use prior to the passage of that legislation. These additives cannot be removed from use until they are proved to be harmful to the public by the FOOD AND DRUG ADMINISTRATION (FDA). Over 600 substances are on the FDA list of GRAS substances. They were placed on this list on the basis of past experience in food use and scientific determination of their safety, as recommended by the FDA and the U.S. Department of Agriculture.

general reporting facility A facility maintaining a quantity of one or more hazardous chemicals in excess of 10,000 pounds, the threshold quantity triggering emergency planning activities, maintenance of material safety data sheets, reporting to state and local emergency planning committees, and listings with local fire departments.

generator 1. A business or industrial facility that produces a HAZARDOUS WASTE. 2. A device used to produce electricity.

genetic assimilation A threat to native species posed by cross-breeding with non-native organisms introduced by humans.

tainer or product to reduce porosity. Glazing containing lead has been a source of lead poisoning when used on earthenware in which acidic foods or beverages are stored.

Global Change Research Program
A unit in the U.S. government established in 1989 to coordinate research on short- and long-term climate change, including the Earth's radiation balance, OZONE LAYER DEPLETION, and marine and continental ecosystem health. Over 300 U.S. and foreign universities and research institutions; offices within the U.S. Departments of Agriculture, Commerce, Defense, Energy, Health and Human Services, Interior, State, and Transportation, and several executive agencies, including the USEPA and the National Aeronautics and Space Administration, participate in the program. See GLOBAL WARMING, UNITED NATIONS ENVIRONMENT PROGRAM, EARTHWATCH. Website: www.usgcrp.gov

Global Environmental Management Initiative (GEMI)
A nonprofit organization of around 40 multinational companies based in the United States "dedicated to fostering environmental, health and safety excellence worldwide." See also RESPONSIBLE CARE, ISO 14000. Website: www.gemi.org

Global Environment Facility
A program begun in 1991 and jointly operated by the United Nations Development Program, the UNITED NATIONS ENVIRONMENT PROGRAM, and the World Bank to provide grants and loans to activities that help combat climate change, ocean pollution, loss of biodiversity, and ozone layer depletion. Website: www.gefweb.org

Global Environment Monitoring System (GEMS)
A part of the United Nations Environment Program's EARTHWATCH, based in Nairobi, Kenya. Purposes of GEMS are to coordinate environmental monitoring and assessment activities worldwide and to provide an environmental data exchange service.

global positioning system (GPS)
U.S. Department of Defense satellite signals are beamed to handheld receivers that calculate and display the user's position on the Earth. Civilian instruments are accurate to about 10 meters.

Global Releaf
A project of American Forests, an American nonprofit conservation organization founded in 1875, to plant 20 million trees for the new millennium. At the beginning of 2000, 11 million trees had been planted in over 500 forests in the United States and 21 other countries. Website: www.americanforests.org

global warming
The predicted excessive warming of the atmosphere resulting from the accumulation of atmospheric CARBON DIOXIDE. The atmosphere is normally warmed when INFRARED RADIATION emitted by the Earth is absorbed by carbon dioxide gas and water vapor in the air. As the amount of carbon dioxide increases as a result of the combustion of fossil fuels and DEFORESTATION, especially of tropical rain forests, the hypothesis is that more heat energy will be retained by the Earth's atmosphere, resulting in a change in rainfall and wind patterns and melting of polar ice, thus raising the global sea level. The change in weather patterns could have devastating consequences for prime agricultural areas. A significant rise in sea level could flood many coastal cities and damage ecologically important coastal wetlands. Other heat-absorbing gases that are increasing in the atmosphere as a result of human activities are methane, nitrous oxide, and the CHLOROFLUOROCARBONS. See GREENHOUSE EFFECT, GREENHOUSE GASES.

global warming potential (GWP)
A weighting of various GREENHOUSE GASES based on their ability to absorb INFRARED RADIATION (heat) relative to that of CARBON DIOXIDE, which is given a GWP of 1. For a 100-year time horizon, the GWP of methane is 21 and for nitrous oxide is 270–310. The HYDROFLUOROCARBONS, substitute compounds for the CHLOROFLUOROCARBONS that threaten the OZONE

LAYER, have GWPs of 1000 to over 10,000.

globe temperature The temperature reading that indicates the level of radiant heat or heat transferred by INFRARED RADIATION. The temperature is read from a thermometer with the bulb set in the center of a black six-inch sphere, which acts to absorb radiant heat in the surroundings.

glove box An airtight container with attached flexible gloves that extend into the box, allowing a worker to manipulate the box contents without exposure to the material(s). Such boxes are used for handling highly radioactive substances and highly infectious materials.

glovebag A device employed to protect workers and the immediate surroundings during the removal of asbestos-containing material, usually pipe insulation. The plastic, baglike enclosure is placed around the area to be cleaned, allowing the removal of the asbestos without unnecessary contamination of the surroundings. See GLOVE BOX.

God committee A group of representatives from U.S. Cabinet offices created by amendment to the ENDANGERED SPECIES ACT (ESA). The committee is allowed to exempt certain projects that would otherwise be blocked by the ESA. The exemption potentially authorizes the extinction of a species, hence the name. See SNAIL DARTER.

Gold Book 1. In water quality management, the USEPA document *Water Quality Criteria 1986,* which contains information on ambient water quality toxicity for acute and chronic exposures to aquatic life and health risks for human exposures to pollutants from water or fish consumption. These criteria are not legally enforceable standards. 2. A guidance document issued by the U.S. BUREAU OF LAND MANAGEMENT and the U.S. Forest Service for oil and gas exploration, drilling, and production from onshore federal lands.

good engineering practice stack height (GEP stack height) A regulatory requirement of the USEPA limiting the height of stacks emitting air pollutants; the prescribed height depends on the date when the stack was constructed, the height or width of nearby structures, and/or the potential ground-level impact.

Good Laboratory Practice Standards (GLP Standards) Standards that must be followed by individuals conducting studies relating to health effects, environmental effects, and chemical fate testing with the support of the USEPA or the Food and Drug Administration. The regulations address the areas of personnel, management, quality assurance, animal care, and substance handling.

Government Printing Office (GPO) The official publisher of U.S. congressional documents and publications of agencies and departments of the executive branch. The office operates bookstores in over 20 cities and distributes government documents to depository libraries across the United States. Website: www.gpo.gov

grab sample Typically, a single air or water sample drawn over a short period, which is not representative of long-term conditions at the sampling site. This type of sampling yields data that provide a snapshot of conditions or chemical concentrations at a particular point in time.

gradient See CONCENTRATION GRADIENT, PRESSURE GRADIENT.

gradient wind Wind that is assumed to move parallel to the curved path of atmospheric ISOBARs with no deviations due to friction between the wind and the ground.

Graham's law Law stating that gases diffuse at a rate proportional to the square root of their density and lighter molecules diffuse faster than heavy molecules. This property of gases can be used to separate uranium-235 hexafluoride (which is lighter) from uranium-238 hexafluoride (which is heavier).

grain 1. A mass unit used in expressions of air concentrations of particulate matter, such as grains per cubic foot. One grain equals 65 milligrams or 0.00014 pound. 2. For jewelers, a unit equaling 0.25 carat, or 50 milligrams.

grain loading An expression of the air concentration of pollutants that are flowing into an emission control device or are being emitted by a smokestack; usually expressed in GRAINS per cubic foot.

gram (g) A unit of mass equal to 10^{-3} kilogram or 1/28 of an ounce. One milliliter (cubic centimeter) of water contains one gram at 4° C.

gram molecular weight (GMW) The mass, in grams, of a substance equal to the MOLECULAR WEIGHT of the substance. For example, the molecular weight of water (H_2O) is 18 (the sum of the atomic weights of two hydrogen atoms and one oxygen atom), so the gram molecular weight is 18 grams. The amount of a material equal to the gram molecular weight of the substance comprises one gram-MOLE of the substance.

gram-negative/gram-positive The response of bacteria to a procedure called the Gram stain. When treated with crystal violet, Gram's iodine, 95% ethanol, and safranin, bacteria usually retain either a red color (gram-negative) or a purple/blue color (gram-positive). In addition to the differences in color upon staining, these two types of bacteria represent bacteria that are fundamentally different in terms of structure, physiological characteristics, ecological features, and pathogenicity. Common gram-negative species are *Escherichia, Salmonella,* and *Pseudomonas;* common gram-positive species are *Bacillus, Staphylococcus,* and *Streptococcus.*

Gram stain See GRAM-NEGATIVE/GRAM-POSITIVE.

grandfather clause In environmental or occupational safety and health laws or regulations, a statement that new rules or strictures apply only to persons or industries beginning business or building new emission sources after a certain date. Operations in existence before that date are "grandfathered" in and do not have to comply with the new rules.

granitic crust That part of the Earth's crust comprising the continents. Derived from the slow cooling of the molten material of the mantle of the Earth, yielding a rock material that is characterized by a density lower than that of BASALTIC crust that underlies the oceans.

granular activated carbon (GAC) A porous carbon material used to remove pollutants from air and water. The extremely large surface area ADSORBS organic compounds (and some inorganics) as the air or water flows through GAC filter beds. See TERTIARY TREATMENT, CARBON FILTRATION, CARBON POLISHING, CARBON TREATMENT, BREAKTHROUGH CURVE.

Graphical Exposure Modeling System (GEMS) A set of interactive computer programs developed by the USEPA that are used to estimate human exposure to environmental contaminants. The programs include fate and transport models for chemicals in air, surface water, groundwater, or soil. Census data for population exposure analysis are incorporated into the package. GEMS is being replaced by the Geographical Exposure Modeling System. Website: www.epa.gov/opptintr/exposure/docs/gems.htm

graphite furnace An ATOMIC ABSORPTION SPECTROPHOTOMETER that uses an electric current, instead of a flame, to heat and atomize the sample. Also called a flameless furnace.

grassed waterway Natural or constructed land area covered with vegetation. Established to allow runoff or effluent to flow to a receiving stream without causing erosion.

grassland A geographical region dominated by shrubs and grasses, receiving 10 to 30 inches of rain annually. Alpine grasslands are in cool, high-elevation areas. Temperate grasslands, called prairie (North America), pampas (South America), steppe (Asia), or veldt (South Africa), are found in regions with moderate temperatures. Tropical grasslands, also called savannas, are found in warmer climates.

grate siftings Material that falls through the openings in the fuel bed of an incinerator that burns solid waste.

gravimetric Pertaining to measurements of the weight (mass) of samples or materials.

gravitational acceleration The change in velocity per unit time of a falling body; for practical purposes, equal to 9.8 meters (32 feet) per second per second.

gravitational constant The constant of proportionality (G) in Newton's equation for GRAVITATIONAL FORCE.

gravitational force The mutual attraction between two physical bodies, as expressed in Newton's law of universal gravitation, the equation describing the gravitational attraction between two bodies with masses m_1 and m_2

$$F_G = G\,\frac{m_1 m_2}{r^2}$$

where F_G is the gravitational force on either of the two bodies and r is the distance between them. G is equal to 6.67×10^{-11} newton-meter^2 kilogram^{-2}.

gravity flow The downhill flow of water or other liquid through a system of pipes, generated by the force of gravity. See GRAVITY SEWER.

gravity sewer Flows in a wastewater system with a free surface exposed to the air space in the pipe and with gravity providing the moving force. Contrast FORCE MAIN.

gravity transport See GRAVITY FLOW.

gray (Gy) The basic SI UNIT of radiation dose absorbed per unit mass of tissue. One gray represents an absorbed dose of one joule of energy per kilogram of tissue. The unit can be used to express the absorption of any type of ionizing radiation and is based on the physical properties of the particular radiation.

gray water Domestic wastewater arising from showers, tubs, sinks, and clothes washers. Domestic wastewater from sources other than toilets. Gray water can be released into the environment after more modest treatment than that applied to sanitary wastewater. Contrast BLACK WATER.

grazing The consumption of live plant biomass by an HERBIVORE.

grazing food chain The feeding pattern of an animal community that is based on the consumption of live plant biomass by the primary consumers. This is contrasted with a DETRITUS FOOD CHAIN, which is based on the consumption of dead plant biomass by the primary consumers.

green belt 1. An area that has been revegetated, especially with trees. Green belts usually represent land areas that have been restored after extensive damage by overgrazing, deforestation, poor agricultural practices, industrialization, or urbanization. 2. An area that has been set aside to preserve natural habitat, vegetation, and open space. Land dedicated as green belts in the United Kingdom to prevent the problems of urban sprawl and development.

Green Cross An international organization founded in 1993 in the wake of the EARTH SUMMIT; dedicated to SUSTAINABLE DEVELOPMENT, water supply issues, lessening of the environmental consequences of warfare, and resource conservation. Chairman since 1993 is Mikhail Gorbachev. Based in Switzerland. Website: www.gci.ch

green-field development 1. Construction of an industrial facility on land that had not previously been used as an industrial site, for example, forested land or agricultural property. 2. Construction of a new industrial facility rather than expansion or modification of an existing site. Contrast BROWNFIELD.

Green Guides *Guides for the Use of Environmental Marketing Claims* issued by the U.S. Federal Trade Commission. The guides require that product labeling or advertising that asserts, for example, that a good is "recyclable" or "degradable" must be substantiated by sound scientific evidence.

Green Lights A voluntary USEPA program that promotes the use of energy-efficient lighting in government offices and large businesses. See also ENERGY STAR.

green parties See GREEN POLITICAL PARTIES.

green political parties Political parties organized around environmental issues and the limitation of industrial society. They offer candidates for election to public office. Green political parties are usually committed to grassroots democracy, environmental protection, dismantling of industrial society, and fundamental revision of the view of industrialization as progress. These parties are allied with a collection of groups arrayed against a culture based on the domination of nature by humankind. The degradation of the natural environment is seen to spring from the same forces responsible for nuclear armaments, dislocations caused by capitalism, and exploitation and mistreatment of women, children, racial minorities, and homosexuals. Such political parties are more active in Europe than in the United States. Website: www.greens.org

green revolution The advances in farming techniques, including increased irrigation and fertilizer use and crop varieties, that have allowed greatly increased crop yields in many areas of the world. See HIGH-YIELDING VARIETIES.

Green Seal An American nonprofit organization that awards its logo, a Green Seal, to products that cause significantly lower harm to the environment than other goods of the same type. Website: www.ns.ec.gc.ca/g7/eco-usa.html

greenhouse effect The natural warming of the atmosphere that results from the INFRARED RADIATION absorbed by CARBON DIOXIDE and water vapor. Rising atmospheric concentrations of carbon dioxide and other GREENHOUSE GASES are predicted to cause GLOBAL WARMING.

greenhouse gases Atmospheric gases or vapors that absorb outgoing infrared energy emitted from the Earth, contributing to the GREENHOUSE EFFECT and potentially contributing to GLOBAL WARMING, an enhancement of the greenhouse effect. The more important gases are carbon dioxide, water vapor, methane, nitrous oxide, and the CHLOROFLUOROCARBONS.

Greenpeace An international organization of environmental activists noted for aggressive and highly visible protest activities, such as members steering small boats in the way of whaling vessels and into waters used for nuclear testing. In 1985, two bombs sank the Greenpeace ship *Rainbow Warrior* in the harbor at Auckland, New Zealand, before it could depart to lead a group of vessels into the French nuclear test site near Tahiti. The sinking, which killed one Greenpeace member, drew worldwide attention. Image events designed to attract media attention have been the central rhetorical activity as the organization attempts to change the way people view the effects of industrialization. The practices and products of industrialism and economic activity based on the domination of nature by humans, the primacy of the scientific method, and the sovereignty of the individual are the special targets as the organization seeks to change the way the public views nature and the environment. In 1999, United States membership was about

171

500,000; worldwide membership was about 3 million. International headquarters is in Lewes, England; U.S. headquarters is in Washington, D.C. Website: www.greenpeaceusa.org, www.greenpeace.org/

grit Sand or fine gravel carried in wastewater.

grit chamber The initial treatment device at a sewage treatment facility; dense material from the incoming wastewater settles to the bottom of the chamber and is removed.

grit removal The process of removing sand and fine gravel from a stream of domestic waste in a GRIT CHAMBER.

grit tank See GRIT CHAMBER.

ground cover Materials, often plants, used to prevent soil erosion or leaching and to protect from freezing.

ground level concentration (GLC) In air pollution modeling, the calculated air concentration of the modeled pollutant downwind of the pollution source at breathing level.

ground-level ozone See OZONE. Compare OZONE LAYER.

ground-penetrating radar Technology employing high-frequency electromagnetic radiation to visualize subterranean geological features and groundwater.

groundwater Water contained in porous strata below the surface of the Earth. These subterranean deposits constitute significant sources of freshwater for use as drinking water, industrial process water, cooling water, and irrigation of crops.

Groundwater Disinfection Rule See GROUNDWATER RULE.

groundwater plume A volume of contaminated groundwater that extends downward and outward from a specific source; the shape and movement of the mass of the contaminated water are affected by the local geological features, materials present in the plume, and flow characteristics of the area groundwater.

groundwater recharge Surface water entering an aquifer by percolation through porous geological deposits.

Groundwater Rule Regulations issued under the SAFE DRINKING WATER ACT focusing on controlling the potential PATHOGENs in public water supply systems delivering groundwater to their customers. The 1989 SURFACE WATER RULE required systems using surface water or groundwater affected by surface water to apply disinfection and filtration treatment but did not apply to most systems supplying groundwater.

groundwater velocity The rate of water movement through openings in rock or sediment. Estimated by using DARCY'S LAW.

ground zero The point on the ground at which a nuclear weapon detonates, or that point on the ground directly under an atmospheric detonation of a nuclear weapon. The center of the area of greatest damage caused by a nuclear weapon.

guidance document A publication of a federal or state agency intended to describe and clarify one of the agency's regulatory programs, including the activities to which the program applies and the requirements for compliance. The program itself is officially implemented through the agency's regulations, which are often difficult to understand.

guideline model Any of the AIR QUALITY DISPERSION MODELS approved by the USEPA for use in the permit review process for new or modified facilities or in the evaluation of a control strategy to solve air quality problems in an area. Each of the models has an appropriate application(s).

guillotine damper A flat grate or plate that can be inserted perpendicular to the flow of a gas within a confined passage to regulate the rate of flow.

gully erosion Severe damage to agricultural land caused by the removal of soils by running water to such an extent that the channels are too large to be repaired by routine plowing.

gully reclamation Projects designed to prevent erosion in gullies by either filling them in or planting vegetation to stabilize the banks.

gypsum Calcium sulfate; a by-product of the reaction between limestone (calcium carbonate) and sulfur dioxide in control devices that reduce sulfur dioxide emissions from stacks. See FLUE GAS DESULFURIZATION.

Haber process The industrial method used for the fixation of atmospheric nitrogen for use as fertilizer. The process forms ammonia by the direct combination of atmospheric nitrogen and molecular hydrogen from natural gas. The reaction requires temperatures of 500–1000° C, pressures of 100–1000 atmospheres, and presence of a catalyst. The ammonia can be chemically combined with carbon dioxide to produce urea or further reacted with oxygen to make nitric acid. Nitric acid added to ammonia forms ammonium nitrate, which, along with urea, is a widely used nitrogen fertilizer.

habitat The specific surroundings within which an organism, species, or COMMUNITY lives. The surroundings include physical factors such as temperature, moisture, and light together with biological factors such as the presence of food or predator organisms. The term can be employed to define surroundings on almost any scale from marine habitat, which encompasses the oceans, to the microhabitat in a hair follicle of the skin.

habitat conservation plan (HCP) A negotiated agreement between the U.S. FISH AND WILDLIFE SERVICE (land use) or the National Marine Fisheries Service (water use) and a nonfederal entity (e.g., a private landowner or a state agency). The provisions of the ENDANGERED SPECIES ACT require an HCP before an INCIDENTAL TAKE permit can be issued. The objective of the plan is to allow lawful economic development or other land and water use that might incidentally harm an endangered or threatened species but to control those activities to minimize the harm to the protected species. An HCP must identify the species of concern, describe the activities that might harm those species, list the measures that the applicant will use to mitigate the possible harm, and describe the alternatives to the project that would prevent harm and why those alternatives are not being used.

habitat indicator A physical or chemical characteristic that can be measured to predict the suitability of a location for the development of a certain biological community. For example, measures of salinity indicate whether a coastal system is suitable for the development of certain flora or fauna, and soil moisture level indicates whether selected species of plants will prosper. High temperature and low moisture indicate a habitat suitable for plants and animals adapted to a desert environment.

Hadley cells One of the three major atmospheric circulation currents (cells) driven by alternate heating and cooling of the air in each hemisphere. Major convection currents in the atmosphere in tropical and subtropical areas. Considering the Northern Hemisphere, as the air at the equatorial region is heated at the surface, rising air currents are created. These ascending currents move upward and northward, gradually cooling in the upper atmosphere. The cooled air becomes denser and subsequently sinks or descends to the surface at approximately 30 degrees north latitude (approximately the Gulf Coast region of North America). The subsiding air moves across the surface of the Earth in a southerly direction, returning to the equatorial region, where heating takes place again, forcing the air again to rise and move northward, forming a continuous circulation. Similar cells operate between about 30 and 60 degrees north and between 60 and 90 degrees north (North Pole). These patterns, which are repeated in the Southern Hemisphere, contribute to mixing of the atmosphere and to the prevailing winds.

hematophagous Describing arthropods such as ticks, fleas, and mosquitoes that feed on blood. Some members of this group have the capacity to transmit infectious diseases, such as Rocky Mountain spotted fever and yellow fever.

half-life The time required for one-half of a radioactive substance to degrade to another nuclear form or to lose one-half of its activity, or for one-half of a chemical material to be degraded, transformed, or eliminated in the environment or in the body. Each radioactive substance has a predictable rate of decay, from millionths of a second to millions of years. Half-lives of chemical materials in the environment vary with the type of chemical and the biological, chemical, and physical conditions present where the chemical is released.

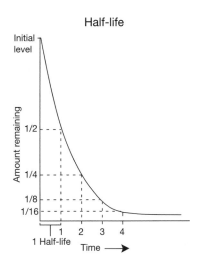

half-value layer (HVL) The shielding material thickness that reduces the quan-

tity of a beam of radiation to one-half strength before the material is traversed. Two half-value layers reduce the radiation to one-fourth of its value. Half-value layers vary with the type of material (lead half-value layers are thinner than those for concrete), and the HVL for a particular material is specific to the type and source of radiation. For example, the HVL for steel against the gamma radiation emitted by cobalt-60 is 0.82 inch; the HVL for steel against the gamma radiation emitted by cesium-137 is 0.64 inch.

haloacetic acids (HAA5) Five DISINFECTION BY-PRODUCTS formed when chlorine is used to remove PATHOGENS from drinking water. Produced by reaction of chlorine with bromide and organic compounds in the treated water. The five haloacetic acids that have federal standards are monochloroacetic acid, dichloroacetic acid, trichloroacetic acid, monobromoacetic acid, and dibromoacetic acid.

halocarbons See CHLOROFLUOROCARBONS.

halocline The boundary between surface freshwater and underlying salt water in a stratified coastal environment. A location where there is a marked change in salinity.

halogen One of the reactive nonmetals: fluorine, chlorine, bromine, iodine, and astatine. These elements are known for their reactivity and are only found combined with other chemicals in nature. At room temperature, fluorine and chlorine are gases, bromine is a liquid, and iodine and astatine are solids. Chlorine is used extensively as an industrial chemical and is a component in many compounds of environmental interest. See CHLORINE, CHLORINATED HYDROCARBONS.

halogenated Describes a chemical compound containing one or more of the HALOGENS, usually chlorine, bromine, or fluorine.

halogenated dibenzofuran (HDF)
See DIBENZOFURANS.

halogenated dibenzo-*p*-dioxin (HDD)
See TETRACHLORODIBENZO-*PARA*-DIOXIN.

halogenated organic compounds (HOCs) Chemical materials containing carbon, hydrogen, and one or more of the HALOGENS, usually chlorine, bromine, or fluorine.

halomethane See TRIHALOMETHANES.

halons Compounds composed of carbon together with fluorine, bromine, and/or chlorine used mainly as fire-extinguishing gases, such as Halon 1211 (CF_2BrCl) and Halon 1301 (CF_3Br). The chemicals are long-lived in the atmosphere and possess high OZONE-DEPLETING POTENTIAL. See OZONE LAYER DEPLETION.

halophytes Plants that require the presence of salt water or salty soil for growth.

hammer mill A machine in which solid material, such as community trash or other solid waste, is pulverized by hammers and rendered into smaller pieces.

hammer provision A provision in an environmental statute that makes effective certain requirements on a specified date if the USEPA or a state agency has not yet taken an action on the statutory requirements. Such provisions are used by Congress in response to common delays by the agencies in issuing regulations or implementing environmental laws. For example, the HAZARDOUS AND SOLID WASTE AMENDMENTS of 1984 contained several hammer provisions concerning the prohibition of certain wastes from land disposal, and the CLEAN WATER ACT called for the USEPA to take action if a state had not implemented an enforceable standard of TOTAL MAXIMUM DAILY LOAD. See LAND DISPOSAL BAN, SOFT HAMMER.

hard pesticide A chemical agent used to kill insects that does not undergo ready biological decomposition. If occurring at all, MINERALIZATION is very slow. Compare SOFT PESTICIDE.

hard water Drinking water that contains sufficient concentrations of metal ions, primarily those of calcium and magnesium, to cause the formation of precipitates with soaps and detergents. Compare SOFT WATER.

Hardin, Garrett (1915–) American Author and professor of human ecology, Hardin is most famous for his essay "TRAGEDY OF THE COMMONS," in which he describes the difficulties inherent in the control of human overpopulation. The concept has been applied to environmental pollution and resource overuse. See EXTERNALITY.

hardness 1. A measure of the amount of calcium and magnesium salts dissolved in water. 2. The relative penetrating ability of X rays.

hardpan A layer on or beneath the soil surface, usually composed of clay particles, sand, gravel, or calcium carbonate, that is compacted and relatively impermeable.

hardwoods Deciduous trees with broader leaves and, usually, slower growth rates than the conifers, or SOFTWOODS. Common temperate-region hardwoods are oak, maple, cherry, walnut, beech, birch, cypress, elm, and hickory.

hazard A physical or chemical agent capable of causing harm to human health or the environment. See also HAZARDOUS SUBSTANCE.

Hazard Analysis Critical Control Point (HACCP) Pronounced "hassip." Food safety regulations and guidelines issued by the U.S. FOOD AND DRUG ADMINISTRATION that identify the aspects (critical control points) of food production that may introduce food-borne hazards (microbial, chemical, or physical) and supply techniques to lessen or eliminate those possible hazards. The system spells out cooking temperatures and times, cleaning techniques, food storage conditions, and worker hygiene standards.

hazard and operability study (HAZOP) A detailed technical review of the potential for failure of a system or system components. The review is usually performed for industrial facilities and seeks to identify all credible scenarios that may lead to a fire, explosion, or chemical release. Steps are then taken to reduce the likelihood of those events.

hazard assessment The rational judgment about the potential problems associated with exposure to some toxic substance arrived at by an evaluation of both the concentration to which individuals are likely to be exposed and the available evidence describing the results of exposure to the chemical. See HAZARD EVALUATION.

Hazard Assessment Computer System (HACS) See CHEMICAL HAZARD RESPONSE INFORMATION SYSTEM (CHRIS).

hazard classes Categories of HAZARDOUS MATERIAL defined by the U.S. Department of Transportation under the HAZARDOUS MATERIALS TRANSPORTATION LAW. A few of the major hazard classes are corrosives, poisons, flammable gases, and FLAMMABLE LIQUIDS.

hazard communication standard (HCS) A program required by the OCCUPATIONAL SAFETY AND HEALTH ACT that includes worker training about the chemicals they use, including labeling, toxicity, signs and symptoms of overexposure, proper safety precautions, correct PERSONAL PROTECTIVE EQUIPMENT, and handling and disposal guidelines. Much of the HCS information is found in a chemical's MATERIAL SAFETY DATA SHEET (MSDS).

hazard evaluation Part of the process of determining the danger to an individual posed by a chemical or other dangerous agent. Gathering and evaluation of information on the types of adverse health effects or injuries that result from exposure to the agent and the conditions of exposure leading to the adverse outcomes.

hazard identification The determination of the possible adverse health effects produced by a chemical or physical agent. The analysis includes an estimation of the amount, frequency, and duration of exposure that may lead to the adverse effects and the identification of any susceptible population subgroups.

hazard index In a RISK ASSESSMENT for exposure to noncarcinogens, the sum of the HAZARD QUOTIENTS. Computed for the contaminants present at a waste cleanup site, for example.

Hazard Information Transmission (HIT) A service of the CHEMICAL MANUFACTURERS ASSOCIATION, which provides a computer printout of emergency information to FIRST RESPONDERS to a hazardous material incident. When given the name or identifying number of the chemical that has been released or is in danger of being released, the system lists hazards of the material, appropriate PERSONAL PROTECTIVE EQUIPMENT, first-aid directions, and methods for controlling any release or fire. The service is free of charge to organizations that register with the Chemical Manufacturers Association. See also CHEMICAL TRANSPORTATION EMERGENCY CENTER.

hazard quotient (HQ) In a RISK ASSESSMENT for exposure to noncarcinogens, the estimated average daily dose of a chemical divided by its REFERENCE DOSE. For multiple chemical exposures, the HQs for all chemicals are summed to produce the HAZARD INDEX.

Hazard Ranking System (HRS) A method for ranking hazardous waste disposal sites for possible placement on the NATIONAL PRIORITIES LIST, as provided for by the COMPREHENSIVE ENVIRONMENTAL RESPONSE, COMPENSATION, AND LIABILITY ACT. The ranking uses information gathered by the PRELIMINARY ASSESSMENT AND

SITE INSPECTION and the LISTING SITE INSPECTION. The need for remedial action is scored on the basis of potential harm to human health resulting from (1) releases into groundwater, surface water, or the atmosphere; (2) fire and explosion; and/or (3) direct contact with hazardous materials. The HRS evaluation assigns to each site an overall numerical value, which determines the priority for cleanup.

Hazardous and Solid Waste Amendments (HSWA) A 1984 federal statute amending the RESOURCE CONSERVATION AND RECOVERY ACT. The amendments expanded and strengthened the regulation of hazardous and solid wastes.

hazardous air pollutants (HAPs) The pollutants listed under the NATIONAL EMISSIONS STANDARDS FOR HAZARDOUS AIR POLLUTANTS (NESHAP) provisions of the CLEAN AIR ACT and approximately 200 chemicals or chemical categories listed as HAPs in the 1990 amendments to the Clean Air Act. See AIR TOXICS.

hazardous material (hazmat) Any chemical designated by the U.S. Department of Transportation under the HAZARDOUS MATERIALS TRANSPORTATION AND LAW.

hazardous material (hazmat) response team A group trained specifically in the actions required to prevent, contain, or clean up a threatened or actual release of HAZARDOUS SUBSTANCES. A hazmat team may be employed by one or more industrial facilities or public agencies, such as fire or police departments. See FIRST RESPONDER(S), HAZWOPER.

Hazardous Materials Transportation Act (HMTA) See HAZARDOUS MATERIALS TRANSPORTATION LAW.

Hazardous Materials Transportation Law Also called the Federal Hazmat Law, the 1994 federal statute governing the transportation of hazardous materials and hazardous waste. Regulations are issued by the United States Department of Transportation (DOT). (The hazardous waste MANIFEST SYSTEM is regulated by the USEPA under the RESOURCE CONSERVATION AND RECOVERY ACT.) The DOT regulations cover container, labeling, and marking standards as well as the use of PLACARDS and routing.

hazardous organic NESHAP (HON) Under the CLEAN AIR ACT, a regulation controlling emissions of listed organic compounds from targeted facilities in the SYNTHETIC ORGANIC CHEMICAL MANUFACTURING INDUSTRY (SOCMI). See also VOLATILE ORGANIC COMPOUNDS, NATIONAL EMISSION STANDARDS FOR HAZARDOUS AIR POLLUTANTS (NESHAP).

hazardous substance Any chemical that poses a threat to human health or the environment if released in significant amounts. Federal statutes and regulations include descriptions and lists of hazardous substances. Materials are regulated as hazardous under Section 311 (b) (2) (A) or Section 307(a) of the CLEAN WATER ACT; Section 112 of the CLEAN AIR ACT; Section 3001 of the RESOURCE CONSERVATION AND RECOVERY ACT; Section 101(14) or Section 102 of the COMPREHENSIVE ENVIRONMENTAL RESPONSE, COMPENSATION AND LIABILITY ACT; or Section 7 of the TOXIC SUBSTANCES CONTROL ACT. Title 40, Section 302.4, of the CODE OF FEDERAL REGULATIONS provides a list of designated hazardous substances. Under OCCUPATIONAL SAFETY AND HEALTH ADMINISTRATION emergency response rules, substances defined at Title 29, Section 1910.120 (a)(3) of the CODE OF FEDERAL REGULATIONS, which includes HAZARDOUS MATERIAL and HAZARDOUS WASTE.

hazardous substance list (HSL) A listing of designated HAZARDOUS SUBSTANCES appearing at Title 40, Section 302.4, of the CODE OF FEDERAL REGULATIONS.

Hazardous Substance Response Trust Fund The original official name of the Superfund; changed in 1986 to the HAZARDOUS SUBSTANCE SUPERFUND.

Hazardous Substance Superfund A federal trust fund for use in the cleanup of spills or sites containing hazardous waste that pose a significant threat to the public health or the environment, also known as the Superfund. The fund, originally called the Hazardous Substance Response Trust Fund, was established by the COMPREHENSIVE ENVIRONMENTAL RESPONSE, COMPENSATION, AND LIABILITY ACT in 1980. Beginning that year, $1.5 billion was to be collected over five years, mainly from taxes on crude oil, petroleum products, petrochemicals, and certain inorganic chemicals. The 1986 reauthorization of the law, which changed the fund's name to the Hazardous Substance Superfund, increased the fund to $8.5 billion and broadened the tax base to include a general corporate Superfund tax. Another $0.5 billion was included to clean up leaks from underground storage tanks.

hazardous waste Any solid waste that is listed as hazardous under the RESOURCE CONSERVATION AND RECOVERY ACT regulations, Title 40, Part 261, of the CODE OF FEDERAL REGULATIONS or that poses a significant threat to human health or safety because the waste is toxic, ignitable, corrosive, or reactive, as determined by specified tests. See LISTED HAZARDOUS WASTE, CHARACTERISTIC HAZARDOUS WASTE.

hazardous waste management facility (HWM facility) All contiguous land and structures used in treating, storing, or disposing of HAZARDOUS WASTE.

hazardous waste operations and emergency response (HAZWOPER) Universally known by its acronym, describing the OCCUPATIONAL SAFETY AND HEALTH ACT regulations (29 CFR 1910.120) that protect workers involved in the cleanup of a HAZARDOUS SUBSTANCE or HAZARDOUS WASTE.

hazardous waste site A location for the treatment, storage, or disposal of HAZARDOUS WASTE.

hazards analysis A systematic evaluation of fixed facilities and transportation sectors as potential sources of an accidental release of HAZARDOUS SUBSTANCES or HAZARDOUS MATERIALS, including probability estimates of releases of different amounts and consequences of the releases under various weather conditions. The exercise identifies vulnerable populations such as those in nursing homes, schools, and hospitals. Different release scenarios can be ranked by probability and consequence.

hazards identification A section of a MATERIAL SAFETY DATA SHEET (MSDS) listing the adverse health effects of a chemical given different exposure routes: inhalation, eye or skin contact, ingestion.

HAZWOPER See HAZARDOUS WASTE OPERATIONS AND EMERGENCY RESPONSE.

head 1. The removable top section of a pressure vessel containing the CORE of a nuclear reactor. The top of the reactor is removed during the refueling of the reactor to provide access to the fuel rods in the reactor core. 2. The pressure exerted by a fluid. See PRESSURE HEAD; HEAD, TOTAL.

head, total In hydraulics, a term for the energy at any point in a hydraulic system. The total head is the sum of the ELEVATION HEAD, PRESSURE HEAD, and VELOCITY HEAD. Head is expressed in length units, such as feet, and is another way of describing water pressure. *Feet of head* refers to the height of a column of water; one foot of water head equals 0.433 pound per square inch.

head loss The loss of energy in a hydraulic system caused by friction between the moving fluid (water) and the pipe as well as by other, smaller factors, such as changes in pipe diameter, bends, and valves.

headspace The zone above the contents in a closed sample container, possibly containing vapors or gases evaporated or released from the sampled material.

health assessment See PUBLIC HEALTH ASSESSMENT.

health effect Adverse effect of substances on the normal functioning of the human body through contact, exposure, and so on.

health physics The scientific principles and practices used to manage the hazards of IONIZING RADIATION.

heap-leach extraction Method of removing valuable minerals (e.g., gold) from low-grade ores. The ore is crushed and formed into large piles. An aqueous solution of a reactive material (alkaline cyanide solution in the case of gold) is sprayed onto the pile and allowed to percolate through it. The migrating liquid dissolves the desirable mineral, which can then be purified from the liquid phase. Because of the reactive nature of the chemicals used to promote the extraction, the pollution of surface water or groundwater and damage to wildlife entering the pools represent potential environmental problems.

hearing In administrative law, a meeting of one or more citizens and one or more representatives of a federal or state environmental agency, for example, the State of Louisiana Department of Environmental Quality. The meeting can range from an informal discussion in an office to a public meeting to a trial-type proceeding before an ADMINISTRATIVE LAW JUDGE. The purpose of the hearing is to allow the citizen(s) to present facts or an interpretation of the facts relative to an administrative agency decision, such as levying a fine or granting a PERMIT. Federal hearing guidelines are found in the ADMINISTRATIVE PROCEDURE ACT.

hearing examiner A representative of an administrative agency at a HEARING. If the hearing is formal in nature, the hearing examiner may be an ADMINISTRATIVE LAW JUDGE.

hearing threshold level At a specific frequency, the SOUND PRESSURE LEVEL a person can barely detect, for each ear.

heat The kinetic energy of a material arising from the molecular motion of the material. Heat is often expressed in CALORIES or JOULES. The heat contained in a body depends on the chemical makeup, mass, and temperature of the body. Heat is transferred from materials at higher temperatures to those at lower temperatures by CONVECTION or CONDUCTION or as RADIANT HEAT.

heat capacity See SPECIFIC HEAT.

heat cramps A heat disorder characterized by painful muscle spasms. Caused by excessive loss of salt and fluids.

heat exchangers Any mechanical device designed to transfer heat energy from one medium to another. The cooling system of an automobile consists of a heat exchange unit consisting of a liquid coolant, pumps, hose, and radiator, all of which transfer heat energy from the engine block through the radiator to the atmosphere. A nuclear reactor has a heat exchange unit designed to transfer heat from the primary coolant in which the core is immersed to the water that is boiled to produce the steam.

heat exhaustion A heat disorder characterized by fatigue, headache, and dizziness. Signs include high pulse rate, pale face, and heavy sweating. Caused by excessive loss of fluids.

heat index An expression combining air temperature and RELATIVE HUMIDITY; an indication of the actual heat stress present. For example, an air temperature of 85° F with a relative humidity of 20% corresponds to a relatively safe heat index of 82, but 85° F and 95% relative humidity translates to a heat index of 105, under which the risk of heat disorders such as HEAT EXHAUSTION is much higher.

heat island See URBAN HEAT ISLAND.

heat of combustion The heat energy liberated by the chemical reaction between an organic fuel and oxygen to form carbon dioxide and water.

heat of condensation The heat released when a vapor changes state to a liquid. See HEAT OF VAPORIZATION.

heat of fusion At the melting point of a solid, the heat required to change it to the liquid state; the value is expressed in calories of heat energy per gram of solid melted. For solid water (ice) at 0° C, the heat of fusion is about 80 calories per gram. This is equal to the heat energy per gram that must be removed to convert water at the freezing point (0° C) to a solid (ice).

heat of vaporization At the boiling point of a liquid, the amount of heat required to cause a change to the vapor state; expressed in calories of heat energy per gram of liquid vaporized. For water at 100° C and standard atmospheric pressure, the heat of vaporization is 540 calories per gram. Conversely, when a liquid condenses, the heat absorbed upon vaporization is lost, giving off HEAT OF CONDENSATION.

heat rate The net efficiency of an electric power generating facility. The measure is calculated by dividing the total heat content of the fuel burned (in BTUs) by the amount of electricity generated (in kilowatt-hours).

heat sink Any material used to absorb heat. In the environment, this is usually air or water that absorbs waste heat produced in the operation of electric power plants or other industrial facilities.

heat stress index (HSI) An expression of the evaporative capacity of an individual to maintain a normal body temperature compared with the maximum possible evaporative capacity under particular environmental conditions. The index is calculated by using the DRY-BULB TEMPERATURE, WET-BULB TEMPERATURE, GLOBE TEMPERATURE, air velocity, and the level of work the individual is performing (metabolic rate).

heat syncope A heat disorder characterized by fainting or near fainting. Caused by blood movement toward the skin surface and pooling in the legs without adequate movement and, consequently, an inadequate supply of blood to the brain.

heat transfer agent A liquid or gas that functions in a HEAT EXCHANGER to facilitate the movement of heat from one location to another. For example, the engine coolant in an automobile serves to transfer heat from the engine block to the atmosphere. Likewise, water facilitates the movement of heat from the reactor core to the outside of a nuclear reactor.

heating degree-day See DEGREE-DAY.

heating value The HEAT OF COMBUSTION per unit mass of a material; typically expressed as joules per kilogram, calories per gram, or BTU/per pound.

heating, ventilation, and air conditioning (HVAC) system The devices and ductwork that deliver air to a building; a poorly functioning HVAC system can contribute to INDOOR AIR POLLUTION.

heatstroke An acute condition resulting from excessive heat exposure, characterized by hot, dry skin and mental confusion. If not treated, the condition can lead to convulsions and death. The disorder stems from the failure of the body's sweating mechanism, which allows a sustained rise in internal body temperature.

heavier-than-air gas A gas with a DENSITY greater than that of the air into which it may be released; air at 25° C, sea level pressure, has a density of 1.18 grams per liter, or 1.18 kilograms per cubic meter.

heavy atom An ISOTOPE of an element that contains more neutrons than the number of neutrons in the most frequently occurring form of that element. For example, the most common isotope of hydrogen has a single proton in the nucleus (atomic weight of 1), whereas the heavy form of hydrogen contains a proton and a neutron in the nucleus (atomic weight of 2).

heavy gas See HEAVIER-THAN-AIR GAS.

heavy hydrogen The ISOTOPE of hydrogen that contains one proton and one neutron in the nucleus rather than the more common one proton and no neutron. Also called deuterium. See HEAVY WATER.

heavy metals The metallic elements of relatively high molecular weight, such as lead, mercury, arsenic, cadmium, chromium, and zinc. Chronic exposure to excessive concentrations of the ionic forms of these elements is associated with a variety of adverse health effects.

heavy oil A thick, viscous CRUDE OIL containing HEAVY METALS and high levels of sulfur; requires special refining to produce lighter hydrocarbon products such as gasoline. Canada and Venezuela have extensive estimated reserves of heavy oil and TAR SANDS.

heavy water Water in which a significant portion of the normal hydrogen atoms, hydrogen with a mass number of 1, have been replaced with the heavy isotope of hydrogen (HEAVY HYDROGEN or deuterium), hydrogen with a mass number of 2. The chemical formula for water is H_2O, and that for heavy water is D_2O. Used in a HEAVY-WATER REACTOR.

heavy-water reactor (HWR) A nuclear reactor in which HEAVY WATER is utilized as both the coolant and the moderator. Heavy water is a much more efficient moderator than normal water and therefore allows the use of naturally occurring, unenriched uranium as the nuclear fuel for the fission process. Compare LIGHT-WATER REACTOR.

hectare A unit of area equal to 10,000 square meters; 1 hectare equals 2.47106 acres.

heliostat A device designed to rotate slowly in order to reflect the rays of the Sun in a fixed direction continuously. Used in some applications of solar energy.

heme One member of a class of compounds derived from porphyrins. These compounds complex with metal ions (such as iron) and proteins to form the central structure in molecules such as hemoglobin and chlorophyll. The iron-porphyrin complex is responsible for the oxygen-carrying ability of hemoglobin. The word *porphyrin* is derived from the Greek word meaning "purple" *(porphyra)*, because porphyrin-containing molecules are highly colored. All porphyrin derivatives have the same central structure but differ in the metal ion included and the structures along the periphery.

hemoglobin The protein in red blood cells that carries oxygen absorbed in the lungs to the body tissues. See HEME.

Henry's law The expression stating that for a gas in contact with water at equilibrium, the ratio of the atmospheric concentration of the gas and the aqueous concentration is equal to a constant (HENRY'S LAW CONSTANT), which varies only with temperature. One form of Henry's law is

$$K_H = P/C_w$$

where K_H is Henry's law constant, P is the equilibrium PARTIAL PRESSURE of a gas in the air above the water, and C_w is the aqueous concentration of the gas dissolved in the water.

Henry's law constant The ratio, at equilibrium and for a given temperature, of the atmospheric concentration of a gas to the aqueous concentration; typical units are ATMOSPHERES per MOLE FRACTION. Henry's law constant is sometimes given as the equilibrium ratio of the aqueous concentration of a gas to the atmospheric concentration; in this case the units are inverted to mole fraction/atmosphere.

hepatitis Inflammation of the liver. A virus-caused disorder transmitted indirectly to humans by the consumption of raw oysters taken from water contaminated with sewage or transmitted directly from person to person.

hepatotoxicity The relative ability of a chemical agent to cause selective damage to the liver.

heptachlor An organochlorine insecticide of the family of chemicals known as the cyclodienes (a family that also includes ALDRIN and DIELDRIN). Heptachlor is persistent in the environment; exposure to the agent has been shown to cause cancer in at least one species of mammals and to result in a reduced immune response in mammals.

herbaceous Describing a nonwoody vascular plant. No parts persist above ground during the winter. Compare EVERGREEN.

herbicide A chemical agent (often synthetic) capable of killing or causing damage to certain plants (usually weeds) without significant disruption of other plant or animal communities.

herbivore An animal that eats only plants, such as a grasshopper, other insects, and cattle. Compare CARNIVORE, OMNIVORE.

herding agent A chemical applied to the surface of water to control the spread of a floating oil spill. Piston film (federal supply system chemical FSN 9G 6810-172-9110) is commonly used.

heritable Describing a property of an organism that is passed on to offspring through GAMETES.

herpetofauna Animals that belong to the amphibian or reptile group, such as snakes, frogs, and lizards.

hertz (Hz) A unit describing the frequency of electromagnetic radiation, wave motion, or sound. One hertz is equal to one cycle per second.

heterotroph An organism that cannot satisfy its carbon nutrition requirements by converting carbon dioxide (inorganic) to organic materials but must derive carbon nutrition from other organic forms of carbon, such as sugars and amino acids. Compare AUTOTROPH.

heuristic routing An approach to the design of routes for solid waste collection vehicles, which uses such basic rules as the minimization of left turns and dead ends and the use of clockwise loops.

hexachlorobenzene (HCB) A fungicide that has been shown to cause severe skin reactions in humans. Chemically the agent is an AROMATIC benzene ring with a chlorine atom substituted for each of the six hydrogens normally present on benzene. The agent is persistent in the environment.

Hexachlorobenzene

high-density polyethylene (HDPE) A low-permeability plastic used in a wide variety of consumer items, including toys, milk and water jugs, containers for laundry chemicals, and plastic garbage bags. The polymer is also used for the construction of landfill LINER. One of the common plastics included in the recycling of household discards.

high-efficiency particulate air filter (HEPA filter) A filter having at least a 99.97% removal efficiency for particles with a diameter of 0.3 micrometer.

183

Required for the control of certain high-hazard dusts, such as asbestos and radioactive materials.

high-level nuclear waste facility A site for the land disposal of very radioactive wastes. These wastes include material from a variety of commercial and governmental sources and commonly consist of used fuel rods from nuclear reactors, products resulting from the recycling of nuclear fuel, and wastes from the production and purification of plutonium. The concept is to encapsulate the waste in a ceramic material, place the ceramic material in corrosion-resistant containers, and then bury the containers in chambers constructed deep within the Earth. The facility must be designed to last for up to 10,000 years or longer. The location selected for a national facility serving the United States is Yucca Mountain, Nevada. See HIGH-LEVEL WASTE, VITRIFICATION.

high-level waste repository See HIGH-LEVEL NUCLEAR WASTE FACILITY.

high-level liquid waste An aqueous solution or suspension containing mixed fission products derived from the recycling of nuclear fuel rods that have been removed from reactors. The waste normally contains more than 100 microcuries per milliliter of fission products. See HIGH-LEVEL WASTE.

high-level radioactive waste See HIGH-LEVEL WASTE.

high-level waste (HLW) Very radioactive material requiring perpetual isolation; waste that includes untreated SPENT FUEL from a nuclear power plant, the residue from the chemical processing of spent fuel, and much of the waste from nuclear weapons production.

high-quality energy Energy useful in doing work. The energy available in natural gas is considered to be of high quality because natural gas can be burned and the released heat energy conveniently used to do work. On the other hand, the energy represented by the warmth of the air in a bathroom after a shower is considered low quality because one cannot conveniently use it to do work. The term is also applied to those energy sources that result in little pollution when converted from one form to another. In this context, electricity would be of high quality, whereas coal would be considered to be of low quality.

high-temperature gas reactor (HTGR) A nuclear reactor that uses enriched uranium as the fission fuel and graphite as the moderator. Helium gas is employed as the coolant. The reactor operates at a CORE temperature in excess of 1000° C, which is about three times the core temperature of water-cooled reactors.

high-volume air sampler (hi-vol) A device for sampling airborne particulate matter. A motor pulls air through a preweighed filter for 24 hours, capturing any airborne particles. The filter is reweighed to determine the mass of particulate in the sampled air. The mass is divided by the volume of air sampled to obtain a mass/volume concentration.

high-yielding varieties (HYVs) Strains of food-crop plants, such as wheat and rice, that have been developed to allow greater production of grain per acre of land cultivated than that of traditional varieties. The new varieties often require a greater input of fertilizer and water to reach full maturity. These strains have contributed to the so-called GREEN REVOLUTION.

highwall A cliff cut into a mountain or hillside during the process of removing coal deposits by surface or strip mining.

holding medium A special fluid employed for maintaining fecal bacteria in a viable state between the time that water samples are processed by filtration and the time that the filters used to remove the bacteria from water can be INCUBATED properly. The medium protects viability between sampling and analysis.

holding pond A reservoir, usually constructed in the ground, designed to retain runoff or effluent prior to treatment or discharge into a receiving stream. Holding facilities are not normally associated with treatment.

holding time The time allowed between the removal of environmental samples and the processing of those samples, such as the time elapsed between the collection of water samples and their bacteriological analysis.

holistic Of or related to a view of the natural environment that encompasses an understanding of the functioning of the complete array of organisms and chemical-physical factors acting in concert rather than the properties of the individual parts. Compare REDUCTIONISTIC.

holothurian A group of marine bottom-dwelling animals related to the sea stars and sand dollars (echinoderms). Unlike their relatives, the holothurians have soft bodies and are long and slender in shape. The sea cucumber is an example.

homeostasis A condition in which the systems of the body act together to maintain a relatively constant internal environment even when external conditions may vary.

homeotherm See ENDOTHERM.

homeowner water system A system, usually a private well, providing running water to an individual household. The system is not connected to a public water supply and usually not monitored by public health agencies.

hood A device that captures or encloses air contaminants. It is usually ventilated to the outside with the aid of a strong exhaust fan. Volatile or irritating chemicals are handled within an enclosure hood to prevent exposure of personnel to the agent.

hopper A container for dusts collected by an ELECTROSTATIC PRECIPITATOR or CYCLONE.

horizon A specific layer of soil that is different from adjacent layers in texture, color, mineral content, and other qualities. Also called soil horizon. See also SOIL PROFILE.

horizontal dispersion coefficient See DISPERSION COEFFICIENTS.

hormonally active agents (HAAs) Because it is less pejorative than "endocrine disruptors," an alternate term for ENVIRONMENTAL HORMONES.

horsepower An engineering unit of power describing the rate at which WORK is performed. The unit is based on the English system of measurements and is equal to 550 foot-pounds per second or 33,000 foot-pounds per minute.

host The larger organism within which a parasite lives. A tick that carries the bacterium that causes Rocky Mountain spotted fever, for example, is said to be the natural host of that pathogen.

hot An informal or colloquial expression meaning "highly radioactive."

hot side The part of a process in a petrochemical facility requiring the heating of a feedstock gas to high temperatures to facilitate desired chemical conversions to make a specific product. Generally requires the use of furnaces heated by natural gas. The products generated by heating are subsequently separated on the COLD SIDE.

hot-side ESP An ELECTROSTATIC PRECIPITATOR located on the upstream side of the AIR PREHEATER, where heat from the exhaust gases has not been partially transferred to incoming boiler air. Air temperatures in the hot-side ESP range from 600° to 800° F (320°–420° C). See COLD-SIDE ESP.

hot soak losses Evaporative emissions of VOLATILE ORGANIC COMPOUNDS from a vehicle after the engine is shut off. Emission control devices are now installed to collect this evaporating gasoline in a canister filled with ACTIVATED CHARCOAL. When the engine is restarted, the gasoline in the canister returns to the engine to be combusted.

hot spot An informal expression designating a specific area as being contaminated with radioactive substances, having a relatively high concentration of air pollutant(s), or experiencing an abnormal disease or death rate.

household hazardous waste (HHW) From a residence, discards that pose a threat to human health or to the environment if released in significant amounts. Common constituents of waste in this category include insecticide residues, paint, used crankcase oil, solvents, ammunition, and gasoline additives.

household waste Discards produced in a residence, commonly MUNICIPAL SOLID WASTE.

human ecology The study of the interactions of people with the natural environment and with each other. Includes the effects that humans have on the natural environment and the effects of the environment on people. The study of these interactions often focuses on human beings as a part of the environment as opposed to separate from the environment. See ENVIRONMENTAL JUSTICE.

human equivalent dose The dose to humans of a chemical or physical agent that is expected to exert the same effect that a certain dose has produced in animals.

human resources People, their abilities, skills, labors, enterprises, and mental capacities.

humic Of or containing HUMUS.

humus Organic matter in soil derived from the partial decomposition of plant and animal remains. Generally the decomposition has proceeded sufficiently to make recognition of the original material impossible. The decomposition products are important contributors to the capacity of soil to hold water, availability of plant nutrients, soil aeration, and texture of the soil. Runoff from land areas rich in humus has a brown or tan color produced by the humic substances.

hydraulic conductivity An empirically derived expression of PERMEABILITY, used as a coefficient in DARCY'S LAW to calculate the velocity of groundwater flow; the units are in length per time. Also called coefficient of permeability.

hydraulic fracturing Any technique involving the pumping of fluid under high pressure into an oil or gas formation to create fissures and openings in the reservoir rock and increase the flow of oil or gas.

hydraulic gradient The change in the elevation of the WATER TABLE per unit horizontal distance, often expressed in feet per mile; an important variable influencing the velocity of groundwater flow. See DARCY'S LAW.

hydraulic head An expression of water pressure in length units; the height to which water will rise in a pipe if one end is inserted into the hydraulic system; hydraulic head *(H)* is given by $H = Z + p/W$, where Z is the ELEVATION HEAD, p is pressure, and W is the SPECIFIC WEIGHT of water. The last term, p/W, is the PRESSURE HEAD. Also called piezometric height.

hydraulic loading For a SAND FILTER wastewater treatment unit, the volume of wastewater applied to the surface of the filtering medium per period. The loading is often expressed in gallons per day per square foot (gpd/ft^2) or cubic meters per square meter per day ($m^3/m^2/d$).

hydraulic radius For a flowing fluid (water), a measure of resistance to flow in pipes or open channels. The radius is expressed in length units and calculated as the flow area divided by the wetted perimeter. For circular pipes with full flow, flow area equals $pD^2/4$, where D is the pipe diameter and the wetted perimeter is the inside circumference $pD;$ therefore the hydraulic radius is $D/4.$

hydraulics The science concerned with water and other fluids at rest or in motion.

hydric Containing an abundance of water.

hydrocarbons Chemical compounds containing carbon and hydrogen as the principal elements. Oil is composed primarily of hydrocarbons.

hydrocracker Process employing a CATALYST and hydrogen for the reduction of high-molecular-weight hydrocarbons to smaller molecules useful in the production of gasoline and other petrochemicals. The technology is especially useful in the removal of nitrogen and sulfur atoms that are a part of the complex molecules composing crude oil, thereby reducing the nitrogen and sulfur content of the gasoline and diesel produced and lowering emissions of FUEL NO$_x$ and SULFUR DIOXIDES when the fuels are combusted. See CRACKING.

hydrogen bonding A type of weak attraction between molecules. The interaction between individual water molecules serves as an example. The water molecule is composed of one oxygen atom and two hydrogen atoms. These atoms are held together by strong covalent bonds that are formed by the sharing of electrons by the two atoms. The covalent bonds are very difficult to break. In addition, there is a weak attraction between the separate water molecules by way of the oxygen atom of one molecule and the hydrogen atom of an adjacent molecule. This attraction, termed a hydrogen bond, constantly forms and breaks among the multitude of water molecules in liquid water. Evaporation of water occurs when individual water molecules at the air-water interface break free of all hydrogen bonds and enter the gas phase. Hydrogen bonds are also found between many organic molecules, for example, between the two strands of deoxyribonucleic acid that constitute individual genes.

hydrogen ion Hydrogen atom without an electron; chemical symbol H$^+$.

hydrogen sulfide A foul-smelling gas (H$_2$S) produced during the ANAEROBIC decomposition of organic material by bacteria and by the metabolism of sulfate-reducing bacteria in the anaerobic sediments of freshwater and marine systems. Also called sewer gas because of the distinctive odor. The gas is responsible for the "rotten egg" odor of some well water. Hydrogen sulfide is very reactive chemically, forming sulfide salts with a host of metals; these metal sulfides are dark in color and are responsible for the black color of many muds. Significant concentrations of the gas are toxic to humans; consequently, the gas represents an occupational hazard to individuals working in industries involved in drilling for, producing, and refining petroleum.

hydrogeology The study of water contained within subsurface geological strata and the connections between subsurface and surface water supplies.

hydrograph A graph of a stream or river DISCHARGE at a certain point over time.

hydrologic cycle The stocks and flows of water in the ecosphere, including the processes of evaporation, precipitation, CONDENSATION, TRANSPIRATION, and surface and subsurface runoff.

hydrology The study of the distribution, movement, and chemical makeup of surface and underground waters.

hydrolysis The chemical reaction of a substance with water, resulting in the splitting of the larger molecule into smaller parts; an important degradation mechanism for pollutants in water or on land.

hydronic Describing a heating or cooling system consisting of heated or chilled water flowing through pipes for the purpose of heat exchange. For example, a building would be cooled by moving heated air over chill-water pipes. The cooled air would be returned to the interior and the water would be piped to an exchange unit for chilling.

hydrophilic Describing a chemical that mixes readily with water. Such substances are often charged molecules and dissolve readily in water. Hydrophilic chemicals are excreted from the body in the urine. Contrast HYDROPHOBIC.

hydrophobic Describing a chemical that mixes well with fatty material but does not mix well with water. Consequently, a chemical that is hydrophobic does not dissolve readily in water. Such materials pass cell membranes and are distributed rapidly throughout the body and are not excreted from the body in the urine.

hydrophyte A plant that lives in water. Compare MESOPHYTE, XEROPHYTE.

hydropower The utilization of the energy available in falling water for the generation of electricity.

hydrosphere That portion of the Earth composed of liquid water, such as lakes, rivers, ponds, and oceans. Together with the atmosphere, lithosphere, and biosphere, constitutes the Earth's ecosphere.

hydrostatic equation See PRESSURE.

hydrostatic pressure Pressure exerted by nonmoving water due to depth alone. Expressed as $P = WH$, where P is the GAUGE PRESSURE at a depth H and W is the SPECIFIC WEIGHT of water.

hydroxide ion (OH⁻) An ion carrying a negative charge that is released when a base or caustic is dissolved in water. Pure water, a neutral molecule, produces an equal number of hydroxide (negatively charged ions) and hydrogen (H^+) ions.

hydroxyl radical A very reactive chemical species consisting of OH carrying an extra unpaired electron. This radical is of paramount importance to the toxic processes of a variety of organic compounds and heavy metals. These molecules are produced when the body is overwhelmed with toxic substances (e.g., excessive doses of certain pesticides or metal ions) causing damage to cell membranes.

hygrometer An instrument used to measure moisture in the air.

hygroscopic Describing a chemical substance with an affinity for water, one that absorbs moisture, usually from the air. Silica gel and zinc chloride are hygroscopic materials that are used as drying agents.

hypersensitive The condition of being much more reactive than normal toward some chemical, drug, allergen, or foreign tissue. Some increased sensitivities reflect uncommon but natural reactions to pollutant chemicals; some are dependent on prior exposure and the development of the overreaction as mediated by a person's immune system.

hypochlorite Reactive ionic species (ClO⁻) produced when chlorine gas or bleach is added to water. The hypochlorite ion is the immediate cause of death of microbes when chlorine compounds are used to disinfect water.

hypolimnion The lower, cooler water layer found in stratified lakes. See EPILIMNION.

hypothesis An informed explanation that best describes a set of available data. The assumption is stated is such a way that subsequent experimentation or observation can test its validity.

hypoxia A condition in which natural waters have a low concentration of DISSOLVED OXYGEN (about 2 milligrams per liter, compared with a normal level of 5 to 10 milligrams per liter). Most game and commercial species of fish avoid waters that are hypoxic.

ideal gas A gas that perfectly obeys the IDEAL GAS LAW. At normal temperatures and pressures, most gases behave similarly to an ideal gas, and the ideal gas law is applied routinely in air pollution calculations.

ideal gas law The equation that describes the relationships among volume, pressure, and temperature for an IDEAL GAS. For n moles of gas, $PV = nRT$, where P = ABSOLUTE PRESSURE, V = total volume, R = UNIVERSAL GAS CONSTANT, and T = ABSOLUTE TEMPERATURE.

identification code (EPA I.D. Number) In hazardous waste management, a unique code assigned to each generator, transporter, and treatment, storage, or disposal facility assigned by the USEPA to facilitate identification and tracking of hazardous materials.

igneous rock Rock formed directly from cooled magma that has erupted from an area deeper in the crust of the Earth. Granite is a common example.

ignitability A characteristic used to define a HAZARDOUS WASTE. An IGNITABLE WASTE is classified as hazardous because of the characteristic of ignitability.

ignitable waste A substance or mixture that meets any of several definitions: (1) a liquid with a FLASH POINT of 60° C or less; (2) a nonliquid that can catch fire through friction, absorption of moisture, or spontaneous chemical change; or (3) ignitable compressed gases or oxidizers, as defined by regulations of the United States Department of Transportation.

illuvial Describing soil material, usually minerals and colloidal particles, that is removed from the upper soil horizon to a lower soil horizon. Illuvial deposits can form a HARDPAN.

Imhoff tank A two-chamber sewage treatment device in which SEDIMENTATION takes place in an upper chamber, after which the solids fall into a lower chamber, where they are digested by microorganisms. Named for its developer, Karl Imhoff.

immediately dangerous to life and health (IDLH) The maximum air concentration of a chemical substance from which a healthy worker could escape within a 30-minute exposure without irreversible adverse health effects or escape-impairing health conditions (dizziness, unconsciousness, and so forth). The level is set by the National Institute for Occupational Safety and Health. This acute air exposure standard is often used to estimate adverse effects on nearby human populations that may be caused by spills or other short-term accidental releases of toxic chemicals.

immigration The movement of individuals into an area or country to assume permanent residence. Compare EMIGRATION.

imminent and substantial endangerment 1. The condition allowing the USEPA to issue an ADMINISTRATIVE ORDER requiring an immediate response to protect human health or the environment from an actual or threatened release of a HAZARDOUS SUBSTANCE. The phrase is from the COMPREHENSIVE ENVIRONMEN-

TAL RESPONSE, COMPENSATION, AND LIABILITY ACT. See POTENTIALLY RESPONSIBLE PARTY, NATIONAL CONTINGENCY PLAN, HAZARDOUS SUBSTANCES SUPERFUND. **2.** The condition allowing the USEPA administrator or a private party using a CITIZEN SUIT to force a responsible party to take actions to ensure the proper management of solid or hazardous wastes to prevent harm to human health or the environment. The phrase is from the RESOURCE CONSERVATION AND RECOVERY ACT.

imminent hazard In pesticide regulation under the FEDERAL INSECTICIDE, FUNGICIDE, AND RODENTICIDE ACT, the designation of a pesticide as likely to cause unreasonable adverse effects on human health or the environment during the time required for normal CANCELLATION proceedings—therefore justifying a SUSPENSION, an expedited ban on use and sale of the pesticide.

immiscible Describing two liquids, neither of which acts as a solvent for the other to any appreciable extent. As a result, the liquids form two layers when mixed and allowed to stand. For example, a gasoline-water mixture separates into two layers with gasoline on the top and water on the bottom (because the bottom layer is denser than gasoline).

immobilization Synonym for STABILIZATION/SOLIDIFICATION.

immunoassay A testing method that uses the bonding specificity of antibodies to detect chemical compounds in the body or in the environment. Immunoassay kits are available for testing pesticide levels in soil and for detecting DIOXIN, PENTACHLORPHENOLS, PETROLEUM products, and POLYCHLORINATED BIPHENYLS (PCBS) in soil and/or water.

immunosuppression The reduction of the capacity of an individual's body to ward off infections or the growth of tumors. The body is equipped with a multifaceted, complex system that removes from it "foreign" materials ranging from viruses, bacteria, and multicellular parasites to transplanted tissue and tumors. Vaccination is a medical procedure intended to enhance the capacity of the immune system. Some medical practices involve suppression or reduce the capacity of the immune system to accomplish such functions as treating acute inflammation and preventing transplant rejection. A variety of environmental toxins, including chlorinated aromatic hydrocarbons, polycyclic aromatic hydrocarbons, arsenic, and organophosphate insecticides, have been implicated as agents that reduce the capacity of the immune system to protect the body.

impact noise A brief, punctuated sound lasting generally less than one-half second and not repeated more than once per second, for example, the sound from a hammering device. Impact noise is difficult to measure accurately.

impaction For inhalation of airborne PARTICULATE MATTER, the collision of particles with the lung surfaces as contaminated air is taken into the lungs. The velocity of the air and the aerodynamic diameter of the particles are the primary variables governing impaction.

impedance An expression of the resistance to an energy flow calculated as the magnitude of the cause of the flow (force, pressure, voltage) divided by the energy flow. For example, Ohm's law states that $Z = E/I$, where Z = impedance, E = voltage, and I = electric current. Energy transfer is most efficient if the impedance of the source is matched to that of the receiver.

impermeability A characteristic of a material that prevents the passage of a fluid through that material, such as the relative impermeability of a LINER at a solid waste landfill. Liners are rated in terms of how slowly a liquid can move through them or how close they are to being truly impermeable.

impingement separator An air pollution control device that removes PARTICU-

LATE MATTER from a gas stream by causing the particles to strike and adhere to plates as the gas direction is changed.

impinger An air-sampling device that collects gases or PARTICULATE MATTER by directing sample air down a glass tube running inside a collection bottle that is filled with an absorbent liquid. The sampled air is mixed with the liquid by the high-velocity air "impingement" near the bottom of the bottle, which increases the collection efficiency.

impoundment Any land area or formation that can hold liquid. Impoundments, which are open to the atmosphere, are used to treat, store, or dispose of waste; they include aeration tanks, HOLDING PONDS, and AERATION LAGOONS. Also called surface impoundment.

in-line filtration In wastewater treatment, a pretreatment process involving the addition of chemical agents coupled with pressure filtration to remove particulate materials from wastewater. The process accelerates wastewater treatment by eliminating the need for FLOCCULATION and SEDIMENTATION to remove particulate matter.

in situ In place. An in situ environmental measurement is one that is taken in the field without the removal of a sample to the laboratory.

in situ conservation Protecting or enhancing the genetic, species, and/or ecological aspects of BIOLOGICAL DIVERSITY within species' natural HABITAT. Compare EX SITU CONSERVATION.

in situ gasification A method of energy extraction that involves igniting and aerating an underground coal seam then capturing a low-energy gas from another well drilled into the coal seam. A portion of the energy of the coal is thereby recovered without physically removing the coal from the ground.

in situ oxidation Technology involving chemical agents to reduce or remove contaminants in groundwater or soil without removing the water or soil from its location.

in situ remediation Treatment technologies that destroy or reduce contaminants in soil or groundwater without removing the water or soil. Underground application of electric current, plasma, microbes, nutrients, and chemical reactants has been shown to be cost-effective at many hazardous waste cleanup sites. See IN SITU OXIDATION, IN SITU STRIPPING.

in situ stripping Technology involving the injection of air into contaminated soil or groundwater to flush volatile organic compounds, commonly hydrocarbon fuels such as gasoline, from the site. The offending chemicals evaporate and are removed from the soil or groundwater with the airstream.

in situ vitrification Technology for the decontamination of soils in place by heating to extremely high temperatures in the range of 3000° F. Offending substances, commonly hydrocarbon-based hazardous chemicals, are destroyed in the process.

in-stream aeration The addition of air to a flowing stream to maintain the DISSOLVED OXYGEN content of the water at an acceptable level.

in-stream use Use of water that is in a stream or channel (e.g., hydroelectric power generation, swimming, boating, fish propagation, and navigation).

in vitro In glass, outside an intact living organism. Refers to experiments that are conducted in petri dishes, test tubes, and like apparatus. Chemical toxicity tests performed in the laboratory using cell or tissue cultures are in vitro tests. Compare IN VIVO.

in vivo In a living organism. Experiments performed inside living organisms, such as chemical toxicity tests that involve the introduction of the tested substance into or on the body of an animal. Compare IN VITRO.

inbreeding The mating of animals that are closely related; the process can result in an increase in the expression of RECESSIVE GENES, which are often maladaptive. The adverse aspects of inbreeding can further harm endangered species, as the small numbers of remaining individuals mate, by necessity, with related individuals. See INBREEDING DEPRESSION.

inbreeding depression The lowering of the quality and vigor of a population of organisms (either plant or animal) due to breeding of individuals that are closely related in a genetic sense. Inbreeding increases the chance that RECESSIVE GENES, which result in the expression of traits that diminish the health, development, or reproduction of the offspring, will become apparent. The lowering of the quality of a captive animal population in a zoo environment due to breeding of closely related animals is a constant concern. The phenomenon is the rationale for the use of hybrid seed in agriculture and the prohibition of marriage to close relatives among humans.

incidence The rate of occurrence of a specific disease or event within a given number of individuals over a standard period. For example, the number of lung cancer cases per 100,000 people per year.

incident command post A facility, mobile or fixed, located a safe distance from the site of an industrial emergency or accident from which the INCIDENT COMMANDER, assisted by key staff and technical representatives, can make decisions and deploy emergency personnel and equipment. Depending on the nature and size of the emergency, the incident command post can be located in a designated and equipped structure on the property of an industry, in a structure off-site equipped with appropriate communications equipment, or in a vehicle designated

for such use. See INCIDENT COMMAND SYSTEM.

incident command system (ICS) The organizational arrangement established to respond to an emergency. The highest-ranking emergency responder designated as the INCIDENT COMMANDER, normally a fire chief, Coast Guard commander, state police officer, or federal official, serves as the leader of an integrated, comprehensive emergency response organization and as the supervisor of the emergency site. The incident command structure in each community is established during planning and preparations for possible emergencies that may possibly occur. For example, the ICS in communities that serve as host to chemical manufacturing facilities typically plans for response to fires, chemical releases, explosions, and transportation accidents.

incident commander Highest-ranking emergency responder (e.g., fire chief, facility safety manager, or federal official) designated to assume control at the site of an emergency. See INCIDENT COMMAND SYSTEM.

incident rate (IR) One of several names applied to the safety record at an industrial facility. See ACCIDENT RATE for description.

incidental take A type of permit giving the holder permission, under the provisions of the ENDANGERED SPECIES ACT, to TAKE (harm or kill) limited numbers of an ENDANGERED or THREATENED SPECIES because the harm is "incidental to, and not the purpose of carrying out otherwise lawful activities." Applicants for an incidental take permit must devise and comply with an approved HABITAT CONSERVATION PLAN.

incineration Burning of organic waste materials. This disposal technique is used to reduce the volume of MUNICIPAL SOLID WASTE and, in the case of WASTE-TO-ENERGY plants, to recover heat for further uses. Incineration technology is often chosen to destroy waste that is otherwise diffi-cult to destroy and is hazardous to public health or the environment. The chemicals are reduced to mineral constituents. For example, chlorinated hydrocarbons are converted to carbon dioxide (CO_2), water (H_2O), and hydrochloric acid (HCl). See MASS BURN.

incipient LC_{50}　The computed concen-tration of some toxic substance in water that would be lethal to 50% of a test pop-ulation of aquatic organisms. The value is extrapolated from experimentally derived data when laboratory exposure to the chemical does not result in a mortality rate exceeding 10% of the test organisms upon exposure to the test pollutant over a 24-hour test period. See also LETHAL CON-CENTRATION—50 PERCENT, LETHAL DOSE—50 PERCENT.

incompatible waste　Hazardous waste that, if mixed, chemically reacts to form hazardous products or excessive thermal discharge. Also, waste that may damage or corrode a container or containment structure, leading to a release.

increment　The allowable increase in ambient air concentrations of certain pol-lutants over the BASELINE level estab-lished under the PREVENTION OF SIGNIFI-CANT DETERIORATION program. Increments are set for SULFUR DIOXIDE, airborne PARTICULATE MATTER (PM_{10}), and NITROGEN DIOXIDE. See INCREMENT CONSUMPTION.

increment consumption　Under the PREVENTION OF SIGNIFICANT DETERIORA-TION air pollution management program, the modeled increase in certain air pollu-tant concentrations over the BASELINE ambient air quality levels. The allowable degradation in air quality is called the INCREMENT. Any air emissions considered to be additional to an area after the base-line date are counted against (consume) the allowable increment.

incubate　To maintain environmental conditions that are optimal for the growth of bacteria. For example, coliforms grow best when held at 37° C.

incubation period　The average time between exposure to an infectious, dis-ease-causing agent and the manifestation of the signs and/or symptoms of the dis-ease.

independent association　In the study of risk factors for a disease, a relationship between a suspected factor and disease risk that remains after adjusting or con-trolling for the influence of other vari-ables.

independent variable　A measurable quantity that can be used to predict the value of a DEPENDENT VARIABLE. In EPI-DEMIOLOGY, an independent variable is an exposure or characteristic that may influ-ence a particular health condition or effect.

indicator　1. In biology, an organism or community of organisms whose character-istics show the presence of specific envi-ronmental conditions or contamination. 2. In chemistry, a substance that shows a vis-ible change, usually a color shift, at a spe-cific point in a reaction. 3. In monitoring, a device that presents the result of some measurement.

indicator organisms　The microorgan-isms that, if present above certain levels in water, indicate contamination by human sewage. The COLIFORM BACTERIA are used commonly as indicators because the test for this class of organism is reliable, rela-tively inexpensive, and produces timely results.

indicator parameters/constituents　Groundwater quality measurements speci-fied in a PERMIT to operate a hazardous waste landfill or surface impoundment TREATMENT, STORAGE, OR DISPOSAL facility. The specified measurements must be made on samples taken at the POINT OF COMPLI-ANCE. The parameter may be an analysis of general water quality, such as SPECIFIC CONDUCTANCE, or measurements of one or more specific chemical constituents, such as arsenic, benzene, or chloroform. The data recovered from the measurements of indicator parameters/constituents are compared statistically to background levels measured at an UPGRADIENT WELL to determine the possibility of adverse influence by the facility on the groundwater.

indigenous   Native; naturally present in an area. Autochthonous. Compare ALLOCHTHONOUS.

indirect association   See SECONDARY (INDIRECT) ASSOCIATION.

indirect discharge   In wastewater management, the introduction of waste from a commercial or industrial facility into the sewer system serving a community. An indirect discharge may be subject to PRETREATMENT requirements.

indirect hot air   A system designed to reheat an exhaust gas that has been cooled by passage through a pollution control device. Ambient air is heated with steam then added to the flue gas to give the exhaust buoyant density.

indirect source   A business, shopping center, or highway, or other source that attracts mobile sources of air pollution. See INDIRECT SOURCE REVIEW.

indirect source review (ISR)   Environmental agency review and possible controls on the siting of shopping centers, office complexes, highways, and airports, among other indirect sources of air pollution, to improve or maintain air quality. The USEPA is not permitted to require an ISR program as part of a STATE IMPLEMENTATION PLAN under current law.

individual lifetime risk   The estimated increase in the lifetime risk of an adverse health effect in a person exposed to a specific amount of a physical or chemical agent for a given period. See UNIT RISK ESTIMATE, POPULATION RISK.

indoor air pollution   The presence of excessive levels of air contaminants inside a home or building from sources such as cigarette smoking, fuel combustion for heating or cooking, certain wallboards, carpets, insulation, air-handling ducts, and the geological characteristics of the area (RADON in soil or rocks beneath the structure). Emissions are more likely to accumulate in structures having limited air exchange with the outside. Many air pollutants typically have higher concentrations indoors than outdoors. See INDOOR/OUTDOOR CONCENTRATION RATIO.

indoor air quality (IAQ)   See INDOOR AIR POLLUTION.

indoor/outdoor concentration ratio ([I]/[O])   The measured or typical concentration of an air pollutant indoors divided by the concentration of the same substance outdoors. If the ratio for a certain pollutant is greater than 1, the indoor air concentration is greater than the outdoor concentration and vice versa for a ratio less than 1. See INDOOR AIR POLLUTION.

Indoor Radon Abatement Act   A 1988 amendment to the TOXIC SUBSTANCES CONTROL ACT that authorizes the USEPA to provide technical and financial help to state agencies for RADON monitoring and mitigation.

induced draft (ID)   The type of ventilation system in which a fan pulls air through a furnace, boiler, or cooling tower, aiding the flow of air out of the smokestack or other exit ducts. See also FORCED DRAFT, NATURAL DRAFT.

induced radioactivity   The creation of an unstable or radioactive atom by the bombardment of a stable isotope of an element with neutrons. For example, the stable isotope of nickel has an atomic mass of approximately 58. Bombardment of that form of nickel with neutrons produces nickel-59, which is a radioactive isotope.

industrial hygiene The study and practice of the recognition, evaluation, and control of chemical, physical, or biological hazards in the workplace.

industrial melanism An increase in the relative abundance of darker strains of insects and spiders in regions around industrial facilities emitting high levels of smoke. There is a natural variation in the pigmentation of the various species, and the survival of the darker pigmented varieties is favored in the surroundings of industrial environments polluted with dark soot. The darker pigmentation provides those strains with camouflage and, therefore, protection from predators. Conversely, the lighter pigmented varieties are easier to see and therefore more susceptible to capture by predators. The results are a gradual enrichment of the entire population for the dark variety and a diminution of the light variety. Air pollution controls shift the selective advantage back to the lighter pigmented strains.

industrial sludge A thick suspension of waste products produced at an industrial facility. See SLUDGE.

Industrial Source Complex model (ISC model) An AIR QUALITY DISPERSION MODEL, approved by the USEPA, for estimating short- and long-term AMBIENT air concentrations in an area using emission rates from smokestacks and/or AREA SOURCES and a compilation of the meteorological data from the area. Annual averages are estimated by the ISCLT (long-term) model and the ISCST (short-term) model concentrations for averaging times of 24 hours or less.

industrial source reduction Practices that reduce the amount of hazardous material, pollutant, or contaminant placed into the waste stream or released from an industrial facility into the environment. The reductions lessen the threat to public health and to the environment associated with the releases. The reductions can come about through upgrades in equipment and technology, changing raw mate-

rials, recycling, and improvements in housekeeping, operations, maintenance, or training. See POLLUTION PREVENTION, POLLUTION PREVENTION ACT, ENVIRONMENTAL MANAGEMENT HIERARCHY.

industrial timber Trees used for lumber, plywood, particleboard, paper, and similar products.

industrial waste Unwanted discards produced in an industrial operation; may include liquids, sludges, solids, or hazardous wastes.

industry category Under the Clean Water Act, the USEPA has divided facilities into industry categories and subcategories for which water EFFLUENT LIMITATIONS, based on the type of process and the technology appropriate to the industry, are written. For example, the battery manufacturing category has seven subcategories, each of which has specific effluent limitations.

inert Describing a substance that does not react chemically with other materials under ordinary conditions. For example, helium does not undergo chemical reactions readily. Compare ACTIVE.

inertial confinement Technology used to promote controlled nuclear FUSION in which a small amount of nuclear fuel is bombarded with high-intensity laser light to increase the temperature to that which promotes the fusion of small atoms.

inertial separator An air pollution control device for PARTICULATE MATTER removal that operates by forcing a dust-containing airstream to change direction abruptly, causing the particulates to collide with a wall, baffle, or louver by the inertia of the particles. The solids fall into a collection HOPPER.

infant mortality rate For a given geographical area and year, the total deaths of infants less than one year of age divided by the number of live births during that year, usually expressed as infant deaths per 1000 live births.

infectious Describing a virus, bacterium, fungus, or protozoan that can invade a host to produce disease. The term is also applied to a disease caused by some pathogenic microbe (e.g., typhoid fever) as opposed to a disease that is not related to the presence of a microbe (e.g., coronary artery disease).

infectious waste A specific class of discarded items from medical or related facilities. The discards do or may contain viable organisms capable of causing INFECTIOUS diseases.

infiltration 1. The wind-induced movement of air into a building through openings in walls, windows, or doors. An important consideration in energy conservation and indoor air quality analysis. 2. The entrance of groundwater into sewer pipes.

infiltration and inflow (I&I) The entrance of groundwater (infiltration) or of surface water (inflow) into sewer pipes. Groundwater can seep through defective pipe joints or cracked pipe sections; roof or basement drains are sources of surface water inflow. Excessive infiltration and inflow can cause sewers to back up or can overload a SEWAGE TREATMENT PLANT, causing a reduction in treatment time or a complete bypass of the treatment process during periods of heavy rainfall.

infiltration gallery A subsurface groundwater collection system constructed to prevent contaminated groundwater from entering streams, ponds, or lakes. These shallow systems are commonly constructed of perforated pipes that drain into a watertight sump from which groundwater is transferred to a wastewater treatment system for processing before release.

inflammatory response A general nonspecific reaction of the body to foreign particles, noxious chemicals, biological or industrial toxins, microorganisms, or similar agents. The characteristic results include redness, fluid accumulation, swelling, pain, and a temperature increase at the location where the agent contacts the body. The reactions are part of the normal response of the body to foreign materials.

inflow General term designating the water or other fluid entering a system.

influent General term designating water, wastewater, or other liquid flowing into a system. Compare EFFLUENT.

informal economy Small-scale business activity operating outside normal business or regulatory constraints, including environmental controls.

information collection request (ICR) A document prepared by a federal administrative agency for approval by the OFFICE OF MANAGEMENT AND BUDGET as required by the Paperwork Reduction Act of 1995. The document contains a description of the information to be gathered, the way it will be collected, the reason why it is needed, and the projected cost of the collection.

information collection rule A nationwide USEPA drinking water program established to gather information about the presence of targeted PATHOGENS in drinking water, including GIARDIA, CRYPTOSPORIDIUM, and viruses and to monitor levels of DISINFECTION BY-PRODUCTS.

information file A collection of documents assembled by the USEPA for each SUPERFUND SITE, describing the types of hazardous substances present and the progress of cleanup activities. The documents are made available to the public at a local school, library, or city hall.

information standard A type of environmental, health, or safety standard that relies on warnings on the container label or on other guidance to prevent or minimize adverse effects from the use or disposal of a material or product; for example, the dosage information on over-the-counter medicines, and application and disposal information on pesticide labels.

infrared radiation (IR) Electromagnetic radiation with wavelengths longer than those of visible light (750 nanometers) and shorter than those of radio waves (3×10^5 nanometers). The Earth, buildings, animals, and people emit this type of radiation. Outgoing infrared radiation from the Earth warms the atmosphere. See GREENHOUSE EFFECT, NEAR INFRARED, INTERMEDIATE INFRARED RADIATION, FAR INFRARED, LONGWAVE RADIATION, INFRARED WINDOW, ABSORPTION BAND.

infrared window The WAVELENGTHS of INFRARED ENERGY between 8 micrometers and 12 micrometers emitted by the Earth that are weakly absorbed by atmospheric gases and therefore escape to space. See GREENHOUSE GASES, GREENHOUSE EFFECT, GLOBAL WARMING.

ingestion The taking in of substances by mouth (eating or drinking); one method of exposure to chemicals.

inhalable diameter The diameter of particles (about 15 micrometers or less) that can be deposited in the human respiratory tract.

initiation/promotion The two-stage theory of carcinogenesis. The first causes the development of premalignant cells, and the second completes the conversion of the cell to the malignant state. See CARCINOGENESIS.

initiator In the two-stage model of carcinogenesis, an initiator produces a change in a cell, commonly called a MUTATION, which leaves the cell in a premalignant condition. In some circumstances, without exposure to a PROMOTER, the changed cell does not become a tumor cell. See CARCINOGENESIS.

injection well A boring through which fluids are injected into the ground for a variety of purposes, including waste disposal, enhanced recovery of crude oil, or solution mining. See CLASS I, II, III, IV, AND V INJECTION WELLS, UNDERGROUND INJECTION CONTROL.

injection zone That geological strata into which fluids are added by an INJECTION WELL.

injury incident rate (IIR) One of several names applied to the safety record at an industrial facility. See ACCIDENT RATE for description.

innocent landowner A designation authorized by the COMPREHENSIVE ENVIRONMENTAL RESPONSE, COMPENSATION, AND LIABILITY ACT (CERCLA). CERCLA regulations allow a purchaser of land that contains hazardous waste a legal defense against liability for the costs of the hazardous waste cleanup if the purchaser can demonstrate that he or she was not aware of the waste when buying the land. The demonstration of innocence is made to a court or to the USEPA. If the legal pleadings of innocence are too burdensome, the landowner may apply for DE MINIMIS status to be freed of the cleanup liability. See PHASE I ENVIRONMENTAL ASSESSMENT.

inoculate 1. To add a viable culture of a virus, bacterium, or fungus to a growth medium or animal host for the purpose of cultivating the microbe. 2. To add specific bacteria to waste disposal ponds, activated sludge systems, or sediments for the purpose of stimulating the decomposition of some special waste. See BIOREMEDIATION.

inoculum The organisms used to INOCULATE a culture medium.

inorganic Of or related to chemicals that do not contain carbon atoms. Mercury ions, sulfur dioxide, lime, copper, and ammonia are examples of inorganic materials. Compare ORGANIC.

insecticide A chemical agent, either natural or synthetic, used to kill or inhibit the growth and development of insects. A type of PESTICIDE.

insolation The interception rate of solar energy on a surface; insolation intensity is greatest when the surface is perpendicular to the solar radiation. Typically expressed in WATTS per square meter.

inspection and maintenance (I&M) An air quality improvement program requiring automobiles to be inspected for the presence and proper operation of emission control devices. The owner may be required to repair or replace the control equipment. Typically implemented locally in areas with air quality problems (NONATTAINMENT AREAS) caused by or related to automobile emissions. See CATALYTIC CONVERTER.

institutional solid waste Discarded items from schools, hospitals, prisons, and other organizations.

instrument detection limit (IDL) The minimal concentration of a substance that is detectable by a specific instrument. Some compounds of environmental concern occur in such small concentrations that an insufficient amount is present to allow detection with available instrumentation. Compare METHOD DETECTION, QUANTITATION LIMIT.

integrated exposure assessment See EXPOSURE ASSESSMENT.

integrated pest management (IPM) The use of a combination of the following to limit pest damage to agricultural crops: (1) agricultural practices, such as field tilling to disrupt insect egg development; (2) biological control agents, such as viruses, fungi, or bacteria; (3) introduction of large numbers of sterile male insects; (4) timed application of synthetic chemical pesticides; (5) application of PHEROMONES and JUVENILE HORMONES. IPM stresses the control of pests at manageable levels rather than their eradication.

Integrated Risk Information System (IRIS) An electronic on-line database provided by the USEPA containing chemical toxicity data and regulatory information for use in health risk assessment and management. Website: www.epa.gov/ngispgm3/iris

integrated waste management Comprehensive plan to direct the collection, processing, and disposal of municipal solid waste, often employing more than one system for each. A municipality might, for example, require residents to separate waste into yard waste (perhaps to be disposed of by COMPOSTING), specific recyclable items (perhaps to be processed through a MATERIALS RECOVERY FACILITY), and other discards (perhaps to be buried in a SANITARY LANDFILL).

Interagency Regulatory Liaison Group (IRLG) A group formed in 1977 to coordinate federal regulation of toxic substances by sharing information, planning agency studies to prevent duplication, and developing consistent federal policy approaches, especially for carcinogens. The group was composed of representatives from the USEPA, the Occupational Safety and Health Administration, the Food and Drug Administration, the Consumer Product Safety Commission, and the Department of Agriculture Food Safety and Quality Service. The IRLG provided important beginnings for the federal government's environmental risk management programs.

interceptor sewer Usually used in describing a bypass mechanism available in municipal sewer systems that carry both water from rainfall and sanitary wastewater from households and commercial establishments. If heavy rainfall creates a runoff of much more water than the system can process, the storm water and a portion of the sanitary wastewater are diverted by interceptor sewers directly to the receiving stream to prevent the backup of sewage into streets and homes.

Intergovernmental Panel on Climate Change (IPCC) An international group of experts formed in 1988 by the WORLD METEOROLOGICAL ASSOCIATION and the UNITED NATIONS ENVIRONMENT PROGRAM to provide scientific advice on the potential for global climate change. The 1990 IPCC report influenced the FRAMEWORK CONVENTION ON CLIMATE CHANGE. See KYOTO PROTOCOL. Website: www.ipcc.ch

interim status Under the RESOURCE CONSERVATION AND RECOVERY ACT, a regulatory designation allowing TREATMENT, STORAGE, OR DISPOSAL facilities in existence on November 19, 1980, to continue to operate under a specific set of regulations (PART A permit) until a final (PART B) permit for the facility is applied for and approved.

intermediate A compound produced during the metabolic conversion of organic compounds by biological organisms. For example, when an organism takes in glucose for use as a nutrient, there is a sequence of biochemical steps between the glucose molecule and the ultimate release of carbon dioxide by the organism. These biochemical steps produce compounds such as fructose-6-phosphate, glyceraldehyde phosphate, pyruvic acid, citric acid, and oxaloacetic acid. Each of these would be considered an intermediate in the metabolic degradation of glucose. Such a series of intermediates could be identified in the degradation or synthesis of almost any organic compound.

intermediate infrared radiation Electromagnetic radiation with wavelengths from 1.5 micrometers to 20 micrometers.

intermittent control system (ICS) The deliberate variation in an air emission rate of a pollutant to maintain a GROUND LEVEL CONCENTRATION of the pollutant below a certain value. The emission rate may be changed to match current weather (dispersion) conditions or to respond to changes in measured ground-level concentrations. New applications of this control technique after December 31, 1970, were not allowed under the CLEAN AIR ACT.

intermittent noise A noise that lasts for at least one second then ceases for at least one second.

internal combustion engine An engine in which the fuel is burned inside a closed cylinder and converted to mechanical energy by a piston, as in gasoline or diesel engines.

internal conversion In nuclear physics, the release of energy from an atomic nucleus to an orbiting electron, ejecting the electron from the atom.

internal dose A measure of the amount of toxic substance or pharmaceutical agent in the body. Expressed as the amount of an agent (1) absorbed after exposure through the respiratory tract, gastrointestinal tract, or skin; (2) stored in internal organs, such as in fatty tissue or bone; or (3) bound to critical sites or tissue of action. See BODY BURDEN, TARGET ORGAN.

internal energy The sum of the KINETIC ENERGY and POTENTIAL ENERGY in the atoms and/or molecules that make up a body or system. Internal energy is affected by changes in temperature and by energy added to or removed from the system as WORK is done on or by the system.

internal radiation Radiation received from radioactive substances located within the body of an organism. For example, strontium-90, absorbed from the atmosphere or from food by an animal or human, can deposit in the bones and become a source of damaging radiation for that organism over a long period.

International Agency for Research on Cancer (IARC) A division of the WORLD HEALTH ORGANIZATION that assigns chemicals to specific categories on the basis of their ability to cause cancer in humans or other animals. The categories are as follows: Group 1, the agent is carcinogenic to humans; Group 2A, the agent is probably carcinogenic to humans; Group 2B, the agent is possibly carcinogenic to humans; Group 3, the agent is not classifiable as a carcinogen for humans; Group 4, the agent is probably not carcinogenic to humans. These categories are then used to formulate regulations concerning industrial and public exposures. Publications include *IARC Monographs on the Evaluation of Carcinogenic Risk of Chemicals to Humans*, studies of more than 800 chemicals. The headquarters of the agency is in Lyon, France. Website: www.iarc.fr

International Atomic Energy Agency (IAEA) A multinational organization that regulates the safety of nuclear power stations, the management of nuclear wastes, and nuclear fuel cycle services; issues radiation protection guidelines; and maintains safeguard agreements with nations that are parties to the Treaty on the Non-Proliferation of Nuclear Weapons. The headquarters of the organization is in Vienna, Austria. Website: www.iaea.org

International Commission on Radiological Protection (ICRP) A group that recommends maximal permissible exposures to IONIZING RADIATION; the commission is financially supported by the WORLD HEALTH ORGANIZATION, the INTERNATIONAL ATOMIC ENERGY AGENCY, and the UNITED NATIONS ENVIRONMENT PROGRAM, among others. Website: www.icrp.org

International Convention for the Prevention of Pollution from Ships Also called the Marpol Convention, a combination of 1973 and 1978 agreements by the members of the International Maritime Organization. The convention controls discharges into the sea, air, or land from ships, focusing on the release of oil, other toxic liquids, radioactive waste, garbage, sewage, and certain air pollutants. Particular provisions of the convention are known by their annex number; for example, MARPOL V governs ocean dumping of garbage.

International Council of Scientific Unions (ICSU) A multinational, multidisciplinary program begun in 1931 to foster research in the natural sciences, including investigations into global environmental change. Started the INTERNATIONAL GEOSPHERE-BIOSPHERE PROGRAMME in 1986. Website: www.icsu.org

International Environmental Information System (INFOTERRA) A part of EARTHWATCH operated by the United Nations Environment Program. The system, based in Nairobi, Kenya, provides a global network of information sources on environmental subjects. Website: www.unep.ch/earthw.html

International Geosphere-Biosphere Programme An interdisciplinary program begun in 1986 by the INTERNATIONAL COUNCIL OF SCIENTIFIC UNIONS that supports research into the physical, biological, and chemical processes that may be related to global change. Headquarters in Stockholm, Sweden. Website: www.igbp.kva.se

International Joint Commission (IJC) A six-member panel established by the 1909 Boundary Waters Treaty between the United States and Canada to cooperate in the management and protection of the rivers and lakes along the boundary between the two countries. Much of the commission's work involves the Great Lakes–St. Lawrence River System, which is the largest freshwater system in the world, containing about one-fifth of the Earth's freshwater. In recent years, the IJC has added agreements on transboundary air pollution issues. Website: www.ijc.org

International Organization for Standardization (ISO) A private group dedicated to the cooperative definition of uniform standards, leading to the promotion of international trade, expansion of technology and communications, and economic development. Each of around 130 countries sends one representative body; the member from the United States is the AMERICAN NATIONAL STANDARDS INSTITUTE. ISO is not an acronym but is the prefix *iso-*, meaning "equal," the standardization sought in the group's name. Established in 1947, it has headquarters in

Geneva, Switzerland. See also ISO 14000. Website: www.iso.ch

International Register of Potentially Toxic Chemicals (IRPTC) A service that provides technical and regulatory information on chemical hazards. Based in Nairobi, Kenya. Part of EARTHWATCH operated by the United Nations Environment Program. Website: http://irptc.unep.ch

international system of units See SI UNITS.

International Union for Conservation of Nature and Natural Resources (IUCN) Known as the World Conservation Union, a nongovernmental agency of international scope that promotes measures to conserve wildlife and natural resources. Based in Gland, Switzerland. Website: www.iucn.org

International Whaling Commission A multinational group organized in 1946 under the International Convention for the Regulation of Whaling. Purposes of the organization include the conservation and development of whale populations, the support of research relating to whales, and the collection of information on current whale population levels. Quotas and restrictions adopted by the commission are not enforceable but rely on self-regulation by member countries. In 1982 the commission adopted a moratorium on all commercial whaling to be implemented by 1986, but Japan and Norway continue to hunt whales, ignoring the international ban. Based in Cambridge, England. Website: http://ourworld.compuserve.com/homepages/iwcoffice/

interplanting An agricultural practice characterized by planting two or more crops, usually in alternating rows, to enhance pest control and land use.

interspecies extrapolation The application of toxicological data obtained from one species to another species, most often from laboratory animals to humans. See SCALING FACTORS.

interstate waters Watercourses (e.g., lakes, streams, or rivers) that either form the boundary between states or flow from one state to another. These bodies of water become important when a downstream state objects to the pollution of upstream waters by entities in an upstream state in excess of the standards in the downstream state.

interstices Pore spaces in soil or rock.

interstitial water Water in the pore spaces in soil or rock.

intertidal zone That area of coastal land that is covered by water at high tide and uncovered at low tide.

intoxication A disorder caused by the presence of some material (toxin) that causes damage to the body. For example, botulism is caused by the presence of a toxin produced by certain bacteria (genus *Clostridium*) that grow in food.

intrinsically safe Describing instruments or machinery designed to be used safely in potentially combustible atmospheres.

inverse square law The relationship describing the reduction in a physical quantity (such as radiation or noise) with increasing distance from a source. The emitted quantity decreases by a factor equal to 1 divided by the square of the distance from the source ($1/d^2$).

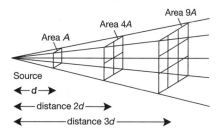

Inverse square law

inversion See TEMPERATURE INVERSION.

invert The bottom of a pipe or duct.

inverted siphon A section of a sewer line that is placed deeper in the ground than normal in order to pass under utility piping, waterways, rail lines, or other obstacles. The sewer line is raised again after passing under the obstacle. Also called a sag pipe.

ion An atom or molecule that carries a net charge (either positive or negative) as a result of an imbalance between the number of protons and the number of electrons present. If the ion has more electrons than protons, it has a negative charge. If the converse is true, the ion has a positive charge. Ions with a positive charge are called cations; ions with a negative charge are called anions.

ion exchange The substitution of one ion for another in certain substances. Either ANION exchange or CATION exchange is possible. In wastewater treatment, exchange sites on a resin column remove ions from the water passing through. The most common cation exchange involves the conversion of hard water to soft water. Hard water contains the divalent ions of calcium (Ca^{+2}) and magnesium (Mg^{+2}), which cause soap and detergents to form gum PRECIPITATES in water. A water softener consists of a resin that is saturated with sodium ions (Na^+). As hard water percolates through the resin, the ions of calcium or magnesium are removed as they attach to the resin, thus releasing (being exchanged for) sodium ions.

ionization The process by which an atom or molecule acquires an electric charge (positive or negative) through the loss or gain of electrons in a chemical reaction, in solution, or by ionizing radiation.

ionization meter A device used to measure dose rates for BETA RADIATION, GAMMA RAYS, and X RAYS. Ions are formed as the radiation passes through a gas-filled chamber and are detected either by the electric current flow, if a constant voltage is applied in the chamber, or by the amount of discharge, if the chamber is charged as a condenser and the ionization drains the charge away.

ionizing radiation Electromagnetic radiation or atomic particles capable of displacing electrons from around atoms or molecules, thereby producing charged atoms or molecules or ions. The more common types of ionizing radiation are X RAYS, GAMMA RAYS, ALPHA PARTICLES, and BETA PARTICLES.

ionosphere The upper layer of the atmosphere above the stratosphere, starting about 80 kilometers from the surface of the Earth. Incoming solar radiation is sufficiently intense to cause the ionization of the sparse gas molecules present at that altitude.

irradiated food Food that is preserved by exposure to gamma RADIATION at levels sufficient to kill bacteria and fungi; used to enhance shelf life. The technology has been applied for decades to the sterilization of medical implements and in microbiology laboratories. Proponents note that irradiation saves energy needed for refrigeration and reduces the use of chemical preservatives. Also called RADIATION STERILIZATION; see GAMMA RAY.

irradiation The use of electromagnetic RADIATION in the ultraviolet, X-ray, or gamma range to promote medical imaging or treatment, preservation of food products, or specific chemical reactions. See IRRADIATED FOOD, RADIATION STERILIZATION.

irrigation The process of supplying water to land that does not receive sufficient rainfall or rainfall that is not sufficiently frequent to allow for the production of agricultural crops. The environmental issues associated with the process include evaporative losses of water leading to the accumulation of salts to such an extent that the land is no longer suitable for agricultural applications, the depletion of groundwater resources, and the destruction of free-flowing surface waters and loss of wildlife habitats resulting from the construction of dams and reservoirs needed to maintain the supply of water.

irrigation return flow Field runoff of agricultural irrigation water; this water is excluded by statute from the definition of solid or hazardous wastes and from the permit requirements of the CLEAN WATER ACT.

irritant Any material that causes an inflammatory reaction in skin or mucous membranes, resulting in reddening, swelling, soreness, or other physical discomfort.

irruptive growth A pattern of growth in a population of organisms characterized by a rapid increase in numbers followed by a sharp decline. The number of individuals oscillates widely through boom and bust cycles.

island biogeography The study of the rates of colonization and extension of species on islands or similarly isolated regions. Because most islands are relatively small and have restricted flora and fauna, numbers of organisms can fluctuate dramatically over time, especially when new species are introduced.

island effects The reduction in species diversity caused by restriction of the size of natural habitat by land development. Natural flora and fauna are restricted to smaller and smaller, unconnected areas.

ISO 14000 The INTERNATIONAL ORGANIZATION FOR STANDARDIZATION guidelines for environmental management, focusing on the activities of an organization and their impact on the environment, including the production, use, and disposal of products. Related to ISO 9000, which is the international standard for quality management in the production of goods and services. Website: www.iso.ch

isobar A line on a map (such as a weather map) connecting points having the same barometric pressure.

isoconcentration Indicating samples that have the same concentration of some chemical. A line on a map or other diagram connecting sampling locations from which equivalent concentration data are obtained as a result of monitoring of the quantity of some constituent in an environmental medium (air, water, soil) across a defined geographical area. See ISOPLETH.

isokinetic sampling In sampling air contaminants inside a stack, the maintenance of the same air velocity in the sample probe inlet as in the exhaust gas carried by the stack. Otherwise, the sample will underestimate or overestimate the pollutant concentration in the stack gas.

isomer A chemical compound (usually organic) that has the same molecular formula as another compound but a different molecular structure and therefore different chemical and physical properties. For example, butane and isobutane both have the molecular formula C_4H_{10}, but they have a different arrangement of the carbon and hydrogen atoms.

isopleth A line on a map connecting points at which a certain variable has the

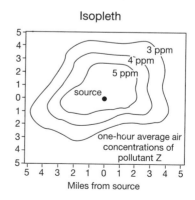

Isopleth

same value, for example, a ring around a smokestack connecting locations with equal estimated ambient air concentrations.

isotherm 1. A line on a map connecting points having the same measured or average temperature. 2. A curve showing the changes in a variable while held at a given temperature. For example, adsorption isotherms show the amount of pollutant adsorbed per mass of adsorbing material compared to the amount (concentration) of the pollutant left in water or air at equilibrium.

isotope Atom of a single element that has the same number of protons (hence the same atomic number) but a different number of neutrons. Consequently, isotopes have different atomic masses and different nuclear properties. For example, carbon-12, carbon-13, and carbon-14 all have six protons (therefore an atomic number of 6) but have six, seven, and eight neutrons, respectively. Each form behaves as carbon in chemical reactions; however, carbon-14 is radioactive (unstable), whereas the other two isotopes are not.

isotropic Displaying equal properties in all directions. For example, isotropic atmospheric pressure at a point six feet above the ground is equal in all directions: up, down, left, right, and all points in between.

itai-itai disease A disease characterized by kidney dysfunction, bone deformities, and bone pain that is caused by long-term excessive CADMIUM exposure. The name is Japanese for "ouch-ouch," because the symptoms were noted in a group of Japanese with high cadmium doses resulting from the consumption of rice and soybeans grown in soil contaminated by cadmium emissions from mining operations.

J curve A graphical representation of a population undergoing EXPONENTIAL GROWTH.

Jackson turbidity unit (JTU) A unit that expresses the cloudiness (turbidity) of water; the measure is related to the distance through water that light can be seen by the unaided eye. The measurement apparatus, a candle turbidimeter, consists of a glass tube with a candle aligned under it. Water is added to the glass tube until the light of the candle cannot be seen; a light path of 2.3 centimeters (cm) equals 1000 JTUs, one of 21.5 cm (clearer water) equals 100 JTUs, and so on (the scale is nonlinear). Named for D. D. Jackson, who, together with G. C. Whipple, published the method in 1900.

jar test A laboratory testing procedure most frequently used in conjunction with the processing of raw water for distribution into a water supply, which also has applications in the treatment of wastewater prior to discharge. Often both systems use a COAGULATION agent to promote the settling of suspended matter in the water. Using drinking water as an example, the laboratory test is done to determine the optimal concentration of the coagulation aid, commonly aluminum sulfate (ALUM). The testing device is a unit that can hold and stir six containers ("jars"). The same amount of water is placed in each container, varying amounts of coagulation agent added, the six mixtures stirred, and the contents in the test containers allowed to settle. The container that demonstrates the most efficient accumulation of particulate material indicates the optimal concentration of coagulant to be applied.

jaundice A condition caused by the presence of abnormal amounts of bilirubin, an end product of hemoglobin metabolism, in the blood and characterized by a yellowing of the skin and eyes. Exposure of humans to toxic substances that damage the liver can result in this condition.

joint and several liability The legal principle, held by the courts to apply to hazardous waste site cleanups under the COMPREHENSIVE ENVIRONMENTAL RESPONSE, COMPENSATION, AND LIABILITY ACT (CERCLA), that allows one party to be held responsible for an entire liability if that party contributed to the liability in part. Thus, a hazardous waste generator with waste at a CERCLA site, a former owner or operator of the disposal facility, or a transporter that chose the facility can be sued by the USEPA for the full cost of cleanup at a hazardous waste site used by many disposers. This principle does not prevent the party held to be jointly and severally liable from legally pursuing other parties for their share of the cleanup. See CONTRIBUTION RIGHTS, POTENTIALLY RESPONSIBLE PARTY, COST RECOVERY, DE MINIMIS, DE MICROMIS, INNOCENT LANDOWNER.

joule (J) A unit of energy. In the SI system, the unit equal to the work done by a force of one newton over a distance of one meter in the direction of the force (1 J = 1 Nm). Also equal to 0.239 calorie, or 9.484×10^{-4} British thermal unit (BTU).

junk science The term applied to the incorrect use or application of scientific data or principles, usually to influence either a judicial outcome or a public policy relating to human health or the environment. For more see Website: www.junkscience.com

juvenile hormones Organic compounds produced by insects undergoing metamorphosis that maintain the insect in a certain stage in its life cycle; once the hormone is no longer secreted, maturation occurs. Synthesized versions of species-specific juvenile hormones are applied to growing crops to prevent a particular pest species from maturing and reproducing, thus controlling the pest.

K

karst A geological formation consisting of porous limestone that may contain caverns and groundwater deposits. The material is sensitive to lowering of the water table in periods of drought and to the withdrawal of groundwater, possibly leading to the collapse of the deposit with the appearance of major sinkholes at the surface above.

Kelvin (K) The SI unit of TEMPERATURE. Zero Kelvin is ABSOLUTE ZERO, and an interval of 1 K is equal to 1 degree on the Celsius scale and 1.8 degrees on the Fahrenheit scale; 0° Celsius equals 273.15° K.

Kepone A chlorinated hydrocarbon insecticide that is persistent in the environment. The agent was banned in 1976 after a number of workers in a Kepone manufacturing unit in Virginia fell ill and fish in the nearby James River were found to contain unsafe levels. Also referred to as chlordecone.

kerogen A solid hydrocarbon embedded in subsurface rock formations. (Sometimes erroneously called oil shale; it is a solid, not liquid oil, and the rock is not shale.) The largest known deposits are in parts of Colorado, Utah, and Wyoming. Although kerogen deposits are estimated to exceed proven oil reserves in the United States by 25–50 times and it is a potential source of fuel, the high extraction and environmental costs have prevented exploitation.

keystone species A single kind of organism or a small collection of different kinds of organisms that occupy a vital ecological NICHE in a given location. The disappearance of these kinds of organisms would have negative consequences within the entire assemblage in a particular locale. Cannot be lost without major consequences to the whole. See COMMUNITY, SPECIES.

kilocalorie (kcal) 1000 calories; 1 kcal = 1 dieter's calorie, the unit applicable to indicate caloric content of food.

kilogram (kg) The SI unit of mass. One kilogram, or 1000 grams, equals about 2.205 pounds.

kilopascal 1000 PASCALs; a common unit for pressure measurements. One kilopascal equals 4.02 inches of water, 0.145 pound per square inch, or 7.5 millimeters of mercury.

kiloton A unit of explosive energy applied to nuclear weapons; equal to 1000 tons of trinitrotoluene (TNT).

kilowatt (kW) A unit of power, which is energy used per time, equal to 1000 WATTs. Since one watt equals one JOULE per second, a kilowatt equals 1000 joules per second. See KILOWATT-HOUR.

kilowatt-hour (kWh) A unit of electrical energy equal to 1000 WATT-hours, or a power demand of 1000 watts (e.g., 10 100-watt incandescent lightbulbs) for one hour. The equivalent of 3.6 million joules. Utility rates are typically expressed in cents per kilowatt-hour.

kinematic viscosity The VISCOSITY of a fluid divided by its mass density (mass per unit volume). Units are length squared divided by time (e.g., square meters per second). The values are also expressed in centistokes, or 0.001 square meter per second.

kinetic energy (*k*)   The energy inherent in a substance because of its motion, expressed as a function of its velocity and mass, or $MV^2/2$.

kinetic rate coefficient   See RATE CONSTANT.

Kirchhoff's law   The law stating that for a given temperature, the ABSORPTIVITY and EMISSIVITY of a substance for a certain wavelength of radiation are equal. Named for the German physicist Gustav Robert Kirchhoff (1824–87).

Kjeldahl nitrogen   The amount of nitrogen contained in organic material as determined by a method based on the digestion of the sample in a sulfuric acid–based reagent that converts the nitrogenous organic material to carbon dioxide, water, and ammonia. Subsequently, the ammonia is quantified. Named for Johann Kjeldahl (1849–1900).

knock-out tray   A device designed to capture the majority of solids, droplets, and mists carried over from a SCRUBBER.

knot   A unit of velocity equal to 1 nautical mile per hour or approximately 1.15 statute miles per hour. A nautical mile equals 1852 meters, or 6072 feet.

known resources   RESOURCES that have been identified as to type, quantity, and location.

Koschmieder's relationship   An equation that gives the visual range (*Lv*) in the atmosphere, using estimates for the scattering and absorption of light by naturally occurring and pollutant gases and particles; expressed as

$$L_v = \frac{3.92}{b_{ext}}$$

where b_{ext} is the extinction coefficient. If the extinction coefficient has units of 1/kilometer, then the visual range is in kilometers.

kraft paper   Paper prepared by the KRAFT PROCESS, which involves the cooking of wood chips (commonly from pine) with a sodium sulfide–sodium hydroxide liquor. The natural-finished product is a brown paper from which corrugated boxes are produced. Substantially different from newsprint and similar grades of paper made by a mechanical pulping process.

kraft process   Technology based on the use of a sodium sulfide–sodium hydroxide liquor to release the cellulose fibers for the preparation of pulp for the manufacture of KRAFT PAPER. Associated with the odor characteristic of paper mills because of the release of reduced sulfur compounds into the atmosphere.

kriging   A technique for the analysis of the spatial variability of data; used in the study of disease INCIDENCE. See CLUSTERING.

krill   Small crustaceans that are abundant and form an important part of the food chain in Antarctic waters; a possible source of human dietary protein, either directly or via animal feed. The effect of large-scale harvesting of krill on ocean food chains is unknown.

K-selected   A type of reproductive strategy of a species that allocates a relatively high proportion of available energy to the maintenance and survival of individual members of the species rather than to reproduction. The number of such organisms in a defined area does not tend to increase beyond the capacity of the environment to support the species. Compare R-SELECTED.

kwashiorkor   Malnutrition caused by protein deficiency although caloric intake is adequate. In the absence of famine, the disorder is usually seen in one- to three-year-old children in very poor areas as a result of premature weaning. Characterized by retarded growth, hair loss, and accumulation of fluids in the abdominal area. Compare MARASMUS.

K waste Material from specific sources defined as LISTED HAZARDOUS WASTE by the USEPA in the *CODE OF FEDERAL REGULATIONS,* Title 40, Part 261.32. See also F WASTE, P WASTE, U WASTE, HAZARDOUS WASTE.

Kyoto Protocol (to the United Nations Framework Convention on Climate Change) A 1997 agreement among 150 nations, including the United States, that requires GREENHOUSE GAS emission reductions from 1990 levels. The European Union agreed to an 8% cut; the United States, 7%; and Japan, 6%. Developing nations' reductions are voluntary. For the full text of the protocol, see www.unfccc.de/resource/protintr.html. See GLOBAL WARMING.

L

L_n An expression indicating how often a certain sound level is equaled or exceeded, when n equals percentage of the measurement time; a summary statistic to describe noise levels. For example, $L_{70} = 55$ dBA means that 70% of the noise measurements in an area were at or above 55 dBA (DECIBELS, A-WEIGHTING NETWORK).

lab pack A container containing laboratory waste, often in numerous individual packages with a wide variety of individual compounds, complicating proper treatment and disposal.

laboratory blank An artificial sample, usually distilled water, introduced to a chemical analyzer to observe the response of the instrument to a sample that does not contain the material being measured. The blank can also detect any contamination occurring during laboratory processing of the sample.

laccolith A geological structure formed when magma is injected between layers of sedimentary rock to form a lens-shaped body that forces the upper layer of rock to bulge upward. The result is a domelike structure in an otherwise flat terrain when viewed from the surface.

lacrimator An air contaminant that causes eye tearing upon excessive exposure; this type of contaminant is often present as an ingredient in PHOTOCHEMICAL AIR POLLUTION (smog), such as PEROXYACETYLNITRATE (PAN).

lacustrine Pertaining to lakes, as in lacustrine environment.

lag phase The period following the inoculation of a growth medium with a culture of bacteria. Some time is required for the organism to adjust to the new environment; during this interval there is no increase in the number of bacteria per unit volume of culture fluid.

lagoon A pond used for the STABILIZATION and DECOMPOSITION of organic materials in wastewater. These ponds allow for the settling and subsequent ANAEROBIC mineralization of particulate matter and for the AEROBIC decomposition of dissolved constituents. Often AERATION devices that agitate the contents of the pond are added to enhance the activities of aerobic bacteria.

laminar flow Smooth, nonturbulent flow of a fluid (water, air) in a pipe or around obstacles. Compare TURBULENT FLOW. See also REYNOLDS NUMBER.

land application A method for the disposal of treated domestic wastewater. The wastewater, which has been subjected to PRIMARY TREATMENT and SECONDARY TREATMENT as well as disinfection to kill or inactivate dangerous microorganisms, is sprayed over the ground to remove plant nutrients and promote the growth of vegetation. A type of TERTIARY TREATMENT.

land ban See LAND DISPOSAL BAN.

land breeze The land-to-sea surface wind that occurs in coastal areas at night. The wind is caused by the rising of the air above the ocean, which is warmer than the land because of the rapid cooling of the land after sunset. Compare SEA BREEZE.

land disposal ban A process initiated by the United States Congress under the HAZARDOUS AND SOLID WASTE AMENDMENTS of 1984. The land disposal of specific hazardous wastes was prohibited automatically unless the USEPA ruled that the disposal ban was unnecessary to protect human health and the environment and issued treatment standards for the wastes. The process included a schedule of phased restrictions on the following types of wastes, in the following order: BULK LIQUID hazardous wastes or hazardous wastes containing FREE LIQUIDS; dioxin-containing wastes and spent solvents; CALIFORNIA LIST WASTES; and all remaining LISTED HAZARDOUS WASTE or known hazardous wastes in three separate groups (known as the first third, second third, and third third). The USEPA has published treatment standards that allow many of these named wastes to be disposed of in landfills. See HAMMER PROVISION, SOFT HAMMER.

land disposal restrictions Rules developed by the USEPA that require HAZARDOUS WASTE to be treated before disposal in a LANDFILL to destroy or immobilize dangerous substances that have the potential to migrate into soil and groundwater. See LAND DISPOSAL BAN.

land farming A technique for the BIODEGRADATION and disposal of organic waste that involves the mixture of waste SLUDGE and soil. Microorganisms in the soil degrade the organic wastes. The biodegradation is enhanced by tilling the soil-waste mixture to ensure adequate oxygen and the control of moisture content, nutrient levels, and soil pH.

land treatment See LAND FARMING.

landfill An area where solid or solidified waste materials from municipal or industrial sources are buried. See SANITARY LANDFILL.

Landsat satellite Short for *land satellite,* a series of Earth resources observation satellites that have collected data since the early 1970s. The Landsats have produced multiwavelength scanner data useful for agricultural management, mapping of soil associations, oil exploration, flood plain mapping, water quality and water resource tracking, air pollutant sensing, weather observation and prediction, and natural resource management.

landscape ecology The study of the patterns of COMMUNITY and ECOSYSTEM distributions within a defined geographical area and of the ecological processes that influence changes in these patterns over time. The ways in which territorial history shapes the features of the land and organisms that inhabit a region.

landscape indicator A measure used to describe spatial patterns of land use within a defined geographic area. Data are derived from maps, REMOTE SENSING information, or other sources to describe such phenomena as forest fragmentation, grassland cover, and wetland loss.

Langelier index An expression of the ability of water to dissolve or deposit calcium CARBONATE scale in pipes. The index is important in industrial water systems, where the formation of scale or sludge can cause equipment or process failure. The index is calculated from direct measurements of the following in the water system: pH, ALKALINITY, calcium concentrations, TOTAL DISSOLVED SOLIDS, and temperature. A positive value indicates a tendency to form scale, and a negative value means the water will dissolve scale and may be corrosive. Named for W. F. Langelier, who devised the index in 1949. Also called the stability index.

langley (ly) A unit expressing the quantity of electromagnetic radiation received or emitted by a unit area of surface. One langley equals one CALORIE per square centimeter. The rate of solar INSOLATION for a point on the surface of the Earth perpendicular to the rays of the Sun is about 1.3 langleys per minute; at the upper reaches of the atmosphere this perpendicular insolation is 2.0 langleys per minute, also called the SOLAR CONSTANT.

lapse rate See ADIABATIC LAPSE RATE, ENVIRONMENTAL LAPSE RATE.

large-quantity generator A facility that produces more than 2200 pounds of HAZARDOUS WASTE per month or more than 2.2 pounds of an acutely hazardous waste (see 40 CFR 261.30) per month. Such locations are subject to all provisions of the RESOURCE CONSERVATION AND RECOVERY ACT. See SMALL-QUANTITY GENERATOR.

latency See LATENT PERIOD.

latent heat See LATENT HEAT TRANSFER.

latent heat transfer The removal or addition of heat when a substance changes state. In the environment, this almost always refers to the release of heat from water upon condensation and the absorption of heat by water upon evaporation. See HEAT OF VAPORIZATION.

latent period The time between exposure to a dangerous substance or radiation and the development of a disease or pathological condition resulting from that exposure.

lateral A municipal wastewater drain pipe that connects a home or business to a branch or MAIN.

lateral expansion Any horizontal expansion of the area that will receive MUNICIPAL SOLID WASTE at an existing landfill; such an expansion is considered a new unit by federal regulations and must meet the (usually more stringent) standards that apply to new facilities.

lateritic soil Land that consists of minerals that are rich in iron and aluminum compounds; other minerals have been removed by LEACHING. The land is hard and unsuitable for agricultural use.

lava Melted rock flowing downslope from fractures or fissures in the crust.

Commonly associated with MAGMA and volcanoes. Lava contains a mixture of solids, liquids, and gases and varies in temperature and viscosity. See BASALTIC ERUPTION, SILICIC ERUPTION.

lava tube A structure formed when free-flowing BASALTIC LAVA cools and solidifies at the margins of the flow while maintaining an inner core of molten liquid material that moves through the solidified exterior.

law of the minimum See LIEBIG'S LAW OF THE MINIMUM.

laws of ecology Axioms for environmental protection proposed by Barry Commoner in his book *The Closing Circle: Nature, Man, and Technology* (1971): (1) everything is connected to everything else; (2) everything must go somewhere; (3) nature knows best; (4) there is no such thing as a free lunch.

Le Châtelier's principle A principle of dynamic equilibrium stating that a change in one or more factors that maintain equilibrium conditions in a system will cause the system to shift in a direction that will work against or adjust to the change(s), with a resulting reestablishment of equilibrium conditions. For example, assume the concentrations of gaseous oxygen in the atmosphere and DISSOLVED OXYGEN in a stream are in equilibrium at a certain temperature. As oxygen dissolves in water, heat is released. If an outside influence (e.g., sunlight) raises the water temperature in the stream, this shifts the equilibrium back in the direction of lower dissolved oxygen and greater atmospheric oxygen, and oxygen escapes from the water. As a result, at higher water temperatures, equilibrium concentrations of dissolved oxygen are lower. Named for the French chemist Henri Louis Le Châtelier (1850–1936).

leachate Water that has migrated through and escaped from a waste disposal site. The fluid contains dissolved and suspended material extracted from the waste and soil.

Leachate collection system

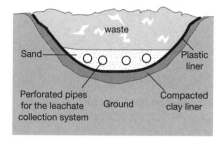

leachate collection system An arrangement of reservoirs and pipes underlying a waste disposal site. The system is designed to accumulate and remove leachate, water that migrates through the waste.

leaching The process by which soluble materials are washed out and removed from soil, ore, or buried waste.

leaching field The area of land into which a SEPTIC TANK drains. The wastewater exiting the tank is dispersed over and percolates through a defined area of land.

lead (Pb) A toxic metal present in air, food, water, soil, and old paint. Overexposure to this metal can cause damage to circulatory, digestive, and central nervous systems. Children less than six years old are considered the most susceptible. Atmosphere levels have dropped sharply with the phaseout of leaded gasoline. Lead in air, water, and food is regulated by the CLEAN AIR ACT; CLEAN WATER ACT; SAFE DRINKING WATER ACT; FOOD, DRUG, AND COSMETIC ACT; and other environmental statutes. See PICA, TETRAETHYL LEAD.

lead agency 1. The federal administrative agency responsible for supervising the preparation of an ENVIRONMENTAL IMPACT STATEMENT when more than one agency is involved in the process. 2. The federal or state agency represented by the ON-SCENE COORDINATOR in a response action to a spill or leak of a hazardous substance.

leaking underground storage tank trust fund (LUST trust fund) Established by a 1986 amendment of the RESOURCE CONSERVATION AND RECOVERY ACT, the fund is used to pay for any required cleanup of leaks from petroleum underground storage tanks for which responsible parties cannot be identified. (If solvent owners can be found, they pay for the cleanup.) The fund was created because the COMPREHENSIVE ENVIRONMENTAL RESPONSE, COMPENSATION, AND LIABILITY ACT excludes petroleum. The money is collected from a 0.1-cent tax on each gallon of petroleum motor fuel sold, generating more than $150 million per year. The fund expired in December 1995 but was reinstated in 1997 to run through 2005. At the end of 1998, the fund balance was $1,250 million. Almost all states have a comparable storage tank remediation fund.

Legionella pneumophila The bacterium that causes legionellosis (or Legionnaires' disease), so named because of the association of the organism with an outbreak of pneumonia during a convention of the American Legion in 1976. The organism can be isolated from many terrestrial and aquatic habitats and is a common inhabitant of cooling towers of air-conditioning systems. The disease is spread to humans via the aerosols generated from cooling towers and other water sources. Legionellosis is a pneumonia characterized by malaise, myalgia, fever, headache, and respiratory illness.

legislative history The committee reports, congressional debates, committee prints, and any other supplemental documents pertaining to an enacted federal statute. The material is used by administrative agencies in formulating regulations based on the law and by courts in reviewing the intent of the law.

legume A plant that produces a seed in a pod, such as a bean plant. This type of plant is important in the symbiotic fixation of atmospheric nitrogen, during which nitrogen gas from the air is converted to a form of nitrogen found in protein. The conversion requires the combined efforts of the plant and bacteria belonging to the genus *Rhizobium*. Legumes are in turn important sources of nitrogen for humans.

lender liability The potential, under the COMPREHENSIVE ENVIRONMENTAL RESPONSE, COMPENSATION, AND LIABILITY ACT, for a bank, if it forecloses on a mortgage or lien or otherwise acts in a defined management capacity over the property, to be held liable for a hazardous waste cleanup on the property.

Lemna gibba The genus and species name of a small, stemless, free-floating plant used in experiments to determine the toxicity of pollutants to aquatic plant life. The species is found in still waters contaminated with sewage. Commonly called duckweed.

lentic water The standing water of ponds, lakes, swamps, or marshes. Compare to LOTIC WATER.

Leopold, Aldo (1887–1948) American Author of *A Sand County Almanac* (1949, posthumous) in which he argued that modern humans' relationship with the land, one of ownership and use, was fatally flawed and should be replaced with a "land ethic" of respect and harmony.

Leq A calculated value of a continuous sound level that equals, in total energy, a combination of different sound levels during a period. The value is expressed as

$$L_{eq} = 10 \log \sum_{i=1}^{N} 10^{L_i/10} t_i$$

where N is the total number of sound measurements, L_i is the sound level in DECIBELs of the ith sample, and t_i is the fraction of the total time for the ith sample. For example, if three 10-minute sound measurements are 50, 40, and 75 decibels, then the equivalent 30-minute continuous sound is 70.2 decibels. Note that the value of the Leq is influenced greatly by high sound levels.

less-developed country Used to describe a country that is characterized by scarce industrial development, low per capita income, high birth rate, rapid population growth, and low levels of technological progress.

lethal concentration—50 percent (LC_{50}) The concentration of a chemical substance that causes death in 50% of the test organisms within a specified period after exposure. The expression usually refers to air concentrations, but the term is used to define water concentrations causing a 50% death rate in aquatic organisms during the testing of chemicals for toxicity in aquatic environments. Usually expressed in units such as parts per million or micrograms per cubic meter if air concentrations. Commonly used water concentration units are milligrams per liter or micrograms per liter. The term can be modified to define a lethal concentration for 25% of the test organisms (LC_{25}), and so on. The LC_{50} is also referred to as the mean or median lethal concentration.

lethal concentration, low (LC_{LO}) The lowest concentration of a chemical substance that has caused death in humans or animals within a certain period after exposure. The term, which usually refers to air concentrations, is also used to define the lowest water concentrations causing death of aquatic organisms during toxicity tests. See LETHAL CONCENTRATION—50 PERCENT.

lethal dose (LD) The absorbed amount of a chemical agent sufficient to cause death. The inclusion of a subscript number with the LD designation indicates the percentage of the exposed population that dies within a certain period as a result of the exposure. For example, LD_{50} indicates a 50% fatality rate. See LETHAL DOSE—50 PERCENT.

lethal dose, low (LD_{LO}) The lowest dose of a chemical introduced by any

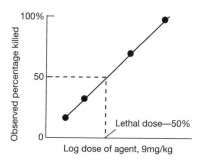

Lethal dose—50 percent

exposure route other than inhalation that has caused death in humans or animals within a certain interval after exposure. A common unit is milligrams of the chemical per kilogram body weight of test subjects. See LETHAL CONCENTRATION, LOW.

lethal dose—50 percent (LD_{50}) The amount of chemical, usually expressed as milligrams of the chemical per kilogram body weight, that causes death in one-half of the exposed organisms. Also called the median lethal dose.

leukemia Cancers of the blood-forming tissues, characterized by the overproduction of white blood cells and their precursors. Excessive exposure to IONIZING RADIATION or BENZENE is associated with an increased leukemia risk.

level of concern (LOC) **1.** In hazards analysis, the air CONCENTRATION of chemical above which acute irreversible health effects may occur over short exposure times. See EXTREMELY HAZARDOUS SUBSTANCE. **2.** In pesticide regulation, a RISK QUOTIENT that, if exceeded, is deemed to be potentially hazardous. The LOC varies with acute or chronic toxicity and the NONTARGET species.

licensed material Any material that contains 0.05% or more of uranium or thorium, enriched uranium, or by-products of nuclear reactors and that must be licensed by the federal government for use, transport, or processing.

lichen Plants that result from a symbiotic relationship between algae and fungi. These composite growths are found on the surfaces of trees, stones, or other structures exposed to the air and represent the most common example of mutualistic relationships of microorganisms. The uptake of air pollutants by lichens has been employed as an indicator of the level of air pollution.

Liebig's law of the minimum The ecological principle that the existence, abundance, or distribution of a population is limited by the essential physical or chemical factor that is in shortest supply relative to the level required by the organisms. The one physical or chemical requirement restricting the growth of a population is called the LIMITING FACTOR. First proposed by Justus von Liebig in 1840. Liebig was a German chemist (1803–73).

life cycle analysis A study of the environmental costs, in terms of resource and energy use, throughout the manufacture, use, and disposal of a product. A system to assess the total impact of some product on the environment.

life cycle assessment See LIFE CYCLE ANALYSIS.

life cycle of a product All stages in the development, manufacture, marketing, use, and ultimate disposal of a product.

life expectancy Usually used as the life expectancy at birth, which is the average age that a newborn can be expected to attain. The period is strongly influenced by the rate of infant mortality. Life expectancy at an age other than birth is the average number of remaining years before death, which is a function of age, geography, sex, race, occupation, diet, health status, and personal habits, inter alia.

life span The longest period of life commonly reached by a type of organism. For example, a dog has a shorter expected life span than a human.

life-span study (LSS) The tracking of individuals and monitoring of their health status throughout their lifetime in an attempt to study long-term effects of exposure to risk factors such as low doses of ionizing radiation.

life table A tabulation of vital statistics on a population in which members are grouped by age. Information on the various age groups includes survivorship, mortality rates, and LIFE EXPECTANCY.

lifetime exposure Either the total dose or the expected constant dose rate per time unit (e.g., two grams per day) of a substance or condition that a person would receive in a lifetime, usually calculated on the basis of a life expectancy of 70 years.

lifetime risk The probability of an individual's dying of a specific cause or contracting a certain disease (such as cancer) as a result of an assumed dose or constant dose rate of a chemical or radiation during 70 years of life.

lifetime risk ratio The probability of a person's dying of a specific cause or contracting a disease as a result of an assumed exposure to a chemical or radiation divided by the probability of an unexposed person's dying of the same cause or contracting the disease.

lift A tier of landfill CELLS.

lift station A pumping facility that raises municipal sewage to a higher elevation to allow for further gravity transport. Such facilities are required in areas with flat topographic features.

light-and-dark bottle technique A method used to determine the extent of photosynthesis in an aquatic ecosystem. Duplicate portions of a water sample are collected. One portion is incubated in a clear bottle, and the other is incubated in a dark, light-tight bottle. After incubation for a prescribed period, the net uptake of carbon dioxide in both is measured and compared. The difference between the two measurements represents a measure of photosynthesis.

light detection and ranging (LIDAR) A technique employing high-intensity laser light to detect and track air pollutants released from industrial facilities. The pulses of laser light are reflected by the aerosols in the atmosphere.

light green technology Applications or innovations with unintentional but real environmental benefits. For example, direct deposit of employee pay is convenient but also reduces traffic congestion on paydays. Compare DARK GREEN TECHNOLOGY.

light nonaqueous phase liquid (LNAPL) A liquid that is IMMISCIBLE with water and with a SPECIFIC GRAVITY less than 1 (less dense than water). Common petroleum fuels and lubricating oils are LNAPLs.

light-water reactor (LWR) A nuclear reactor that uses ordinary water as the coolant and MODERATOR. The CORE of the reactor, which contains the uranium fuel and CONTROL RODS, is totally immersed in the water. Heat generated by the FISSION of the uranium fuel raises the temperature of the water, which is then pumped to HEAT EXCHANGERS for the production of steam and subsequent generation of electricity. The process results in a continuous transfer of heat from the reactor to the outside. The water also functions as a moderator to reduce the energy level of neutrons released by the fission process in order to allow some of the neutrons to promote additional fission events. The light-water reactor is the most common type of nuclear reactor operated in the United States. Compare HEAVY-WATER REACTOR.

lignin A major macromolecular polymer contained in the woody structure of plants. The polymer is a random arrangement of phenylpropane subunits. BIODEGRADATION of lignin is slow and takes place only in aerobic environments. Lignins are important in the formation of HUMIC materials and the building of topsoil.

lignite   A type of coal, with the lowest energy content of coals that are widely used. Also called brown coal.

lime treatment   1. The addition of technical-grade calcium oxide powder (lime) to acidic soils or surface water bodies to raise the pH. See ACID DEPOSITION. 2. The addition of lime to chemical toilets, the pit of an outhouse, or animal yards to "improve" the sanitary quality by controlling odor, reducing fly infestations, and lowering the microbial load.

limestone   A sedimentary geological deposit formed by the consolidation of CARBONATE minerals that were formerly shells of living organisms. Heating of this mineral results in the formation of lime. The mineral is used in the control of sulfur dioxide releases in industrial exhausts. See FLUE GAS DESULFURIZATION.

limestone scrubbing   See FLUE GAS DESULFURIZATION.

liming   See LIME TREATMENT.

limit of detection (LOD)   The lowest concentration of some substance that can be assayed by using the technology available.

limited degradation   1. Slow or minimal biological, chemical, or physical breakdown of a complex material into simpler components. 2. Environmental policies that restrict the increase in environmental concentrations of pollutants to certain levels over an area-specific baseline level. See PREVENTION OF SIGNIFICANT DETERIORATION, BASELINE CONCENTRATION, ANTIDEGRADATION POLICY.

limited   water-soluble   substances   Water pollution chemicals that are soluble in water at less than one milligram of substance per liter of water.

limiting factor   The dominant factor that restricts the continued reproduction or spread of a particular species. The factor may be a physical constraint, such as light, or a chemical resource, such as an essential nutrient.

limits of tolerance   Minimal and maximal amounts of a required physical or chemical environment factor within which a particular species can exist; an optimal level, supporting the maximal number of organisms, lies within the tolerance limits.

limnetic zone   The open water of a pond or lake supporting PLANKTON growth. Compare to PROFUNDAL ZONE.

limnology   The study of inland bodies of freshwater and the flora and fauna in those waters.

Lindane   A commercial CHLORINATED HYDROCARBON insecticide that consists of several isomers of hexachlorocyclohexane, with the gamma isomer the primary active agent. The chemical is a six-carbon, cyclic compound; however, lindane is not AROMATIC. Lindane is persistent in the environment

line source   A roadway source of vehicular air pollutants. In the analysis of air contamination, pollutants are assumed to be released at a constant rate along each segment of the roadway.

linear accelerator   A device for accelerating charged subatomic particles in a straight line. The device is used to study the structure of atoms.

linear alkyl sulfonate (LAS)   A common SURFACTANT used in detergents. BIODEGRADABLE LAS replaced ALKYLBENZENE SULFONATE, another surfactant, when the latter, which is not readily biodegradable, caused foaming discharges from municipal sewage treatment plants.

linear dose model   A projection that postulates that the increase in risk associated with exposure to a dangerous substance or radiation is in direct proportion to the dose of that agent.

linear energy transfer (LET) A measure of the loss of energy as IONIZING RADIATION passes through tissue, expressed as the frequency of ionizing events, or energy locally imported to the tissue, per unit length of biological material. X RAY and GAMMA RAY are characterized by low LET, since ionizing events are widely spaced as the radiation passes through cells, whereas ALPHA PARTICLES and fast neutrons are described as having high LET, since ionizing events are spaced closely together as they pass through cells. See also QUALITY FACTOR, RELATIVE BIOLOGICAL EFFECTIVENESS.

linear growth See ARITHMETIC GROWTH.

linearized multistage model A mathematical method that uses the observed probability of tumor formation in animals following very high doses of a chemical to estimate the expected incidence of cancer for humans who will receive much lower doses. The model incorporates the assumption that a clinical cancer results after progressing through several stages and that cancer risk is proportional to dose (a linear relation) at low exposures. One form of the model, for a fixed time of exposure, is

$$P(d) = 1 - \exp\{-\sum_{i=0}^{n} q_i\, d^i\}$$

where $P(d)$ is the cumulative probability of cancer at dose d, n is the number of stages, and q is and empirically derived nonnegative coefficient

liner 1. In waste management, a barrier, usually consisting of a combination of clay and plastic sheeting, installed in an excavation or lagoon to prevent the movement of contaminants into groundwater. 2. In pollution prevention, an insert or sleeve placed inside old or damaged sewer pipes to prevent the loss of wastewater.

lipid A diverse collection of organic substances that have a common property of solubility in fat or organic solvents such as methanol, acetone, chloroform, and benzene but are insoluble in water. Often used as a synonym for fatty substance. Lipids are important constituents of the CELL.

lipid solubility The tendency of substances to dissolve in fat solvents and conversely to be insoluble in water. Substances that are lipid soluble tend to be absorbed readily and retained in the body, whereas water-soluble substances tend to be eliminated in the urine. See BIOACCUMULATION.

lipophilic Describing chemicals that are FAT SOLUBLE and therefore less likely to be excreted from the body. WATER-SOLUBLE chemicals are eliminated more readily.

liquefaction The conversion of a substance in the solid state into the liquid state.

liquefied natural gas (LNG) Natural gas, mainly methane and ethane, condensed to a liquid state to reduce its volume greatly for international transport in specially designed refrigerated tanker ships. At the destination, the LNG is revaporized and introduced into a pipeline network. The facilities for liquefication, transport, and revaporization are industrial operations, and the public does not have direct access to LNG. Compare LIQUEFIED PETROLEUM GAS.

liquefied petroleum gas (LPG) A liquid hydrocarbon fuel composed of condensed propane and butane gases; the fuel is stored and transported in pressurized tanks for distribution to public consumers that are not serviced by natural gas facilities. LPG is derived from hydrocarbon fractions recovered in the refining of crude oil and during some petrochemical operations. Compare LIQUEFIED NATURAL GAS.

liquid chromatography A technique for the separation and analysis of higher-molecular-weight organic compounds. See CHROMATOGRAPHY.

liquid injection incinerator A device used to burn liquid wastes. DESTRUCTION AND REMOVAL EFFICIENCY is improved by using high pressures to create tiny droplets that are introduced into the incineration chamber.

liquid-metal fast breeder reactor (LMFBR) See BREEDER REACTOR.

liquid scintillation counter See SCINTILLATION COUNTER.

liquor A liquid solution containing dissolved substances. A concentrated solution of process chemicals or raw materials added to an industrial process. Compare SLURRY.

list 1. In environmental regulation, to add a chemical to a regulatory program. For example, the USEPA would list a chemical as a CRITERIA POLLUTANT before writing a NATIONAL AMBIENT AIR QUALITY STANDARD for that pollutant. See DELISTING. 2. A slang term for the list maintained by the USEPA of facilities or firms prohibited from participating in government contracts because they have excessive violations of environmental regulations. See DEBARMENT.

listed hazardous waste Chemical substances or processes that produce chemical substances defined as hazardous waste by the USEPA and published in the CODE OF FEDERAL REGULATIONS, Title 40, Parts 261.31–261.33. Listed wastes are identified in four categories: F WASTE, K WASTE, P WASTE, AND U WASTE. A facility producing hazardous waste is then regulated as a "generator"; these rules appear in the *Code of Federal Regulations,* Title 40, Part 262. Special provisions apply for SMALL-QUANTITY GENERATORS.

listed waste See LISTED HAZARDOUS WASTE.

listing site inspection (LSI) A more extensive investigation at a potential SUPERFUND SITE after the PRELIMINARY ASSESSMENT AND SITE INSPECTION (PA/SI). The results of the PA/SI and the LSI are used to calculate a site score by using the HAZARD RANKING SYSTEM. The score determines whether the site is included on the NATIONAL PRIORITIES LIST.

liter (l) A metric volume unit equivalent to 1000 cubic centimeters. One liter equals 1.057 quarts or 0.0353 cubic feet.

lithology The study of soil, sediment, or rocks employing physical properties such as grain size and texture.

lithosphere The solid portion of the crust of the Earth. Constitutes, together with the ATMOSPHERE, HYDROSPHERE, and BIOSPHERE, the ecosystem of the Earth.

lithotroph A type of bacteria capable of obtaining metabolically useful energy from the oxidation of inorganic chemicals, chiefly ammonium, nitrite, iron, and various forms of sulfur. These bacteria obtain their carbon from carbon dioxide, as do the green plants. Compare AUTOTROPH, HETEROTROPH.

litter fence A movable fence used at sanitary landfills to catch blowing debris.

littoral An interface region between the land and a lake or the sea.

littoral zone The area of a lake or pond close to the shore; includes rooted plants. Compare PROFUNDAL ZONE.

ln See NATURAL LOGARITHM.

load The amount of chemical material or thermal effluent released into a receiving stream by human or natural sources. See TOTAL MAXIMUM DAILY LOAD.

load allocation (LA) The portion of the pollution LOAD of a stream attributable to human NONPOINT SOURCES. The amount of pollution from each point source is the WASTELOAD ALLOCATION. See TOTAL MAXIMUM DAILY LOAD.

loading 1. The air concentration of a pollutant in a gas duct before entry into an air pollution control device. 2. Synonym for the pollution LOAD of a stream.

loading, acute toxicity test In the laboratory, testing of the toxicity of pollutants to aquatic biota, the ratio of the biomass of the test species, in grams of wet weight, to the volume (liters) of solution used.

loading capacity See TOTAL MAXIMUM DAILY LOAD.

local education agency (LEA) Under the provisions of the ASBESTOS HAZARD EMERGENCY RESPONSE ACT, the public or private school or school management unit that is required to inspect the school(s) for ASBESTOS-CONTAINING MATERIAL and to prepare asbestos management plans. See ASBESTOS.

local emergency planning committee (LEPC) A committee formed under the STATE EMERGENCY RESPONSE COMMISSION as required by TITLE III of the SUPERFUND AMENDMENTS AND REAUTHORIZATION ACT. The duties of the committee include the preparation of an emergency response plan for the local emergency-planning district, typically a county, which includes the emergency response to accidental releases of toxic chemicals that affect the public.

locally undesirable land use (LULU) Any project or activity that has actual or potential negative effects on residents in its vicinity, such as a landfill, hazardous waste incinerator, smelting operation, airport, or freeway. See EXTERNALITY.

loam A type of rich soil consisting of a mixture of clay, silt, sand, and organic material.

lodging The bending over of a cereal crop that has grown too tall, damaging the grain, reducing yields, and/or making the crop difficult to harvest. This problem is exhibited by native plant varieties given large fertilizer applications. The new HIGH-YIELDING VARIETIES possess dwarf genes that prevent lodging.

lofting Describing a PLUME from a smokestack that is being emitted above a TEMPERATURE INVERSION; the inversion prevents dispersion of the stack gases downward. Compare CONING, FANNING, FUMIGATION, LOOPING, TRAPPING.

log-normal distribution In air pollution data analysis, a distribution of airborne particles that is a normal, bell-shaped curve if the logarithm of the midpoint of various particle size ranges is plotted against the percentage of the total mass in that size range (mass function). A log-normal distribution also exists if the graph of the particle sizes is a straight line when plotted against the mass percentage of the total particulate that is less than (or greater than) that size, on log-probability paper.

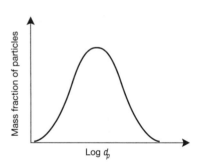

A log-normal particle distribution

log phase That period during which growth (cell division) of bacteria is occurring at a maximal rate. When bacteria are placed in an environment suitable for their growth, a pattern of increase in cell numbers that is typical of bacteria is observed. The various phases of growth are LAG PHASE (a pause in cell division), log phase, STATIONARY GROWTH PHASE (cells stop dividing), and DEATH PHASE (cells are dying). Same as exponential phase.

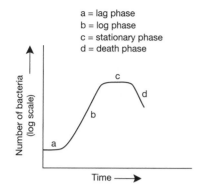

a = lag phase
b = log phase
c = stationary phase
d = death phase

Bacterial phases (log scale)

logarithm (log) The value of the exponent that a fixed number (the base) must have to equal a given number. The logarithm is calculated as $b^x = y$, where b is the base and x is the logarithm. The base for the common logarithm is 10. For example, the logarithm of 100 is 2 since $10^2 = 100$. Also written as $\log_{10} 100 = 2$. The NATURAL LOGARITHM base is approximately equal to 2.718282.

logistic curve A graphic representation resembling a flattened S (sigmoid curve), obtained when certain phenomena (e.g., population size for a species) are plotted against time. The growth rate is high at the beginning, reaches a maximum during the early stages, starts to slow near the center, and declines to zero at the end.

London Dumping Convention The Convention on the Prevention of Marine Pollution by Dumping of Wastes and Other Matter, an international agreement designed to control pollution of the sea by restricting dumping from oceangoing vessels. Primary attention is given to hydrocarbons, wastes generated during the operations of a ship at sea, and radioactive materials.

long-range transport The movement of air contaminants as far as hundreds of miles downwind before their removal by physical or chemical processes, especially precipitation; such movement is considered to be a significant contributor to ACID DEPOSITION.

long-term sampling A sample taken over an extended period to determine the average concentration of some constituent over the period represented by the sampling process. See COMPOSITE SAMPLE.

longitudinal study See COHORT STUDY.

longwave radiation Electromagnetic energy with WAVELENGTH equal to or greater than that of INFRARED RADIATION; often refers only to infrared radiation emitted from the surface of the Earth.

looping Describing the behavior of a PLUME from a smokestack being emitted into unstable, turbulent atmospheric conditions; causes short-term high concentrations downwind when the looping plume hits the ground. Compare CONING, FANNING, FUMIGATION, LOFTING, TRAPPING.

loss-of-coolant accident (LOCA) Potentially one of the more serious accidents involving a nuclear reactor. Water used to transfer heat from, and therefore cool, the reactor core is lost from the vessel housing the reactor, resulting in melting of the core contents. See MELTDOWN, CHINA SYNDROME.

lost-time accident Any injury occurring on the job that requires the employee to miss the next scheduled work shift.

lotic water The flowing water of rivers and streams. Compare LENTIC WATER.

loudness The perceived intensity of a sound, which depends on the SOUND PRESSURE LEVEL (DECIBEL level) and the FREQUENCY of the sound. Both factors are required in the determination of loudness because the human ear is more sensitive to higher frequencies.

Love Canal The site of an extraordinary contamination case involving hazardous waste. The episode represents a major hinge in history in that the outcome has profoundly altered the way all waste is handled in the United States as well as the relationship between private citizens and government at all levels. Love Canal Chemical Waste Landfill site was located on the eastern side of Niagara, New York, close to the Niagara River. The canal was constructed in the 1890s by William T. Love but never finished. The excavation was later used by the city of Niagara for placement of municipal solid waste and by Hooker Chemical for disposal of industrial waste from 1942 until 1953, at which time the property was transferred to the Niagara Falls Board of Education, which built a school on the property. The resurfacing of the waste, especially in the mid-1970s, led to the appearance of adverse health effects among the residents, a long-running emotional controversy involving conflict between local residents and the state of New York, and ultimately the closing of the school and the buyout of local residents. The nature of the conflict contributed to the passage of amendments to the RESOURCE CONSERVATION AND RECOVERY ACT, the COMPREHENSIVE ENVIRONMENTAL RESPONSE, COMPENSATION, AND LIABILITY ACT, and the creation of the HAZARDOUS SUBSTANCES SUPERFUND. See GIBBS, LOIS MARIE.

low-density polyethylene (LDPE) A plastic resin used in film plastics such as grocery sacks and bread bags.

low-emission vehicle (LEV) An automobile or light truck that meets emission restrictions more stringent than normal emission standards. The low-emissions standard requires that a vehicle demonstrate a release of hydrocarbons that is 70% lower than that in standard allowable emissions and a release of nitrogen oxides that is 50% lower than that of standard allowable emissions. National LEV standards have been drafted with the cooperation of the states belonging to the OZONE TRANSPORT COMMISSION, the USEPA, environmental protection groups,

oil companies, and automobile manufacturers.

low-emissivity (low-E) window Window design and installation intended to increase energy efficiency in homes and commercial buildings by lowering the amount of energy loss or gain through windows. The windows inhibit the transmission of radiant heat while allowing passage of sufficient light.

low fence A hearing damage criterion that equals a 25-decibel threshold for the sound frequencies 500, 1000, and 2000 hertz.

low-flow augmentation The release of reservoir water into a stream during periods of low flow to maintain acceptable DISSOLVED OXYGEN levels. This practice is not permitted as a water quality maintenance technique in the United States.

low-head hydropower Small-scale systems designed to capture energy from water retained behind dams that retain a relatively small amount of water. There is little difference in the water level behind the dam and the water level in front of the dam. This type of technology causes much less ecological damage to the environment than large dam projects.

low-level radioactive waste (LLRW) Radioactive waste material with a radiation intensity of less than 10 nanocuries per gram. Such materials are commonly produced by hospitals and research laboratories (examples include glassware, tubing, paper products, and gloves) and are not as dangerous as waste generated in nuclear reactors. See HIGH-LEVEL WASTE.

low-level waste (LLW) See LOW-LEVEL RADIOACTIVE WASTE.

low-quality energy Widely dispersed forms of energy, characterized by high disorganization and lack of usability, as in the heat lost from a mechanical operation.

lower detectable limit See DETECTION LIMIT.

lower explosive limit (LEL) The air concentration of a gas or vapor below which the chemical will not burn or explode in the presence of a flame or other ignition source. Compare UPPER EXPLOSIVE LIMIT.

lower flammable limit (LFL) Same as LOWER EXPLOSIVE LIMIT.

lowest achievable emission rate (LAER) The emission rate required for an air pollutant emitted by a new or modified source in an area that does not meet the national ambient standard for that pollutant (NONATTAINMENT AREA). The CLEAN AIR ACT requires that the emission rate be set as the most stringent emission limitation in use for that pollutant and type of facility.

lowest-observed-adverse-effect level (LOAEL) The lowest dose of a chemical or physical agent that causes a significant increase in the frequency or severity of an adverse effect in an exposed population compared with an unexposed (control) population.

lowest-observed-effect level (LOEL) The lowest dose of a chemical or physical agent that causes a significant biological change in an exposed population compared with an unexposed (control) population. The observed change is not necessarily deleterious. Compare LOWEST-OBSERVED-ADVERSE-EFFECT LEVEL.

Luddite Term applied to an individual who resists technological progress in favor of what he or she considers protection of the individual and of local environments. One who challenges the legitimacy of science and technology and who opposes technological determinism. The term stems from the battle against power looms in England during the early stages of the Industrial Revolution in the late 1790s and early 1800s. The hero of the movement was Ned Ludd, variously described as a feebleminded man who smashed weaving machinery and an imaginary hero who led the battle to protect skills and livelihoods of craftsmen displaced by industrial development. In the modern context, the forebears of radical environmental and environmental justice groups that question the exploitation of labor and the environment in the pursuit of profits. Often used as a label for those who are wary of computers and other technological advances or generally opposed to modern industrial society.

Lurgi process A method for COAL GASIFICATION developed in the 1930s that mixes pulverized coal, oxygen, and steam under high pressure, producing methane gas. The net effect is the conversion of a solid fossil fuel to a gaseous form, making distribution and use easier for the consumer.

lymph nodes Small glands, located throughout the body (as in the neck, armpits, and groin), containing large numbers of lymphocytes, which are white blood cells used to defend the body against bacteria and viruses. Tumor cells that have migrated from an original site of growth are frequently observed in these structures.

lymphoma Cancers of the lymphatic tissue; such tumors are characterized by the overproduction of cells that compose LYMPH NODES.

lysimeter A device for measuring loss of water vapor from soil covered with vegetation. See EVAPOTRANSPIRATION.

M

macroconsumer A class of heterotrophic organism that ingests other organisms or PARTICULATE ORGANIC MATTER; includes HERBIVORES, CARNIVORES, and OMNIVORES. Compare MICROCONSUMER.

macroion See POLYELECTROLYTE.

macronutrient Chemical compound or element that an organism must obtain from the environment in relatively large quantities to sustain life functions. Plant macronutrients include nitrogen, potassium, phosphorus, calcium, sulfur, and magnesium. Compare MICRONUTRIENT.

macrophage See ALVEOLAR MACROPHAGE.

macropore Large, usually natural channel in the Earth that allows the rapid movement of material from the surface to the subsurface without the need to migrate through the soil structure. Examples include channels through geological strata left from the rotting of roots from long-dead trees or cracks in the Earth from drying or from seismic activity. These channels are of significance in the contamination of groundwater.

macroscopic Capable of being seen with the naked eye. Items that can be seen only with the aid of a microscope are termed MICROSCOPIC.

Mad Hatter's disease A nervous system disorder resulting from chronic exposure to inorganic mercury compounds. Affected individuals may experience exaggerated emotional responses, tingling sensations of the skin, mental disturbances, or hallucinations. The term derives from the occupational exposure syndrome caused by overexposure to the mercuric nitrate used to soften fur in the making of felt hats.

magma Molten rock found in the mantle, beneath the crust of the Earth. When forced toward the surface, magma cools and solidifies to become IGNEOUS ROCK.

magnet A naturally occurring or treated object almost always containing iron that exhibits the elemental natural force of magnetism. Magnets exert a MAGNETIC FIELD around them and can affect objects within the field. The ends of magnets are called the north and south poles, because if a bar magnet is allowed to turn freely, the magnet will align itself with the magnetic field of the Earth (from the iron core). The like poles of magnets repel each other; a north pole and south pole attract each other. The passage of an electric current through a conducting coil of wire surrounding an iron or steel core causes the core to become a magnet while the current is flowing; this is the phenomenon used to convert electrical energy to mechanical energy in electric motors. If a magnet is put in motion near a conducting coil of wire, then a current (ELECTRICITY) is induced in the coil; that is the operating principle of an electric generator.

magnetic field The area around a magnetic pole or moving charge under the influence of the forces exerted by the magnet.

magnetic separation Use of magnetic devices to segregate ferrous metal from other waste material so that the metal can be used in a recycling operation.

magnetohydrodynamic generation A theoretical concept for the recovery of a greater amount of electricity from power plants that burn coal to generate electric energy. Potassium would be added to the hot gases produced by the burning of coal. The extremely high heat (about 2000° C) would result in the ionization of the potassium, and the ions would then be forced to move through a magnetic field, generating an electric current. If successfully developed, the design could double the efficiency of currently used coal-burning power plants.

main A relatively large pipe in a distribution system for drinking water or in a collection system for municipal wastewater.

major source Under the CLEAN AIR ACT, a source with the POTENTIAL TO EMIT more than 100 tons per year of a named air pollutant or, in the case of certain HAZARDOUS AIR POLLUTANTS (HAPS), 10 tons per year for one, 25 tons per year for any combination of HAPs. Also, sources of CRITERIA POLLUTANTS in NONATTAINMENT AREAS for the emitted pollutant with the potential to emit as little as 10 tons per year.

major source facility See MAJOR SOURCE.

major stationary source See MAJOR SOURCE.

majors A regulatory designation applied by the USEPA to PUBLICLY OWNED TREATMENT WORKS (municipal sewage treatment plants) serving at least 10,000 people, receiving industrial wastewater and having approved PRETREATMENT programs, or processing at least 1 million gallons of wastewater per day. (The average amount of wastewater produced by each person is 100 gallons per day.) Compare MINORS.

makeup water Water added to the flow of water used to cool condensers in electric power plants. This new water replaces condenser water lost during passage of the cooling water through cooling towers or discharged in BLOWDOWN.

malathion An insecticide of the ORGANOPHOSPHATE class. The agent is known as a soft pesticide because of the relatively quick degradation of malathion in the environment. Likewise, malathion has been shown to be less toxic to mammals than some of the more dangerous pesticides such as DDT and dieldrin.

malignant Describing a tumor that produces cells that can migrate to new sites in the body where additional tumors can subsequently develop.

malignant neoplasm A fast-growing tumor that invades other tissues and undergoes METASTASIS.

Malthusian An environmental philosophy taking as a central dogma the analysis developed by the English economist and clergyman Thomas Malthus in the late 18th century. Malthus proposed that the human population would soon exceed available agricultural resources, with a resulting catastrophic famine and loss of human life as population levels adjusted to the available food supply. The basic philosophy has been adopted by those who believe that scarcity in natural resources will ultimately lead to catastrophic failures in the world economy. Generally a core belief grounded in pessimism concerning environmental issues. Compare CORNUCOPIAN, CORNUCOPIAN FALLACY.

mammalian selectivity ratio (MSR) A parameter employed to judge the relative toxicity of insecticides by comparing the amount needed to kill rats with the amount needed to kill houseflies. The proportion is calculated as the ratio of the oral dose (milligrams per kilogram body weight) required to kill 50% of experimental rats to the topical dose (in the same units) needed to kill 50% of exposed houseflies. A ratio less than 1 indicates that the agent kills rats more effectively than it kills houseflies, and vice versa.

Man and Biosphere program (MAB) A United Nations Educational, Scientific, and Cultural Organization project dedicated to the SUSTAINABLE DEVELOPMENT and protection of terrestrial natural resources. The MAB works with the UNITED NATIONS ENVIRONMENT PROGRAM, INTERNATIONAL UNION FOR CONSERVATION OF NATURE AND NATURAL RESOURCES, and WORLD METEOROLOGICAL ASSOCIATION, among many other organizations. See AGENDA 21, MAN AND BIOSPHERE PROGRAM RESERVE. Website: www.unesco.org/mab/

Man and Biosphere program (MAB) reserve A design initiated and encouraged by the United Nations for future economic development of nature preserves. The concept involves the division of protected areas into multizone use; a highly protected core area is surrounded by a buffer zone in which modest ECOTOURISM is allowed, all of which are surrounded by a peripheral (transition) zone where human habitations and sustainable resource removal are permitted. The development of a forest under this philosophy would consist of a center section, where the presence of humans is not apparent and natural species are allowed space and a surrounding zone, into which some human intervention would be permitted for specified purposes that do not involve the removal of trees; all surrounded by a zone where sustainable forestry practices are allowed along with the development of compatible modest housing. Over 350 biosphere reserves have been developed in over 90 countries. Website: www.unesco.org/mab/wnbr.htm

mandatory recycling Requiring by law or regulation the separation of recyclable items from the waste stream. The objective of such programs is to reduce the amount of discards going into landfills.

mangrove swamp A tidal swamp forest populated by plant species capable of growth and reproduction in areas that experience periodic tidal submergence in ocean water with a resulting increase in saline conditions. These forests develop along coastal regions in tropical climates. More than 50 species of plants may be present; however, mangrove swamps are dominated by trees referred to as red mangrove, *Rhizophora mangle;* black mangrove, *Avicennia germinans;* and white mangrove, *Laguncularia racemosa.* Typically, these trees have large, exposed root systems.

manifest system The regulations applicable to transporters of HAZARDOUS WASTE under the RESOURCE CONSERVATION AND RECOVERY ACT; found in Title 40, Part 263, of the CODE OF FEDERAL REGULATIONS. Transporters are required to use a set of shipment forms to track wastes in transit from their origin to their final site of disposal.

man-made mineral fibers (MMMFs) Fibrous materials prepared from substances such as glass wool, rock wool, slag wool, and ceramic fibers. These materials have been used as replacements for ASBESTOS in insulation and building materials. Although questions remain concerning the relationship of these fibers to human health, they have coarser diameters and do not become airborne as readily as asbestos fibers, which have been implicated in lung diseases.

Manning formula An equation describing the volumetric flow rate (Q) of water in an open channel as follows:

$$Q = \frac{1.0 \text{ or } 1.49}{N} AR^{2/3} S^{1/2}$$

where A is the cross-sectional area of the water, S is slope of the channel sides, N is a channel roughness coefficient, and R is the HYDRAULIC RADIUS. Use $1.0/N$ for SI units, $1.49/N$ for American units. Also known as the Chezy-Manning formula.

manometer An instrument for measuring pressure, consisting of a transparent tube filled with a liquid such as water or mercury. The level of the liquid in the tube indicates the pressure.

mantle The division of the interior of the Earth between the core and the crust. The mantle is composed mainly of silicate rock and is around 2900 kilometers (1800 miles) thick.

manual separation Hand sorting of recyclable material (steel cans, aluminum beverage containers, cardboard, newsprint, to name a few) from other discards or the separation of commingled recyclable materials (plastic containers into various resin types or glass containers into different colors) collected from households. This separation is done to improve the quality of recyclable items before sale to a buyer. Some facilities have mechanical devices to accomplish a similar sorting. See MAGNETIC SEPARATION, MATERIALS RECOVERY FACILITY.

maquiladoras Industrial facilities located near the U.S.-Mexican border that are allowed to ship goods into the United States without import duties. Inadequate waste disposal practices and toxic emissions to air and water from the *maquiladoras* are ongoing issues between the United States and Mexico. See NORTH AMERICAN FREE TRADE AGREEMENT (NAFTA), NORTH AMERICAN AGREEMENT FOR ENVIRONMENTAL COOPERATION (NAAEC).

marasmus The deficiency condition caused by inadequate intake of calories and protein. In the absence of famine, the condition is usually seen in poor areas in infants less than one year old, after weaning and/or gastrointestinal infection that is accompanied by severe or prolonged diarrhea. Compare KWASHIORKOR.

margin of exposure (MOE) The ratio of the NO-OBSERVED-EFFECT LEVEL to the estimated dose received. For example, if the no-observed-effect level is 20 milligrams per kilogram body weight per day and the dose received is 5 milligrams per kilogram body weight per day, then the MOE is 4. The higher the ratio, the safer the exposure.

margin of safety (MOS) In setting an environmental standard, a factor applied to the maximal dose of a chemical substance showing no adverse effect in toxicity tests; the factor accounts for uncertainty in the toxicology data. The factor lowers the allowable amount of exposure or dose, which lowers the standard in terms of the acceptable concentration in an environmental medium (air, water, food). Also used to describe the difference between the NO-OBSERVED-EFFECT LEVEL and the estimated dose received by a population.

mariculture The cultivation of marine organisms for use as a food resource.

marine Associated with the sea.

Marine Protection, Research, and Sanctuaries Act (MPRSA) A 1972 law that includes provisions requiring citizens of the United States to obtain a permit from the USEPA before disposing of materials in the oceans. Subsequent amendments to the act have limited the types of waste that may be permitted for ocean disposal. For example, sewage sludge and industrial waste dumping was ended in 1992. Commonly called the OCEAN DUMPING ACT. See OCEAN DUMPING PERMIT.

marine sanitation device Equipment installed on vessels or drilling platforms to prevent the discharge of raw, untreated SEWAGE into the oceanic environment. The devices collect, store, treat, or discharge wastewater.

Marpol convention See INTERNATIONAL CONVENTION FOR THE PREVENTION OF POLLUTION FROM SHIPS, MARPOL V.

Marpol V The 1988 annex to the INTERNATIONAL CONVENTION FOR THE PREVENTION OF POLLUTION FROM SHIPS, which governs the discharge of garbage into the oceans. Ships must be a required distance from land and are under severe restrictions in designated areas. It also includes a complete ban on the ocean dumping of plastic.

marsh A coastal region where the soil has a high moisture content because of periodic flooding caused by the tides. The vegetation is normally dominated by grasses.

Marsh, George Perkins (1801–82) American Statesman and naturalist. Author of *Man and Nature* (1864), in which he warns against the human misuse of natural resources.

marsh gas Gas produced during the decomposition of organic material buried in wetland soils. The primary gas is methane.

mass The quantity of matter in a substance. The SI base unit of mass is the kilogram; mass is expressed in milligrams, grams, kilograms, and similar units.

mass balance An approach used to estimate pollutant releases to the environment, based on the conservation of matter—that matter cannot be created nor destroyed. If the input mass to a process and the output, or product, mass are known and if the chemical reactions within the process are taken into account, the difference between the known mass in and known mass out is considered to be the amount lost to the environment.

mass burn The INCINERATION of MUNICIPAL SOLID WASTE without prior material recovery or processing but often accompanied by an energy recovery system, such as steam and electricity production. See WATERWALL INCINERATOR, WASTE-TO-ENERGY.

mass burn waterwall incinerator See MASS BURN, WATERWALL INCINERATOR, WASTE-TO-ENERGY.

mass density Mass per unit volume; commonly expressed as kilograms per cubic meter or pounds per cubic foot.

mass flow rate For a liquid or gas, $m = Qp$, where m is the mass flow rate, Q is the VOLUMETRIC FLOW RATE, and p is the fluid DENSITY. The unit is mass per unit time.

mass flux See MASS VELOCITY.

mass number The number of protons plus the number of neutrons in the nucleus of an atom. This number is also the approximate atomic weight of an atom. See ATOMIC WEIGHT.

mass spectrometer (MS) A sensitive device used for the analysis of organic materials in environmental samples. Extracts prepared from water, sediment, biota, or other materials are injected into the instrument, and organic compounds extracted from the sampled material are fragmented by a stream of electrons. The positive ions that are produced as a result are separated on the basis of their mass. A detector within the system responds to each ion and produces a corresponding peak in a computer graphics file or on chart paper. The height of each peak indicates the relative abundance of each element in the original organic compound of interest. The identity of organic materials recovered from environmental samples can be determined by a comparison of the pattern of peaks (mass spectrum) produced by known compounds with the spectrum obtained from the unknown chemicals extracted from samples collected from within the environment. The comparison is done by computer.

mass spectrum See MASS SPECTROMETER.

mass velocity The MASS FLOW RATE in a duct or other enclosure per unit area. Expressed as
$$M_v = m/A$$
where M_v is the mass velocity, m is the mass flow rate, and A is the cross-sectional

area of the enclosure; used in gas absorber design. Typical units are kilograms per second per square meter or pounds per hour per square foot. Also called mass flux.

matching The selection of persons for an epidemiological study of a disease so that the study groups are similar with respect to characteristics that are not being tested but that are related to the disease in question. The selection process is an attempt to allow the study results to be influenced only by the variable being studied. All such potential CONFOUNDING VARIABLES cannot be eliminated, however. For example, in a study of the effects of exercise on the risk of heart disease, in which a frequently exercising group is compared with a nonexercising group, one important variable related to heart disease that must be equally distributed in the study groups is age. When this is accomplished, the groups are said to be matched on age, and any difference in heart disease noted in the groups cannot be attributed to age. Although the exercising group may show a lower risk of heart disease, frequency of exercise may not be the CONTROLLING VARIABLE if, for example, the groups are not matched on smoking history.

material balance See MASS BALANCE.

material exchange A step in the management of waste in which a facility takes temporary possession of waste prior to the transfer to a disposal location.

material flow chart Part of the waste management/waste reduction plan for an industrial facility. A complete survey of a facility in terms of all raw materials, process chemicals, desired end products, by-products, and wastes of all types is conducted, and the losses and emissions quantified. The data allow the operation to be evaluated in terms of when and where wastes are generated (and possible reductions made).

material safety data sheet (MSDS) A written summary of information about chemical substances required by the United States Occupational Safety and Health Administration to be distributed to all employees who may be exposed to the chemicals and to purchasers of the chemical materials. The required information for each chemical substance includes the address and telephone number for the manufacturer; common and scientific names for the chemical; important physical and chemical characteristics; fire, explosion, and reactivity data; adverse health effects that may be expected after overexposure; methods for safe handling, use, and disposal; and recommended methods for workplace control of exposure. One of many MSDS listings on the Internet is http://siri.uvm.edu/msds/.

materials recovery The removal of economically reusable items or substances from waste material. Used frequently to describe the recovery of items usable for recycling from MUNICIPAL SOLID WASTE (e.g., glass and plastic containers, aluminum beverage cans, and newsprint from household garbage).

materials recovery facility (MRF) Pronounced "murf." A building designed to facilitate the separation, sorting, and preparing of recyclable items derived from households so that the items are acceptable to companies that convert the resources back into commercial items. A clean MRF normally handles recyclable items that have been separated from the normal household garbage by individual citizens. At a dirty MRF, the sorting of all household garbage for removal of items that can be recycled is done at a central facility.

matrix The material in which an environmental sample is embedded or contained; the material can be soil, water, dried biomass, or other substances.

matrix interference The adverse influence of the environmental sample MATRIX on the ability to detect the presence or amount of a chemical substance in the sample.

maximum acceptable toxicant concentration (MATC) The highest concentration at which a pollutant can be present and not exert an adverse effect on the biota; used experimentally to determine the toxicity of the chemical.

maximum allowable concentration (MAC) The predecessor workplace air standard to the THRESHOLD LIMIT VALUES.

maximum achievable control technology (MACT) The level of air pollution control technology required by the 1990 amendments to the CLEAN AIR ACT for sources of certain AIR TOXICS. The USEPA bases MACT on the best-demonstrated control technology or practices of the industry.

maximum contaminant level (MCL) The maximal permissible concentration of a drinking water contaminant set by the USEPA under the authority of the SAFE DRINKING WATER ACT. MCLs are set for inorganic and organic chemicals, turbidity, coliform bacteria, and certain radioactive materials.

maximum contaminant level goal (MCLG) A nonenforceable drinking water concentration defined by the USEPA for several listed contaminants that cause no known or anticipated adverse human health effects, including a margin of safety. MCLGs are set for certain organic contaminants and for fluoride.

maximum exposed individual (MEI) See MAXIMUM INDIVIDUAL RISK, MOST EXPOSED INDIVIDUAL.

maximum holding time The longest period that water samples can be retained between the taking of the sample and the laboratory analysis for a specific material before the results are considered invalid. The times vary from none in the case of the test for residual chlorine levels to six months for the testing of radioactivity. Some types of analyses require that preservatives be added to the sample, and some require storage of samples at refrigerated temperatures. See HOLDING TIME.

maximum individual risk (MIR) The excess lifetime health risk (the risk above that of a nonexposed individual) for a certain disease (e.g., cancer) in an individual estimated to receive the highest exposure to a particular environmental pollutant. See MOST-EXPOSED INDIVIDUAL.

maximum permissible concentration (MPC) The amount of toxic chemical or radioactive material in air, water, or food that would be expected to result in a MAXIMUM PERMISSIBLE DOSE at the normal rate of consumption by an individual.

maximum permissible dose (MPD) The amount of exposure to toxic chemicals or radioactivity below which no adverse health effects are anticipated.

maximum sustainable yield The greatest amount of a renewable natural resource (e.g., forests or wildlife) that can be removed without diminishing the continuing production and supply of the resource.

maximum tolerated dose (MTD) Used in tests for carcinogenicity, the highest lifetime dose a test animal can receive without experiencing significant adverse effects other than cancer. Such large doses can overwhelm the normal metabolic detoxification and excretion responses of animals and not be representative of the risk posed by lower doses. The cell division in response to the chronic internal wounding occurring at the MTD could itself be the cause of tumors. The MTD, however, is used to overcome the problem of finding positive results in studies of cancers with perhaps a 1 in 100,000 risk in humans, in which only around 100 test animals are used for each exposure (dose) level.

maximum total trihalomethane potential (MTTP) The highest combined concentration of chloroform, bromodichloromethane, dibromochloromethane, and bromoform produced in water containing chlorine, chlorine dioxide, or chloroamines when held for seven days at 25° C or above. The measure is employed to assess the impact of chlorine addition on raw water supplies. See TRIHALO-METHANE.

MCF   One thousand cubic feet.

McPhee, John (1931–) American Writer. Many of McPhee's works have environmental themes, among them *The Pine Barrens* (1968), a portrait of natural and human community in New Jersey; *Encounters with the Archdruid* (1971), a profile of the environmental activist DAVID BROWER; *Coming into the Country* (1977), on Alaska wilderness; *Rising from the Plains* (1986), on Wyoming wilderness; and *The Control of Nature* (1989), on the vanity of human efforts to control large-scale natural phenomena.

mean free path   The average distance that a molecule in a fluid (air or water) moves before colliding with another molecule.

mean relative growth rate (RGR)   A measure used under test conditions to document the effects of pollutants on the growth rate of aquatic algae. Control and treatment conditions are compared during the active growth of the algae by the formula $RGR = (\log n_2 - \log n_1)/(t_2 - t_1)$, where n_1 is the number of algal cells at time 1 (t_1) and n_2 is the number of algal cells at time 2 (t_2).

measure of effect   See END POINT.

measure of exposure   See EXPOSURE, EXPOSURE ASSESSMENT.

measurement end point   Numerical data that can be compared with similar information collected at reference or control sites to judge the effectiveness of efforts to clean up a contaminated location. These numerical data are related to some desired ecological outcome that is to result from the cleanup operation. For example, in the cleanup of a location contaminated with POLYCHLORINATED BIPHENYLS (PCBs), the desirable ecological outcome might be the protection of the reproductive potential of the hawk. Since the measurement of PCB concentration in wild hawks would be difficult and potentially dangerous to the animal one is trying to protect, the concentration of PCBs in field mice, a common food source in this example, would be used as the measurement end point. The information collected from field mice would be used in association with data on prey ingestion rates to determine whether the PCB contamination is likely to cause reproductive problems in the hawk. The cleanup would continue until the concentration of PCB in the mice reached a level considered to be safe for the hawk, the measurement end point. See ASSESSMENT END POINT.

mechanical aeration   The use of agitation devices or the injection of compressed air into treatment basins to increase the DISSOLVED OXYGEN level in wastewater. The increase in the amount of dissolved oxygen accelerates the decomposition of organic materials and prevents the degradation of receiving streams when wastewater is released into the environment. See AERATION, AEROBIC DECOMPOSITION.

mechanical integrity   The demonstrated absence of significant leaks in the outer wall of the borehole (well casing), tubing, or packer of an underground INJECTION WELL. Also, the absence of the downward vertical migration of the injected fluids through underground pathways adjacent to the borehole of the well.

mechanical separation   The use of air classification, MAGNETIC SEPARATION, or other mechanized methods to separate solid waste into recyclable categories. See AIR CLASSIFIER.

mechanical turbulence   Randomly fluctuating air motion caused by air moving over a rough surface or past objects. The eddies produced by this erratic motion help dilute polluted air with unpolluted air by a process called EDDY DIFFUSION. Usually less significant than the effects of THERMAL TURBULENCE.

media Air, surface water, groundwater, food, and soil that are regulated by federal or state environmental and health agencies. For example, air represents a medium into which pollutants are released, with the potential for exposure of individuals to the pollutant through skin contact or breathing. See MEDIUM for applications of the term to microbiology.

median lethal concentration (LC_{50}) See LETHAL CONCENTRATION—50 PERCENT.

median lethal dose (MLD) See LETHAL DOSE—50 PERCENT.

Medical Literature Analysis and Retrieval System (MEDLINE) A computer database containing toxicological and medical information from over 3000 research journals. The system is operated by the National Library of Medicine, Bethesda, Maryland. Website: www.ncbi.nlm.nih.gov/PubMed/ or http://igm.nlm.nih.gov/

medical surveillance 1. Attention to the health status of workers through periodic examination and screenings for indicators of adverse health effects associated with employment or exposure to potentially hazardous conditions. 2. Periodic examination and screenings of residents in the vicinity of a hazardous waste site for adverse health effects that might be linked to chemical releases, often the result of a court order.

medical waste Discarded paraphernalia associated with the practice of medicine, such as bandages, needles, syringes, test tubes, and specimen containers. The CLEAN AIR ACT amendments of 1990 require the USEPA to set NEW SOURCE PERFORMANCE STANDARDS for medical waste incinerators, the OCCUPATIONAL HEALTH AND SAFETY ADMINISTRATION has

issued blood-borne pathogen regulations (29 CFR 1910.1030), and the U.S. Department of Transportation has shipping regulations (49 CFR 173.196). State laws further control the disposal of medical wastes.

Medical Waste Tracking Act (MWTA) A 1988 law attached to the RESOURCE CONSERVATION AND RECOVERY ACT that started a two-year medical waste management demonstration program; the act requires sources generating certain wastes at a rate greater than 50 pounds per month, excluding incinerator ash from on-site burning, to keep detailed records of the movement of those wastes from their facilities to the final disposal site. Information from the MWTA contributed to the current federal regulation of medical waste disposal. See MEDICAL WASTE.

Mediterranean climate area An area with a climate similar to that of some regions close to the Mediterranean Sea: namely, wet and cool winters alternating with hot, dry summers. Plants that are adapted to this type of climate (such food crops as wheat, oats, apples, grapes, and melons) have seeds or root structures that ensure survival over the hot, dry periods.

medium In microbiology, a sterile mixture, either solid or liquid, of nutrients used for the cultivation of bacteria or fungi. Solid media (plural form) are prepared by using agar as the agent to solidify the liquid solution containing required nutrients.

mega- A prefix meaning 1 million. For example, 1 megawatt equals 1 million watts.

megacity A city with a population of 10 million or more.

megagram (Mg) One million grams, 10^6 grams.

megahertz (MHz) One million HERTZ; a unit of FREQUENCY.

megalopolis An area in which separate cities have expanded to the point that they have fused to create a continuous urban environment with about 10 million or more people.

megawatt A unit of electrical power equal to 1 million WATTS. Nuclear power plants are commonly rated with a capacity of 1000 megawatts.

megaton An expression of explosive energy, applied to nuclear weapons, equal to 1 million tons of trinitrotoluene (TNT).

meltdown A theoretical worst-case accident at a nuclear reactor. A meltdown is caused by a loss of coolant with a resulting melting of the reactor core and fuel elements. If the molten core of a reactor escaped from the containment structure housing the reactor, the potential for catastrophic consequences would be great. Large amounts of highly radioactive materials would be released into the air, water, and soil. The event that most closely followed a theoretical worst-case meltdown was the explosion, with subsequent melting of the damaged reactor core, at CHERNOBYL. See LOSS-OF-COOLANT ACCIDENT, CHINA SYNDROME.

melting point The temperature at which a solid changes to a liquid. The temperature varies and is consistent at equal temperatures and pressures for each element or solid.

membrane filter A flat paperlike disk typically made of modified cellulose. The filters have a small pore size that allows for the retention of bacteria. These units are employed to remove bacteria from liquid samples for examination for public health purposes as well as for sterilization of liquids. These membranes are available in a variety of sizes, with a diameter of 47–50 millimeters the most common.

membrane filter method A procedure used to recover and count bacteria in samples of liquid substances, such as water. The liquid is drawn through a MEMBRANE FILTER by using a slight vacuum, and the bacteria in the liquid is retained on the filter. The filter disk is then transferred to a MEDIUM suitable for the growth and incubation of the bacteria.

mercaptans A group of compounds that are sulfur analogs of alcohols and phenols. For example, the chemical formula of methyl alcohol is CH_3OH, and that of methyl mercaptan is CH_3SH. The compounds are noted for their odor and are used to impart odor to natural gas in public gas supplies. Air emissions from chemical manufacturers and pulp and paper mills may contain mercaptans.

mercury One of a group of elements classified as heavy metals. Mercury has an atomic number of 80 and an atomic weight of approximately 200, and the element is the only metal that is a liquid at room temperature. Unlike most other metals, mercury tends to form COVALENT bonds with other inorganic materials as well as organic compounds. Elemental (metallic), inorganic (ionic), and organic forms of mercury are released into the environment through natural and anthropogenic sources. The latter include mining, smelting, and industrial discharges. The chloralkali and paper industries were major sources at one time. The burning of coal and the refining of petroleum continue to be important sources. Once released into the environment, the element can undergo many transformations and conversions among the three major forms given. The most common human exposures to elemental mercury are by the inhalation of vapor. The element is rapidly absorbed and distributed by the circulatory system. Absorption of mercury from the gastrointestinal tract is very limited for elemental mercury but very high for organic forms of the element. Mercury is excreted slowly from the body, with about one-half of an absorbed dose lost in 30 to 90 days. That part of the human body most sensitive to mercury poisoning is the central nervous system. Symptoms can range from a mild tingling of the skin to death. See also

METHYL MERCURY, MINAMATA SYNDROME.

meromictic lake A lake in which the waters are only partly mixed during the course of a year so that there is permanent STRATIFICATION of the water. Compare DIMICTIC LAKE.

mesh size A measure of particle size determined by the size of a sieve or screen (mesh) that particles can pass through.

mesophyll cells Chlorophyll-containing leaf tissue.

mesophyte A plant that requires only moderate amounts of water for growth. Compare HYDROPHYTE, XEROPHYTE.

mesoscale The scale of air motion smaller than the SYNOPTIC (the scale of a weather map) and larger than the microscale (smaller, turbulent airflow). Appropriate horizontal distances are several tens of kilometers; the time scale is about one day. Air STAGNATION is a mesoscale phenomenon.

mesosphere The division of the atmosphere above the STRATOSPHERE; this layer begins at about 50 kilometers in altitude and extends to about 80 kilometers.

mesothelioma Cancer of the membranes lining the abdominal and chest cavities.

mesotrophic Describing a body of water with a moderate nutrient content. Compare EUTROPHIC, OLIGOTROPHIC.

metabolism A general term describing all of the chemical reactions that occur in the body. Taken collectively, these are the life processes. Some of the reactions involve the utilization of nutrients as sources of energy or as substances used to build mass. Some of the reactions involve the breakdown and utilization of macromolecules in the body (e.g., fat). See ANABOLISM, CATABOLISM.

metabolite A chemical substance produced by the metabolic reactions of an organism. Because the body metabolizes most absorbed materials, tests for exposure to many materials are analyzed for their metabolites rather than the original chemical substances absorbed. For example, the test for exposure to benzene is the presence of phenol, a benzene metabolite, in the urine. See METABOLISM, BIOLOGICAL MONITORING, BIOLOGICAL EXPOSURE INDEX.

metal fume fever An acute condition caused by short-term high exposure to the FUMES of zinc, magnesium, or their oxides. The condition is characterized by fever, muscle pain, and chills. Symptoms begin several hours after the high dose is received.

metalimnion Same as THERMOCLINE.

metamorphic rock Rock formed by the exposure of SEDIMENTARY or IGNEOUS material to high temperatures, high pressures, and chemical processes deep beneath the surface of the Earth. For example, limestone, a sedimentary rock, is converted to marble, a metamorphic rock.

metastasis The migration of tumor cells to new sites in the body where additional cancerous growth can arise. See CARCINOGENESIS.

meteorology The study of the atmosphere and weather conditions in the atmosphere. Knowledge of this science is required for an understanding of the movement and activities of pollutants released into the atmosphere.

methane A gaseous hydrocarbon that is the main component of NATURAL GAS; the molecule contains a single carbon atom bonded to four hydrogen atoms. ANAEROBIC DECOMPOSITION produces methane in a LANDFILL. Also called landfill gas, swamp gas, BIOGAS. See GREENHOUSE GAS, FOSSIL FUEL, FIREDAMP.

methane hydrate Natural formations consisting of mounds of icelike material on or just below the sea floor containing large amounts of METHANE trapped within a lattice of icelike crystals. These stable structures are formed at temperatures and pressures characteristic of oceanic depths exceeding 300 to 500 meters. The formations, which vary greatly in size, are especially common in areas along the CONTINENTAL SHELF experiencing rapid accumulation of organic BIOMASS or sediments. The formations can occur at shallower depths at the polar regions. The methane appears to be of BIOGENIC origin, with the methane produced by microbes caged within an icelike lattice instead of dissipating in the water. The hydrates also form in pipelines transporting natural gas from offshore wells, plugging the pipes. Natural accumulations of methane hydrate, mostly in oceanic environments, appear to hold more hydrocarbons than the combined world inventory of all coal, oil, or conventional natural gas. One cubic centimeter of methane hydrate produces up to 164 cubic centimeters of natural gas upon thawing. May be an important source of natural gas fuel in the future.

methanol A one-carbon alcohol (CH_3OH), which has multiple uses. Methanol is used as an OXYGENATED FUEL additive to reduce certain air pollutants, as a drying agent in gasoline, as a solvent in cleaners, and as an antifreeze. Methanol is a poison when ingested. Also known as wood alcohol.

methemoglobin A hemoglobin molecule with the central iron atom in an oxidized (ferric) state. Methemoglobin cannot combine with and carry oxygen to the tissues, and thus, a high level of red blood cells in the methemoglobin state can lead to HYPOXIA, a deficiency of tissue oxygen. Environmental, dietary, or workplace exposure to NITRITE, NITRATE, aniline, and nitrobenzene, among other materials, produces methemoglobin in the blood. See METHEMOGLOBINEMIA.

methemoglobinemia The disease state characterized by inadequate tissue oxy-genation caused by excessive levels of blood METHEMOGLOBIN.

method detection limit (MDL) The minimal concentration of a chemical substance present in a sample that can be measured and reported with a 99% probability (confidence level) that the measured concentration is above zero.

methoxychlor A CHLORINATED HYDRO-CARBON insecticide belonging to the dichlorodiphenylethanes, the class of agents that includes dichlorodiphenyl-trichloroethane (DDT), but has a relatively short HALF-LIFE in the environment; its human toxicity is not well-known. In animals, high doses cause adverse effects on the nervous system, and studies indicate the potential to interfere with reproductive capacity. See DICHLORODIPHENYL-TRICHLOROETHANE (DDT).

methyl bromide A low-molecular-weight compound belonging to the halogenated hydrocarbons (CH_3Br) that has been used as a fumigant to enhance the keeping quality of grains. Because methyl bromide's OZONE-DEPLETING POTENTIAL is excessive, it is being phased out under provisions of the CLEAN AIR ACT to protect the OZONE LAYER.

methyl mercury One of the common ALKYL MERCURY compounds in which an atom of mercury is bonded to a methyl (CH_4) group. Methyl mercury is produced by a variety of organisms (chiefly bacteria) in the natural environment after contamination by ionic forms of inorganic mercury. Methyl mercury is absorbed by a variety of animals and undergoes BIOAC-CUMULATION within the food chain. Humans exposed to significant levels of this compound can experience symptoms ranging from mild neurological disorders to paralysis or death. The compound has also been linked to birth defects.

methyl tertiary butyl ether (MTBE) An additive to REFORMULATED GASOLINE and OXYGENATED FUEL that increases the octane rating while reducing emissions of

H
|
H−C−H
H   H
| H |
H−C — O — C — C−H
| | |
H   H−C−H   H
|
H
|
H

Methyl tertiary butyl ether

VOLATILE ORGANIC COMPOUNDS and CARBON MONOXIDE. Unlike traditional gasoline, the agent is water soluble and disperses quickly in water. Acute effects of overexposure include headaches; eye, nose, and throat irritation; and dizziness. MTBE is an animal carcinogen, but its potential to cause cancer in humans is not known. Its toxicity together with the strong taste and odor of the additive have combined to heighten public awareness of the contamination of groundwater by leaking underground storage tanks at retail gasoline outlets. See OXYGENATE.

methylchloroform Also named 1,1,1-trichloroethane. A colorless, nonflammable liquid. The primary use is for the cleaning of metal surfaces. This agent is not likely to cause environmental damage; however, methylchloroform has an excessive OZONE-DEPLETION POTENTIAL, and it is being phased out under provisions of the CLEAN AIR ACT to protect the OZONE LAYER.

metric ton (t) A unit of mass equal to 1000 kilograms or 2204.62 pounds.

Metropolitan Statistical Area (MSA) A measure adopted by the U.S. Bureau of the Census to track the growth of cities in the United States. A geographical area (may be an entire county, for example) with a minimum population of 100,000 with at least 50,000 of the populace residing in a center city. More than three-fourths of the population of the United States lives in 280+ such locations.

microhydro generator Small turbine device used for generating small amounts of electricity. This technology is intended for use in small streams and in the supply of electrical power to one or a small number of homes without interfering with the ecological or hydrological features.

microbar One-millionth of a BAR. A unit used to express sound pressures. Standard atmospheric pressure equals about 1 million microbars, or 1 bar.

microbe Short for microorganism. Small organisms that can be seen only with the aid of a microscope. The term encompasses viruses, bacteria, yeast, molds, protozoa, and small algae; however, *microbe* is used most frequently to refer to bacteria. Microbes are important in the DEGRADATION and DECOMPOSITION of organic materials added to the environment from natural and human sources. Also called germs.

microbial growth The increase in mass of microscopic organisms, most often reflected in an increase in the number of such organisms. Also, commonly associated with the biochemical transformations that these organisms are capable of (for example, the production of wine or the souring of milk). Refrigeration and the use of common preservatives are intended to prevent the growth and metabolism of spoilage organisms, and the conditions set for the production of compost from yard waste are designed to encourage the growth of such organisms. See BACTERIA, FUNGI.

microbial load The total number of bacteria and fungi in a given quantity of water or soil or on the surface of food. The presence of the bacteria and fungi may not be related to the presence of disease-causing organisms. The greater the microbial load, the faster food spoils.

microbiology The study of organisms that can be seen only with the aid of a microscope. The science deals with the structure and chemical composition of various microbes, the biochemical changes within the environment that are caused by members of this group, the diseases caused by microbes, and the reaction of animals, including humans, to their presence.

microbiota The plants, animals, and microorganisms that can be seen only with the aid of a microscope.

microconsumer A class of heterotrophic organisms that utilizes waste material from other organisms or the tissues of dead animals or plants. Mainly composed of BACTERIA and fungi. Compare MACROCONSUMER.

microcosm A laboratory model of a natural ecosystem in which certain environmental variables can be manipulated to observe the response. The model test results are not always applicable to an actual ecosystem because the microcosm is, of necessity, a simplified collection of selected physical, chemical, and biological ecosystem components.

microenvironment **1.** Extremely small habitats occupied by microorganisms: a break in the skin, a small pore on the surface of a leaf, or a part of the surface of a grain of sand. These locations may present apparently paradoxical circumstances, for example, tiny depressions on the surface of normal teeth that provide an environment suitable for the growth of bacteria that are unable to live in the presence of oxygen, even though the mouth is strongly aerobic. **2.** An area in a home, office, or automobile that can be considered uniform in terms of the concentration of a toxic chemical or other agent that can have negative consequences on individuals.

microfauna Animals invisible to the naked eye, such as copepods and mites.

microflora Plants invisible to the naked eye, such as diatoms and algae.

microgram (μg) A mass unit equal to one-millionth GRAM.

micrograms per cubic meter (μg/m³) An expression of the air concentration of a solid, liquid, or gaseous substance. The concentration of airborne solids or liquids must be expressed as a mass per unit volume, such as micrograms per cubic meter, but gas concentrations can also be expressed as volume/volume ratios, such as PARTS PER MILLION (PPM). At 25° C and one atmosphere of pressure, a gaseous concentration in parts per million can be converted to micrograms per cubic meter by multiplying the number of parts per million of the substance by its MOLECULAR WEIGHT then multiplying by 40.9. For example, a carbon monoxide concentration (by volume) of 9 ppm is equal to 9 times the molecular weight of carbon monoxide (28) times 40.9, or about 10,300 micrograms per cubic meter.

microinjection Process used for introducing genes into cells. With the aid of a microscope, a tiny needle is used to inject deoxyribonucleic acid (DNA) into a recipient cell. The process does not cause the rupture or destruction of the cell.

microliter (μl) A volume unit equal to one-millionth LITER.

micrometer (μm) A unit of length equaling one-millionth meter.

micron Synonym for MICROMETER. Airborne particle diameters are commonly expressed in micrometers.

micronutrient Chemical nutrient required in very small amounts by an organism. Plant micronutrients include copper, manganese, iron, zinc, vanadium, molybdenum, cobalt, boron, chlorine, and silicon. Compare MACRONUTRIENT.

microscopic Describing an object or organism visible only with the aid of a microscope. Objects that can be seen with the naked eye are called MACROSCOPIC.

microwave The area of the ELECTROMAGNETIC SPECTRUM between the INFRARED region and radio waves; energy with wavelengths between about 0.003 and 0.3 meter or with frequencies between

100,000 and 1000 megahertz. Lower-frequency microwaves (< 3000 megahertz) are absorbed by internal body tissues. Home microwave ovens operate at 2450 megahertz. Overexposure can raise the body temperature and thus can could lead to burns or adverse effects on reproduction.

midnight dumping   The deliberate disposal of hazardous waste at a site other than a permitted disposal facility, often taking place at night.

migration route   The environmental medium, air, water, or land, through which waste material can be released.

migration velocity   See DRIFT VELOCITY.

mil   a length unit equaling 0.001 inch. Often used to express the thickness of a plastic LINER used in the construction of landfills rated to receive MUNICIPAL SOLID WASTE or HAZARDOUS WASTE.

Milankovitch cycles   See MILANKOVITCH THEORY.

Milankovitch theory   An explanation of climate changes based on changes in the orbit of the Earth around the Sun. Based on the work of Milutin Milankovitch, a Serbian mathematician and astronomer who in the 1920s and 1930s determined that the orbit of the Earth around the Sun is neither circular nor constant. These "wobbles" influence global temperature, migration of polar ice caps, and climate. He described three basic cycles: changes in the tilt of the axis of the Earth between 21.5 and 24.5 degrees relative to the Sun, with a period of 41,000 years; variation in the shape of the orbit around the Sun between circular and elliptical, with a period of 100,000 years; and variation of the axis of the rotation of the Earth from usually pointing directly at the North Star (Polaris), with a period of 23,000 years.

millfeed   Mineral ores, such as uranium ore, that enter the refining process.

milliequivalents per liter (meq/l)   An expression of the concentration of a material dissolved in water; the expression is calculated by dividing the concentration, in milligrams per liter, by the EQUIVALENT WEIGHT of the dissolved material. For example, the equivalent weight of aluminum is 9.0. A water concentration of aluminum of 1.8 milligrams per liter equals an aluminum concentration of 0.2 milliequivalent per liter.

milligram (mg)   A unit of mass equal to one-thousandth of a GRAM.

milligrams per liter (mg/l)   An expression of water concentration of a dissolved material; one milligram per liter is equal to one PART PER MILLION (PPM).

millilter (ml)   A unit of volume, equal to one cubic centimeter (cm^3 or cc). One thousand milliliters equals one liter.

Millipore filter   A thin membrane of modified cellulose that is used as a filter in the bacteriological examination of water or wastewater. The filter is typically used to filter a given quantity of aqueous sample, followed by transfer of the filter to the surface of a special medium to allow for the growth of the bacteria that have been retained by the filter. The filters are also used to filter sterilize aqueous solutions. The only significant commercial source of the filters was the Millipore Corporation for many years. Although the filters are currently available from a variety of sources, workers in the microbiology area refer to all such filters as Millipore filters.

millirem (mrem)   A unit of IONIZING RADIATION dose equal to one-thousandth of a REM.

Minamata syndrome   Insidious neurological disorders resulting from the consumption of fish and shellfish contaminated by ORGANOMERCURIALS. The most famous case involved a substantial number of people near Minamata Bay, Japan. Fish and shellfish in the bay accumulated ALKYL MERCURY discharged into the bay by an industrial facility. (Although inorganic mercury can be converted to organic mercury in sediments, the Minamata Bay contamination resulted from the direct discharge of alkyl mercury compounds.) Consumption of the contaminated seafood resulted in several hundred poisonings during the 1950s through the 1970s, ranging from mild neurological disorders to paralysis and deaths. See METHYL MERCURY.

mine drainage See ACID MINE DRAINAGE.

Mineral Lands Leasing Act The 1920 federal statute authorizing the secretary of the interior to issue leases for the extraction of coal, oil, natural gas, phosphate, sulfur, and other minerals from public lands. The leasing program is administered by the BUREAU OF LAND MANAGEMENT. Later statutes required that the environmental impact of mining on public lands be considered before a lease is issued or that specific controls be applied. These laws include the Mining and Mineral Policy Act of 1970, the FEDERAL LAND POLICY AND MANAGEMENT ACT, and the SURFACE MINING CONTROL AND RECLAMATION ACT of 1977.

mineral water A category of bottled water containing at least 250 parts per million total dissolved solids (consisting primarily of dissolved inorganic salts). The dissolved inorganic salts are present in the water when drawn from the aquifer and carried to the surface. Inorganic salts are not added artificially. Most mineral waters are carbonated with the characteristic effervescence.

mineralization The conversion of an organic material to an inorganic form by microbial DECOMPOSITION.

minimization Usually used to refer to a process or plan to reduce the amount of waste material produced or released by a facility.

minimum moisture content The amount of water in soil during the driest time of the year.

minimum tillage farming A farming technique that reduces the degree of soil disruption. Crop residues are not plowed under after harvest, and special planters dig narrow furrows in the crop residue when new seeds are sown. Advantages of the technique include reductions in energy consumption by farm equipment, less soil erosion, and lower soil moisture losses during the fallow season. Disadvantages include the possibility of encouraging insect pests by leaving the crop residue in the field and the use of herbicides to control weeds in the place of mechanical cultivation.

mining of an aquifer The overuse of an aquifer by withdrawing groundwater at a rate that exceeds the replenishment flow of water into the aquifer's RECHARGE ZONE.

mining waste Rock, dirt, and other debris displaced in the course of operations to remove mineral deposits from the Earth. The disposal and ultimate disposition of these materials are not subjected to the same regulatory control as municipal or industrial solid wastes.

minors A regulatory designation applied by the USEPA to PUBLICLY OWNED TREATMENT WORKS (municipal sewage treatment plants) serving fewer than 10,000 people, not requiring a PRETREATMENT program, or processing less than 1 million gallons of wastewater per day. (The average amount of wastewater produced by each person is 100 gallons per day.) Compare MAJORS.

minute volume The amount of air moving through the lungs each minute as determined by the product of the breathing rate and the TIDAL VOLUME.

miscible liquids Two different liquids (e.g., water and ethanol) that mix and remain mixed when placed together. Compare IMMISCIBLE.

mist Liquid AEROSOL; small droplets suspended in air.

mist eliminator A device placed downstream from a SCRUBBER to remove particles that were introduced to the airstream by the turbulent gas-liquid contact in the scrubber. Also called an entrainment separator.

mitigation Actions taken to lessen the actual or foreseen adverse environmental impact of a project or activity.

mitigation banking The creation, enhancement, or restoration of WETLANDS to offset loss of wetlands by development. See NO NET LOSS, SECTION 404 PERMIT.

mixed funding agreement Under the COMPREHENSIVE ENVIRONMENTAL RESPONSE, COMPENSATION, AND LIABILITY ACT, an agreement to clean up a waste site that involves payments from both the HAZARDOUS SUBSTANCES SUPERFUND and companies that are held responsible for the hazardous waste at the site. See CASH-OUT.

mixed glass Glass containers of different colors that have been placed in the same receptacle for recycling. Mixed glass has relatively little recycling value; maximal value in recycled glass is achieved when glass containers are separated into three categories: clear, green, and amber.

mixed layer See MIXING HEIGHT.

mixed liquor The liquid/ACTIVATED SLUDGE mixture undergoing decomposition in a wastewater treatment plant's AERATION tank.

mixed liquor suspended solids (MLSS) A measurement of solid material, mainly organic compounds and ACTIVATED SLUDGE, in the AERATION tank of a wastewater treatment plant.

mixed liquor volatile suspended solids (MLVSS) That portion of MIXED LIQUOR SUSPENDED SOLIDS that vaporizes when heated to 600° C; this volatile fraction is mainly organic material and thus indicates the BIOMASS present in the aeration tank. The material that does not vaporize in this test, mostly inorganic substances, is said to be fixed.

mixed metals Metal items that have been collected for recycling but not separated into recyclable categories (e.g., aluminum versus tin cans).

mixed municipal waste Recyclable items separated from the remainder of the garbage but not sorted into categories. All the items, for example, newsprint, plastic containers, bottles, beverage cans, and boxboard, collected for recycling from a residence.

mixed paper Postconsumer paper not sorted into the common categories of newsprint, boxboard, cardboard, white office paper, magazines, and so forth. Mixed paper has less value as a recyclable item than paper sorted into types.

mixed plastic Postconsumer plastic containers that have not been sorted by resin type. Plastic items intended for recycling have greater value when separated by resin type. See PLASTIC RECYCLING.

mixed waste Solid waste containing diverse materials (e.g., routine household garbage).

mixing height The altitude, at a particular time, below which atmospheric dilution of pollutants can occur; the value is determined by the ENVIRONMENTAL LAPSE RATE present that day (or time of day). A low mixing height for an extended period allows air contaminant concentrations to increase, possibly to unhealthy levels. See TEMPERATURE INVERSION.

mixing ratio The concentration of water vapor in the atmosphere, commonly expressed as grams of water vapor per kilogram of dry air. See also SATURATION MIXING RATIO, RELATIVE HUMIDITY.

mixing zone In water quality management, the area around a POINT SOURCE OUTFALL (discharge pipe) within which

Mixing zone

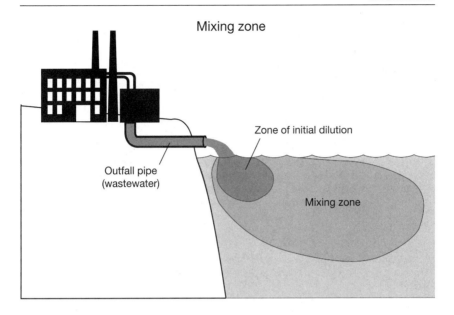

Zone of initial dilution

Outfall pipe
(wastewater)

Mixing zone

mixing of the EFFLUENT water and the receiving water occurs; the water quality standards in a NATIONAL POLLUTANT DISCHARGE ELIMINATION SYSTEM (NPDES) permit do not apply to this dilution volume. Mixing zone exceptions to NPDES standards are prohibited if the receiving (diluting) water does not attain minimal water quality standards. Mixing zones may also be eliminated for certain toxic water pollutants that can undergo BIOACCUMULATION in the aquatic environment. See ZONE OF INITIAL DILUTION.

mixture Two or more elements or compounds present in various proportions. A mixture can be separated into the constituent parts by physical or mechanical means, and the materials composing the mixture retain their individual physical and chemical properties. Compare COMPOUND.

mixture rule The USEPA regulatory provision stating that (1) with certain exceptions, any mixture of LISTED HAZARDOUS WASTE with nonhazardous solid waste is a HAZARDOUS WASTE, and (2) any mixture of CHARACTERISTIC HAZARDOUS

WASTE with nonhazardous solid waste must be tested to determine whether it constitutes hazardous waste.

MMCF Million cubic feet.

MMCFD Million cubic feet per day; used to express natural gas production, transport, or combustion rates.

mobile incinerator system A movable furnace used for the burning or destruction of hazardous waste. Commonly moved from place to place to participate in the cleanup of abandoned hazardous waste disposal sites. See HAZARDOUS SUBSTANCES SUPERFUND.

mobile sources Moving emitters of air pollutants such as automobiles, trucks, boats, and airplanes. See STATIONARY SOURCE, AREA SOURCE.

mobilization The introduction to the air or water of a chemical material that formerly was not circulating in the environment. For example, certain metals can be chemically bound to soil or clay until they are dissolved into and then move with acidic groundwater. The carbon in fossil fuel resources can be said to have been mobilized dramatically since the 19th century, with much of the carbon in the fossil fuels previously stored underground emitted to the atmosphere as carbon dioxide when the fossil fuel is burned.

model 1. A simplified representation of an object or natural phenomenon. The model can take many possible forms: a set of equations or a physical, miniature version of an object or system constructed to allow estimates of the behavior of the actual object or phenomenon when the values of certain variables are changed. Important environmental models include those estimating the transport, dispersion, and fate of chemicals in the environment and evaluation of the potential long-term effects of adding carbon dioxide to the atmosphere. 2. The action of using a model, for example, to model the atmospheric dispersion of the emissions from a smokestack.

moderator A material employed in a NUCLEAR REACTOR to decrease the energy level of fast, high-energy NEUTRONS released in FISSION reactions and to increase thereby the possibility of a sustained fission reaction. When slow, low-energy neutrons collide with nuclei of the fissionable fuels used in nuclear reactors (e.g., uranium-235), the neutrons are captured by those nuclei, causing the production of an inherently unstable atom that undergoes fission. When fission takes place, fragments of the uranium are produced (fission products) along with a large amount of heat. The fission releases or ejects fast, high-energy neutrons. These neutrons must be slowed in order to increase the chance that they will collide with another atom of fissionable material to create a sustained nuclear chain reaction. The slowing of the neutrons is the function of a moderator. Regular (or light) water, HEAVY WATER, and graphite are the most common moderators in nuclear reactors.

modular incinerator A prefabricated MASS BURN incinerator unit that is trans-ported to a municipal waste incineration site. Each modular unit typically has a capacity of 5 to 120 tons per day.

moisture content The mass of water contained in a material divided by the total dry mass of the material, usually expressed as a percentage.

molal volume The MOLECULAR WEIGHT of a substance divided by the DENSITY of that same substance; common units are cubic centimeters per mole (grams per mole divided by grams per cubic centimeter).

molar absorptivity For a given chemical substance, the proportionality constant used in the calculation of ABSORBANCE for a given wavelength of light by a solution of the chemical. The BEER-LAMBERT LAW defines absorbance of a chemical solution as the product of the molar absorptivity of the absorbing chemical, the length of the light path, and the aqueous concentration of the chemical. The units for molar absorptivity are typically expressed per molar concentration per centimeter, where molar concentration is the concentration of the chemical in water and centimeters indicate the length of the light path.

molar concentration The number of MOLES of a chemical substance per unit volume of a medium, for example, 0.2 mole (8 grams) of sodium hydroxide per liter of water.

molarity The number of MOLES of a dissolved chemical substance per liter of solution.

mole The SI unit for the amount of a substance that contains AVOGADRO'S NUMBER (6.02×10^{23}) of atoms or molecules. Frequently expressed as a gram-mole, equal to the molecular mass of a substance in grams. For example, the molecular mass of water (H_2O) is 18. Thus, 1 gram-mole of water equals 18 grams and contains 6.02×10^{23} molecules.

mole fraction An expression of the concentration of a chemical substance in a solution (water) or mixture (air). Calculated by dividing the number of MOLES of the chemical substance by the total number of moles of the various substances making up the mixture or solution.

molecular pharming A technique using TRANSGENIC LIVESTOCK (such as cows, goats, or sheep) that have been modified by the introduction of genes that direct the production of medically important molecules. These products are released in the milk produced by the altered livestock. The usual practice involves the placing of the desired gene from some outside source adjacent to genes that normally promote the production of milk by the lactating animal. As a result, the medical agent is released in the milk of the livestock and is subsequently purified for use for medical purposes. Human hormones and agents that dissolve blood clots and treat emphysema, hemophilia, and diabetes are manufactured by this technology.

molecular sieve A crystalline aluminosilicate material with uniform pore spaces that can be used to separate molecules by size. Technical applications include drying of gases, ion exchange, catalysis, and gas chromatography.

molecular weight The sum of the ATOMIC WEIGHTs of the ATOMs in a molecule. For example, the molecular weight of water (H_2O) is 18, the sum of the atomic weights of two hydrogen atoms (1 + 1 = 2) and oxygen (16).

molecule A group of atoms held together by chemical bonds. They may be either atoms of a single element (O_2) or atoms of different elements that form a compound (H_2O). The smallest unit of a compound that retains the chemical properties of that compound.

molten salt reactor A device that thermally oxidizes organic wastes in a bed of fluid carbonate salts with a very high DESTRUCTION AND REMOVAL EFFICIENCY.

monitored natural attenuation In the REMEDIATION of contaminated groundwater, if natural processes are chosen to accomplish a cleanup, the responsible party agrees to a long-term monitoring period to ensure that human health and the environment are protected and that the cleanup is actually working. See NATURAL ATTENUATION.

monitoring Sampling and analysis of air, water, soil, wildlife, and so on to determine the concentration(s) of contaminant(s).

monitoring well A well drilled in close proximity to a waste storage or disposal facility to check the integrity of the facility or to keep track of leakage of materials into the adjacent groundwater.

monoclonal antibody A specific protein globulin produced by a CLONE of genetically identical cells derived from a single ANTIBODY-producing cell, termed a lymphocyte. The cell line therefore produces a single type of antibody. The cell line is produced by fusing an antibody-producing cell (each antibody-producing cell releases a single kind of antibody that reacts with only one specific antigen) with a myeloma cell (a type of tumor cell). A large amount of homogeneous, highly specific antibody can be produced. Monoclonal antibodies are important tools in medical research, diagnostic and therapeutic medicine, as well as cancer diagnosis and treatment. See BIOTECHNOLOGY.

monoculture The growing of a single plant species over a large area.

monofill A solid waste disposal facility containing only one type or class of waste.

monolith A solid mass of waste-containing material that has undergone SOLIDIFICATION.

monomer A compound that under certain conditions joins to other compounds of the same type to form a molecular chain called a polymer. For example, vinyl

Monomer

Vinyl chloride → Polyvinyl chloride

chloride monomers can polymerize to polyvinyl chloride (PVC).

monomictic lake A lake that experiences OVERTURN once a year, usually in the fall. Such lakes are usually deep, are located in temperate climates, and do not freeze over during the winter. Compare DIMICTIC LAKE.

monsoon Seasonal winds that develop in response to the differing rates of heating of the land and the ocean. The most typical examples are found along the Indian subcontinent, where the tropical and subtropical summer leads to the heating of the land during that season. As the land heats, the air in contact with it heats and rises, drawing cooler, moisture-laden air from the ocean (which does not heat as rapidly) over the land by strong onshore winds. As the oceanic air is in turn forced to rise by a combination of land temperature and geographic features (e.g., mountains), the entrained water condenses, producing torrential rainfall.

montane A forest ecosystem in mountainous areas of the tropics. The montane forest has far fewer plant species than does the TROPICAL RAIN FOREST, which is found at lower elevations below the mountains.

Monte Carlo method A method that produces a statistical estimate of a quantity by taking many random samples from an assumed probability distribution, such as a normal distribution. The method is typically used when experimentation is not possible or when the actual input values are difficult or impossible to obtain.

Montreal Protocol A 1987 international agreement, subsequently amended in 1990, 1992, 1995, and 1997 that establishes a schedule for the phaseout of CHLOROFLUOROCARBONS and other substances with an excessive OZONE-DEPLETING POTENTIAL in participating countries. The full name is the Montreal Protocol on Substances That Deplete the Ozone Layer. See OZONE LAYER DEPLETION. Website: www.unep.org/ozone/

moral extensionism Expansion of the rights generally ascribed to humans to nonhuman entities. Holding a position that animals or plants, the land, or the environment in total be considered worthy of moral protection normally given to humans. See ECOCENTRIC; ECOFEMINISM; LEOPOLD, ALDO.

morbidity Statistics related to illness and disease.

morbidity rate See INCIDENCE.

more-developed country (MDC) Modern industrial nation. A nation that is characterized by high personal income and rates of consumption of natural resources, low death rates, low birth rates, a high standard of living, significant urbanization, and massive dissemination of formal education.

mortality The loss of members of a population through death; NATALITY results in an addition to the number of individuals in a population.

mortality rate   The number of deaths in a given area during a specified period, usually one year, divided by the number of persons in the area, multiplied by a constant, typically 1000. For example, if 200 deaths occurred in 1999 in Jonesville and if the population of Jonesville was 25,000, the 1999 mortality rate was 8 per 1000.

most exposed individual (MEI)   In a risk assessment of the off-site impact of a pollutant released by a facility, a hypothetical person receiving the highest dose of the pollutant. For example, in an analysis of air toxics risk, an individual assumed to spend 24 hours per day outside at the location that is predicted to have the maximal concentration of the modeled pollutant emissions. After the MEI is identified, the health risk attributable to the exposure is estimated. See MAXIMUM INDIVIDUAL RISK.

most probable number (MPN)   A statistical estimate of the levels of COLIFORM bacteria in a water sample, expressed as the number of coliforms per 100 milliliters. See MULTIPLE-TUBE FERMENTATION TEST.

mottling, of teeth   Discoloration of the teeth, which can be caused by extended intake of excessive FLUORIDE. The condition is seen in certain populations who use groundwater containing naturally occurring high fluoride levels.

mousse   A water-oil emulsion formed when higher-density CRUDE OIL is spilled into surface waters.

mucociliary escalator   The mechanism that sweeps particles from the air-conducting tubes (bronchi and the smaller bronchioles) in the lungs. The bronchi and bronchioles are lined with hairlike projections called cilia, which move in unison to force a constantly supplied sheet of mucus upward from just above the ALVEOLAR REGION toward the pharynx (throat).

Trapped particles move upward with the mucus.

mucosa   A mucous membrane lining those parts of the body communicating with the exterior. The mucosa in the gastrointestinal and respiratory tracts can therefore be irritated by certain ingested or inhaled environmental pollutants.

muck soil   Poorly consolidated soil high in organic content resulting from the partial decay of large amounts of plant mass. Such soil is seldom dry and usually consists of very soft mud. Marsh soils.

muffle furnace   A device used to determine the organic content of a soil sample. A preweighed soil sample is heated in the furnace at temperatures sufficient to vaporize the HUMUS, and the mass lost (the organic content) is calculated by subtracting the sample weight after the furnace treatment from the pretreatment weight. The device is also used to determine that fraction of PARTICULATE ORGANIC MATTER that can be volatilized by heating to high temperatures.

Muir, John (1838–1914)   American Naturalist, preservationist, and writer. First president of the SIERRA CLUB, influential promoter of the U.S. national park system. Advocated PRESERVATION. Compare CONSERVATION; PINCHOT, GIFFORD.

mulch   Natural or synthetic material spread over the ground to protect soil, conserve water, prevent erosion, or limit weed growth. Items used for mulch include compost, stabilized manure, wood chips, straw, leaves, plastic, and paper.

multiclone   Individual CYCLONES connected in parallel to control air emissions of particulate matter.

multimedia inspection   A major government audit of the environmental compliance of an industrial facility. A team from federal and state regulatory agencies examines permits, laboratory methodologies, records, training procedures, monitoring practices, emissions, releases, waste management, planning documents, and other data for all media (air, water, and ground). The team usually arrives unannounced and may remain at a facility for a week or more.

multiple chemical sensitivity syndrome (MCS) A condition in which some individuals are HYPERSENSITIVE to a variety of chemical agents; those thought to have MCS express discomfort at exposure levels that normally have no effect on most people. The usual symptoms include nasal congestion, headaches, lack of concentration, fatigue, and loss of memory. Controlled studies have tended to eliminate an immunologic mechanism. Proposed mechanisms underlying the expression of the symptoms involve neurological responses associated with inflammation of the nervous system or behavioral and psychological reactions.

multiple-tube fermentation test A method used to estimate the number of specific types of bacteria, such as COLIFORM BACTERIA, in a water sample. Three different quantities of a sample, for example, 10 ml, 1 ml, and 0.1 ml, are placed into three sets of five tubes each containing lactose broth. The 15 tubes are allowed to incubate at 35° C for 24 hours, and the number of tubes in each set of five tubes showing positive results (gas production) is determined. Most-probable-number tables are then consulted to provide a statistical estimate of the number of coliform bacteria in the water sample. The population levels of different types of bacteria can be determined by altering the media used and the incubation conditions. See STANDARD METHODS.

multiple use The policy of allowing public land to be used for varied purposes, such as timber production, camping and hiking, animal grazing, mineral extraction, and/or wildlife preservation. The national forests and the federally owned rangelands in the western United States and Alaska are managed under the multi-

ple-use principle. The U.S. Forest Service manages the national forests (the 15% of these lands that are WILDERNESS AREAS do not have multiple-use management), and the BUREAU OF LAND MANAGEMENT administers the rangelands.

Multiple Use and Sustained Yield Act A 1960 statute directing the U.S. secretary of agriculture to manage the national forest system for MULTIPLE USE and SUSTAINED YIELD, that is, for recreation, wildlife habitat, and timber production.

multistage cancer risk model The most frequently used model of carcinogenesis; the model assumes that a normal cell or cells pass(es) through two or more stages before becoming a detectable tumor. See CARCINOGENESIS.

municipal sewage Wastewater originating in residences and businesses. On average, each person produces 100 gallons of wastewater per day.

municipal solid waste (MSW) Solid waste, including GARBAGE and TRASH, that originates in households, commercial establishments, or construction/demolition sites. Nonhazardous SLUDGE from municipal sewage treatment plants or nonhazardous industrial waste can also be placed in this category of solid waste.

muon An elementary, subatomic particle with either a positive or a negative charge and a mass 207 times the mass of an electron.

musculoskeletal disorders (MSDs) Impairments of the nerves, muscles, tendons, and supportive structures of the body, for example, back pain, tendinitis, and carpal tunnel syndrome. Many MSDs have been associated with the workplace, and the OCCUPATIONAL SAFETY AND HEALTH ADMINISTRATION has given MSDs regulatory attention.

muskeg Large boggy area found in Canada and Alaska, part of the North American boreal forest BIOME.

mustard gas A group of gases related to 2,2-dichlorodiethyl sulfide. The gas was used extensively during World War I as a vesicant (blistering gas).

mutagen Any agent that has the capability of causing a permanent change in the deoxyribonucleic acid (DNA) of a cell. See MUTATION.

mutagenicity The ability of an agent to cause permanent changes in the genetic material of a cell. See MUTATION.

mutation A significant change in the genetic material of a cell. These changes can be reflected in the physical or biochemical properties of the cell and can be transferred to offspring of that cell. Most mutations are deleterious but persist in a population because they are not expressed (are recessive) and thus are carried in the GENOTYPE without affecting the appearance or viability of the organism. Excessive environmental exposures to a variety of agents, including X rays, ultraviolet radiation, and an array of chemical compounds, can cause mutations.

mutualism An interaction between two or more distinct biological species in which the members benefit from the association. Mutualism describes both symbiotic mutualism (a relationship requiring an intimate association of species in which none can carry out the same functions alone) and nonsymbiotic mutualism (a relationship between organisms that is of benefit but is not obligatory: that is, the organisms are capable of independent existence). Compare NEUTRALISM.

mycorrhizal An association between plant roots and fungi. There is an integration between the roots and the fungal mycelium (filaments) to produce a distinct morphological unit. The fungal mycelium can either form an external layer on the root or actually invade living cells of the root structure.

mycotoxin Natural toxic materials produced by molds. AFLATOXIN, a natural carcinogen, is a mycotoxin commonly found in rice, peanuts, wheat, and corn.

nanometer (nm) An SI unit of length equal to 10^{-9} meter.

nappe The stream of water flowing over a dam or WEIR; from the French for "sheet."

narrative standards Water quality standards that use descriptions to define acceptable levels of quality; for example, water quality shall be "suitable for wildlife habitat" or should "allow safe recreational use." Also called descriptive criteria and criticized as being too vague. The CLEAN WATER ACT also provides for numerical water quality standards, which define allowable pollutant levels in terms of certain chemical-specific CONCENTRATIONS, for example, phosphorus levels of less than five parts per million, daily average.

natality The addition of members to a population by the reproductive process of birth or hatching. Compare MORTALITY.

National Acidic Precipitation Assessment Program (NAPAP) A $500+ million 10-year research effort (1980–90) established by the U.S. Congress to study the ACID DEPOSITION problem. The 1990 report did not fully support the addition of Title IV to the 1990 amendments to the CLEAN AIR ACT, a nationwide ACID RAIN control program. In 1990, the mandate for NAPAP to study acid rain was extended indefinitely, and research reports are still issued. Website: http://www.nnic.noaa.gov/CENR/NAPAP/

National Air Monitoring System (NAMS) A national network of air monitoring stations designed by the USEPA and individual state environmental agencies to assess the ambient air quality in major urbanized areas and used by the EPA to track long-term air quality trends. About 1100 stations monitor for PARTICULATE MATTER, SULFUR DIOXIDE, NITROGEN DIOXIDE, CARBON MONOXIDE, OZONE, and LEAD. The NAMS stations are part of the STATE AND LOCAL AIR MONITORING SYSTEM.

national ambient air quality standards (NAAQS) The AMBIENT air concentration standards set for PARTICULATE MATTER, SULFUR DIOXIDE, NITROGEN DIOXIDE, OZONE, CARBON MONOXIDE, and LEAD to protect human health (primary standards) or welfare (secondary standards).

National Audubon Society A large American environmental interest group that encourages natural resource and wildlife conservation. Named in honor of John James Audubon (1785–1851), who was one of the first American conservationists and who gained recognition for his paintings of birds. Founded in 1905. In 1999 600,000 members. Website: www.audubon.org

National Cancer Institute (NCI) A federal agency under the National Institutes of Health, United States Department of Health and Human Services, that conducts and supports research on cancer and the identification of carcinogens. Based in Bethesda, Maryland, the NCI is a participant in the NATIONAL TOXICOLOGY PROGRAM. Website: www.nci.nih.gov/

National Contingency Plan (NCP) The outline of procedures, organization, and responsibility for responding to spills and releases of hazardous substances and oil into the environment. Prepared by the USEPA as required by sections of the COMPREHENSIVE ENVIRONMENTAL RESPONSE, COMPENSATION, AND LIABILITY ACT and the CLEAN WATER ACT. The plan applies to sudden, accidental releases and to nonsudden, gradual leaks. SUPERFUND SITE cleanups are performed in accordance with the NCP. Officially called the National Oil and Hazardous Substances Pollution Contingency Plan, the NCP is codified at Title 40, Code of Federal Regulations, Part 300.

National Council on Radiation Protection and Measurements (NCRPM) A private group of scientists who recommend safe occupational and public exposure levels to IONIZING RADIATION. Formerly the National Committee on Radiation Protection. Based in Bethesda, Maryland. Website: www.ncrp.com

National Emission Standards for Hazardous Air Pollutants (NESHAP) National technology-based limits set by the USEPA for air emissions of pollutants determined by the agency to pose a significant risk of death or serious illness upon long-term exposure. The NESHAPs have been superseded by the AIR TOXICS provisions of the 1990 amendments to the CLEAN AIR ACT. See HAZARDOUS AIR POLLUTANT, MAXIMUM ACHIEVABLE CONTROL TECHNOLOGY.

National Environmental Performance Partnership System (NEPPS) A program of the USEPA and the ENVIRONMENTAL COUNCIL OF THE STATES begun in 1995 that gives states greater flexibility in the management of environmental quality within their jurisdictions, with an emphasis on self-assessment of progress using selected indicators and goals tailored to each state, as detailed in Performance Partnership Agreements (also called Environmental Performance Agreements). Website: www.epa.gov/ocirpage/

National Environmental Policy Act (NEPA) A 1969 statute that requires all federal agencies to incorporate environmental considerations into their decision-making processes. The act requires an ENVIRONMENTAL IMPACT STATEMENT for any "major Federal action significantly affecting the quality of the human environment." See COUNCIL ON ENVIRONMENTAL QUALITY, ENVIRONMENTAL ASSESSMENT.

National Estuary Program (NEP) Authorized by the 1987 amendments to the Clean Water Act, a program that involves local, state, and federal agencies; citizen groups; educators; and other local stakeholders in the development of plans to manage, protect, and, if needed, restore the margins between the land and the sea. The NEP seeks to go beyond basic water quality protection to include appropriate economic and recreational activities and maintenance of the chemical, physical, and biological integrity of these wetland areas. Grants are available from federal sources to formulate plans, conduct research, and operate restoration projects. Website: www.epa.gov/owow/estuaries/nep.htm

National Fire Protection Association (NFPA) An international organization with voluntary membership whose functions are to promote and improve fire prevention and to establish safeguards against the loss of life or property by fire. The organization has produced the National Fire Code, which lists standards for recommended practices and materials handling. Based in Quincy, Massachusetts. Website: www.nfpa.org

National Forest Management Act (NFMA) A 1976 statute that amended and expanded the Forest and Rangeland Renewable Resources Planning Act of 1974. The act requires the secretary of agriculture to develop a management program for national forest lands based on MULTIPLE USE, sustained-yield principles and to implement a resource management plan, including appropriate timber harvesting rates, methods, and locations, for each unit operated by the NATIONAL FOREST SYSTEM. The NFMA is the basic law by which U.S. national forests are managed. See MULTIPLE USE AND SUSTAINED YIELD ACT.

National Forest System The 155 national forests totaling 192 million acres managed by the U.S. Forest Service.

National Institute for Environmental Health Sciences (NIEHS) A government research organization attached to the National Institutes of Health, United States Department of Health and Human Services, located in Research Triangle Park, North Carolina. The purpose of the NIEHS is to conduct or sponsor research on the adverse effects of environmental agents on human health. The research results are used by federal environmental regulatory agencies in their prevention and control programs. Website: www. niehs.nih.gov/

National Institute for Occupational Safety and Health (NIOSH) An agency of the United States Public Health Service that recommends occupational exposure limits for chemical and physical agents and certifies respiratory and air-sampling devices. Based in Cincinnati, Ohio, and Washington, D.C. Website: www.cdc.gov/niosh/

National Institute of Standards and Technology (NIST) An office within the U.S. Department of Commerce established by Congress "to assist industry in the development of technology . . . needed to improve product quality, to modernize manufacturing processes, to ensure product reliability . . . and to facilitate rapid commercialization . . . of products based on new scientific discoveries." The four major programs of the NIST are measurement and standards laboratories, the advanced technology project, the manufacturing extension partnership, and the Malcolm Baldrige quality award. Website: www.nist.gov

National Oceanic and Atmospheric Administration (NOAA) A federal agency within the Department of Commerce responsible for mapping and charting of the oceans, environmental data collection, monitoring and prediction of conditions in the atmosphere and oceans, and management and conservation of marine resources and habitats. Based in Suitland, Maryland, and Rockville, Maryland. Website: www.noaa.gov

National Oil and Hazardous Substances Pollution Contingency Plan (NCP) The full name of the NATIONAL CONTINGENCY PLAN.

National Parks and Conservation Association (NPCA) A private nonprofit group dedicated to protecting the American system of national parks; the NPCA uses education, advocacy, and litigation to accomplish its goal. Membership of 400,000 in 1999. Website: www.npca.org

National Pollutant Discharge Elimination System (NPDES) The program established by the CLEAN WATER ACT that requires all POINT SOURCES discharging into any "waters of the United States" to obtain a permit issued by the USEPA or a state agency authorized by the federal agency. The NPDES permit lists permissible discharge(s) and/or the level of cleanup technology required for wastewater.

National Primary Drinking Water Regulations (NPDWR) Regulations for public drinking water supply systems that include health-based standards for various contaminants and monitoring and analysis requirements. Issued by the USEPA under authority of the SAFE DRINKING WATER ACT. See MAXIMUM CONTAMINANT LEVEL, MAXIMUM CONTAMINANT LEVEL GOAL. Compare NATIONAL SECONDARY DRINKING WATER REGULATIONS. Website: www.epa. gov/safewater/mcl.html

National Priorities List (NPL) A list of the hazardous waste disposal sites most in need of cleanup; the list is updated annually by the USEPA, based primarily on how a site scores using the HAZARD RANKING SYSTEM. Also called the Superfund list. Website: www.epa. gov/superfund/sites/npl/npl.htm

National Research Council (NRC)
A group of volunteer professionals supported by the National Academy of Sciences, National Academy of Engineering, and Institute of Medicine who, working through study committees, conduct independent research for the U.S. government on public policy issues in science and technology. Based in Washington, D.C. Website: www.nationalacademies.org/nrc/

National Response Center (NRC)
The U.S. Coast Guard unit that receives reports of hazardous chemical spills and is responsible for notifying other agencies that help plan, coordinate, and respond to the release. Based in Washington, D.C. The 24-hour telephone number of the center is 800-424-8802. See SPILL CLEANUP INVENTORY, EMERGENCY RESPONSE NOTIFICATION SYSTEM. Website: www.nrc.uscg.mil

National Response Team (NRT)
A group of 16 federal agencies that coordinate the emergency response to releases of oil or hazardous substances. The NRT, under the leadership of the USEPA, serves as an umbrella organization at the federal level; its functions include evaluating methods to respond to discharges or releases; recommending needed changes in the response organization; making recommendations related to the training, equipping, and protection of response teams; evaluating response capabilities; reviewing regional responses to discharges; and coordinating the activities of federal, state, and local governments as well as private organizations in response to discharges. See REGIONAL RESPONSE TEAM, NATIONAL CONTINGENCY PLAN. Website: www.nrt.org

National Science Foundation (NSF)
An independent U.S. government agency based in Washington, D.C., that supports basic and applied research in science and engineering. Website: www.nsf.gov

National Secondary Drinking Water Regulations (NSDWR)
Regulations governing the operation of public water supply systems under the Safe Drinking Water Act. The regulations define secondary maximum contaminant levels, that is, the maximal concentrations of certain substances in drinking water that affect aesthetic quality. The NATIONAL PRIMARY DRINKING WATER REGULATIONS set standards intended to protect public health. Website: www.epa.gov/safewater/mcl.html

National Stream Quality Accounting Network (NASQAN)
A data system operated by the U.S. Geological Survey that, since 1995, compiles measurements of water pollutant concentrations in four major rivers in the United States: the Mississippi (which includes the Missouri and Ohio), the Columbia, the Colorado, and the Rio Grande. Between 1973 and 1995, NASQAN collected water quality data from over 600 watersheds. Website: http://water.usgs.gov/nasqan/

National Strike Force (NSF)
Units within the U.S. Coast Guard that respond to spills of oil or hazardous substances as part of the NATIONAL RESPONSE TEAM following the NATIONAL CONTINGENCY PLAN. Force units are based along the Atlantic, Gulf, and Pacific coasts. They provide, among other services, communication support, advice, and assistance in the event of discharges; shipboard damage control; containment and removal of discharges; and diving activities related to damage assessment and surveys. Website: www.uscg.mil/hq/nsfcc/nsfweb/

National Technical Information Service (NTIS)
An agency that sells reports from government-funded studies to the public. Located in Springfield, Virginia. Website: www.ntis.gov

National Toxicology Program (NTP)
An organization within the U.S. Department of Health and Human Services (DHHS) charged with coordinating toxicology research. The NTP *Report on Carcinogens* lists chemicals to which large numbers of American are exposed that can (or may) cause cancer in humans. The

agencies involved are the Food and Drug Administration (National Center for Toxicological Research), the National Cancer Institute, the National Institute for Occupational Safety and Health, and the National Institute for Environmental Health Sciences. The NTP executive committee also includes the heads of other federal health research groups and regulatory agencies. Website: http://ntp-server.niehs.nih.gov/

National Wildlife Federation (NWF) An international conservation organization headquartered in Vienna, Virginia, that promotes sustainable management of wildlife and other natural resources. Largest U.S. conservation organization; 1999 membership of 4.4 million. Website: www.nwf.org

natural attenuation In the REMEDIATION of contaminated groundwater, the reduction in concentration or mobility of toxic materials that occurs by natural physical, chemical, and/or biological means, which can include ADSORPTION to soil and clay particles and BIODEGRADATION. See MONITORED NATURAL ATTENUATION.

natural draft A gas flow created by the difference in pressure between hot gases and the atmosphere, such as the draft operating in a fireplace chimney, incinerator stack, or NATURAL-DRAFT TOWER. See also FORCED DRAFT, INDUCED DRAFT.

natural-draft tower A cooling tower that is designed to remove waste heat from a heated effluent. The air that receives the heat from the water rises and exits the tower by convection currents without the aid of blowers or fans.

natural experiment An unplanned situation in which parts of a human population are exposed to an environmental risk factor and others similar in other ways to the exposed group are not, much as if the exposed and control groups were part of a laboratory experiment. Differences in disease INCIDENCE in the two groups can be used to judge the strength of the risk factor.

natural gas A colorless, odorless, flammable mixture of METHANE, ethane, propane, butane, and other hydrocarbons. Often found with PETROLEUM deposits. A clean-burning fuel widely used for heating and cooking.

natural history The study of nature (e.g., plants; animals; geological, chemical, or other physical aspects of the Earth) in a nontechnical manner. Description of the distribution of a plant or animal as well as the interactions of organisms within the natural environment.

natural increase A positive change in the number of individuals in a community as a result of a larger number of births than deaths.

natural logarithm (ln) The value of the exponent that the base, e, must have to equal a given number. The natural logarithm of y is the value of x in the expression $e^x = y$. For example, the natural logarithm of 5 is the power (x) to which e (approximately 2.718282) must be raised to equal 5, or $e^x = 5$; therefore x is about 1.60944.

natural minor See TRUE MINOR; compare SYNTHETIC MINOR.

natural organic matter (NOM) In drinking water treatment, ORGANIC CARBON present in the source water. CHLORINATION or OZONATION can react with the NOM to produce DISINFECTION BY-PRODUCTS.

natural radioactivity Ionizing radiation from sources that are not related to human activities, for example, cosmic rays and radiation emitted by RADIOISOTOPES found naturally in the crust of the Earth.

natural resources damages Monetary penalties assessed by a court against a source that spills or otherwise discharges excessive amounts of hazardous substances or oil into the environment. As authorized by the COMPREHENSIVE ENVIRONMENTAL RESPONSE, COMPENSATION,

AND LIABILITY ACT and the OIL POLLUTION ACT, the assessment is based on the expense required to clean up and restore the natural resources (estuary, shoreline, surface water body, and so forth) harmed by the discharge. State and federal administrative agencies are TRUSTEES who receive the monetary award for use in restoration.

Natural Resources Defense Council (NRDC) A private U.S. environmental organization. A participant in numerous lawsuits that have shaped national environmental policies. For example, NRDC suits resulted in the listing of lead as a CRITERIA POLLUTANT (*NRDC v. Train*, 1976) and the adoption of best-available technology effluent standards for toxic water pollutants (*NRDC v. Train*, 1976; see FLANNERY DECREE). Membership in 1999: 350,000. Headquarters in New York City. Website: www.nrdc.org

natural selection A natural process by which certain members of a POPULATION that are well adapted to prevailing environmental conditions survive and reproduce at greater rates than those organisms not suited to that particular environment. Expressed by Charles Darwin, the process is often referred to as the survival of the fittest (i.e., the genetic makeup that best fits the environment is most successfully passed to offspring) and is considered to be the selection pressure that drives EVOLUTION.

natural sink 1. A habitat that serves to trap or otherwise remove chemicals such as plant nutrients, organic pollutants, or metal ions through natural processes. For example, a river that enters a swamp may carry a substantial amount of dissolved plant nutrients but the water exiting the swamp will have lower nutrient concentrations. The swamp has then served as a sink to trap the nutrients that are no longer available for subsequent plant growth downstream from the swamp. 2. A natural process whereby pollutants are removed

from the atmosphere. A sink process can be physical (particulates removed by rain), chemical (the reaction of ozone with nitric oxide to form nitrogen dioxide and oxygen), or biological (the uptake of airborne hydrocarbons by soil microorganisms). Also called a scavenging mechanism.

natural water Any bottled water obtained from a groundwater source, including spring water, well water, artesian water, or mineral water. Water not obtained from a municipal water supply.

naturally occurring radioactive material (NORM) Radioactive material in fluids moved to the surface during the production of oil or gas, usually in the PRODUCED WATER. The material can contaminate drilling pipe by forming radioactive scale or, if discharged into surface waters, can potentially accumulate in an aquatic ecosystem.

Nature Conservancy, The An international, private, nonprofit organization dedicated to the conservation of plant and animal species through the protection of natural habitats. The organization purchases and manages land in association with private landowners and government organizations for the protection of diverse land and aquatic habitats. The conservancy oversees the largest system of private nature sanctuaries in the world, with holdings in 50 states and Canada, and the group has partnerships with like-minded organizations in Latin America, the Caribbean, and Asia. Membership in 1999: 1 million. Headquartered in Arlington, Virginia. Website: www.tnc.org

navigable waters Water to which the CLEAN WATER ACT applies; such waters include "the waters of the United States," that is, any body of water with any connection to interstate waters or commerce, including almost all surface water and wetlands. There is no requirement, despite the name, that vessels be able to navigate these waters.

near field   The area very close to a noise source in which the sound pressure level does not drop with the inverse square of the distance from the source (INVERSE SQUARE LAW). Compare FAR FIELD.

near infrared   Electromagnetic radiation with wavelengths from 0.75 micrometer to 1.5 micrometers. See INFRARED RADIATION.

near miss   Any unintentional or unplanned event in an industrial setting that could have resulted in a violation of health, safety, or environmental standards. The event is characterized as a near miss regardless of its consequences. These occurrences are recorded and studied, and corrective action is taken to prevent future accidents that may have unfavorable outcomes related to health, safety, or environmental standards.

necrosis   Localized death of body tissue that results in the development of a lesion characterized by inflammation and pus accumulation. A festering sore.

negawatt   Negative watt. Describing the electrical energy saved by conservation or efficiency measures.

negligible residue   An amount of pesticide remaining in or on raw agricultural commodities that would result in a daily intake of the agent regarded as toxicologically insignificant. See PESTICIDE RESIDUES.

negotiated rulemaking   In administrative law, a process sometimes used to draft RULES (regulations) by which administrative agency staff consult industrial representatives, members of environmental activist groups, and any other STAKEHOLDER as the regulations are being developed. The interaction, which may include the assistance of FACILITATORS, is designed to produce consensus and prevent litigation.

nekton   Animal in an aquatic system that is free-swimming, independent of currents or waves. Compare BENTHOS, PLANKTON.

nematode   Roundworm. One of the most common kinds of animals, with about 10,000 species known. Members of this group live in almost all known habitats, ranging from polar regions to the tropics and from soil to the deep ocean. Depending on the species, nematodes can be parasitic to humans, livestock, and plants. Some common human pathogens are hookworm, pinworm, intestinal roundworm, and whipworm.

neo-Luddite   See LUDDITE.

neo-Malthusian   See MALTHUSIAN.

neoplasm   The growth of new or abnormal tissue that has no prescribed physiological function. A tumor. See BENIGN NEOPLASM, MALIGNANT NEOPLASM.

nephelometer   A device that measures the scattering of light by particles (or bacteria) suspended in air or water compared with a reference suspension. The instrument consists of a light source aimed at a sample cell and a detector placed at right angles to the light path through the sample. The light scattered at right angles is measured by a nephelometer in NEPHELOMETRIC TURBIDITY UNITS; the decrease in light transmitted directly through the sample cell is a measure of TURBIDITY in JACKSON TURBIDITY UNITS.

nephelometric turbidity unit (NTU)   A unit used to express the cloudiness (turbidity) of water as measured by a NEPHELOMETER. Nephelometric turbidity units are approximately equal to JACKSON TURBIDITY UNITS.

neritic   Of the shallow regions of a lake or ocean that border the land. The term is also used to identify the biota that inhabit the water along the shore of a lake or ocean.

Nessler reagent   An aqueous solution of mercury and potassium salts of iodine and sodium hydroxide that is used to test for the presence of ammonia in water or reaction mixtures.

net community productivity (NCP)
The gain of BIOMASS within a defined region over time. The total amount of carbon dioxide fixed by the photosynthetic plants within the area (PRIMARY PRODUCTION) minus that amount of carbon dioxide lost through metabolism at all TROPHIC LEVELS within the same region.

net energy yield A measurement used to compare different energy-conversion systems, calculated as the ratio between the output of useful energy (available to do useful work) and the energy cost of construction, fuel extraction, energy conversion, energy transmission, and disposal of waste produced by the conversion. The ratio for hydroelectric power has been estimated to be about 20:1, indicating that the yield of useful energy is high relative to the energy that must be expended to recover the hydroelectric power, whereas the ratio for BIOGAS (METHANE produced from the ANAEROBIC decomposition of BIOMASS) is estimated to be on the order or 2:1, indicating that it takes the expenditure of almost as much energy to utilize the resource as the amount of useful energy recovered.

net precipitation A factor used to evaluate the potential for LEACHATE generation at a waste disposal site. The factor is computed for a specific location by subtracting the annual evaporation from lakes in the region from the normal annual rainfall.

net primary productivity (NPP) The number of grams of carbon dioxide fixed by photosynthesis of plants per unit area or volume of water minus the number of grams of carbon dioxide produced during the respiration of those plants. See also PRIMARY PRODUCTIVITY.

net reproductive rate (R_0) In population ecology, the average number of surviving offspring produced by each individual during its lifetime. A population with an R_0 of 1 is experiencing zero growth.

netting See BUBBLE POLICY, NETTING OUT, EMISSION REDUCTION CREDITS, EMISSIONS TRADING.

netting out The exemption of certain facility modifications from a more-detailed air pollution permit process (a NEW SOURCE REVIEW) if the emissions of a particular pollutant from the proposed modification and emission reductions of the same air pollutant within the same source result in no net increase in facility emissions for that pollutant.

neural tube defects Malformation or other harm to the tissue from which the CENTRAL NERVOUS SYSTEM develops; results from excessive exposure to toxic compounds during early embryonic development.

neurotoxin A substance that can damage or destroy nerve tissue. BOTULISM toxin represents a common example.

neuston Small particles or microorganisms found in the surface film that covers still bodies of water.

neutralism A circumstance describing the absence of interactions between two species when those organisms are living in close association. Compare MUTUALISM.

neutralization Moving the pH of a matrix to 7 or neutral. If the water is acidic, a caustic or a basic material is added to absorb or otherwise remove some of the protons so that neutrality will be achieved. On the other hand, if water is too alkaline (pH above 7), acid is added to lower pH until neutrality is achieved. Frequently, the pH of wastewater must be adjusted to achieve neutrality before treatment or release into the environment.

neutrino A very small subatomic particle released from unstable atoms that emit BETA PARTICLES as they undergo nuclear decay. The particle does not carry a charge as do the protons and electrons of an atom.

neutron One of the elementary particles in the nucleus of all atoms except hydrogen. A neutron does not have a charge, and the atomic mass is approximately 1, the same as the mass of the proton.

new source performance standards (NSPS) **1.** Pollutant-specific national uniform air emission standards for new or modified stationary sources, set by the USEPA by facility type, based on available emission control technology. **2.** Effluent limitations set by the USEPA for new POINT SOURCES of water pollution. The standards are applied to an INDUSTRY CATEGORY, such as a petroleum refinery or a phosphate manufacture.

new source review (NSR) The procedural steps defined by an environmental regulatory agency for the issuance of a permit for a new facility (or major modification of an existing facility) that will emit significant quantities of air pollutants. The review includes specification or approval of air pollution control devices or methods and AIR QUALITY DISPERSION MODELING of the estimated emissions of a facility to assess their impact.

new towns Planning models that attempt to construct environments that combine urban living with a close association with nature. Planned communities providing for clusters of residential and commercial development separated by extensive green areas. See GREEN BELT.

newton (N) The SI unit of force equaling a mass of one kilogram accelerated at one meter per second per second, Expressed as $1 N = 1$ kg m s^{-2}.

niche In ECOLOGY, a term that includes both the HABITAT and role (functional status) of an organism within an ecosystem.

night soil Solid human excrement.

nitrate A chemical compound having the formula NO$_3$⁻. Nitrate salts are used as fertilizers to supply a nitrogen source for plant growth. Nitrate addition to surface waters can lead to excessive growth of aquatic plants. High groundwater nitrate levels can cause METHEMOGLOBINEMIA in infants. See NITRITE, EUTROPHICATION, CULTURAL EUTROPHICATION.

nitric acid A strong mineral acid having the formula HNO$_3$. This acid is one of the constituents of ACID RAIN.

nitric oxide (NO) The gas formed by heating air to high temperatures (THERMAL NO$_x$) or by the oxidation of organic nitrogen contaminants in a fuel during combustion (FUEL NO$_x$).

nitrification The oxidation of ammonia to NITRATE by bacteria in soil or water.

nitrilotriacetic acid (NTA) Chemical considered a substitute for SODIUM TRIPOLYPHOSPHATE in detergents. The polyphosphate is added to detergents to act as a BUILDER, or agent to bind with ions of calcium, iron, and magnesium, thus preventing the accumulation of deposits of detergent on clothing. The water pollution resulting from the large amounts of phosphorus in streams and lakes stimulated the search for a substitute. The NATIONAL TOXICOLOGY PROGRAM has determined that there is sufficient evidence that NTA is an animal carcinogen. Limited use of NTA as a builder indicated that the agent was effective; however, concerns about lawsuits and product boycotts resulted in cancellation of plans to use the agent. See EUTROPHICATION, CULTURAL EUTROPHICATION.

nitrite An oxidized nitrogen molecule with the chemical formula NO$_2$⁻. Nitrite can be formed from NITRATE (NO$_3$⁻) by microbial action in soil, water, or the human digestive tract. Excessive nitrate levels in rural well water, caused by fertilizer application, can cause METHEMOGLOBINEMIA, typically in infants. Sodium nitrite preservative added to bacon, lunch meats, hot dogs, ham, and other foods can react with dietary amines (compounds found in cereals, fish, cheese, beer, and others) to form carcinogenic NITROSAMINES. However, relative to nitrites formed by normal body metabolism and dietary intake of natural nitrates (in many vegetables), the health risk of nitrite preservatives is small.

nitrogen cycle A model illustrating the conversion of nitrogen from one form to another through a combination of biological, geological, and chemical processes. The process is continuous, with N_2 in the atmosphere (constituting about 80% of the gas in the atmosphere) converted to forms usable by biota and then ultimately returning to the atmosphere as N_2.

nitrogen dioxide (NO_2) A brownish colored gas that is a major ingredient in PHOTOCHEMICAL SMOG. This oxide of nitrogen is readily produced in the atmosphere from nitric oxide by the addition of an oxygen atom ($NO + \frac{1}{2}O_2 \rightarrow NO_2$). Nitrogen dioxide can be converted by atmospheric reactions to PEROXYACYL NITRATE or to NITRIC ACID, an ingredient in ACID DEPOSITION. See NITROGEN OXIDES.

nitrogen fixation The conversion of nitrogen in the atmosphere (N_2) to a reduced form (e.g., amino groups of amino acids) that can be used as a nitrogen source by organisms. The process is important since all organisms require a source of nitrogen for nutrition, and N_2 cannot be used by the great majority of the biota to satisfy that need. Biological nitrogen fixation is carried out by a variety of organisms; however, those responsible for most of the fixation are certain species of blue-green algae, the soil bacterium *Azotobacter,* and the symbiotic association of plants of the legume variety and the bacterium *Rhizobium.* In industry, the HABER PROCESS is used to fix atmospheric nitrogen for use as fertilizer. See also SYMBIOSIS.

nitrogen oxides (NO_x) Gases containing nitrogen and oxygen; they include NO, NO_2, NO_3, N_2O, N_2O_3, N_2O_4, and N_2O_5. The first two, NITRIC OXIDE (NO) and NITROGEN DIOXIDE (NO_2), are the primary NO_x air pollutants. About 60% of U.S. NO and NO_2 emissions are from stationary sources (smokestacks), and most of the remaining 40% is from transportation exhaust, mainly from automobiles. NITROUS OXIDE is a significant GREENHOUSE GAS. See THERMAL NO_x, FUEL NO_x.

nitrogenous BOD The amount of molecular oxygen required for the microbial oxidation of ammonia and nitrite contaminants in a specified volume of wastewater. This type of oxygen demand can complicate the interpretation of data obtained from the determination of the BIOCHEMICAL OXYGEN DEMAND (BOD) of treated sewage, although a chemical can be added to the BOD test to prevent ammonia oxidation. Ammonia and nitrite are oxidized by chemoautotrophic bacteria. See CHEMOAUTOTROPHS.

nitrogenous waste Wastewater that contains organic or inorganic forms of nitrogen, including ammonia and nitrate.

nitrosamines A large and diverse family of synthetic and naturally occurring compounds having the general formula $(R)(R')N—N=O$. Almost all are CARCINOGENIC, with biochemical activation to a cancer-causing intermediate taking place in tissue fluids. Different members of the family of compounds and different concentrations of the same nitrosamine result in cancer development in different tissues: liver, lungs, esophagus, kidney, pancreas, among others. Several nitrosamines produced from nicotine by bacterial activity during the tobacco curing process are responsible, in part, for the cancer-causing potential of smoking and use of smokeless tobacco products.

nitrous oxide (N_2O) A colorless gas introduced to the atmosphere by microbial metabolism; used as an anesthetic (laughing gas). N_2O strongly absorbs INFRARED RADIATION, and rising atmospheric concentrations have implicated it as an important GREENHOUSE GAS (contributor to GLOBAL WARMING).

no further remedial action planned (NFRAP) The designation applied by the USEPA when an assessment of a hazardous waste disposal site indicates that the site does not pose a significant health risk to the public or risk for damage to the environment. Sites rated as NFRAP are now removed from the USEPA active list of waste sites, COMPREHENSIVE ENVIRONMENTAL RESPONSE, COMPENSATION, AND LIABILITY INFORMATION SYSTEM (CERCLIS). No further activity is anticipated under the COMPREHENSIVE ENVIRONMENTAL RESPONSE, COMPENSATION, AND LIABILITY ACT. Waste site listings are available at www.epa.gov/enviro/.

no migration 1. A demonstration by a municipal solid waste LANDFILL that there is no potential for hazardous substances to migrate from the landfill into an AQUIFER. USEPA approval of a no migration petition exempts the landfill from groundwater monitoring at the site. 2. For a hazardous waste TREATMENT, STORAGE, OR DISPOSAL facility, a demonstration that a waste that would otherwise be subject to the LAND DISPOSAL BAN will not migrate into an AQUIFER; a successful demonstration allows the waste to be buried.

no net loss In WETLANDS conservation, the policy of identifying replacement wetland for any wetland habitat that is destroyed by development. See SECTION 404 PERMIT, MITIGATION BANKING, CONSTRUCTED WETLANDS.

no-observed-adverse-effect level (NOAEL) See NO-OBSERVED-EFFECT LEVEL.

no-observed-effect level (NOEL) In a toxicology study, the highest dose at which no adverse effect is observed. Also called the NO-OBSERVED-ADVERSE-EFFECT LEVEL.

noise Unwanted sound. Noise differs from most environmental insults in that if the source is stopped, it disappears immediately and has no residual accumulation. Noise also has important subjective aspects. No direct measurements can quantify the loudness of a noise or the disturbance quality. The Occupational Safety and Health Administration workplace standard for an eight-hour exposure is a limit of 90 DECIBELS, A-WEIGHTING NETWORK (dBA). There are no enforceable federal community noise standards, but the Federal Aviation Administration and the Department of Urban Development have established a guideline, which is a DAY-NIGHT SOUND LEVEL of 65 dBA. See NOISE-INDUCED HEARING LOSS and other entries that start with the word *sound*.

noise-induced hearing loss (NIHL) A PERMANENT THRESHOLD SHIFT attributable to excessive noise exposure; hearing loss beyond the normal decline with age.

noise reduction coefficient The arithmetic average of the SABIN ABSORPTION COEFFICIENTS for a certain material at the frequencies of 250, 500, 1000, and 2000 hertz, rounded to the nearest 0.05. The coefficient is used for comparing the sound absorption characteristics of materials.

nominal variable A sample characteristic expressed as a class or descriptive category, such as marital status, ethnic group, socioeconomic class, and religious preference. These are qualitative, not quantitative, data, and only NONPARAMETRIC TESTS are appropriate for data collected as nominal variables. See also ORDINAL VARIABLE, PARAMETRIC TESTS.

nomograph A graphical solution to a multivariable equation. Parallel vertical scales, one for each variable, are arranged such that a straight line across the scales produces values for all variables that together solve the equation.

non-time-critical removal action Under the provisions of the NATIONAL CONTINGENCY PLAN, a chemical release that does not require an emergency (immediate) response; however, the response must begin within six months.

See REMOVAL ACTION, TIME-CRITICAL REMOVAL ACTION (TCRA).

nonaqueous phase liquid (NAPL) Organic liquid that is relatively insoluble in water and less dense than water. When it is mixed with water or when an aquifer is contaminated with this class of pollutant (frequently hydrocarbon in nature), these substances tend to float on the surface of the water. Compare DENSE NON-AQUEOUS PHASE LIQUID (DNAPL).

nonattainment area (NAA) A geographical area that does not meet a NATIONAL AMBIENT AIR QUALITY STANDARD for a particular pollutant. The extent of the area is defined by air quality monitoring data, air dispersion modeling, and/or the judgment of the state environmental agency and the USEPA. The CLEAN AIR ACT requires a STATE IMPLEMENTATION PLAN to contain the steps necessary to become an ATTAINMENT AREA.

nonbinding preliminary allocation of responsibility (NBAR) An allocation by the USEPA of percentages of total cleanup costs for a hazardous waste site to each POTENTIALLY RESPONSIBLE PARTY under the authority of the COMPREHENSIVE ENVIRONMENTAL RESPONSE, COMPENSATION, AND LIABILITY ACT. The system is used to encourage settlements.

nonbiodegradable Describing organic compounds, usually synthetic, that are not decomposed or mineralized by microorganisms. This term is somewhat archaic and has been replaced by RECALCITRANT in most circumstances

noncommunity water system As defined in SAFE DRINKING WATER ACT (SDWA) regulations, a drinking water supply and distribution system that serves at least 15 connections or 25 or more people but not on a year-round basis, for example, the system at a summer camp. The SDWA applies different requirements depending on a water system's classification.

noncontact cooling water Under CLEAN WATER ACT regulations, COOLING WATER that does not have direct contact with raw material, products, by-products, or waste. NATIONAL POLLUTANT DISCHARGE ELIMINATION SYSTEM permits for noncontact cooling water control THERMAL POLLUTION, and, if necessary, set limits for chemical additives used to inhibit corrosion or to control pipe SCALE.

nonconventional pollutants Under the CLEAN WATER ACT, water pollutants not listed as CONVENTIONAL POLLUTANTS, toxic pollutants, or thermal discharges. The nonconventional agents include chloride, iron, ammonia, color, and total phenols.

nondestructive testing (NDT) In geophysical surveying, methods used to detect subsurface water, subsurface containers, or the areal extent of groundwater contamination without soil borings. The testing involves the use of acoustic sounding, infrared radiation, X rays, magnetic field perturbation, electrical resistivity, and other methods.

nondiscretionary (duty) See CITIZEN SUIT PROVISION.

nondispersive infrared analysis (NDIR) An analytical method that uses the molecular absorption of INFRARED RADIATION to measure the concentration of certain chemical compounds in the ambient air. A broad (nondispersive) band of infrared radiation is used, in contrast to the particular wavelength tuning applied in methods like ULTRAVIOLET PHOTOMETRY. NDIR is the USEPA reference method for ambient measurements of CARBON MONOXIDE.

nonferrous metal Metal that is not attracted by a magnet, such as aluminum, copper, and lead. Recycling operations cannot use MAGNETIC SEPARATION for materials made of nonferrous metals.

nonfissionable Describing an atom that is not capable of undergoing nuclear

FISSION, or breaking into two or more pieces, when bombarded by NEUTRONS. Commonly used to refer to a specific ISOTOPE of some ELEMENT, other isotopes of which normally undergo nuclear fission. For example, uranium-235 is an isotope of uranium that is capable of undergoing fission, whereas uranium-238 is an isotope of uranium that is nonfissionable.

nongovernmental organizations (NGOs) Excluding commercial businesses, the array of private research groups, foundations, advisory agencies, professional societies, and political organizations holding an interest in environmental affairs and resource utilization (and a host of other causes).

nonhazardous oil field waste (NOW) Waste generated by drilling of and production from oil and gas wells that is not classified as a HAZARDOUS WASTE by regulations of the USEPA. Typical NOW wastes include drilling muds, cuttings, drilling fluids, and PRODUCED WATER.

nonionizing radiation Electromagnetic radiation that contains less than 10–12 ELECTRON VOLTS per PHOTON, which is not sufficiently energetic to produce ions when absorbed by matter, in particular water. Examples are ULTRAVIOLET light, visible light, and radio waves. See IONIZING RADIATION.

nonmethane hydrocarbons (NMHC) Hydrocarbon compounds present in ambient air that participate in photochemical reactions leading to an accumulation of PHOTOCHEMICAL OXIDANTS, especially OZONE. Methane is excluded because it is relatively nonreactive and does not contribute to the atmospheric reactions forming ozone. Compare NONMETHANE ORGANIC COMPOUNDS.

nonmethane organic compounds (NMOC) Airborne NONMETHANE HYDROCARBONS, plus any oxygenated hydrocarbons, such as aldehydes and ketones, that participate in photochemical reactions producing OZONE or other PHOTOCHEMICAL OXIDANTS.

nonparametric statistics See NONPARAMETRIC TESTS.

nonparametric tests Statistical procedures that yield information about populations but not about population PARAMETERS. This type of test uses data from NOMINAL VARIABLES or ORDINAL VARIABLES and does not require the assumptions about sample population distributions that must be made or acknowledged using PARAMETRIC TESTS.

nonpersistent pollutant A substance that can cause damage to organisms when added in excessive amounts to the environment but is decomposed or degraded by natural biological communities and removed from the environment relatively quickly. Compare PERSISTENT POLLUTANT, RECALCITRANT.

nonpoint source A diffuse, unconfined discharge of water from the land to a receiving body of water. When this water contains materials that can potentially damage the receiving stream, the runoff is considered to be a source of pollutants. Runoff from city streets, parking lots, home lawns, agricultural lands, individual septic systems, and construction sites that enters lakes and streams constitutes an important source of SEDIMENTS, oil and grease, and nutrients causing EUTROPHICATION. Nonpoint sources are the largest contributors to the nation's remaining water pollution problems. See STORM WATER RUNOFF.

nonpolar solvent A solvent with no positive POLARITY or negative polarity. This type of solvent is a good dissolver of other nonpolar materials. For example, oils dissolve in benzene because both are nonpolar materials. Compare POLAR SOLVENT.

nonpotable Describes water that is undrinkable because it may contain excessive levels of infectious agents, hazardous chemicals, or other substances that render it unpalatable.

nonreactive See INERT.

nonrenewable energy A source of energy such as oil or natural gas that is not replaceable after it has been used. In contrast, firewood and hydroelectricity production are sources of RENEWABLE ENERGY.

nonrenewable resource A natural resource such as coal or mineral ores that is not replaceable after its removal. In contrast, food crops and timber forests are RENEWABLE RESOURCES.

nonroad emission Air pollutant released by internal combustion engines that do not power cars or trucks (e.g., farm vehicles, lawn and garden equipment, stationary construction equipment, and water recreation vehicles).

nontarget organism Plant or animal harmed by PESTICIDE use other than the TARGET ORGANISM.

nonthreshold pollutant Any chemical or physical agent for which any human exposure is assumed to produce an increased risk of an adverse effect. Ionizing radiation and chemical compounds that are suspected to be human carcinogens are considered by the regulatory policy of the United States to be nonthreshold pollutants. See THRESHOLD DOSE, THRESHOLD EFFECT, THRESHOLD HYPOTHESIS.

Nonthreshold pollutant

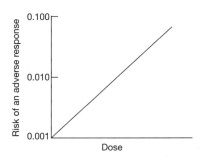

normal solution A solution containing one gram EQUIVALENT WEIGHT per liter.

North American Agreement for Environmental Cooperation (NAAEC) A parallel agreement to the NORTH AMERICAN FREE TRADE AGREEMENT (NAFTA) that focuses on environmental enforcement and environment/trade issues; the NAAEC created the COMMISSION FOR ENVIRONMENTAL COOPERATION of Canada, Mexico, and the United States. Website: http://naaec.gc.ca/

North American Free Trade Agreement (NAFTA) The 1993 agreement liberalizing trade among the United States, Mexico, and Canada. Environmental concerns prompted some, but not all, U.S. environmental groups to oppose NAFTA. The main issues were the relatively less stringent environmental regulations in Mexico, looser environmental compliance standards causing the possible relocation of U.S. plants to Mexico, and possible relaxation of U.S. environmental standards attendant to the new free trade rules, especially those related to PESTICIDE RESIDUES on Mexican produce. The environmental issues were at least partly solved by the NORTH AMERICAN AGREEMENT FOR ENVIRONMENTAL COOPERATION, a supplemental side agreement on the environment, not part of NAFTA, that implements an array of environmental protection measures.

North American Industry Classification System (NAICS) A numerical code system that classifies businesses by type that is replacing the STANDARD INDUSTRIAL CLASSIFICATION (SIC) codes. The NAICS is being used by the United States, Canada, and Mexico as an outgrowth of the NORTH AMERICAN FREE TRADE AGREEMENT (NAFTA) and puts a greater emphasis on services and technology than the SIC codes.

North American Water and Power Alliance (NAWAPA) A controversial scheme to channel water from Alaska and western Canada through the Canadian

Rocky Mountains to the western United States and upper Mexico. The project has never moved from the planning stage thanks to the excessively high economic and environmental costs involved.

nosocomial Describing a disease or infection contracted in a hospital because of either excessive exposure to infectious organisms or the general compromised condition of the patient

not in my backyard (NIMBY) An expression of public opinion acknowledging that waste materials must be treated and disposed of, but asserting that this should be done somewhere else. See also BUILD ABSOLUTELY NOTHING ANYWHERE NEAR ANYONE (BANANA); NOT IN MY TERM OF OFFICE (NIMTOO); YES, IN MY BACKYARD, FOR A PRICE (Y,IMBY,FAP)

not in my term of office (NIMTOO) The slogan sometimes attributed to politicians voting against new landfills or other potentially controversial projects. See also BUILD ABSOLUTELY NOTHING ANYWHERE NEAR ANYONE (BANANA), NOT IN MY BACKYARD (NIMBY), YES, IN MY BACK-YARD, FOR A PRICE (Y,IMBY,FAP).

notice-and-comment rule making The process used by the USEPA to write rules (regulations or standards) that implement federal environmental laws. The rules are proposed in the *FEDERAL REGISTER*, followed by a public comment period and, often, public hearings at locations across the country. The agency responses to public comments appear in another *Federal Register* and the rules may be revised and reproposed for further comments or are issued (promulgated) as final with only one round of comments.

notice letter 1. A formal notice from the USEPA to a POTENTIALLY RESPONSIBLE PARTY that a REMEDIAL INVESTIGATION/ FEASIBILITY STUDY or a cleanup action is to be undertaken at a site at which hazardous substances have been released or pose a substantial threat of release in accordance with the COMPREHENSIVE ENVIRONMENTAL

RESPONSE, COMPENSATION, AND LIABILITY ACT. 2. A letter written to an environmental administrator under the CITIZEN SUIT PROVISIONs of various environmental laws, informing the administrator that the plaintiff alleges a failure to perform a nondiscretionary act. The citizen suit cannot begin until 60 days after the notice letter (also called a 60-day letter).

notice of deficiency A communication from the USEPA or a state agency to an applicant for an environmental PERMIT indicating that additional information is required from the applicant before the agency can complete its regulatory review.

notice of intent to cancel An announcement by the USEPA that it is starting CANCELLATION proceedings for a pesticide.

notice of intent to deny A communication from the USEPA or a state agency to an applicant for an environmental PERMIT indicating that the agency has made a preliminary decision to deny (not issue) a permit.

notice of intent to suspend An announcement by the USEPA that it will issue a SUSPENSION order for a pesticide.

notice of noncompliance A communication from a state agency or the USEPA that a facility is in violation of environmental protection regulations. In many cases, a facility that is not complying with regulations is issued a monetary penalty for noncompliance based on the economic advantage enjoyed by not controlling emissions. See BEN MODEL.

notice of violation (NOV) An official communication from the USEPA or a state enforcement department that a facility has violated a regulation or exceeded the discharge limits or other provisions of the operating PERMIT. See COMPLIANCE ORDER, COMPLIANCE SCHEDULE.

notice of violation—potential penalty (NVPP) Official communication issued by a governmental regulatory agency to an industrial facility warning of the possible imposition of a financial penalty for a violation of environmental protection regulations. Facility management may explain what happened to cause the violation and what steps are being taken to decrease the likelihood of future violations to convince the regulatory agency to reduce or waive the penalty.

nuclear fission   See FISSION.

nuclear fusion   See FUSION.

nuclear reactor   A device to promote and control nuclear FISSION for producing heat and then steam to generate electricity or for producing a certain RADIOISOTOPE. All reactors have a CORE containing nuclear fuel, which serves as the energy source, and CONTROL RODS, which regulate the rate of fission. The fuel is usually a mixture of URANIUM-238 (about 97%) and URANIUM-235 and/or PLUTONIUM-239 (about 3%). The fuel is formed into pellets and packed into metal tubes called FUEL RODS. The core is immersed in COOLANT (commonly water). The system is energized by removal of the control rods. Heat generated by fission is used to make steam, which is used to turn generators that produce electricity. Some reactors operated for the U.S. government produced plutonium, tritium, and other RADIONUCLIDES used in nuclear weapons; WEAPONS-GRADE production ended by 1990. Other than their safe operation, one of the primary environmental concerns related to nuclear reactors is the accumulation of waste materials that are very radioactive and remain dangerous for hundreds of years. These radioactive wastes accumulate in the fuel rods as the reactor operates and within the reactor structure itself. Consequently, used fuel rods and reactors that are decommissioned are sources of large amounts of dangerous radioactive waste.

Nuclear Regulatory Commission (NRC)   A five-member U.S. government commission with supporting staff responsible for issuing licenses for the construction and operation of nuclear power plants; the commission succeeded the Atomic Energy Commission in January 1975. Website: www.nrc.gov

Nuclear Waste Policy Act (NWPA)   A 1982 federal statute that established a schedule to identify a site for and construct an underground repository for SPENT FUEL from nuclear power reactors and HIGH-LEVEL RADIOACTIVE WASTE from federal defense programs. Initially three sites were recommended: Deaf Smith County, Texas; Hanford Reservation, Washington; and Yucca Mountain, Nevada. A 1987 amendment to the NWPA provided for further studies at the YUCCA MOUNTAIN site only. The NWPA is being implemented by the Office of Civilian Radioactive Waste Management (OCRWM) within the DEPARTMENT OF ENERGY. Website: www.rw.doe.gov

nuclear winter   Predicted consequences of a war involving the use of a large number of nuclear weapons. The detonation of the weapons and the resulting fires have been predicted to cause high concentrations of dust and smoke in the atmosphere. The high particulate levels are predicted to remain long enough to reduce the amount of solar radiation reaching the surface of the Earth dramatically, lowering surface temperature to harmful levels worldwide. These theoretical predictions have been questioned strongly and remain a subject of much debate.

nucleic acid   See DEOXYRIBONUCLEIC ACID.

nucleon   Common name applied to a particle in the nucleus of an atom. The most common are the PROTON and the NEUTRON.

nucleus   1. In terms of atomic structure, the corelike center of an atom containing PROTONS and NEUTRONS and surrounded by the electron cloud. The nucleus contains almost all of the mass of an atom. 2. In terms of the structure of biological cells, the largest of the structures within a cell, containing the genetic material, that is, the GENES, CHROMOSOMES, and DEOXYRIBONUCLEIC ACIDS, that determines the properties of a cell and directs the hereditary transmission of features to daughter cells when the original divides.

nuclide A general term for the various configurations of PROTONS and NEUTRONS in atomic nuclei. On the basis of the number of protons in the nucleus, over 100 different ELEMENTs have been identified. An element often has several ISOTOPES with different numbers of neutrons. The elements and their various isotopes number about 1000 nuclides.

null hypothesis The hypothesis to be tested; the statement that there is no difference between two populations in terms of a measured or observed variable. Statistical tests are used to determine whether a significant difference exists, that is, whether the null hypothesis should be rejected.

nutrient Any substance that an organism obtains from the environment for use as an energy source, growth factor (such as a vitamin), or basic material for the synthesis of BIOMASS. The term is often used to identify substances used for growth by plants.

nutrient cycle The cyclic conversions of nutrients from one form to another within the biological communities. A simple example of such a cycle would be the production and release of molecular oxygen (O_2) from water (H_2O) during PHOTOSYNTHESIS by plants and the subsequent reduction of atmospheric oxygen to water by the respiratory METABOLISM of other biota. The cycle of nitrogen is much more complex, as the nitrogen atom undergoes several changes in oxidation state (N_2, NO_3^-, $R\text{-}NH_2$, and NH_4^+) during the cycling of this element through the biological community and into the air, water, or soil, and back.

nutrient sink See NATURAL SINK.

O

Occupational Safety and Health Act (OSHAct) The 1970 federal statute that established the OCCUPATIONAL SAFETY AND HEALTH ADMINISTRATION for the purpose of ensuring, to the extent feasible, a safe and healthy workplace. The act has produced an extensive body of health and safety regulations, a detailed record-keeping system of employee exposure to potentially harmful agents and of workplace injuries and illnesses, periodic workplace inspections, and the creation of the NATIONAL INSTITUTE FOR OCCUPATIONAL SAFETY AND HEALTH to conduct studies of occupational hazards and recommend standards.

Occupational Safety and Health Administration (OSHA) An agency of the U.S. Department of Labor responsible for issuing and enforcing regulations to protect the safety and health of workers. Authorized by the OCCUPATIONAL SAFETY AND HEALTH ACT. Website: www.osha.gov

Ocean Dumping Act See MARINE PROTECTION, RESEARCH, AND SANCTUARIES ACT.

ocean dumping permit Under the Ocean Dumping Ban Act of 1988, which amended the 1972 MARINE PROTECTION, RESEARCH, AND SANCTUARIES ACT, a permit from the USEPA is required to discharge waste into the oceans, for example, the disposal of dredged material. Permits for the dumping of sewage sludge or industrial waste are not allowed after 1991 except under emergency conditions that threaten health and safety.

ocean floor sediment Unconsolidated material that settles and accumulates on the floor of the deep ocean. These materials can be fine muds and clays, quartz grains, dust, glacial debris derived from the landmasses, oozes that comprise microscopic shells of plants or animals, and substances precipitated directly from seawater.

ocean thermal energy conversion (OTEC) The use of the solar energy absorbed by the ocean to produce electricity. The temperature difference between the warm surface water and the cooler, deeper water is as much as 20° C in the tropics, and the movement caused by this thermal gradient can drive an evaporation-condensation cycle of a fluid to turn a turbine generator. Low efficiency and saltwater corrosion are two current technical problems with OTEC. Website: www.nrel.gov/otec/

oceanic crust One of the two basic types of materials forming the outermost layer of the surface of the Earth. The oceanic crust is the thinner but denser of the two types of CRUST; is composed primarily of basalt rocks; and is the material underlying the ocean basins.

oceanic island Landmass formed from features arising from the ocean basins, particularly from volcanic activity. An island that is not derived from continental or granitic material. Some oceanic islands of the Pacific region have been formed through the growth of coral.

oceanic trench Very deep regions of the ocean, commonly close to the continents, where the denser basaltic OCEANIC CRUST is being forced downward by the overriding action of the less dense, but more massive, granitic continental crust.

The resulting down-folding of the oceanic crust results in the formation of deep trench-shaped features on the ocean floor.

octane number A measure of the tendency of gasoline to knock, or ignite prematurely, in an INTERNAL COMBUSTION ENGINE. A fuel with a low knocking potential, isooctane, is arbitrarily given an octane number of 100, and a higher-knocking fuel, *n*-heptane, is given a value of 0. The octane number of a fuel is determined by equating the knocking tendency of the fuel to the knocking of a mixture of isooctane and *n*-heptane. The octane number is the volume percentage of the isooctane in the equated isooctane/*n*-heptane mixture. The higher the octane number, the greater the antiknock property and the more smoothly an internal combustion engine operates.

octanol-water partition coefficient (K_{ow}) A ratio derived from the laboratory measurement of the solubility of a chemical compound in water relative to solubility in *n*-octanol. The ratio is expressed as micrograms of substance per milliliter of *n*-octanol divided by the micrograms of the same substance per milliliter of water. The coefficient can be related to the water solubility, the SORPTION to soil organic matter, and the BIOCONCENTRATION factors of certain organic compounds.

octave bands A series of frequency intervals used to analyze the makeup of a sound. Each interval is the range represented from a given frequency to twice the given frequency. Often, 11 octave bands are used, starting at 22 hertz (Hz) and ending at 44,800 Hz. Therefore, the first octave band is 22–44 Hz, the next, 44–88 Hz, and so on, up to the 11th band, represented by 22,400–44,800 Hz. The center band frequency for each octave band is the GEOMETRIC MEAN of the frequencies at each end of the range; for example, the center band frequency of the 22–44 Hz octave band is the square root of the product of 22 and 44, or approximately 31.5 Hz.

odds ratio An estimate of disease risk used in epidemiological CASE-CONTROL STUDIES. The ratio is calculated as the rate of exposure (to the suspected causative agent) in the diseased group divided by the rate of exposure in the nondiseased group. Compare RELATIVE RISK.

odor fatigue The loss of odor sensitivity that occurs after a period of continuous exposure to an odor.

odor threshold The lowest concentration of a vapor or gas that can be detected as an odor by a stated percentage of a panel of test individuals.

off-gas The normal gas emissions from any process vessel or equipment.

off-site facility A destination for the TREATMENT, STORAGE, OR DISPOSAL of an industrial waste that is not at the location of the generation of the waste. The TOXICS RELEASE INVENTORY compiles off-site transfers of waste.

Office of Civilian Radioactive Waste Management (OCRWM) See NUCLEAR WASTE POLICY ACT.

Office of Information and Regulatory Affairs (OIRA) The executive branch agency within the OFFICE OF MANAGEMENT AND BUDGET that reviews the REGULATORY IMPACT ANALYSIS documents submitted by federal administrative agencies for all major regulations and standards. The OIRA, following EXECUTIVE ORDER 12866, seeks to maximize the benefits and minimize the burdens of new federal rules.

Office of Management and Budget (OMB) An agency in the executive branch with responsibility for drafting the annual budget of the federal government. The OMB also reviews major regulatory proposals, in accordance with EXECUTIVE ORDER 12866. See OFFICE OF INFORMATION AND REGULATORY AFFAIRS, EXECUTIVE ORDER 12866. Website: www.whitehouse.gov/OMB

Office of Solid Waste and Emergency Response (OSWER) The administrative unit of the USEPA responsible for solid and hazardous waste management, including responses to chemical releases. Website: www.epa.gov/swerrims/

Office of Technology Assessment (OTA) An agency established in 1972 to provide Congress with background and overview reports on various technically complex issues. OTA reports have influenced a wide range of legislation, including key environmental protection statutes. In 1995, Congress voted to close the agency. An archive of OTA publications is kept at Princeton University. Website: www.wws.princeton.edu/~ota/

office paper A class of recyclable paper consisting of uncoated high-grade paper characteristic of that used for copiers, typing, and office printers. White paper made with chemical pulping processes. Differing from newsprint and slick, coated stock used to print magazines.

offset An air quality management rule that requires a facility locating or expanding in an area that does not meet a NATIONAL AMBIENT AIR QUALITY STANDARD for an air pollutant to reduce emissions or obtain emission reductions from other facilities in an amount greater than the additional emissions of the nonattainment pollutant associated with the new or expanded facility. The emission reductions must "offset" the new emissions for the pollutant that violates an ambient standard. Emission reductions from existing sources within the facility are an application of the BUBBLE POLICY, sometimes called an internal offset. See NONATTAINMENT AREA, NETTING OUT, EMISSION REDUCTION CREDITS, EMISSIONS TRADING.

oil and gas waste Drilling muds, brines recovered along with oil and gas in production wells, and other liquid and sludge wastes associated with the drilling for and recovery of crude oil and gas. These wastes are typically contaminated with metal ions and high-molecular-weight hydrocarbons. See also NATURALLY OCCURRING RADIOACTIVE MATERIAL, NONHAZARDOUS OIL FIELD WASTE.

Oil and Hazardous Materials Technical Assistance Data System (OHMTADS) A system developed by the USEPA to provide technical information to units responding to spills or releases of hazardous substances. The database includes more than 1,400 chemicals.

oil desulfurization The removal of sulfur from oil fuels to reduce the amount of SULFUR DIOXIDE released into the atmosphere when the fuel oil is burned. Reaction with hydrogen gas is a common step. See HYDROCRACKER.

oil fingerprinting A method for establishing responsibility for oil discharged illegally into the environment. Oil recovered from each deposit has a characteristic mixture of hydrocarbons. By employing instruments capable of measuring the suite of hydrocarbons present in oil discharged into the environment then comparing the results with previously determined characteristics of oil from known locations (comparing fingerprints), the source of spilled oil may possibly be determined. Complicating the comparison are evaporation, photochemical reactions, biodegradation, and dissolution changes in the oil after it enters the environment. Oil fingerprinting is also used by exploration companies to determine the underground flow of hydrocarbons by comparing samples taken from different wells.

Oil Pollution Act of 1990 (OPA90) The federal statute that requires oil spill contingency plans establish liability for oil spill cleanups similar to the liability for hazardous substance cleanups under the COMPREHENSIVE ENVIRONMENTAL RESPONSE, COMPENSATION, AND LIABILITY ACT. OPA90 was enacted after the oil spill from the EXXON VALDEZ in 1989. Responses must be in accordance with the NATIONAL CONTINGENCY PLAN. New vessels operating in U.S. waters must meet additional design standards, including DOUBLE BOTTOMS.

oil shale   See KEROGEN.

oil skimmer   A device that collects and removes oil from a water surface. Ropes, belts, rotating drums, and similar devices are used as adhering surfaces for the oil, which is pressed out or scraped off into a holding tank.

oil spill   Release of crude oil or of petroleum products prepared from crude oil into the environment. See SHEEN RULE.

old field   Cropland that is no longer used to produce an agricultural crop and that has been allowed to revert to natural plant cover.

old growth   Forests that either have never been cut or have not been cut for many decades. Forests characterized by a large percentage of mature trees.

olfactometer   A device that allows a selected group of evaluators (odor panel) to compare the odor from a sample of ambient air to a series of different concentrations of a reference odorant. Each panel member inhales air delivered through a mask or sample port of the olfactometer then selects the reference dilution that best matches the strength of the sample odor. The olfactometer-odor panel approach attempts to introduce a degree of objectivity to the identification and control of odors in community air.

oligotrophic   Describing a body of water with low content of plant nutrients. Compare EUTROPHIC, MESOTROPHIC.

omnivore   An animal that eats both plants and animals, such as a human being. Compare CARNIVORE, HERBIVORE.

on-scene coordinator (OSC)   A federal official designated by the USEPA or the U.S. Coast Guard to direct and coordinate federal responses to oil spills and the releases of HAZARDOUS SUBSTANCES. The OSC is responsible for removal actions after spills. The prevention or minimization of the release of oil or hazardous substances is the responsibility of the REMEDIAL PROJECT MANAGER.

on-site   A location on the same or geographically continuous property or on properties connected by a private right-of-way.

on-site disposal   See ON-SITE FACILITY.

on-site facility   In waste management, a hazardous waste disposal facility located on the same property with the plant that generates the waste. Also used to designate any operation to dispose of nonhazardous solid waste on geographically continuous property occupied by the plant that produces the solid waste.

on-site handling, storage, and processing   In waste management, activities associated with the management of waste at a location that is the same as or is geographically continuous with the property on which a waste is generated.

on-site release   The discharge of pollutants into the air, water, ground, or an injection well on the same or geographically continuous property occupied by an industrial facility. The TOXICS RELEASE INVENTORY compiles on-site releases annually.

onboard refueling vapor recovery   The capture of gasoline vapors (VOLATILE ORGANIC COMPOUNDS) when a vehicle is refueled. The vapors are stored in an activated carbon canister under the hood then routed into the engine when the vehicle is started, preventing loss of the vapors to the atmosphere. Also called onboard controls. Compare STAGE II VAPOR RECOVERY.

onboard control   See ONBOARD REFUELING VAPOR RECOVERY.

once-through cooling   See OPEN-CYCLE COOLING.

oncogene A gene that is involved in cell division through the regulation of the cell cycle. Alteration or damage to oncogenes caused by exposure to chemicals or ionizing radiation is associated with the development of certain cancers.

oncogenic Describing any virus, chemical, or radiation that has the capacity to give rise to tumors. The term is most frequently used to describe viruses. See CARCINOGENIC.

one-hit theory The assumption that a single molecular interaction between a chemical or IONIZING RADIATION and a cell, or "hit," at the cellular level can cause irreversible changes in the cell, leading to a tumor.

one-hundred-year floodplain The land bordering a river or stream that one could reasonably expect to flood once every 100 years on the basis of the historical records of rainfall in the area of land that drains into the waterway. See FLOODPLAIN.

opacity For smokestacks, the degree of light transmission through the plume indicating the concentration of PARTICULATE MATTER in the exhausting air. See RINGELMANN CHART, SMOKE READER.

open access system A resource that is held in common and for which there are no ownership, management, or regulatory rules. The resource is depleted or damaged through excessive use unless the community has unarticulated controls that ensure the continuing availability of the resource. See TRAGEDY OF THE COMMONS.

open canopy A forest where the tree foliage covers less than 20% of the ground. The forest floor in such a system receives enough sunlight to support the growth of brush.

open dump A landfill operated without the environmental safeguards required by current law. All open dumps were forced by the RESOURCE CONSERVATION AND RECOVERY ACT of 1976 to upgrade to SANITARY LANDFILL status or to close.

open range Natural grazing land, frequently federal land on which cattle owners place stock.

open system An environment or defined area that is characterized by an ongoing input of chemical elements used to support the growth of plant and animal communities and a corresponding loss of biomass or chemical elements from the area to the outside. Compare CLOSED SYSTEM.

open water loop Any process in which water is routed through a facility, then not reused, but discharged into a surface water body after any appropriate treatment. Compare CLOSED WATER LOOP.

open-cycle cooling The practice of withdrawing surface or well water to cool the condensers of an electric power plant or other industrial equipment, followed by release of the heated water to the ocean, a river, or a lake.

open-pit mining See STRIP MINING.

operable unit In the cleanup of a HAZARDOUS SUBSTANCES SUPERFUND site, each of the discrete activities associated with the cleanup (e.g., removal of contaminated earth or transport of drums stored on site). The REMEDIAL INVESTIGATION/FEASIBILITY STUDY defines and evaluates the operable units.

opt in To adopt provisions of environmental control statues and their accompanying regulations although the measures do not legally apply. For example, some states require the use of REFORMULATED GASOLINE (RFG) in certain urban areas although these areas are not among those where its use is mandated.

optical coefficient An expression of the fraction of light energy striking a surface that is absorbed, is reflected, or passes through the material.

oralloy Uranium that has been enriched with the fissionable isotope uranium-235. The mixture is capable of supporting a sustained CHAIN REACTION.

order of magnitude A factor of 10 times.

ordinal variable A characteristic expressed by rank or order. For example, a sample group might be classified by shades of the color red, with the number 1 being the lightest red and 10 the darkest red. Although the ranking is often in numerical order, the data do not represent actual quantities; there is no indication of the measured difference between the ranked samples. Used only in NONPARAMETRIC TESTS. See also NOMINAL VARIABLE.

organ of Corti The cells in the ear that translate vibrations into nerve impulses to be sent to the brain. Different cells react to different sound frequencies. The aging process and excessive noise can damage these cells and cause permanent hearing impairment. The cells responding to higher-frequency sounds are almost always harmed first.

organic Of or related to a substance that contains CARBON atoms linked by carbon-carbon bonds. All living matter is organic. The original definition of the term *organic* related to the source of chemical compounds; organic compounds were those carbon-containing compounds obtained from plant or animal sources, whereas INORGANIC compounds were obtained from mineral sources. We now know that compounds containing bonds between carbon atoms can be made in the laboratory and industrially.

Organic Act 1. The 1916 federal statute establishing the National Park Service to manage national parks, monuments, and reservations so as "to conserve the scenery and the natural and historic objects and the wild life therein and to provide for the enjoyment of the same in such manner and by such means as will leave them unimpaired for the enjoyment of future generations," providing a strong stewardship impetus for U.S. public lands. **2.** Any document establishing or providing the organizational framework for a unit of government.

organic acid A compound consisting of carbon atoms that are linked with carbon-carbon covalent bonds that release hydrogen ions when put into solution in water. A simple organic acid is vinegar, chemically known as acetic acid. Organic acids are responsible for the bitter taste of citrus fruits, and some large-molecular-weight organic acids are responsible for some body odor. See ORGANIC, ACID.

organic carbon 1. Carbon atoms in an organic compound that are linked to other carbon atoms by a COVALENT BOND. Used to distinguish such compounds from inorganic forms of carbon, such as those found in the carbonates and cyanides. **2.** The amount of organic material in soil or water.

organic carbon partition coefficient (K_{oc}) A measure of the extent to which an organic chemical is ADSORBED to soil particles or sediment. The measure is expressed as the ratio of the amount of adsorbed carbon per unit mass of total organic carbon (milligrams of carbon adsorbed per kilogram organic carbon) to the equilibrium concentration of the chemical in solution (milligrams dissolved carbon per liter, or kilogram, of solution). Higher values of K_{oc} indicate greater organic carbon adsorption and greater retention of the organic chemical in the soil. Lower values indicate less retention and greater transport of the chemical in the ground or surface water.

organic compounds See ORGANIC.

organic load The amount of organic material added to a body of water. The quantity of material, usually added by human activities, that must be mineralized or degraded within a particular environment

organic matter Normally used to refer to the remains of plant or animal matter or to FECAL MATERIAL.

organic nitrogen Nitrogen that is bound to carbon-containing compounds. This form of nitrogen must be subjected to MINERALIZATION or decomposition before the nitrogen can be used by the plant communities in aquatic and terrestrial environments. By contrast, inorganic nitrogen is in the mineral state and more readily utilized by plant communities. See ORGANIC WASTE.

organic phosphorus Phosphorus bound to carbon-containing compounds. See ORGANIC NITROGEN.

organic solvent A liquid hydrocarbon such as methyl ethyl ketone or toluene used to dissolve paints, varnishes, grease, oil, or other hydrocarbons.

organic waste Carbon-containing materials that are discarded into the environment. The term is often used as a euphemism for domestic SEWAGE.

organismal cloning Technology that produces two or more genetically identical animals. The successful pioneering experiment was the creation of Dolly, the lamb that was the first cloned animal. A mammary cell was removed from a donor sheep and grown in laboratory tissue culture. An unfertilized egg was removed from a second sheep and the nucleus removed. The mammary cell from the laboratory culture and the egg from which the nucleus was removed were fused and then induced to divide in a laboratory setting. The product was then transferred to a surrogate ewe, in which normal embryonic development took place. The lamb that was born to the surrogate ewe was genetically identical to the donor sheep from which the original mammary cell was removed. The end product is a mammal created by a process that does not require the participation of a male sperm donor.

organochlorine See CHLORINATED HYDROCARBONS.

organomercurials Common name applied to organic molecules complexed with MERCURY. These compounds are toxic to humans at low doses; the TARGET ORGAN is the nervous system. The most well-known member of the group is METHYL MERCURY.

organophosphates A family of organic insecticides that includes tetraethyl pyrophosphate, diazinon, malathion, and parathion. The compounds, which are esters of phosphoric acid, contain phosphorus as an integral component of the molecules. Members of this family of insecticides interfere with nerve transmission through the inhibition of cholinesterase. In humans, they cause headache, weakness, and dizziness in

Organophosphate

Parathion

lower doses and paralysis, convulsions, and coma in higher doses. Members of the family react readily with clay particles, ions, and other components of soil. They are relatively nonpersistent in the environment. See CHOLINESTERASE INHIBITORS.

orifice meter   A device for measuring gas flow rates in which the gas flows through a hole (orifice) in a plate positioned in a pipe. The difference in static pressure measured upstream and downstream of the plate is used to compute the gas flow rate.

orographic lifting   The upward movement of air when currents in the atmosphere encounter mountains. The rising air expands and cools, condensing moisture in the air as clouds and resulting in precipitation. The air descending on the other side of the mountains has lost moisture, and, as the air descends, it is compressed and warms.

orphan site   See ABANDONED SITE.

Orsat analysis   A determination of the relative amounts of selected gases in a smokestack by using chemical-specific solvent absorption.

osmosis   The diffusion of a solvent (typically water) across a membrane (either natural or artificial) separating two solutions of different concentrations. The semipermeable membrane allows the passage of water but prevents the passage of substances dissolved in the water. The water movement is from the more dilute solution toward the more concentrated solution and continues until the two solutions are equal in concentration. If pressure is applied to the more concentrated side, the flow of water reverses, from concentrated side to the more dilute side. See REVERSE OSMOSIS.

osmotic lysis   The rupture of a cell placed in a dilute solution. For example, when a red blood cell is placed in distilled water, water tends to move into the cell because of the osmotic pressure generated

as a result of the concentration of the materials inside the cell. As the amount of water increases within the cell, the cell membrane can no longer withstand the pressure and ruptures. The process is not unlike puffing air into a balloon until it pops.

osmotroph   An organism that obtains nutrients through the active uptake of soluble materials across the cell membrane. This class of organism, which includes the bacteria and fungi, cannot directly utilize particulate material as nutrients. Compare to PHAGOTROPH.

outage   1. The difference between the volume capacity of a container and the actual content. 2. The period when industrial equipment is shut down for routine maintenance. 3. An interruption of electric power delivery, resulting from a power plant failure or a break in the distribution system.

Outer Continental Shelf Lands Act   The federal statute authorizing the Department of Interior to sell leases on the federal portion of the CONTINENTAL SHELF. The lease auctions are mainly for exploration and production of oil and gas resources. The act requires the secretary of the interior to manage, with coastal states' participation, the development of outer continental shelf RENEWABLE RESOURCES and NONRENEWABLE RESOURCES to minimize adverse environmental, economic, and social effects, recognizing that marine, coastal, estuarine, and onshore areas all could be harmed by the federal leasing.

outfall   The location where wastewater is released from a POINT SOURCE into a receiving body of water.

outgassing   The loss of vapors or gases by a material, usually as a result of raising the temperature of and/or reducing the pressure on the material.

Outstanding   Natural   Resource   Waters (ONRW)   Specific bodies of water designated by the states under provisions of the CLEAN WATER ACT for preservation, protection, reclamation, or enhancement because of wilderness, aesthetic, and ecological properties.

outwash Sand and gravel deposited by meltwater from glaciers. Individual deposits tend to be of uniform particle size.

overburden The rock and dirt that overlie a mineral deposit and that must be removed before the mineral deposit can be extracted by surface mining.

overdraft The sustained extraction of groundwater from an aquifer at a rate greater than that at which the aquifer is recharged, resulting in a drop in the water table. Also called MINING OF AN AQUIFER.

overfishing The removal of a sufficiently large number of certain fish from a body of water such that breeding stocks are reduced to levels that will not support the continued presence of the fish in desirable quantities for sport or commercial harvest.

overland flow The discharge of wastewater in such a way that the water flows over a defined land area prior to entering a receiving stream. The movement over vegetated land fosters the removal of plant nutrients from the wastewater. See TERTIARY TREATMENT.

overpack An external, secondary container used to enclose a packaged hazardous material, hazardous waste, or radioactive waste.

overshoot The extent to which the number of animals in the population of a specific species exceeds the number that a particular environment can support over an extended period. See CARRYING CAPACITY.

oversight In U.S. federal environmental law, congressional review of an administrative agency's implementation and management of the regulatory programs for which it is responsible. The USEPA is subject to oversight by 7 House of Representatives committees, 6 Senate committees, and over 30 subcommittees of the two legislative branches. Some of the statutes identifying the USEPA as the federal administrative agency are the CLEAN AIR ACT; CLEAN WATER ACT; TOXIC SUBSTANCES CONTROL ACT; RESOURCE CONSERVATION AND RECOVERY ACT; COMPREHENSIVE ENVIRONMENTAL RESPONSE, COMPENSATION, AND LIABILITY ACT; POLLUTION PREVENTION ACT; and NATIONAL ENVIRONMENTAL POLICY ACT, each creating many separate programs requiring congressional oversight.

overturn See FALL TURNOVER, SPRING TURNOVER, TURNOVER.

oxidant See PHOTOCHEMICAL OXIDANT.

oxidation An array of reactions involving several different types of chemical conversions: (1) loss of electrons by a chemical (the most common definition), (2) combination of oxygen and another chemical, (3) removal of hydrogen atoms from organic compounds during biological metabolism, (4) burning of some material, (5) biological metabolism that results in the decomposition of organic material, (6) metabolic conversions in toxic materials in biological organisms, (7) stabilization of organic pollutants during wastewater treatment, (8) conversion of plant matter to compost, (9) decomposition of pollutants or toxins that contaminate the environment. See REDUCTION.

oxidation pond A pond into which an ORGANIC waste (sewage) is placed to allow decomposition or MINERALIZATION by aerobic microorganisms.

oxidation-reduction reaction A coupled reaction in which one atom or molecule loses electrons (through OXIDATION) and another atom or compound gains electrons (through REDUCTION).

oxides of nitrogen (NO_x) See NITROGEN OXIDES.

oxides of sulfur (SO$_x$) See SULFUR OXIDES.

oxidizing agent Any material that attracts electrons, thereby oxidizing another atom or molecule. The oxidizing agent is reduced: that is, gains electrons. The material donating the electrons is the REDUCING AGENT. Chlorine and oxygen are good oxidizing agents. See OXIDATION.

oxygen cycle The use and production of molecular oxygen (O$_2$) within the biosphere. The oxygen in the atmosphere is in DYNAMIC EQUILIBRIUM in that diatomic oxygen is continually produced during PHOTOSYNTHESIS and is continually utilized in AEROBIC metabolism as well as in various chemical reactions.

oxygen demand The requirement for molecular oxygen (O$_2$) of biological and chemical processes in water. The amount of molecular oxygen that dissolves in water is extremely limited; however, the involvement of oxygen in biological and chemical processes is extensive. Consequently, the amount of oxygen dissolved in water becomes a critical environmental restraint on the BIOTA living in water. The METABOLISM of large organisms such as submerged plants and fish, the microorganisms engaged in DECOMPOSITION, and spontaneous chemical reactions all require (demand) a portion of a limited resource, molecular oxygen. See BIOCHEMICAL OXYGEN DEMAND, CHEMICAL OXYGEN DEMAND, DISSOLVED OXYGEN.

oxygen-demanding waste Any organic material that, when discharged into a natural waterway, stimulates the METABOLISM of bacteria with a corresponding use of DISSOLVED OXYGEN. Often used as a euphemism for domestic SEWAGE.

oxygen depletion The removal of DISSOLVED OXYGEN from a body of water as a result of bacterial metabolism of degradable organic compounds added to the water, typically by human activities.

oxygen sag curve A graph of the measured concentrations of DISSOLVED OXYGEN in water samples collected (1) upstream from a significant POINT SOURCE of readily degradable organic material, (2) from the area of the discharge, and (3) from some distance downstream from the discharge, plotted by sample location. The amount of dissolved oxygen is typically high upstream, diminishes at and just downstream from the discharge location (causing a sag in the line graph), and returns to the upstream levels at some distance downstream from the source of pollution.

oxygenate The primary additives to REFORMULATED GASOLINE. A group of

Oxygen sag curve

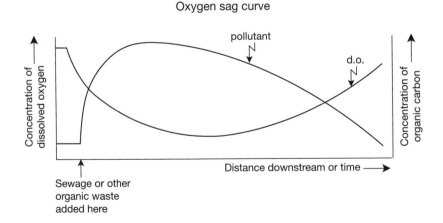

compounds that contain oxygen as part of the structure. The addition of these additives (in amounts up to 15% by volume) increases the OCTANE NUMBER of the final product and, by promoting more complete fuel combustion, decreases the emission of VOLATILE ORGANIC COMPOUNDS and CARBON MONOXIDE. Some of the compounds serving as oxygenates are METHANOL, ETHANOL, METHYL TERTIARY BUTYL ETHER (MTBE), ethyl tertiary butyl ether (ETBE), and tertiary amyl methyl ether (TAME). See OXYGENATED FUEL.

oxygenated fuel 1. Gasoline containing OXYGENATE. 2. For tailpipe CARBON MONOXIDE emission reductions, gasoline containing 2.7% oxygen by weight. Federal regulations require oxygenated fuel to be used during the winter months in areas not meeting the carbon monoxide air quality standards. See also REFORMULATED GASOLINE, OXYGENATE.

oxygenated solvent A solvent containing oxygen as part of the chemical structure, for example, alcohols and ketones used in paints.

oxyhemoglobin The hemoglobin of red blood cells that is bound to molecular oxygen. Hemoglobin of red blood cells that is functioning normally.

ozonation The use of OZONE gas (O_3) as a disinfectant to reduce microbial load and to kill dangerous pathogenic bacteria. The treatment can be applied to a public drinking water supply before the water enters the distribution system or to wastewater before discharge into a receiving stream.

ozone Triatomic oxygen (O_3). A very reactive gas produced by PHOTOCHEMICAL reactions or lightning in the troposphere and by the absorption of ultraviolet radiation in the lower stratosphere. In sufficiently high concentrations at ground level, the gas acts as an irritant to the eyes and respiratory tract. See OZONE LAYER,

PHOTOCHEMICAL AIR POLLUTION, PHOTOCHEMICAL OXIDANTS.

ozone-depleting potential (ODP) Describing the relative damage to the OZONE LAYER caused by CHLOROFLUOROCARBONS and certain brominated compounds. It is rated in terms of the ODP of CFC-11, which equals 1.

ozone hole A seasonal drop in stratospheric ozone concentration over a large area of Antarctica. The sharp drop in ozone concentration is linked to wintertime POLAR STRATOSPHERIC CLOUDS in the POLAR VORTEX, a region above the South Pole with very strong winds and isolated from the rest of the atmosphere. The polar stratospheric clouds convert chlorine atoms from CHLOROFLUOROCARBONS into chemical forms that destroy ozone. The increasing solar input during the Southern Hemisphere's spring provides energy that accelerates the ozone destruction. The ozone hole abates by late spring. See OZONE LAYER DEPLETION.

ozone layer An area of the stratosphere, about 12 to 30 miles in altitude, where the intensity of short-wavelength ULTRAVIOLET light from the sun (between 130 and 200 nanometers) is sufficiently high to convert normal diatomic oxygen (O_2) to ozone (O_3). The ozone thus formed provides a measure of protection to plant and animal life on the surface of the Earth because the substance absorbs ultraviolet radiation at wavelengths that are mutagenic (about 250 to 270 nanometers). See OZONE LAYER DEPLETION.

ozone layer depletion The destruction of ozone molecules in the OZONE LAYER of the stratosphere by chemical reactions with materials released by human activities. The main ozone-consuming chemicals are the CHLOROFLUOROCARBONS (CFCs) and the HALONS, both of which are extremely stable in the troposphere, with typical atmospheric lifetimes of 60 to 100 years. If the CFCs or halons migrate to the stratospheric ozone layer, the ULTRAVIOLET radiation there is strong enough to break the molecules apart, releasing chlorine atoms (CFCs) or bromine atoms (halons), which react with and destroy ozone. See MONTREAL PROTOCOL, OZONE HOLE.

ozone threshold value The ambient air quality standard for GROUND-LEVEL OZONE used by the European Environment Agency.

Ozone Transport Assessment Group (OTAG) A 1995–97 united effort by the USEPA, the ENVIRONMENTAL COUNCIL OF STATES, and industrial and environmental representatives that studied the transport of GROUND-LEVEL OZONE or chemical PRECURSORS to ozone across state boundaries, especially focusing on the northeastern U.S. corridor from Washington, D.C., to Boston. OTAG was prompted by the concern that the STATE IMPLEMENTATION PLAN for controlling ozone formation within the boundaries of each state was not working and that a coordinated response was required.

Ozone Transport Commission A group of environmental officials from 12 Northeast and mid-Atlantic states, the District of Columbia, and the USEPA authorized by the CLEAN AIR ACT to assess the interstate movement of GROUND-LEVEL ozone and ozone PRECURSORS and to coordinate their ozone control measures. The work of the OZONE TRANSPORT ASSESSMENT GROUP provided important technical background to the commission. Website: www.sso.org/otc/

Ozone Transport Rule A USEPA program requiring 22 states to revise their STATE IMPLEMENTATION PLANS to control emissions of NITROGEN OXIDES, which are PRECURSORS to GROUND-LEVEL OZONE formation. The work of the OZONE TRANSPORT ASSESSMENT GROUP concluded that interstate movement of ozone or its precursors required a multistate approach. Each state is given a fixed level of nitrogen oxide emissions and is free to choose which mix of controls it will use to reduce emissions to the state limit. The program includes a market-based emissions trading program that allows sources that achieve emission reductions beyond their requirements to sell the EMISSION REDUCTION CREDITS to another source for which installation of control devices would cost more than the cost of the credits. This trading program is modeled after the successful use of the TRADABLE EMISSION ALLOWANCE program in the control of ACID RAIN.

P

p (*p*-value)   A value, calculated by various statistical tests, used to evaluate the relationship between two sets of data. The p statistic provides a mechanism for deciding whether the differences between two sample populations in regard to some characteristic or disease incidence are due to sampling variability alone or whether the observed differences are real. For example, consider a study done to investigate the association between smoking and bladder cancer. The NULL HYPOTHESIS is that cigarette smoking is not a significant risk factor for the development of bladder cancer. Observations from the study reveal that the incidence of bladder cancer among smokers is 16 per 10,000 individuals and the incidence among the nonsmokers is 8 per 10,000 individuals. The question is, Is this observed difference real? The p-value computed for the study is 0.01. This p-value means that, *if the two groups are actually part of the same population* (there is no relationship between bladder cancer and smoking and the null hypothesis is true), then there is 1 chance in 100 that a random sampling of that population will produce a difference as large as was actually observed, that is, 16 per 10,000 in the smokers and 8 per 10,000 in the nonsmokers. Consequently, if we are willing to accept a 0.01 chance that we could be wrong, the null hypothesis would be rejected and the conclusion would be that cigarette smoking increases one's risk of developing bladder cancer. The conclusion is that the observed difference is real based on our acceptance of the p-value.

P waste   Acutely hazardous chemicals defined as LISTED HAZARDOUS WASTE by the USEPA in the *Code of Federal Regulations,* Title 40, Part 261.33. See also F WASTE, K WASTE, U WASTE, HAZARDOUS WASTE.

packaging   Container or wrapping used to protect, store, or display a commodity. The covering material may be made of cardboard, plastic, paper, boxboard, wood, or similar material. The use of excess material has been criticized because of the contribution to the solid waste that must be managed, although packaging maintains freshness, prevents unsanitary contamination, allows efficient stacking and display of goods, prevents damage during shipping: all of which also permit goods to be shipped long distances at reasonable costs. The printed material on the packaging conveys important consumer information such as nutritional value, proper storage conditions, and directions for safe use and disposal. "Excess" packaging is often used to discourage shoplifting of small items. In these ways, packaging has also contributed to the democratization of consumption.

packed bed absorber   See PACKED TOWER.

packed tower   An air pollution control device in which contaminated air is moved through a tower containing materials (PACKING) that have a large surface area; the air contaminants are then absorbed into a liquid flowing over the packing materials. The flow is usually countercurrent; that is, the liquid falls downward and the air is forced upward through the tower. See also SPRAY TOWER, TRAY TOWER. Compare PACKED TOWER AERATION.

packed tower aeration   A process for the removal of organic contaminants from groundwater. The groundwater flows downward inside a tower filled with materials (PACKING) over a large surface area. Air is introduced at the bottom of the tower and is forced upward past the falling water. Individual organic contaminants are transferred from the water to the air, according to the gas and water equilibrium concentration values of each contaminant. See also AIR STRIPPING, HENRY'S LAW CONSTANT.

packer A device lowered into a well to produce a fluid-tight seal.

packing Corrosion-resistant materials, such as porcelain, stainless steel, or polypropylene, used in a PACKED TOWER to maximize gas-liquid contact for air pollutant removal efficiency. Packing types vary by shape and include Berl saddles, Lessing rings, Pall rings, Raschig rings, and Intalox saddles.

pahoehoe flow Hawaiian term (pronounced "pa ho'e ho'e") used to describe a type of basaltic lava flow from a volcano. The lava is characterized by molten material containing much more gas than normal, resulting in a thin, fast-moving flow of lower viscosity.

paint filter liquids test (PFLT) A test used to determine the presence of FREE LIQUIDS in an untreated waste sample or in a sample of a waste that has undergone STABILIZATION/SOLIDIFICATION. The test is performed by placing the sample in a supported funnel lined by a paint filter; if no liquids pass through the filter and flow from the funnel within five minutes, no free liquids are present.

palatable water Drinking water that is free of tastes, odors, color, or turbidity that would render it unacceptable to the public. Despite the factors that make the water unacceptable for drinking, it does not represent a danger to human health.

palisade parenchyma Along with the spongy parenchyma, chlorophyll-containing cells between the upper and lower epidermis, or outer layers, of a leaf. These cells can be damaged by excessive exposure to some air pollutants.

palustrine Describing marsh or wetlands.

paper tape sampler An air-sampling device for suspended PARTICULATE MATTER. Sampled air is directed toward a filter tape. The tape is moved periodically, and the series of small circular dust spots produced is evaluated by measuring the transmittance of white light through the spots compared with the transmittance through an unsoiled filter. See COEFFICIENT OF HAZE.

paper trail See ADMINISTRATIVE RECORD.

parabolic mirror Curved reflective surfaces that, when facing the Sun, reflect light energy on a central point. The mirror is used in systems termed high-temperature solar energy projects. A long array of such mirrors are positioned so that the energy is concentrated on a tube containing a heat-absorbing liquid that is elevated to a high temperature. The heat collected in the tube can be coupled to a power-generating system.

paraffins Straight- or branched-chain hydrocarbons with carbon-carbon single bonds; the members of this group are relatively nonreactive. A large component of CRUDE OIL. Also called ALIPHATIC hydrocarbons. When the carbon atoms are arranged in a ring structure, they are called cycloparaffins.

paralytic shellfish poisoning (PSP) A pathological condition in humans caused by the consumption of certain marine mussels or clams that have fed on planktonic dinoflagellates of the genus *Gonyaulax*. The mussels or clams become contaminated with a NEUROTOXIN produced by the dinoflagellates, and subsequent consumption by humans results in respiratory or gastrointestinal distress. The condition occurs in conjunction with the phenomenon known as the RED TIDE, a bloom of dinoflagellate populations in marine waters.

parameter The true value of a population characteristic. The estimate of a parameter, called a statistic, is a measurement of a sample of the population. For example, the average height of 12-year-old girls, based on measuring the height of 10 girls, may be stated as 54 inches, a statistic. The parameter corresponding to the true average height of all 12-year-old girls is unknown but is approximated by the sample results.

parametric tests Statistical tests that attempt to draw inferences about population PARAMETER, for example, the mean or standard deviation. These tests use quantitative data (measurements) and require certain assumptions about the sample population data distributions. Compare NONPARAMETRIC TESTS.

paraquat The best-known example of the bipyridyl herbicides, 1,1'dimethyl-4,4'-bipyridylium. Large exposures to the agent result in damage to the lungs, liver, and kidneys. For a fatal exposure, the cause of death is usually associated with respiratory distress even when the herbicide is taken by mouth. Evidence of lung damage can become apparent up to two weeks after the exposure, at which time the agent has been eliminated from the body. Paraquat adsorbs strongly to soil particles and resists microbial degradation.

parasite An organism (the parasite) whose natural habitat is either on or inside another, larger organism (the host). The presence of the parasite results in damage to the host.

parasitism A biological interaction between species in which a PARASITE gains nourishment by living on or inside a host organism. The interaction is beneficial to the parasite and detrimental to the host.

parathion One of the early ORGANO-PHOSPHATE insecticides developed in the 1940s and used as a replacement for dichlorodiphenyltrichloroethane (DDT). The agent, O,O-diethy-O-p-nitrophenyl phosphorothioate, is a stable, AROMATIC compound that binds tightly to sediments. The insecticide has little specificity in terms of toxicity to target and nontarget species. The relatively high toxicity to humans associated with the inhibition of nerve function resulted in many bizarre accidents when the agent was introduced in the 1950s. Workers carelessly handled this new insecticide as they did DDT-related agents because they did not appreciate the immediate toxicity of this compound as compared with the relatively benign nature of ORGANOCHLORINE insecticides.

parent material The rock from which a soil is derived; bedrock.

parenteral In animal toxicology studies, the administration of the test chemical by any route other than oral; most often by skin absorption or via inhalation.

Parshall flume A device commonly used to measure flow rates of water entering a wastewater treatment facility. As the untreated wastewater enters a treatment facility by way of an open channel, the water is forced to flow through a constricted section built into the channel. The geometry of the constricted area allows for the calculation of the flow rate on the basis of the height of the water as it enters the constricted region. Unlike most other devices used to measure flow rates, the flume is designed to allow for monitoring the flow rate of water that carries a large amount of SUS-PENDED PARTICULATE MATTER, which would foul most other devices used to monitor liquid flows

Part A Under the RESOURCE CONSERVA-TION AND RECOVERY ACT, the first part of the two-part permit application for a hazardous waste TREATMENT, STORAGE, OR DISPOSAL facility; the form asks for general information about the facility. INTERIM STATUS facilities are said to operate under a Part A permit. See PART B.

Part B The detailed, technical second part of the permit application for a hazardous waste TREATMENT, STORAGE, OR DISPOSAL facility under the RESOURCE CONSERVATION AND RECOVERY ACT. Approval of Part B constitutes a final permit to operate the facility. See PART A.

partial closure In hazardous waste management, the closure of one or more parts of a TREATMENT, STORAGE, AND DISPOSAL facility, such as a tank, surface IMPOUNDMENT, or waste pile, while the remainder of the facility remains in use. The closure must follow the requirements found in Title 40 of the *Code of Federal Regulations,* Parts 264 and 265, as appropriate.

partial pressure The pressure exerted by an individual gas in a gaseous mixture. If the mixture of gases is assumed to behave as an IDEAL GAS, the sum of the individual partial pressures in a mixture is equal to the total pressure of the mixture of gases. For example, the atmosphere contains roughly 78% nitrogen, 21% oxygen, and many other gases, which total the remaining 1%. Of a total atmospheric pressure of 760 millimeters of mercury (mm Hg), the partial pressure of the nitrogen is about 593 mm Hg (78% of 760 mm Hg), the oxygen partial pressure is about 160 mm Hg (21%) and the other gases total about 7 mm Hg. See DALTON'S LAW.

particle size distribution A presentation of the relative masses of airborne particles in different size ranges present in an air sample. For example, the air might contain particles 0–2 micrometers in diameter, 15 milligrams (8% of total); 2–4 micrometers, 20 milligrams (11% of total); and so on.

particulate control device Any of several types of air pollution control equipment that remove PARTICULATE MATTER from exhaust gases. See ELECTROSTATIC PRECIPITATOR, BAGHOUSE, SCRUBBER, PACKED TOWER, TRAY TOWER, SPRAY TOWER.

particulate loading The mass of solid materials present in a given amount of air or water. A measure of air or water pollution.

particulate matter 1. In water pollution, material in either the solid or dissolved state. Insoluble particulate matter includes particulate substances that either settle from water that is allowed to stand or are removed by passing the water through a filter. Sand, clay, and some organic matter constitute insoluble particulate matter. Dissolved substances that neither settle if water is allowed to stand nor are removed by passage through a filter but will be recovered if the water is allowed to evaporate are called dissolved particulate matter or DISSOLVED SOLIDS. Salt is an example of this type. 2. In air pollution, solid particles or liquid droplets suspended in air or carried by a stream of air or other gas through a duct, for example. See PARTICULATE MATTER, 10-MICRON DIAMETER; PARTICULATE MATTER, 2.5-MICRON DIAMETER; PARTICULATE ORGANIC MATTER; PARTICULATE PHOSPHATE.

particulate matter, 10-micron diameter (PM_{10}) Airborne particles with diameter equal to or smaller than 10 micrometers. The size range is defined in the ambient air quality standard for particulate matter established by the USEPA. Particles of this size can penetrate deep into the respiratory system and can potentially damage the gas-exchange surfaces of the lung or be absorbed into the blood, with possible systemic toxic effects. See ALVEOLAR REGION; PARTICULATE MATTER, 2.5-MICROMETER DIAMETER ($PM_{2.5}$)

particulate matter, 2.5-micron diameter ($PM_{2.5}$) Airborne particles with diameters equal to or smaller than 2.5 microns, small enough to penetrate to the gas-exchange area (ALVEOLAR REGION) of the lungs. The USEPA has proposed the addition of a NATIONAL AMBIENT AIR QUALITY STANDARD for $PM_{2.5}$. The existing standard is for PARTICULATE MATTER, 10-MICROMETER DIAMETER.

particulate organic matter (POM)
Material of plant or animal origin that is suspended in water. The amount of this type of material suspended in water can be estimated by first removing the suspended material from water by filtration then either directly measuring the amount of carbon retained on the filter or estimating the amount of carbon present from the weight lost upon heating of the filter in excess of 500° C. Generally, the greater the amount of particulate matter present, the more severe the water pollution problem.

particulate phosphate   That portion of the total amount of phosphate (PO_4^{-3}) suspended in water that is attached to particles and does not pass through a filter. The aggregates can be either inorganic or organic. This form of phosphate must be solubilized before use as a plant nutrient.

particulates   See PARTICULATE MATTER.

partition coefficient   A measure of the distribution of some chemical between two IMMISCIBLE solvents. See OCTANOL-WATER PARTITION COEFFICIENT.

partitioning   The division of a chemical into two or more COMPARTMENTS in an ecosystem or body of an organism.

parts per billion (ppb)   A unit of measure commonly employed to express the number of parts (e.g., grams) of a chemical contained within a billion parts of gas (air), liquid (water), or solid (soil). Conversions of various parts per billion expressions to metric units are given in the Appendixes. See CONCENTRATION.

parts per million (ppm)   A unit of measure commonly employed to express the number of parts (e.g., grams) of a chemical contained within a million parts of gas (air), liquid (water), or solid (soil). Conversions of various parts per million expressions to metric units are given in the Appendixes. See CONCENTRATION.

pascal (Pa)   The SI unit of pressure equal to one NEWTON per square meter. Expressed as 1 Pa = 1 N m^{-2}.

Pasquill-Gifford stability class   One of six atmospheric stability categories used to estimate DISPERSION COEFFICIENTS for a GAUSSIAN PLUME MODEL. For a particular stability class at a given distance downwind from a source of air pollution, a dispersion coefficient is input, with other variables, to the model to estimate the air concentration of the released pollutant. The classes represent different combinations of solar insolation and wind speed present during a pollutant release.

passive smoking   The involuntary inhalation of ENVIRONMENTAL TOBACCO SMOKE (ETS) from another person's cigarette, cigar, or pipe. Studies have shown an increased risk of respiratory illnesses in children and adults exposed to ETS. The USEPA has classified ETS as a human carcinogen. Also called second-hand smoke.

passive solar system   A design that captures sunlight to heat a structure. Solar energy is captured directly through the use of large windows, special structural materials, or greenhouses, and that heat is distributed within the structure without the aid of mechanical devices such as fans. Compare ACTIVE SOLAR SYSTEM.

patchiness   In ecology, describing the uneven distribution of organisms within a habitat or particular environment.

pathogen   A microorganism, such as a bacterium or fungus, that has the capacity to cause disease under normal conditions.

pathway   1. Physical course traveled by a pollutant from a source to an exposed organism. The route can be through air, surface water, groundwater, sediment, dust, or food. 2. The series of chemical conversions occurring as a substrate or toxic material is metabolized in an organism. Each step is catalyzed by a specific ENZYME.

pay-as-you-throw pricing A MUNICI-PAL SOLID WASTE management program under which residents pay collection and disposal fees based on the amount of discards collected by the city rather than a flat fee regardless of the amount discarded. The payment can be based on either weight or volume.

peak load The maximal electricity demand in an area, expressed on a daily, seasonal, annual, or other time basis.

peak shaving The use of supplemental electricity-generating units to meet large but short-term power demands.

peat The residue of partly decomposed plant material in which various plant parts such as stems can easily be discerned. The residue is mined for use as a soil builder in gardening or as a low-grade fossil fuel. Peat is considered to be an early stage in the conversion of plant residues to coal.

pelagic Describes marine biota that live suspended in the water. Compare BENTHIC.

penetration In air pollution control, an expression of the fraction of particulate matter in a gas stream that is not removed by a collection device.

pentachlorophenol (PCP) A synthetic compound very similar to hexachloroben-zene; the only difference is the substitution of one of the chlorine atoms (Cl) with a hydroxyl group (OH). (See HEXA-CHLOROBENZENE for a diagram.) The substance has been used in the United States since 1936 as an insecticide, herbicide, and fungicide. The treatment of wood to prevent attack and decomposition by fungi and insects is one of the major uses for this chemical. Because it is widely used and because the material is persistent in the environment, PCP can routinely be recovered from soil and water samples at concentrations less than one part per billion. Such concentrations do not represent a health hazard. Adverse reactions to the agent have been reported as a result of accidental exposures to relatively high concentrations. Commercial preparations of the chemical have been found to be contaminated with other POLYCHLORI-NATED compounds, such as DIBENZO-PARA-dioxin and DIBENZOFURANS. PCP is now a RESTRICTED-USE PESTICIDE, meaning that it is not available to the general public.

percent destruction See DESTRUCTION AND REMOVAL EFFICIENCY, FOUR NINES, SIX NINES.

percent error An expression of the error of a measurement as a fraction of the measurement itself. For example, a measurement could be expressed as 50 centimeters, plus or minus 2%, meaning that the true value is thought to be within 49 ($50 - 0.02 \times 50$) and 51 ($50 + 0.02 \times 50$).

percent saturation The amount of a solid or liquid that is dissolved in a solution compared with the total amount of material that could be dissolved. If 14 grams of sodium chloride (salt) is dissolved in a total volume of 100 ml of water, the saturation is about 50% because a total of 28 grams of salt will dissolve in that volume of water.

perched water Groundwater held above the WATER TABLE because the water is positioned on a layer of impermeable stone or other material.

percolation The natural movement of water through subsurface soil as driven by gravity. The process allows for the migration of water and entrained pollutants from the surface into groundwater.

percutaneous Describing the absorption of chemicals through the intact skin.

perennial species Plants that do not die back over winter: herbs, shrubs, and trees. This type of plant has the capacity to store carbon derived from the atmosphere for long periods.

perfluorocarbon A class of synthetic compounds consisting of fluorine and carbon atoms. Examples include perfluoropropane (three carbon atoms), perfluorobutane (four carbon atoms), perfluoropentane (five carbon atoms), and several more complex structures. These compounds are exceptionally stable, colorless, odorless, noncorrosive, nonflammable, and essentially nontoxic. They do not lead to the destruction of the OZONE LAYER but they have high GLOBAL WARMING POTENTIAL. Their applications range from replacements for CHLOROFLUOROCARBON coolants, insulating fluids, and direct contact cooling to uses in the cosmetics industry.

perforated tray absorber A mechanical device designed to remove unwanted materials from an exhaust gas. A series of plates containing holes are arranged so that a liquid that reacts with or traps the pollutant flows over the trays as the exhaust gas is forced through the holes. This arrangement provides for maximal contact between the exhaust gas and the liquid.

performance curve A graph depicting the performance of a pollution control device for various conditions, for example, the collection efficiency of a cleaning device for various sizes of airborne particulate matter.

performance data In environmental regulation, information collected during the operation of a process or for a proposed treatment technique that is used to determine whether the process or technique meets design, regulatory, or emission criteria. May be used to make decisions on the issuance or renewal of a PERMIT for operations.

Performance Partnership Agreement See NATIONAL ENVIRONMENTAL PERFORMANCE PARTNERSHIP SYSTEM.

performance standards Environmental or workplace health and safety standards written in terms of the result to be achieved but without detailed requirements for specific actions. Compare SPECIFICATIONS STANDARDS.

period For electromagnetic radiation, the inverse of FREQUENCY, or the time required for one wave cycle to pass a given point. Frequency is given in units of inverse time (1/second); a period in time units (seconds).

periodic table An arrangement of the chemical elements by ATOMIC NUMBER (increasing number of protons) to show the relationships among elements. Horizontal rows are called periods; vertical rows are called groups.

peripheral nervous system The part of the nervous system exclusive of the brain and spinal cord.

periphyton Microscopic plant species attached to submerged objects in aquatic ecosystems. Compare NEKTON, PLANKTON.

permafrost Soil layers that remain frozen throughout the year. The upper layers of soil in the TUNDRA region of the Arctic thaw during the summer months; however, the lower layers remain frozen.

permanent hardness Water hardness that cannot be reduced or removed by heating the water, a reflection of the presence of dissolved calcium, magnesium, iron, and other divalent metal ions. These ions react to form insoluble precipitates. See HARDNESS. Compare TEMPORARY HARDNESS.

permanent retrievable storage A waste disposal option considered for nuclear HIGH-LEVEL WASTE. The material would be placed in containers and the containers put into a secure location, a salt mine or a special warehouse, from which they could be recovered in the future should the need arise. Burial of waste in a landfill would not be considered retrievable storage.

permanent threshold shift (PTS) An irreversible reduction in hearing ability for a certain sound frequency. Exhibited by an enduring increase in the HEARING THRESHOLD level. Normal hearing loss with age (PRESBYCUSIS) and NOISE-INDUCED HEARING LOSS begin with a permanent threshold shift in the higher sound frequencies. See TEMPORARY THRESHOLD SHIFT.

permeability The ease with which water and other fluids migrate through geological strata or landfill LINERs. Compare IMPERMEABILITY, POROSITY.

permeant An animal that can move freely among different types of environments, or an animal that can move freely from one group of animals to another group.

permissible exposure limit (PEL) An employee's allowable workplace exposure to a chemical or physical agent; the levels are set and enforced by the OCCUPATIONAL SAFETY AND HEALTH ADMINISTRATION.

permit Required under many federal and state environmental protection statutes, the official document containing the limitations applicable to a source's discharge into the air, water, or land. Sometimes called the "license to pollute." Discharge permits are required under provisions of the CLEAN AIR ACT, CLEAN WATER ACT, and RESOURCE CONSERVATION AND RECOVERY ACT, among many others.

permit by rule Permission to perform certain activities that does not require an individual, detailed permit application. Environmental protection agencies define activities that, if performed as stipulated, will have minimal impact on human health or the environment. If the activity falls under the permit by rule program, then a facility, contractor, or business is required to notify the agency of the activity, then abide by the regulatory requirements, to avoid expensive and lengthy permit review.

permit requirements The performance standards of a PERMIT.

peroxyacetylnitrate (PAN) A component of PHOTOCHEMICAL AIR POLLUTION with the chemical formula $CH_3C(=O)NO_2$; excessive exposure to this compound can cause eye irritation and injuries to vegetation. See PEROXYACYL NITRATE.

peroxyacyl nitrate (PAN) Air pollutant created by photochemical reactions involving sunlight, reactive hydrocarbons, and oxides of nitrogen. The general chemical formula is $RC(=O)NO_2$; R stands for a hydrocarbon group. The most common species is PEROXYACETYLNITRATE.

persistence The relative ability of a chemical to remain chemically stable after release into the environment. Persistent chemicals resist BIODEGRADATION and thus are of greater concern in the treatment of wastes.

persistent See PERSISTENCE.

persistent pesticide A chemical used to control unwanted plants or insect pests and only slowly decomposed, often by physical rather than biological processes, after application.

persistent pollutant Some substance that does not undergo ready decomposition or degradation when added to the environment. The term is usually applied to materials such as dieldrin, dichlorodiphenyltrichloroethane (DDT), and polychlorinated biphenyls (PCBs), which remain in the environment for years after their release.

person-gray A unit that reflects the total radiation exposure of a defined group (population) of people. A population dose, in person-grays, is found by multiplying the number of persons within each subgroup of the population receiving a certain radiation dose by that dose (in GRAYs) and summing the products obtained from each of the other subgroups. For example, a population of 8 individuals, 2 of whom receive a dose of 4 grays (8 person-grays) and 6 of whom receive a dose of 5 grays (30 person-grays), is exposed to a total of 38 person-grays.

person-rem Same as PERSON-GRAY except that the amount of radiation is given in REMS.

personal air samples Air samples taken with a small device carried by a worker and with a filter or other collection apparatus placed in his or her BREATHING ZONE.

personal protective equipment (PPE) Items worn to protect against workplace hazards, such as AIR-PURIFYING RESPIRATORS, SUPPLIED-AIR RESPIRATORS, a SELF-CONTAINED BREATHING APPARATUS, steel-toed boots, nonflammable clothing, earplugs, gloves, or safety glasses.

persons at risk The number of individuals who can possibly experience a health-related event during a period. See PERSON-YEARS-AT-RISK, POPULATION AT RISK.

person-years-at-risk (PYAR) A unit that reflects both the number of individuals who have been exposed to potentially toxic substances or radiation and the length of time during which adverse side effects have been expected from that exposure. The unit is computed by multiplying the total number of individuals exposed times the number of years since the exposure began. PYAR is used as the denominator in an INCIDENCE rate or MORTALITY RATE.

perturbation Some action or event that disrupts the normal functioning of the environment.

pest Any organism that degrades a resource; usually used to refer to insects that damage food crops or cause human misery.

pest resurgence A dramatic reappearance of a specific strain of insect pest in an environment that has been treated with an agent to kill or destroy the insect in ques-

tion. When related to pesticide use or overuse, the circumstance is most frequently related to the development of an insecticide-resistant strain of pest or to the depletion of other animals that kept the insect population under control.

pesticide A chemical agent used to kill an unwanted organism. Most of the agents are not highly selective in their action; however, they can be placed into categories depending on their target species. The most common pesticides are the insecticides (chemical agents applied to the environment to kill insects). Another class of materials used in great quantities are the herbicides (chemical agents applied to the environment to kill unwanted plants). Fungicides, which are frequently used as preservatives to prevent rotting, and rodenticides, which are used to control mice, rats, and similar animals, represent two lesser-used classes of chemicals. The pesticides were introduced to control vector-transmitted diseases, to increase agricultural productivity, to exercise urban pest control, and to preserve materials such as wood.

pesticide rain Deposition of airborne chemicals that were applied to agricultural operations to control unwanted insect pests. Some agents can be carried by air currents and carried to the ground by rain at a considerable distance from the field on which the agent was applied. During the first part of a rainfall, the raindrops can have excessively high concentrations of various chemicals being washed out, including pesticides.

pesticide residues Small amounts of insecticides, fungicides, or herbicides remaining in or on food when consumed.

pesticide tolerances See TOLERANCES, PESTICIDE.

pesticide treadmill The need for the application of an insecticide at an increasing rate or amount as a result of the gradual increase in insecticide resistance within the target group. Or the ongoing requirement for the development of new agents to combat insect strains that are resistant to previously used chemical agents.

Peterson dredge   A device used to collect sediment samples for the identification of bottom-dwelling animals in lakes and streams. The device has the appearance of a closed metal cylinder sectioned in half through the long axis. The two sections are hinged together in such a way that they resemble large jaws when locked open for deployment by wire. When the device is on the bottom and the line is pulled to retrieve the apparatus, the apparatus is levered in such a way that the two halves close to trap a sample of sediment. Weights can be attached to the outside of the cylinder sections to provide a deeper bite into the sediment. The Peterson dredge is useful in the sampling of sediments that have a high content of sand and gravel. Also called the Peterson grab.

Petri dish   The customary device for the isolation and/or cultivation of bacteria in the laboratory. The standard dish is 100 × 15 millimeters and consists of two overlapping halves. A solid medium (agar) is placed in the bottom portion of the dish for INOCULATION of cultures of bacteria.

petroleum   A liquid FOSSIL FUEL consisting of various HYDROCARBONS. Also called oil. The mixture is found in various geological deposits and can be refined to produce such products as gasoline, fuel oil, kerosene, and asphalt. See CRUDE OIL, CRUDE OIL FRACTION, HEAVY OIL.

petroleum derivatives   Chemical compounds derived from HYDROCARBONS, either products from the refining of CRUDE OIL or chemicals released when oil or an oil product evaporates or is degraded, such as gasoline, kerosene, various organic solvents, or other VOLATILE ORGANIC COMPOUNDS.

petroleum exclusion   The exclusion of oil and natural gas from the definition of a HAZARDOUS SUBSTANCE under the COMPREHENSIVE ENVIRONMENTAL RESPONSE, COMPENSATION, AND LIABILITY ACT. More specifically, the exclusion applies to CRUDE OIL and CRUDE OIL FRACTIONS (except certain fractions listed as hazardous substances), NATURAL GAS, natural gas liquids, LIQUEFIED NATURAL GAS, and SYNTHETIC NATURAL GAS.

Pfiesteria piscicida   A very small (about seven micrometers) single-celled marine dinoflagellate (alga) capable of producing a powerful toxin. The organism has been detected in coastal waters, especially those subjected to pollution with organic material. *Pfiesteria* species can assume more than 20 different shapes depending on the life cycle, including a difficult-to-detect cyst stage, an amoeboid stage, and a very toxic vegetative stage. Normally an inhabitant of coastal sediments, the organism produces a parasitic stage when the overlying water contains a multitude of fish, especially fish suffering from exposure to low dissolved oxygen. The parasitic stage swims with the aid of flagella and infects the hapless fish, producing a sore or lesion and eventually contributing to the death of the fish. The organism also appears to be capable of infecting humans handling the fish or swimming in the water. Laboratory workers, swimmers, and fishers infected with the organism have experienced a distinctive clinical syndrome characterized by skin lesions, headaches, lightheadedness, and chronic difficulties with learning and memory.

pH   A unit used to express the strength of an acidic or basic solution; calculated as the negative logarithm of the hydrogen ion concentration. Values commonly range from 0 to 14: less than 7.0 is acidic and greater than 7.0 is basic. A pH of 7.0 is considered neutral. Because the units are derived from common logarithms, a difference of 1 pH unit indicates a 10-fold (10^1) difference in acidity; a difference of 2 units indicates a 100-fold (10^2) difference in acidity.

phagocytic   Describing a cell capable of engulfing particles. For example, white blood cells and cells that reside in lymph nodes can engulf and destroy bacteria and other microorganisms that enter the body.

phagotroph An organism that obtains nutrients through the ingestion of solid organic matter. This class of organism includes all animals from the simplest, single-celled animal (for example, protozoa) to the higher forms. Organisms have some type of device to ingest particles, a digestive system, and a system to discard waste products.

pharmacokinetics The study and mathematical description of the ABSORPTION, DISTRIBUTION, PARTITIONING, METABOLISM, and excretion of a drug or pollutant in an organism.

Phase I Under the RESOURCE CONSERVATION AND RECOVERY ACT (RCRA), the first phase of the assumption of responsibility by a state for administering the RCRA hazardous waste (SUBTITLE C) program. Phase I state authorization refers to regulations identifying hazardous wastes and the standards for generators, transporters, and TREATMENT, STORAGE, OR DISPOSAL facilities with INTERIM STATUS. Compare PHASE II.

Phase I environmental assessment Performed by purchasers of real estate to document any required INNOCENT LANDOWNER defense under the liability provisions of the COMPREHENSIVE ENVIRONMENTAL RESPONSE, COMPENSATION, AND LIABILITY ACT. The assessment usually involves a title search, an examination of public records of activities on the site, aerial photographs, an inspection of surrounding land use, and a site visit.

Phase II Under the RESOURCE CONSERVATION AND RECOVERY ACT, the second phase of the assumption of responsibility by a state for the hazardous waste (SUBTITLE C) program, which covers the detailed, technical requirements for issuing final permits to TREATMENT, STORAGE, OR DISPOSAL facilities for hazardous wastes.

phenol An ingredient or chemical intermediate for many plastics, drugs, and explosives; the chemical formula is C_6H_5OH. Commonly known as carbolic acid. Environmental exposures primarily are through industrial employment. The compound has an adverse effect when discharged into sewage treatment systems.

phenology The scientific study of the changes in biological organisms that coincide with the yearly seasons, such as the timing of plant emergence and the progression of fall tree colors from north to south.

phenotype The genetically based traits of an organism that are actually observable as some morphological, behavioral, or biochemical characteristic of the organism. Because some traits are not expressed (are recessive), they may not be reflected in the appearance of the plant or animal. The total genetic makeup of the organism is termed the GENOTYPE.

phenyl group The chemical group C_6H_5 that is contained in many organic compounds, both natural and synthetic. For example, POLYCHLORINATED BIPHENYLS contain two phenyl groups.

pheromone A substance produced by an animal that serves as a chemical signal to other members of the same species. For example, the sex attractant produced by many animals and a chemical substance used to mark territory are pheromones.

phon A unit used to express the perceived LOUDNESS of a sound. For a particular sound, the unit is equal to the DECIBEL level of a 1000-hertz sound evaluated by a group with healthy hearing to be of the same loudness as that particular sound. FLETCHER-MUNSON CONTOURS are expressed in phons. Compare SONE.

phosphates The general term used to describe phosphorus-containing derivatives of phosphoric acid (H_3PO_4). The chemical containing the phosphate group (PO_4^{-3}) can be either organic or inorganic and either particulate or dissolved. An important plant nutrient.

phosphogypsum pile A special waste generated by the processing of phosphate ore, primarily calcium phosphate, using sulfuric acid to produce phosphoric acid, which is used in the fertilizer and chemical industries. The by-product is calcium sulfate or gypsum. An estimated 1 billion tons is stored in Florida, the location of over 90% of the phosphate ore in the United States. This solid waste contains acidic residues and may contain radioactive elements (NATURALLY OCCURRING RADIOACTIVE MATERIAL). Rainwater runoff from these piles of waste constitutes a pollutant source that must be controlled and treated prior to release. Also called phosphogypsum stacks in some regulations.

phosphorus (P) An essential chemical element in the growth of plants and animals. Phosphorus often acts as the plant nutrient that limits the growth of aquatic and terrestrial vegetation. Fertilizers frequently contain a source of phosphorus. The addition of excess levels of this element to freshwater environments causes the overgrowth of algae. See EUTROPHICATION, LIMITING FACTOR.

phosphorus cycle A series of integrated biological, geological, and chemical reactions (termed biogeochemical reactions) involved in the conversion of phosphorus from a solid mineral form in rocks, to a soluble form, to organic BIOMASS, and to and from soluble to insoluble form in a continuous cyclic manner.

photic Related to either the presence or the effects of light.

photic zone That area of a body of water into which light penetrates. The upper portion of a lake or sea within which light is sufficiently abundant to support the growth of PHYTOPLANKTON. Compare APHOTIC.

photochemical Describing a chemical reaction that is driven by the energy available in sunlight.

photochemical air pollution A type of air pollution resulting from the production of chemicals in the atmosphere by reactions between sunlight and airborne substances released by automobiles and industrial facilities. The most important pollutants that undergo these reactions are VOLATILE ORGANIC COMPOUNDS and NITROGEN OXIDE. These are converted by sunlight to NITROGEN DIOXIDE, OZONE, and PEROXYACYL NITRATE (PAN). These gases can mix with PARTICULATE MATTER and produce concentrations of pollutants high enough to impart a tint to the air, popularly called smog. See PHOTOCHEMICAL OXIDANTS, PHOTOCHEMICAL CYCLE.

Photochemical Assessment Monitoring Stations (PAMS) Urban air quality monitoring sites established to assist a state in efforts to lower the intensity and frequency of OZONE (smog). The PAMS record weather conditions and measure concentrations of VOLATILE ORGANIC COMPOUNDS, OZONE, and NITROGEN OXIDE. See PHOTOCHEMICAL AIR POLLUTION, PHOTOCHEMICAL OXIDANTS.

photochemical cycle A complex series of chemical reactions leading to an accumulation of OZONE or other PHOTOCHEMICAL OXIDANTS in the troposphere. Generally, the cycle involves the absorption of sunlight by NITROGEN DIOXIDE (NO_2), causing the conversion of the dioxide into NO and [O] (atomic oxygen). The atomic oxygen is very reactive and can combine with a diatomic oxygen molecule (O_2, the usual form of oxygen in the atmosphere) to form ozone (O_3). The O_3 can react with the NO to form NO_2 and O_2 (the PHOTOLYTIC CYCLE). If reactive hydrocarbons are present, atmospheric reactions can convert them to FREE RADICALS, which can oxidize the NO formed by the initial splitting of NO_2. If oxidized to NO_2, NO cannot react with and destroy the O_3, thus allowing ozone to accumulate with a resulting rise in the concentration of ozone.

photochemical oxidants Components of PHOTOCHEMICAL AIR POLLUTION that include OZONE, PEROXYACYL NITRATE (especially PEROXYACETYLNITRATE), and oxygenated hydrocarbons, such as the ALDEHYDES. The photochemical oxidant present in the highest concentrations in the ambient air is ozone. The USEPA initially set a NATIONAL AMBIENT AIR QUALITY STANDARD for all photochemical oxidants but now uses a single standard for OZONE to represent the group.

photochemical smog See PHOTOCHEMICAL AIR POLLUTION.

photodegradable plastic A plastic polymer that decomposes or depolymerizes under the influence of sunlight. At one time, proposed as a partial solution to litter and to the nonbiodegradable nature of plastic.

photoelectric effect An electric current induced when light strikes certain metals, the principle behind the operation of automatic exposure meters in cameras and PHOTOVOLTAIC devices.

photoionization detector (PID) An analytical instrument used to quantify the presence of organic compounds. Ultraviolet light with an energy content equal to or greater than a chemical-specific ionization potential removes the outermost electrons from a molecule of the target substance. The resulting current flow is proportional to the number of ionized molecules.

photolysis The breakdown of a material by sunlight. Photolysis is an important degradation mechanism for air pollutants and for contaminants in surface water and in the terrestrial environment.

photolytic cycle An atmospheric reaction that forms OZONE in the troposphere. The cycle involves the splitting, by the absorption of energy available in sunlight, of a nitrogen dioxide molecule (NO_2) into NO + [O] (atomic oxygen). The atomic oxygen atom [O] can then combine with a diatomic oxygen molecule (O_2) to form ozone (O_3). In the absence of reactive hydrocarbons, the ozone that is formed is destroyed by reacting with the NO to form NO_2 and O_2. See also PHOTOCHEMICAL CYCLE.

photomultiplier tube An electronic device in some analytical instruments that enhances the PHOTON signal emitted by a reaction or interaction used to detect the presence of a particular chemical substance. Photons are emitted by certain chemical reactions or by an interaction between a specific wavelength of electromagnetic energy and the substance being measured.

photon A quantum of electromagnetic radiation. A unit of intensity of electromagnetic radiation, including light. A photon has properties that relate to both particles and waves. A photon has no charge or mass; however, it does have momentum. Photon energy equals the product of PLANCK'S CONSTANT and FREQUENCY.

photoprocessing pollution Liquid waste from facilities that process photographs or X-ray images of various types. The discards include spent chemicals for developing and fixing of negatives and prints.

photosensitization The increased susceptibility of the skin to ULTRAVIOLET light caused by exposure to photosensitizing agents such as COAL TAR, derivatives of coal tar, hexachlorophene, and certain plant products.

photosynthate Carbohydrates and other organic molecules produced and often released by algae during the process of PHOTOSYNTHESIS.

photosynthesis A process in green plants and some bacteria during which light energy is absorbed by CHLOROPHYLL-containing molecules and converted to chemical energy (the light reaction). During the process, carbon dioxide is reduced and combined with other chemical elements to provide the organic intermediates that form plant biomass (the dark reaction). Green plants release molecular oxygen (O_2), which they derive from water during the light reaction.

photovoltaic Producing an electric current as the result of light striking a metal; the direct conversion of radiant energy into electrical energy.

photovoltaic cell See SOLAR CELL.

phreatophyte A plant with roots extending to the water table and with a relatively high TRANSPIRATION rate. Such plants can move a significant amount of groundwater to the atmosphere.

phthalates Agents added to plastics to improve their flexibility. Over 25 different compounds are produced for commercial use; di(2-ethylhexyl) phthalate (DEHP), diisononyl phthalate (DINP), and di-*n*-butyl phthalate (DBP) are the most common. Because they are used in every major category of consumer products and because they are only slowly degradable, phthalates are distributed throughout the environment. The level of acute toxicity of the compounds is very low. Chronic exposure can cause adverse reproductive effects and excess tumors in rodents, but evidence indicates that the toxicity is species specific and that humans are not a sensitive group.

phylogeny The evolutionary development of a group or species of organisms.

physical factors Those nonbiological factors that influence the growth, development, and survival of organisms (e.g., temperature, light, climate, chemicals, and water).

phytoplankton Microscopic plants, such as algae, suspended in aquatic environments.

phytotoxicant A chemical that can damage or kill plants in aquatic environments.

phytotoxicity The ability of chemicals to damage or kill plants in aquatic environments.

phytotreatment Biological technology used to treat wastewater, typically that which contains excessive plant nutrients. As rooted plants or algae are allowed to grow, the nutrient is removed from the wastewater and converted into plant mass. Adaptations of the technology can also be applied to the removal of hazardous substances from contaminated sediment. See BIOREMEDIATION, TERTIARY TREATMENT.

pica The tendency of children to eat nonfood items, such as chips of old paint. See LEAD.

pico- (p) An SI unit prefix meaning 10^{-12}.

picocurie A unit of radioactivity equal to 1×10^{-12} CURIE.

picocuries per liter (pCi/l) 1×10^{-12} CURIE of RADIOACTIVITY per liter of air. The unit used to express RADON levels in indoor air.

piezometer An instrument used for the determination of water table elevation or the water level in a tightly cased well, the latter called the POTENTIOMETRIC SURFACE.

piezometric height See HYDRAULIC HEAD.

piezometric surface See POTENTIOMETRIC SURFACE.

pig 1. A container used to store or ship radioactive materials. The container is constructed of materials that act as shielding to prevent the transmission of high-energy radiation. 2. A device forced through unlined cast-iron water pipes by hydraulic pressure to scrape iron oxide scale from the inside surfaces.

pillow lava The cylinder-shaped (picture a structure formed by toothpaste being pushed from a tube as it is squeezed vigorously) mass of lava produced when lava from a BASALTIC ERUPTION is extruded under water. The lava cools rapidly, yielding the semitube appearance.

Pinchot, Gifford (1865–1946) American Conservationist, first head of the U.S. Forest Service. Promoter of sustainable use of natural resources. See CONSERVATION. Compare PRESERVATION; MUIR, JOHN.

pioneer community The first plant species to colonize a land area that is essentially free of plant life. The nonvegetated areas may be present because of previous farming, surface mining, or severe fires. These initial plant communities usually give way to different species that arrive later. See SUCCESSION, PRIMARY SUCCESSION.

pitchblende A mineral ore containing commercial quantities of the elements uranium and radium.

pitot tube A device used to measure the velocity of a gas or liquid. Two concentric tubes are oriented in the same axis as the material flow, and the tubes measure the total pressure and static pressure of the flowing fluid. The difference between the total pressure and static pressure is the VELOCITY PRESSURE, which can be related to the velocity of the fluid.

pK_a The negative logarithm of the acid DISSOCIATION CONSTANT. Lower pK_a values indicate stronger acids.

pK_b The negative logarithm of the base DISSOCIATION CONSTANT. Lower pK_b values indicate stronger bases.

pK_s The negative logarithm of the SOLUBILITY PRODUCT CONSTANT for a chemical compound dissolving in water.

placard An identifying sign required on trucks, railcars, and large containers carrying hazardous materials. The diamond-shaped placard will usually display a four-digit identification number. See HAZARDOUS MATERIAL TRANSPORTATION LAW.

Planck's constant (*h*) The constant relating the energy content of electromagnetic radiation to its frequency; the constant is equal to 6.626×10^{-34} joule-second.

Planck's law The radiation energy emitted per unit area by a blackbody, by wavelength (λ) at absolute temperature T is expressed by

$$E_\lambda = \frac{2\pi h c^2 \lambda^{-5}}{\exp{(hc/\lambda kT)} - 1]}$$

where h is PLANCK'S CONSTANT, c is the speed of light, and k is Boltzmann's constant. Planck's law also expresses the energy value *(E)* of a photon of electromagnetic radiation by $E = h\nu$, where ν is the radiation frequency.

planetary boundary layer The atmosphere between the surface and the altitude at which the GEOSTROPHIC WIND occurs.

plankton Microscopic plants and animals that live suspended in water. Generally, these organisms move along with currents. See PHYTOPLANKTON, ZOOPLANKTON. Compare BENTHOS, NEKTON.

plant nutrients The primary mineral ingredients of fertilizer: phosphate (PO_4^{-3}), nitrate (NO_3^-), ammonium (NH_4^+), and potassium (K^+) together with an extensive array of chemical elements used in lesser amounts to support the growth of plants. See TRACE ELEMENTS, EUTROPHICATION, LIMITING FACTOR.

plasmid A general term used to describe a small circular piece of deoxyribonucleic acid (DNA) that exists independently of the nuclear chromosomes that control most traits of a cell or organism. An extrachromosomal piece of genetic material. Plasmids are useful in GENETIC ENGINEERING. See Ti PLASMID.

plastic recycling The conversion of POSTCONSUMER plastic containers into items that are suitable for other consumer uses. The diversion of plastic containers from the solid waste stream. The recycling of plastic is complicated by the number of resins used to manufacture consumer plastic; consequently, plastic must be sorted into resin type to facilitate reentry of the material into the commercial market. On containers and many other consumer items, the resin is identified by a number code that appears in a triangle composed of three arrows. The codes, developed by the Society of the Plastic Industry, together with resin types and common products are listed in the Appendixes.

plasticizers Chemical additives (often PHTHALATES) used to increase the flexibility of plastics.

plate boundaries According to the theory of PLATE TECTONICS, the locations where the rigid plates that compose the crust of the Earth meet. The plates move slowly on the molten material beneath in the process called CONTINENTAL DRIFT. As the plates meet, the boundaries can be classified as divergent (places where the plates are moving apart, as at the midocean ridges of the Atlantic Ocean), convergent (places where the plates are colliding, as at the Himalayas), and transform (places where the plates are sliding past each other, as in California).

plate count A method used to determine the microbial population of a sample of soil, water, or biological material. The source material is suspended or mixed with a sterile solution, and a portion of the mixture is applied to the surface of a suitable agar medium in a PETRI DISH. Theoretically, each bacterium present in the sample applied to the dish divides until a visible colony develops. By counting the colonies after incubation and applying the necessary mathematical manipulations, the number of bacteria in the original sample can be determined.

plate tectonics A concept stating that the crust of the Earth is composed of crustal plates moving on the molten material below. Seven major and many minor

plates are recognized. The seven major plates constitute the great majority of the crust and are named for the continents or oceans (Pacific, Eurasian, African, Australian, North American, South American, and Antarctic). These and the minor plates are moving slowly but relentlessly. The boundaries of the plates are the foci for earthquakes and volcanic activity. Deep ocean ridges are formed where plates diverge or move apart; transform faults like those in California are formed when plates slide past one another; ocean trenches and volcanic activity are produced when one plate overrides another and pushes the underlying plate into the mantle. The latter boundary is characteristic of areas like the region bordering the western Pacific Ocean.

plate tower scrubber See SCRUBBER, PLATE TOWER.

plenum A chamber in a ventilation system used for air distribution via connecting ducts.

plug flow See CONTINUOUS-FLOW SYSTEM.

plugged and abandoned Describing an unsuccessful exploration well for oil or natural gas; the well borehole is sealed and the drilling equipment removed from the site.

plugging Stopping the flow of water, gas, or oil into or out of a well. Also, cementing or otherwise blocking the casing of a dry oil, gas, or water well.

plume A relatively concentrated mass of emitted chemical contaminants spreading in the environment. In surface water, the effluent added to a receiving stream near a point source. For example, when a heated-water discharge is added to a stream, the heated water does not mix immediately with the stream water. The mass of hot water remains detectable for some distance downstream. In groundwater, the leachate leaking downgradient from a site of buried waste material. In air pollution, visible or invisible gases, vapors, or particulate matter emitted from a smokestack or moving downwind from an urban area. See URBAN PLUME.

plume reflection The assumption in the GAUSSIAN PLUME MODEL, used for air quality dispersion estimates, that an expanding plume carried downwind from an elevated source (stack) will, when the plume reaches the ground, reflect back upward instead of being absorbed.

plume rise The movement of an identifiable mass of gas (plume) upward for a period after its exit from a stack; the movement is caused by (1) the vertical momentum of the exiting gas, which, when spent, allows the plume to bend over with the wind; and (2) the buoyancy effect of the lower-density hot gas mixture. Plume rise is estimated to calculate EFFECTIVE STACK HEIGHT (physical stack height plus plume rise), which is a required input to models that estimate air concentrations downwind from the stack.

plutonium-239 (^{239}Pu) A radioactive element, not found in natural ores, produced in nuclear reactors by the bombardment of uranium-238 with FAST NEUTRONS. The atomic number of plutonium is 94. The 239 isotope is a fissionable material used as the primary nuclear fuel in nuclear weapons. Plutonium-239 is a long-lived isotope and is extremely toxic to humans.

pneumoconiosis A general term for diseases of the lung resulting from chronic overexposure to dusts. See ASBESTOSIS, COAL WORKERS' PNEUMOCONIOSIS, SILICOSIS.

Pogo principle "We have met the enemy and he is us," the tag line in the "Pogo" cartoon appearing on Earth Day 1971, by Walt Kelly, showing Pogo and Porkypine gingerly stepping through a garbage dump.

poikilotherm See ECTOTHERM.

point of compliance (POC) For a hazardous waste TREATMENT, STORAGE, OR DISPOSAL facility, the location, specified by the operations permit of the facility, for a DOWNGRADIENT WELL. Each well is placed to detect the presence of any contaminants released from the facility into groundwater that will move into the uppermost aquifer in the area.

point-of-entry treatment device An apparatus connected to the drinking water supply entering a residence (or other building) to remove particulate matter, dissolved calcium, or other metals (see HARD WATER), or organic contaminants from the water supply. All of the water entering the home from a municipal supply would be treated.

point-of-use/point-of-entry (POU/POE) An approach to the management of the quality of drinking water that locates a water-cleaning device at the entrance of the water to the residence (POINT-OF-ENTRY TREATMENT DEVICE) or at an individual faucet (POINT-OF-USE TREATMENT DEVICE) in a household; sometimes used in homes supplied by a private well that does not meet drinking water standards.

point-of-use treatment device An apparatus connected to a single tap supplying drinking water at a residence or other facility to remove particulate matter or organic contaminants from the water supply. Only the water from one tap would be treated.

point source An identifiable and confined discharge point for one or more water pollutants, such as a pipe, channel, vessel, or ditch. Compare NONPOINT SOURCE.

poise (P) A unit, equal to 1.0 gram per centimeter-second, of VISCOSITY.

polar solvent A solvent, with a slight negative charge on one part of the molecule and a slight positive charge at another position, that dissolves other polar materials. The most common polar solvent is water. See POLARITY; compare NONPOLAR SOLVENT.

polar stratospheric clouds Water–nitric acid mixtures on which reactions occur that release chlorine from the compounds $ClONO_2$ and HCl. When chemically bound, the chlorine does no harm to the ozone layer, but when released, it reacts with and destroys stratospheric ozone molecules. See OZONE LAYER, OZONE HOLE, OZONE LAYER DEPLETION.

polar vortex A region of extremely cold air that forms over Antarctica during the winter months in the Southern Hemisphere. The strong winds move in a circle, and the air is isolated from the rest of the atmosphere. (Although there is an Arctic polar vortex also, it is not as strong or as isolated.) POLAR STRATOS- PHERIC CLOUDS form in the vortex and result in the seasonal OZONE HOLE over Antarctica.

polarity A property of a molecule that causes one side or end of the molecule to have a slight negative charge and the other a slight positive charge. When two atoms are joined with a COVALENT BOND, they share a pair of electrons. For example, the water molecule is composed of one oxygen atom joined to two hydrogen atoms by covalent bonds. The oxygen atom exerts a slightly stronger attraction for the shared electrons than do the hydrogen atoms. Consequently, the oxygen atom is slightly negative and the hydrogen atoms are slightly positive in character. This creates a molecule with a positive pole and a negative pole. Solvents composed of molecules that exhibit polarity (water) are better able to dissolve ionic materials than are solvents consisting of molecules that do not exhibit polarity (NONPOLAR SOLVENTS, e.g., carbon tetrachloride).

polarized light microscopy Use of a light microscope, equipped with polarizing filters (which screen out all light waves not vibrating or moving in the same plane), to observe specific optical characteristics of geological materials. This type of microscopy is useful in the identification of asbestos.

polishing The removal of low concentrations of dissolved, recalcitrant organic compounds from either water intended for human consumption or wastewater that has been subjected to PRIMARY TREAT- MENT and SECONDARY TREATMENT. The passage of water through ACTIVATED CHARCOAL is a frequently employed polishing technique. Also called CARBON FIL- TRATION, CARBON TREATMENT. See CAR- BON POLISHING.

pollen Male reproductive structure of plants. Pollen release in the spring often triggers hay fever or allergic responses in sensitive individuals, and the presence of large amounts of pollen can cause a yellow discoloration in the air and a yellow dust on cars and in homes. Pollen transfer is an important component of plant reproduction and fruiting.

pollutant A chemical or physical agent introduced to the environment that may lead to POLLUTION.

Pollutant Standards Index (PSI) A one-number representation of the air quality of an urban area. The index (number) is obtained by combining data from air-sampling stations for different air pollutants using a standardized method. PSI values falling within certain numerical ranges are then described by standard terminology: 0–50, good; 51–100, moderate; 101–200, unhealthy; and so on. The index was developed to communicate summary air quality information to the public.

polluter pays principle The environmental management theory under which the (potential) emitter of a pollutant pays for controls to reduce emissions and/or pays a fee to compensate for residual harm resulting from the emissions. In practice, these two could be used in combination to minimize any EXTERNALITY from the polluting activity.

pollution   The addition of one or more chemical or physical agents (heat, electromagnetic radiation, sound) to the air, water, or land in an amount, at a rate, and/or in a location that threatens human health, wildlife, plants, or the orderly functioning or human enjoyment of an aspect of the environment.

pollution exclusion clause   A feature of a COMPREHENSIVE GENERAL LIABILITY POLICY since approximately 1970; the clause excludes bodily injury or property damage coverage for routine emissions or gradual leaks of pollutants from a facility. See ENVIRONMENTAL IMPAIRMENT LIABILITY POLICY.

pollution indicator organism   A plant or animal species that is not normally present in an aquatic environment unless the body of water has been subjected to damage by pollution. For example, *Escherichia coli* is a bacterium that is not found in the aquatic environment unless the system has been contaminated by the addition of fecal material. The organism signals the presence of pollution.

pollution offset   See OFFSET.

pollution prevention (P$_2$)   Concept that moves away from regulations requiring the treatment of exhausts and effluents to remove offending substances, or away from the protection of human health and the environment by the cleanup of contaminated land or water. Rather, pollution prevention means eliminating the source of the pollutant, recycling process waste, converting waste material to products, and eliminating the toxicity of process residuals or by-products. For example, eliminating the source of the pollutant can involve modification of the process, change of raw materials, internal recovery of solvents for reuse, and substitution of process chemicals. One of the Codes of Management Practice of the RESPONSIBLE CARE initiative of the Chemical Manufacturers Association. See SOURCE REDUCTION.

Pollution Prevention Act   A 1990 federal statute that requires facilities to emphasize actions that reduce pollution at its source; the reduction can be in volume and/or in toxicity and should follow the ENVIRONMENTAL MANAGEMENT HIERARCHY. See POLLUTION PREVENTION.

polyampholytes   A class of POLYELECTROLYTES that contain both positively charged (cationic) and negatively charged (anionic) functional groups in the same polymer.

polycentric   In urban planning, describing large urban areas characterized by a collection of interconnected urban centers, each of which has a complete set of services and supporting facilities. A single, concentrated downtown center does not serve the entire area, but each "satellite" has a full complement of services. Such arrangements describe old cities plus outlying communities in which the centralized activities in the downtown area of the old city are now dispersed.

polychlorinated biphenyls (PCBs)   A diverse mixture of AROMATIC compounds that in the past were used extensively as insulating and cooling agents in electrical transformers, as plasticizers in waxes, and in the manufacture of paper and inks. The compounds are very stable, are widely distributed in the environment, and undergo BIOACCUMULATION in mammals. Excessive exposures cause a severe acnelike eruption (chloracne) in humans, and the material has been shown to induce cancer development in mammals. As required by the TOXIC SUBSTANCES CONTROL ACT (1976), the USEPA has written extensive regulations for the manufacture, use, marking, storage, disposal, and cleanup of PCBs. They are found under Title 40, Part 761, of the CODE OF FEDERAL REGULATIONS.

polychlorinated dibenzofurans (PCDFs)   Contaminants found in commercial preparations of POLYCHLORINATED BIPHENYLS and PENTACHLOROPHENOL that may be responsible for some of the physiological effects ascribed to those compounds. Exposure to the furans results in severe acnelike eruptions (chloracne) in humans; they are less toxic than the DIOXINS, which also contaminate the two classes of chemicals indicated. See DIBENZOFURANS.

polycyclic Describing chemical compounds composed of multiple units of the six-carbon AROMATIC nucleus BENZENE. Anthracene and related compounds are polycyclic. See POLYCYCLIC AROMATIC HYDROCARBONS.

polycyclic aromatic hydrocarbons (PAH) A group of AROMATIC ring compounds that are derivatives of anthracene, which consists of three benzene rings in a row. Other aromatic rings or organic groups are attached to the anthracene. They are found in coal, tar, and petroleum and are emitted by combustion-related activities. Many different compounds can be formed through metabolic conversions involving the basic aromatic nucleus within biological systems, as well as during chemical syntheses, because of the reactive nature of anthracene and derivatives of this compound. This class of compounds appears to be responsible in part for the cancer-causing properties of cigarette smoke.

polycyclic hydrocarbons See POLYCYCLIC AROMATIC HYDROCARBONS.

polyelectrolyte Synthetic or natural polymer containing many positive or negative charges. This material is water soluble, promotes changes in the fluid properties of aqueous suspensions and slurries, is strongly reactive with particles or ions carrying the opposite charge, and is frequently resistant to biological DECOMPOSITION. These polymers have wide applications as FLOCCULATION agents, thickeners, COAGULANTS, dispersants, water conditioners, soap and detergent additives, soil conditioners, and superabsorbers in disposable diapers and in oil recovery processes. Also called macroions and

polyions. See POLYAMPHOLYTES, PRIMARY TREATMENT.

polyethylene (PE) Plastic resin consisting of polymerized or linked ethylene (a two-carbon hydrocarbon) that is used in numerous construction, industrial, and consumer products. HIGH-DENSITY POLYETHYLENE is used in the manufacture of containers for water or milk (natural color) or household chemicals, juices, and other liquid (pigmented). LOW-DENSITY POLYETHYLENE is commonly used in applications requiring film plastic.

polyethylene terephthalate (PET) A plastic used for beverage containers and collected by many RECYCLING programs. The most commonly recycled plastic resin. Coded with a number 1 on the RECYCLING SYMBOL.

polyions See POLYELECTROLYTES.

polymer Macromolecule composed of repeating units of some smaller molecule. Many natural materials (cellulose, proteins, starches, nucleic acids) and synthetic materials (nylon, plastic, rubber) are made by linking smaller units. Cellulose is a natural polymer composed of glucose molecules; polyethylene plastic is an artificial polymer composed of ethylene molecules.

polynuclear aromatic hydrocarbons (PAH) See POLYCYCLIC AROMATIC HYDROCARBONS.

polynuclear organic matter (POM) See POLYCYCLIC AROMATIC HYDROCARBONS.

polyvinyl chloride (PVC) A strong synthetic POLYMER plastic used in pipes, toys, electrical coverings, and many other products. Produced from vinyl chloride MONOMER. See VINYL CHLORIDE.

population All of the members of a single SPECIES that inhabit a defined geographical area. The snow goose population of a marsh would consist of every member of the species *Chen caerulescens* found in that marsh.

population at risk (PAR) The number of persons who can potentially contract or develop a disease, adverse health effect, or condition; the proper denominator for use in calculating the rate of a disease or condition, which is the number of diseased persons divided by the population at risk.

population crash A sudden and dramatic decline in the numbers of individuals of a specific species within an area of interest. The reasons are varied and include loss of food source, introduction of a predator species, disease, environmental toxins, and extreme weather events or natural disasters.

population dose The sum of the radiation doses received by individuals in a given population, expressed as PERSON-GRAYS or PERSON-REMS. A measure of total population exposure to ionizing radiation.

population equivalent (PE) A way to express the strength of industrial waste in terms of the comparable amount of BIOCHEMICAL OXYGEN DEMAND (BOD) in the household wastewater produced by one person. An industrial waste that has a PE of 1000 is equivalent to the BOD of waste produced by 1000 people.

population explosion A rapid increase in the number of individuals of a specific species in a defined location. Commonly, the number of individuals exceeds the capacity of the environment to provide resources, leading to a POPULATION CRASH. The result is a marked swing in numbers from high to low, back to high, and so on.

population momentum The impetus for continued expansion of the number of people in a country when the age structure is characterized by a large number of children. Even if birth control efforts are effective in the adult community and the number of new births per person decreases, the number of people in the country expands as the large population of children reach reproductive age.

population risk A risk estimate equal to the product of INDIVIDUAL LIFETIME RISK and the size of the population exposed. Population risk is typically expressed as the number of excess cases of a disease, such as cancer, per year attributable to a given exposure. See UNIT RISK ESTIMATE.

porosity A description of the total volume of soil, rock, or other material that is occupied by pore spaces. A high porosity does not equate to a high PERMEABILITY in that the pore spaces may be poorly interconnected.

portal-of-entry effect Adverse consequences that are experienced at the location where the body encounters or is exposed to a toxin. If a dangerous substance causes chemical burns, then the skin is damaged as a direct result of contact with the chemical. This is different from damage to a TARGET ORGAN, the liver, for example, is damaged when a chemical gains entry into the system at some location (or portal of entry) and subsequently is transported via the circulatory system to the liver, where the damage is done. See PRIMARY IRRITANT, SECONDARY IRRITANT.

positive association The direct relationship between two variables, the values of which fluctuate together in the same direction. For example, as the strength of incoming solar radiation increases seasonally, the atmospheric temperature increases.

positive crankcase ventilation (PCV) The routing of automobile crankcase emissions back into the cylinders for combustion. The process lowers the amount of hydrocarbons released into the atmosphere.

postclosure plan A document prepared by a hazardous waste TREATMENT, STOR-

AGE, OR DISPOSAL facility outlining the groundwater monitoring and reporting, waste containment provisions, and security arrangements for the 30-year period after CLOSURE.

postconsumer recycling Recovery of materials that have served their intended final use for remanufacture into commercial items. Reuse of materials that have passed through the hands of the intended consumer (e.g., newsprint from discarded newspapers or glass from beer bottles). Contrast PRECONSUMER RECYCLING.

postconsumer waste See POSTCONSUMER RECYCLING.

postmaterialist values A personal philosophy that rejects modern materialism based on the accumulation of possessions. Generally rejects continued industrial expansion and advocates a return to natural surroundings and an emphasis on living "in harmony with nature."

postmodernism An outlook that does not accept the concept of progress, industrial development based on scientific reasoning and technology, domination of nature by human beings, rational planning of ideal social order, or the standardization of knowledge. Indeed, truth cannot be known at all. Postmodernists would rewrite many words in this definition with quotation marks, as "progress," "development," "scientific reasoning," "knowledge." For environmental issues, this philosophy at least partially underlies the social movements known as ENVIRONMENTALISM, ECOFEMINISM, and ENVIRONMENTAL JUSTICE.

potable water Water suitable for human consumption. Compare NONPOTABLE.

potency See CANCER POTENCY FACTOR.

potency slope factor See CANCER POTENCY FACTOR.

potential energy The energy available in a substance because of position (water held behind a dam) or chemical composition (hydrocarbons). This form of energy can be converted to other, more useful forms (for example, electrical energy).

potential temperature The temperature of a dry air parcel if compressed or expanded, without loss or gain of heat from the surroundings, to a pressure of 1000 millibars. The potential temperature is sometimes used to express ATMOSPHERIC STABILITY. If potential temperature increases with altitude, the atmosphere is stable; if it decreases with altitude, the atmosphere is unstable.

potential to emit Under the CLEAN AIR ACT, the amount of a pollutant a facility would emit when operating at the full rated capacity, taking normally operating control devices into account. This can be lowered by an agreement between the source and the regulatory agency stating that the source operations will be limited to a certain level. The potential of a source to emit pollution is used to classify the facility as a MAJOR SOURCE, a TRUE MINOR source, or a SYNTHETIC MINOR source.

potentially responsible party (PRP) Any hazardous waste generator, owner, or operator of a hazardous waste disposal facility or transporter identified by the USEPA as potentially liable for the cost of cleaning up a site under the COMPREHENSIVE ENVIRONMENTAL RESPONSE, COMPENSATION, AND LIABILITY ACT. All companies that disposed of hazardous wastes in a disposal facility are potentially responsible if problems relating to public health or environmental pollution develop in the future as a result of the presence of the disposal facility. See CONTRIBUTION RIGHTS, COST RECOVERY, DE MICROMIS, DE MINIMIS, INNOCENT LANDOWNER, JOINT AND SEVERAL LIABILITY.

potentiation The enhancement of the toxic effects of one chemical by another when two agents or drugs have a greater effect than would be expected by the additive effects of the two drugs. For example, the drug disulfiram is not toxic when taken alone in therapeutic doses; however, this medication is used to discourage alcohol consumption by alcoholics. Intake of alcohol after the administration of disulfiram causes the person to become violently ill because the agent interferes with the metabolism of ethanol. The occupational exposure to carbon tetrachloride produces much more pronounced liver toxicity when the individual is also exposed to isopropyl alcohol.

potentiometric surface The water level in a tightly cased well, or the level to which water would rise if a well were sunk at a particular point. For an artesian well, for example, the potentiometric surface is above ground level.

pounds per square inch (absolute) (psia) Common units for the expression of ABSOLUTE PRESSURE exerted by gases relative to zero pressure. Absolute pressure units are required in IDEAL GAS LAW calculations.

power (*P*) The rate at which work is done, expressed as work/time. Units of power are the WATT and HORSEPOWER. One watt of power is equal to one JOULE per second. One horsepower equals about 33,000 FOOT-POUNDS per minute or 746 watts.

pozzolanic Certain materials containing silicates or aluminosilicates that solidify when combined with cement or lime. This type of material is used to trap (immobilize) certain hazardous waste contaminants before land disposal. Common sources of pozzolanic materials are fly ash, lime kiln dusts, and slag from blast furnaces.

precautionary principle In environmental management, the belief that if a technology, chemical, physical agent, or human activity can be reasonably linked to adverse effects on human health or the environment, then controls should be implemented even if the problem or the cause-effect relationship is not fully understood; to wait for scientific certainty (or near certainty) is to court disaster.

preconsumer recycling Using or reusing materials before they have reached the consumer. The items include raw materials, by-products, mill waste, shavings, or other items that result from the fabrication process, for example, wastepaper generated during the paper-making process that is put back into the pulping process for reuse. Contrast POSTCONSUMER RECYCLING.

precipitant An agent added to a liquid mixture to encourage the formation of solid materials that will settle from the mixture. For example, ALUM (aluminum sulfate) is added to sewage to promote the formation of FLOC, which facilitates the removal of organic materials from the wastewater. See PRIMARY TREATMENT, COAGULANT.

precipitate The solid that settles from a liquid suspension. The solid produced by a chemical reaction involving chemicals that are in solution.

precipitation scavenging The removal of particles or certain gases from the atmosphere by rain or snow.

precision The repeatability of a series of test results; whether the testing method gives the same answer under the same set of circumstances. The answer is not necessarily accurate. Compare ACCURACY.

precursor In air pollution, an air contaminant that reacts with sunlight and/or other compounds to produce new chemical materials, for example, the production of OZONE from the atmospheric reactions involving NITROGEN OXIDES and VOLATILE ORGANIC COMPOUNDS.

precycling Describes the practice of making purchases on the basis of the potential to recycle the product after use. Adherence to precycling means that if items cannot be recycled, they are not purchased because they will contribute to the accumulation of solid waste.

predation The process of one animal's killing and eating another.

predator An animal that kills and eats other animals.

predictive maintenance A system for the protection of fixed equipment or vehicles through the monitoring of a specific characteristic that can provide an indication that certain mechanical problems are looming if corrective action is not taken. For example, one might maintain a surveillance of the chemical components of crankcase oil to determine when oil changes would be appropriate. Contrast with PREVENTATIVE MAINTENANCE which would require the crankcase oil be changed every three months regardless of the conditions.

preheat The heating of a raw material or reactant before it is added to some process, or the heating of a gas exhaust before release into the atmosphere.

preliminary assessment and site inspection (PA/SI) The first data collection and evaluation at a site containing hazardous waste that may require remediation under the COMPREHENSIVE ENVIRONMENTAL RESPONSE, COMPENSATION, AND LIABILITY ACT. Depending on the assessment results, no further action may be recommended or the site may be scheduled for a LISTING SITE INSPECTION and possible inclusion on the NATIONAL PRIORITIES LIST.

preliminary assessment information rule (PAIR) A requirement of the regulations implementing the TOXIC SUBSTANCES CONTROL ACT under which the USEPA gathers certain information about a chemical substance, including the location and quantity manufactured, an estimate of worker exposure, and consumer uses. The agency uses this information for a preliminary assessment of the risk posed by the chemical to human health and the environment. This rule can be supplemented by the COMPREHENSIVE ASSESSMENT INFORMATION RULE (CAIR).

preliminary remediation goals (PRGs) In the cleanup of a contaminated waste site, estimates of maximal chemical concentrations in air, water, and soil that can be present without adverse effects on human health, including on susceptible groups. The concentrations are estimated by using assumed EXPOSURE factors and toxicity data. For example, if chemical X has shown no adverse effects for a daily DOSE of two milligrams and the only exposure to chemical X is via drinking water pumped from a well drawing groundwater from beneath the contaminated site, then using a standard factor of two liters of tap water consumed per day yields a preliminary remediation goal of one milligram of X per liter in the groundwater at the site, assuming the chemical is not a carcinogen and without accounting for absorption rates, uncertainty factors, and other necessary assumptions. USEPA guidance emphasizes that PRGs are only screening concentrations, and a measured concentration for a chemical in air, water, or soil above a PRG at a site only leads to further evaluation and is not necessarily the cleanup target concentration. See APPLICABLE OR RELEVANT AND APPROPRIATE REQUIREMENT.

premanufacturing notice (PMN) Under provisions of the TOXIC SUBSTANCES CONTROL ACT, toxicity test results and/or other appropriate data required to be submitted to the USEPA by a producer of any new chemical or existing chemical proposed for a significant new use. The act requires the agency to determine whether the new chemical or new use will pose an "unreasonable risk" to human health or the environment. See SIGNIFICANT NEW USE RULE.

presbycusis The normal loss of hearing ability in the higher frequencies due to age.

prescribed burning A forest management technique that uses planned fires to maintain a desirable ecological balance. Also termed fire ecology.

preservation The natural resources policy that stresses the aesthetic aspects of forests, rivers, wetlands, and other areas and tends to favor leaving such areas in an undisturbed state. Compare CONSERVATION.

preservative Chemical additive intended to prolong the useful life of a product or food. Those items added to food to assist in maintaining quality, without causing damage to human health.

pressed-wood products Materials, used in the construction and furniture industry, composed of wood chips, particles, veneers, or other fiber that are bonded together by adhesives applied with heat and pressure. Particleboard is an example of such a product. The environmental issues associated with these products are the reuse of a potentially discarded waste material to make a new product and potential indoor air pollution, caused by the volatilization of chemicals from the wood, for example, FORMALDEHYDE.

pressure (p) Force per unit area. Enclosed fluids exert a force perpendicular to the surface of the containing vessel. The shape of the container does not affect the fluid pressure. At equilibrium, the general expression for the relationship between pressure at a point in a fluid and the elevation (depth) z is

$$\frac{dp}{dz} = -\rho g$$

where ρ is the fluid density and g is gravitational acceleration. This is called the hydrostatic equation. This expression indicates that an increase in elevation is accompanied by a drop in pressure.

For pressures p_1 and p_2 at the elevations z_1 and z_2, $p_2 - p_1 = -\rho g(z_2 - z_1)$. Elevation z_2 is higher than elevation z_1.

For stationary incompressible fluids (liquids), density is constant and the ABSOLUTE PRESSURE p_{abs} at a certain depth h below the point z_2 at the surface of the liquid is expressed as $P_{abs} = \rho g h + p_a$, where $h = z_2 - z_1$ and p_a is ATMOSPHERIC

PRESSURE, equal to p_2 at point z_2.

For stationary compressible fluids (gases), because density varies with elevation z, the pressure change between two elevations is expressed as an integral:

$$p_1 - p_2 = \int_{z_1}^{z_2} \rho g \, dz$$

Pressure is typically expressed in pounds per square inch, newtons per square meter, or PASCALS.

pressure filter A device used to remove fine particulate matter from water; the filter consists of a filter medium, such as sand or anthracite coal, packed in a watertight vessel.

pressure gradient The change in pressure with distance, from lower to higher pressure, or vice versa. See ISOBAR, PRESSURE GRADIENT FORCE.

pressure gradient force The force causing horizontal air flow from areas of relatively high pressure to areas of relatively low pressure. Air flow is perpendicular to the lines of equal pressure (ISOBARS). Air moving under the pressure gradient force (wind) is deflected to the right by the CORIOLIS FORCE in the Northern Hemisphere. Centrifugal force and friction also affect wind direction and velocity.

pressure head The hydraulic energy represented by the water pressure per unit weight. The numerical value is expressed in length units: water pressure in pounds per square foot divided by the unit weight of the water, in pounds per cubic foot, equals the pressure head, in feet. The sum of the pressure head and ELEVATION HEAD is the HYDRAULIC HEAD. See also HEAD, TOTAL.

pressure ridges Geological formations associated with lava flows from volcanic regions. The lava flows cool more quickly on the surface, producing a crust. The hot gases trapped beneath the surface push up and crack the surface, forming permanent ridges on it.

pressure sewer A type of wastewater collection system used in rural areas as a less expensive alternative to the GRAVITY SEWER used in urban areas. The polyvinyl chloride (PVC) pipe in pressure sewers is of smaller diameter (two to four inches) and is buried just below the frost line, not as deep as the clay or concrete pipe systems in urban areas needing a downward flow between LIFT STATIONS. The pumps in a pressure sewer are located at each household or are placed at nodes in the collection system. Pressure sewers are classified by the way they prevent solids from entering and fouling the sewer system. In grinder-pump systems, solids are ground into a slurry before being pumped into the collection system. Others use a SEPTIC TANK at each household to remove the solids before the sewage is pumped to the sewer. A pump is located in each septic tank effluent pump (STEP) unit; septic tank effluent filter (STEF) units drain the sewage by gravity to a larger collection tank serving several households, from which it is pumped into the main collection system. See VACUUM SEWER.

pressurized-water reactor (PWR) A type of nuclear reactor in which water is employed to transfer heat produced in the reactor CORE. The water that is in contact with the reactor core, called the primary loop, is maintained under high pressure to prevent boiling. Compare BOILING-WATER REACTOR.

presumptive test The first of three steps in the analysis of water or wastewater for the presence of bacteria of fecal origin. Portions of a water sample are inoculated into lactose broth and incubated for 24 hours at 35° C. The presence of acid and gas after that time is a positive test result, and the water is presumed to be contaminated. See CONFIRMED TEST, COMPLETED TEST.

pretreatment Under the CLEAN WATER ACT, the required alteration and/or reduction of certain water pollutants in a waste stream before the wastewater is discharged into a PUBLICLY OWNED TREAT-MENT WORKS (POTW). The purpose of this requirement is to prevent discharges that will reduce the efficiency of the water treatment facility or to treat materials that are not treated or are inadequately treated by the POTW.

pretreatment standards for existing sources (PSES) See PRETREATMENT.

pretreatment standards for new sources (PSNS) See PRETREATMENT.

prevalence The number of cases of a disease or pathological condition present in a given population at a certain time, expressed as a rate; for example, 3 cases of measles per 1000 persons during the month of April.

prevalence study See CROSS-SECTIONAL METHOD.

preventative maintenance A system for the protection of fixed equipment or vehicles through upkeep on a set schedule. For example, one might change crankcase oil every three months regardless of the level of use. Contrast with PREDICTIVE MAINTENANCE, which would require the crankcase oil be monitored for the presence of agents that indicate when an oil change is necessary.

prevention of significant deterioration (PSD) A Clean Air Act regulatory program under which air quality in an area can only worsen by a fixed amount for particular pollutants above a defined baseline, even if the NATIONAL AMBIENT AIR QUALITY STANDARD for the pollutant is met. The United States is divided into Class I, II, and III areas, and the classification determines how much the air quality in the area can deteriorate; Class I allows the smallest deterioration. See INCREMENTS, INCREMENT CONSUMPTION.

prey An animal that is potentially or actually killed and consumed by another animal (the PREDATOR).

primacy State agency authority to enforce federal environmental regulations; granted by the USEPA if the state meets certain criteria.

primary air pollutant A pollutant that is emitted directly into the air from a source and found in the ambient air in the same form, such as ambient carbon monoxide that was emitted by an automobile. Compare SECONDARY AIR POLLUTANT.

primary clarifier In a WASTEWATER treatment plant, the CLARIFIER that is used early in the treatment process, after the water passes through the BAR RACKS and GRIT CHAMBER.

primary consumer In a food chain, a HETEROTROPHIC organism that feeds on plants. An herbivore. An animal that eats other animals is a SECONDARY CONSUMER.

primary industry categories The 34 types of facilities requiring the BEST-AVAILABLE TECHNOLOGY ECONOMICALLY ACHIEVABLE (BAT) for toxic water pollutants under the Clean Water Act. They are listed as Appendix A to Title 40, Part 122, of the CODE OF FEDERAL REGULATIONS. See FLANNERY DECREE.

primary irritant A chemical that, upon overexposure, causes irritation to the skin, eyes, or respiratory tract at the site of contact but does not cause an adverse SYSTEMIC effect. For example, ammonia inhalation causes irritation of the upper respiratory tract but no other systemic effect in the body. Compare SECONDARY IRRITANT.

primary pollutant A general term for a pollutant in the environment that is in the same chemical form as it was when released. Compare SECONDARY POLLUTANT and PRIMARY AIR POLLUTANT.

primary producer See PRODUCER, AUTOTROPH.

primary productivity The weight of plant BIOMASS accumulated as a result of PHOTOSYNTHESIS, per unit area or volume of water over a specific interval. The value can be expressed as GROSS PRIMARY PRODUCTIVITY or NET PRIMARY PRODUCTIVITY. See also SECONDARY PRODUCTIVITY.

primary recovery The oil and gas produced from a well that is forced to the surface by natural pressure. See SECONDARY RECOVERY, ENHANCED OIL RECOVERY.

primary settling tank A holding tank where raw sewage or other wastewater is retained to allow the settling and removal of particulate material. The material that separates from the suspension is often termed SLUDGE. See PRIMARY TREATMENT.

primary sludge The sludge produced by PRIMARY TREATMENT in a wastewater treatment plant.

primary standards Standards set by the USEPA for the maximal amount of pollutants that can be present in air and water without adverse health effects on humans. The six air pollutants with primary standards are particulate matter, sulfur dioxides, carbon monoxide, nitrogen dioxide, ozone, and lead. The primary standards for drinking water are set for about 80 chemical, biological, or physical agents, ranging from arsenic to fluoride and from pesticides to radionuclides and microorganisms. Compare SECONDARY STANDARDS.

primary succession The development of plant and animal communities in a land area that does not contain topsoil, for example, in an area covered by lava that has solidified. This type of succession contributes to and depends on the slow weathering of rock. Compare SECONDARY SUCCESSION.

primary treatment The removal of particulate materials from domestic wastewater, usually by allowing the materials to settle as a result of gravity; setting can be hastened by the addition of ALUM or POLY-

ELECTROLYTES that increase FLOCCULA-TION and COAGULATION. Typically the first stage of treatment encountered by domestic wastewater entering a treatment facility. The wastewater is allowed to stand in large tanks, termed clarifiers or settling tanks. Also, any process used for the decomposition, stabilization, or disposal of sludges produced by settling. The water from which the solids have been removed is then subjected to SECONDARY TREATMENT.

principal organic hazardous constituents (POHCs) Chlorinated organic compounds found in certain chemical waste mixtures used to determine the DESTRUCTION AND REMOVAL EFFICIENCY of a hazardous waste incinerator. See TRIAL BURN.

prior informed consent (PIC) procedure A voluntary program run by the United Nations FOOD AND AGRICULTURAL ORGANIZATION and the UNITED NATIONS ENVIRONMENT PROGRAM that obtains formal consent from a country's DESIGNATED NATIONAL AUTHORITY (DNA) before that country receives an international shipment of certain hazardous chemicals that have been banned or their use severely restricted in the exporting country.

priority pollutants A list of 129 chemicals in 65 classes of chemical materials defined as toxic pollutants by Section 307 of the 1977 CLEAN WATER ACT, which also requires TECHNOLOGY-BASED standards for the control of these chemicals. Also see FLANNERY DECREE.

probabilistic risk assessment (PRA) The analytical estimation of the probability of an undesired consequence using the probabilities of events that will lead to the consequence, such as a study of the likelihood of a catastrophic release of radiation from a nuclear power plant. See EVENT TREE, FAULT TREE.

procarcinogen A chemical that is converted to an active CARCINOGEN through a variety of biotransformations that follow the ingestion and metabolism of the chemical by animals, including humans. Most chemicals known to induce cancer fall into this category. For example, aflatoxin produced by the mold *Aspergillus* in grain stored under moist conditions and subsequently ingested by an animal is converted to aflatoxin-2,3-epoxide within the tissues of that animal. The epoxide form of the toxin then combines with deoxyribonucleic acid (DNA) to cause the MUTATION needed to induce cancer development.

procaryotic (prokaryotic) Describing an organism composed of cells that do not contain membrane-bound organelles such as nuclei, mitochondria, or chloroplasts. The genetic material of this type of cell is not associated with large chromosomes within the nucleus, as is characteristic of higher plants or animals. The bacteria and blue-green algae constitute this group. Compare EUCARYOTIC.

procedure violations A program used by industrial facilities to reduce the incidence of injuries and environmental releases. The program maintains surveillance to determine when a prescribed procedure to protect the safety of workers or to ensure environmental protection was not followed. In these events, neither injuries nor environmental contamination occurred; however, the failure to follow prescribed procedures could have caused a problem and must not be repeated to prevent future problems. A proactive program to prevent future accidents.

process chemical Chemical that is a raw material, intermediate, or product. Some process chemicals or their by-products may be released as pollutants. See FEEDSTOCK.

process hazards analysis (PHA) An exercise performed by an industrial facility as part of the PROCESS SAFETY MANAGEMENT program to identify failures in equipment, piping, storage vessels, and so on that could cause toxicity, fire, or explosion hazards to workers or the public. See HAZARD AND OPERABILITY STUDY, EVENT TREE, FAULT TREE.

process safety management A regulatory program of the OCCUPATIONAL SAFETY AND HEALTH ADMINISTRATION designed to minimize the likelihood and consequences of the accidental release of chemicals that pose toxicity, fire, and explosion hazards. See SEVESO DIRECTIVE, TITLE III, CHEMICAL AWARENESS AND EMERGENCY RESPONSE, RISK MANAGEMENT PLAN.

process variable A measured characteristic, such as temperature, pressure, flow rate, or a chemical quantity such as carbon dioxide level, used to monitor a process. The process is adjusted to maintain the chosen variables within the correct limits and is used to maintain product quality, to control by-product (pollutant) releases, and to monitor the performance of air pollution control devices and wastewater treatment units.

process vent Opening through which gases from an industrial operation are continually or periodically discharged, usually without emissions control.

process waste The discards of an industrial operation. Something that is a by-product or discard produced as a result of an industrial operation. A SLUDGE collected in a device to remove solids from crude oil at a refinery would be a process waste, and the facility must determine whether it must be managed as a HAZARDOUS WASTE. Office trash collected in the management office of the refinery would be classed as a commercial solid waste and would rarely meet the definition of a hazardous waste.

process water Any water that comes in contact with a raw material or product. The water is often released as wastewater after use.

produced water As crude oil is extracted from a well, the water that

rises to the surface with the oil. The produced water can comprise a large fraction of the total fluids extracted and is either pumped back into an underground formation via an INJECTION WELL or treated and discharged to surface water. See NATURALLY OCCURRING RADIOACTIVE MATERIAL.

producer Any organism that is involved in the fixation of carbon dioxide that results in the gain of organic BIOMASS within an environment, primarily green plants. The green plants, together with a few types of bacteria, assimilate carbon dioxide and other inorganic nutrients into organic material, which, in turn, serves as food for consumers. The activity of producers is an important process removing carbon dioxide from the atmosphere. See CONSUMER.

product stewardship One of the six codes of management practice under the RESPONSIBLE CARE initiative organized by the Chemical Manufacturers Association. Basically, the code requires a manufacturer of a chemical to be responsible for the health, safety, and environmental consequences of a product from the time of product design through manufacture to ultimate disposal. Under the provisions of this code, wastes are considered a product for which the generating facility is responsible.

productivity The rate of BIOMASS accumulation within an environment as a result of PHOTOSYNTHESIS. The rate of plant growth. See PRIMARY PRODUCTIVITY, GROSS PRIMARY PRODUCTIVITY, NET PRIMARY PRODUCTIVITY.

products of incomplete combustion (PICs) Potentially hazardous organic materials formed during the incineration of wastes containing chlorinated organic compounds. PICs in the exhaust gases of an incinerator could include POLYCYCLIC AROMATIC HYDROCARBONS, DIBENZOFURANS, AND TETRACHLORODIBENZO-PARA-DIOXIN.

profundal zone In deep lakes, the deep-water stratum found beneath the LIMNETIC ZONE, or the shallow portion of the lake where light is sufficient for plant growth. This dark, low-oxygen zone often experiences conditions that support the activity of DECOMPOSERS.

progeny The offspring of animals or plants. The term is also used to describe the products of RADIOACTIVE DECAY. See DAUGHTER.

programmatic impact statement An ENVIRONMENTAL IMPACT STATEMENT prepared for a broad action or a series of cumulative actions to be taken by a federal agency, such as the development of a new technology or a series of mineral leases on public lands.

Project XL An initiative of the USEPA to give states and the regulated industrial community the flexibility to develop comprehensive strategies as alternatives to the heterogeneous collection of current regulations. The pilot program strives for excellence, leadership, and innovative, cost-effective regulatory approaches, in contrast to the prevailing system of COMMAND AND CONTROL REGULATION. Numerous projects are found at www.epa.gov/ooaujeag/projectxl/.

Promethean environmentalism Used as a term to ridicule and scorn those who are optimistic about the future on the basis of scientific rationality and the potential for human ingenuity and enterprise to find cures for societal problems. Term inspired by Prometheus, a character in Greek mythology who suffered torment for his sin of pride. See CORNUCOPIAN, CORNUCOPIAN FALLACY, MALTHUSIAN.

promoter In the two-stage model of carcinogenesis, a substance that converts an initiated cell to a tumor. Many promoters stimulate cell growth. Some hormones can serve this function. See CARCINOGENESIS.

promulgate To announce, propose, or issue in final form any RULES or regulations by an administrative agency. See FEDERAL REGISTER, CODE OF FEDERAL REGULATIONS, ADMINISTRATIVE PROCEDURE ACT.

pronatalist Describing an individual or a policy that encourages an increase in the human birth rate in order to achieve a larger human population. Compare ANTI-NATALIST.

propagule 1. Seed, cutting, or other plant parts capable of separately and individually starting a new plant. 2. A piece of fungal biomass that is capable of starting the development of a fungal colony if the proper nutritional and environmental conditions are met. The piece can be a viable fragment of fungal hyphae or a spore.

proportional counter An instrument used to measure IONIZING RADIATION. The device usually consists of a gas-filled cylinder with a wire running through the center. A voltage is applied to the wire. Any ionizing radiation ionizes the gas in the cylinder, allowing a current to flow. The output voltage is proportional to the number of ionizing events.

proportionate mortality ratio (PMR) The fraction of deaths attributable to a specific disease in an exposed population divided by the fraction of deaths caused the same disease in a control (nonexposed) population.

Proposition 65 The Safe Drinking Water and Toxic Enforcement Act of 1986, a California state law that requires larger businesses to warn any individuals who may be exposed to certain chemicals discharged by that business into water or land (i.e., into routes that lead to potential sources of drinking water, either surface water or groundwater). Warnings (on labels, for example) are required for chemicals that have been listed by the state as carcinogens or reproductive toxins and that will result in an exposure that poses a "significant risk." The act also requires government employees to inform local government officials if an illegal discharge of a hazardous waste that occurs in their jurisdiction may be harmful to public health or safety. Website: www.oehha.ca.gov/prop65.html

prospective study See COHORT STUDY.

protective factor (PF) In industrial hygiene, an expression of the efficiency of a RESPIRATOR used to protect the wearer from harmful air concentrations. The factor is calculated as the ratio of the air concentration of a particular air contaminant outside the respirator to the concentration of the contaminant inside it. A higher PF means greater protection from that material.

protein One of the major macromolecules in biological systems. Composed of long chains of amino acids linked by peptide bonds, covalent chemical bonds between the amino group of one amino acid and the carboxy group of the adjacent amino acid. Proteins differ on the basis of the kinds, number, and linear sequence of amino acids. Enzymes are protein molecules. One gene directs the synthesis of one protein.

proteinuria Excessive protein in the urine; a sign of kidney disease or toxicity.

proton One of the elementary nuclear particles. The particle is located in the nucleus of atoms, has a mass number of 1, and carries a positive charge. The number of protons in the nucleus of an atom is referred to as the atomic number.

protoplast An intact bacterial cell or plant cell from which the cell wall has been removed. Protoplasts are produced in the laboratory and must be protected from OSMOTIC LYSIS.

protozoa Unicellular animals, most of which lack cell walls, ingest food particles, and move about freely. Most are harmless to humans and are important members of aquatic and soil communities involved in the DECOMPOSITION of organic materials. They are important members of the FOOD

WEBS in natural environments and in wastewater treatment facilities. A few species cause diseases, including malaria, encephalitis, amoebic dysentery, and GIARDIASIS.

proximate analysis A chemical analysis that determines the fractions of volatile carbon, FIXED CARBON, water, and noncombustibles in a waste. This crude analysis is an attempt to classify a waste for possible incineration without undertaking a more extensive analysis of the precise chemical composition of the material. Compare ULTIMATE ANALYSIS.

proxy climate indicators Direct measurements of temperature and precipitation spanning the last several hundred years are available for only a few locations. Past climatic conditions available in the "fossil record" are termed proxy climate indicators because close examination reveals information about climate long before direct measurements of temperature and precipitation began. Some of the indicators are to be found in mud deposits left by floods, annual layers in the Greenland and Antarctic ice sheets, pollen and vegetation analysis in sediments, archaeological remains of mammals, tree rings, and ocean bed deposits.

psychrometer An instrument used to determine RELATIVE HUMIDITY that consists of two thermometers, one dry and the other with a bulb surrounded by a wetted wick. Relative humidity can be read from a psychrometric chart by using the two thermometer readings.

psychrometric chart A graph showing the relationships among DRY-BULB TEMPERATURE, WET-BULB TEMPERATURE, ABSOLUTE HUMIDITY, RELATIVE HUMIDITY, and DEW POINT.

public comment period The time allowed for the public to relay views concerning a planned action as announced in the *Federal Register* by an administrative agency (e.g., the USEPA). The ADMINISTRA-

TIVE PROCEDURE ACT requires that agencies follow the procedures of NOTICE-AND-COMMENT RULE MAKING.

public health The organized pursuit of good community health, including physical and mental well-being. The primary focus of public health research and activity is on the *prevention* of human disease, which involves the maintenance of safe and healthy physical, chemical, and biological conditions for human populations. This includes controls on and monitoring of the food supply (fresh, packaged, and restaurant food), surface water, groundwater, indoor air, outdoor air, workplace health and safety, farm and ranch animals, pets, human immunization status, and exposure to ionizing and nonionizing radiation. Public health is also involved in case tracking of communicable diseases, design and operation of consumer products, accident prevention (especially vehicles), and promotion of safe and healthy behavior, such as campaigns against tobacco use and excessive alcohol consumption, encouragement of vehicle seat belt use and a balanced diet, and prevention of sexually transmitted diseases.

public health approach In environmental management, an emphasis on the reduction of risks of human exposures and potential adverse health effects. Top priority is assigned on the basis of providing for the protection of the largest number of individuals and preventing the most severe impacts on people.

public health assessment A study performed to identify the concentrations and distribution of toxic materials in air, water, or soil and to estimate the exposure and possible adverse effects on human health of the contamination. The AGENCY FOR TOXIC SUBSTANCES AND DISEASE REGISTRY performs a public health assessment for all hazardous waste sites on the NATIONAL PRIORITIES LIST. See HAZARDOUS SUBSTANCES SUPERFUND.

public hearing See HEARING.

public notice On certain occasions, the USEPA and other regulatory agencies are required to inform the public of significant regulatory or pollution events. For example, the public must be informed when the USEPA issues a draft PERMIT or schedules a public HEARING. Under the provisions of the SAFE DRINKING WATER ACT, violations of the national PRIMARY STANDARDS for drinking water must be announced to the public.

public trust doctrine First seen in the law of the Roman Empire, the doctrine that reserves, as a public trust, the public's use of certain natural resources, in particular common resources such as oceans, lakes, rivers, and the atmosphere. Ongoing public use implies that the resources must be protected by public TRUSTEES, such as state and federal regulatory agencies. In common law, public trust resources were originally defined as submerged land along waterways, but the public trust has been increasingly applied more widely to include the resources Roman law protected: aquatic life, terrestrial wildlife, parklands, and possibly archaeological and historic artifacts.

Public Utility Regulatory Policies Act of 1978 (PURPA) Section 210 of PURPA is designed to encourage alternate forms of electric power generation by requiring public utilities to purchase power from small generators classified as qualifying facilities by the FEDERAL ENERGY REGULATORY COMMISSION. The public utility must pay the small generator a rate equal to the incremental cost (also called the avoided cost) that it would incur to generate the additional power. In 1997, about 7% of U.S. electric power generation was from small generators participating in PURPA, such as stations that burn BIOMASS (e.g., rice hulls), small-scale HYDROPOWER units, SOLAR CELLS, landfill-generated METHANE, and WIND FARMS. See QUALIFYING FACILITY.

public water system A utility that provides piped water for human consumption to at least 15 service connections or 25 individuals.

publicly owned treatment works (POTW) A SEWAGE TREATMENT PLANT owned and operated by a public body, usually a municipal government.

pug mill A mechanical device used to blend dry solids with waste material in order to improve handling characteristics of the waste.

pull factors In urban planning, used to categorize conditions in urban areas that draw people from rural areas: available housing, city utilities, educational institutions, cultural events, obtainable employment, and modern medical facilities, to name a few. See PUSH FACTORS.

pulmonary edema An abnormal accumulation of fluids within the lungs. The condition can result from a bacterial infection or from an irritation caused by exposure to certain chemical agents.

pump and treat A cleanup technique for contaminated groundwater; the water is pumped to the surface, contaminants removed, and the cleaned water pumped back underground. See AIR STRIPPING.

pumped storage Water that is pumped upgrade into a reservoir or lake during periods of low electrical power consumption. The water added to the reservoir is later released through the hydroelectric facility to generate electricity during times of high power demand. See PEAK SHAVING.

purgeable organic A volatile organic compound that has a boiling point less than or equal to 200° C and that is less than 2% water soluble.

purgeable organic carbon (POC) The amount of carbon in the PURGEABLE ORGANICS that can be removed from soil or water.

purgeable organic halogens (POX) Those organic derivatives containing chlo-

rine, bromine, or fluorine that are found within the PURGEABLE ORGANIC materials that can be removed from soil or water.

purging Cleaning pipes or containers by removing stale air or water from equipment before use or sampling.

push factors In urban planning, used to categorize conditions in rural areas that stimulate the movement of people into the city environment: poverty, unemployment, lack of medical care, dilapidated housing, and isolation, to name a few. See PULL FACTORS.

putrefaction The partial degradation of organic materials in the presence of an insufficient oxygen supply. The result is the release of noxious decomposition products and gases.

putrescible Describing any substance that is likely to result in the production of a rotten, foul-smelling product when decomposed by bacteria; examples are the plant and animal waste material in municipal solid waste.

pyramid of biomass A visual representation obtained by depicting the total weight of organisms residing in each TROPHIC LEVEL in a given area as a horizontal bar. The mass of plants within the area are placed on the bottom, and the mass represented by the top carnivores is placed on the top, with other forms of biota placed at intervening levels. Once arranged in an order representative of the feeding structure in the community, the drawing resembles a pyramid.

pyramid of energy A drawing that is the same as the PYRAMID OF BIOMASS except that the total caloric content of the organisms at each TROPHIC LEVEL is used to construct the representation.

pyranometer An instrument used to measure direct solar radiation and indirect sky radiation, the incoming solar energy that is scattered downward to the surface.

Pyramid of biomass

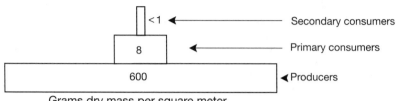

Grams dry mass per square meter

pyrethroid A class of natural insecticides developed from pyrethrum extracted from plants belonging to the genus *Chrysanthemum*. Agents in this group of materials are nonpersistent in the environment.

pyrite A mineral consisting of iron sulfide. Burning of the mineral (frequently done when high-sulfur coal is burned) results in the release of SULFUR OXIDES (SO_x) into the atmosphere, which contributes to ACID RAIN.

pyroclastic flow See ASH FLOW.

pyroclastic texture The result of ASH FLOW associated with a volcanic eruption characterized by the release of very hot gases, fragments of rock, and ash. As the material settles, the temperature is so high that the ashlike fragments fuse into a dense mass.

pyrolysis The thermal destruction of some material (e.g., coal, oil, wood, or other organic substance) in the absence of molecular oxygen. Also termed destructive distillation.

pyrophoric Describing a material that can ignite spontaneously in air at less than 130° F.

Q

q1* See CANCER POTENCY FACTOR.

quad (q) An energy unit representing 1 quadrillion (a million billion, or 1×10^{15}) BTUs that is sometimes used to discuss national or global energy production or consumption. Compare QUINT.

qualifying facility A business that is certified as a small power production unit or a COGENERATION facility by the FEDERAL ENERGY REGULATORY COMMISSION in accordance with the provisions of the PUBLIC UTILITY REGULATORY POLICIES ACT OF 1978 (PURPA).

qualitative analysis The examination of a substance or sample to determine what chemical compounds or elements are present irrespective of the amounts of those compounds or elements. Compare QUANTITATIVE ANALYSIS.

qualitative variable See CATEGORICAL VARIABLE.

quality assurance/quality control (QA/QC) All methods and procedures used to obtain accurate and reliable results from environmental sampling and analysis. Includes rules for when, where, and how samples are taken; sample storage, preservation, and transport; and, during the analysis, the use of BLANKs, duplicates, and SPLIT SAMPLES.

quality factor (QF) A factor by which a dose of IONIZING RADIATION is multiplied to obtain a quantity that corresponds to the biological effect of the radiation. The factor adjustment reduces radiation exposures to a common scale, weighing them by their ionizing potential. See RELATIVE BIOLOGICAL EFFECTIVENESS.

quantitation limit The minimal amount of some chemical that can be detected and measured with a suitable degree of reliability with currently available instruments. Usually three to five times the INSTRUMENT DETECTION LIMIT.

quantitative analysis The examination of a substance or sample to determine the precise amounts of certain chemicals or elements that are present in air, water, soil, or other media. Compare QUALITATIVE ANALYSIS.

quantitative variable A characteristic that can be measured and expressed in units that describe the quantity present, for example, length, mass, or time. Compare CATEGORICAL VARIABLE.

quantum A unit of energy conveyed by an electromagnetic wave. The magnitude of the unit is proportional to the frequency of the wave; for example, a unit of ultraviolet light (relatively high frequency) conveys more energy than a unit of infrared light (relatively low frequency). Also referred to as a PHOTON.

quarternary ammonium compounds A class of chemicals related to ammonium chloride in that the hydrogen atoms attached to the nitrogen are replaced by various organic compounds. These compounds are of low toxicity to people, are readily soluble in water, and are strongly antimicrobial. They are found in soaps and disinfectants that are widely used as cleaning and sanitizing agents in hospitals, nursing homes, and other public places.

quench tank Tank holding water used to cool industrial processes.

quencher The inlet section of a vessel designed to remove particles from an exhaust gas. The hot exhaust gas is cooled before particles are removed.

quint (Q) An energy unit representing 1 quintillion (1×10^{18}) BTUs that is sometimes used to discuss national or global energy production and consumption. See QUAD.

R

rabbit A container used for substances to be subjected to exposure in a nuclear reactor.

rad See RADIATION ABSORBED DOSE.

radiant heat INFRARED ENERGY emitted by a surface; the amount of heat is proportional to the fourth power of the ABSOLUTE TEMPERATURE of the radiating body, according to the STEFAN-BOLTZMANN LAW.

radiation Emitted energy, as particles or as electromagnetic waves. Based on energy content, the two major divisions are IONIZING RADIATION and NONIONIZING RADIATION.

radiation absorbed dose (rad) The non-SI unit expressing the amount of radiation absorbed by any medium: 1 rad is equal to an energy absorption of 100 ergs per gram of medium (usually human tissue). The equivalent SI unit is the GRAY: 1 gray equals 100 rads.

radiation chemistry The study of the effects of alpha and beta particles and X rays and gamma rays on the structure and properties of matter.

radiation inversion An atmospheric temperature INVERSION (cooler air beneath warmer air) caused by the loss of heat, or INFRARED ENERGY, from the ground on a cool, clear night. This radiation heat loss causes the air closest to the ground to be cooler than the air above it. By midmorning, incoming solar energy warms the air near the ground sufficiently to reestablish the normal atmospheric temperature profile within the atmos-

phere. During inversion conditions, the dispersive capability of the atmosphere is lowered, and air pollutants typically increase in concentration. See also SUBSIDENCE INVERSION.

radiation shielding Material, such as lead, placed between people and a source of IONIZING RADIATION to reduce human radiation exposure.

radiation sickness Acute adverse health effects, such as gastrointestinal upset, decreased blood cell production, and loss of hair, caused by overexposure to IONIZING RADIATION.

radiation sterilization The use of IONIZING RADIATION, such as GAMMA RAYS, either to render a plant or animal incapable of reproduction or to kill all microorganisms associated with some material or product.

radiative forcing Factors contributing to a change in the amounts of energy absorbed or the rate at which energy is emitted by the Earth. Cyclic variations in solar energy output and changes in the GREENHOUSE GAS concentrations in the atmosphere are examples of positive radiative forcing. See SUNSPOT CYCLE, GREENHOUSE EFFECT, GLOBAL WARMING.

radical In chemistry, an atom or group of atoms possessing an odd (unpaired) electron. Such structures are highly reactive and short-lived. Also referred to as FREE RADICALS. Radicals are of immense importance in the toxicity of organic chemicals in the body, and the generation and control of radicals influence the development of tumors and resistance to infectious diseases. Some of the nutritional supplements taken as vitamins influence the generation and removal of radicals in the body.

radioactive Describing an unstable atom that undergoes spontaneous decay, releasing IONIZING RADIATION.

radioactive decay The spontaneous emission of mass or energy from the nucleus of an unstable element. The emission results in the production of a new element or a different ISOTOPE of the same element. These new products may in turn undergo further decay. Each step in the decay process takes place at a predictable rate normally expressed as a HALF-LIFE. See RADIOISOTOPE, IONIZING RADIATION, RADIOACTIVE SERIES.

radioactive isotope An unstable atom of a single element that differs from other forms of the same element in terms of the number of neutrons in the nucleus. Undergoes spontaneous decay, releasing IONIZING RADIATION. See ISOTOPE, RADIOACTIVE.

radioactive series A series of ELEMENTs produced by RADIOACTIVE DECAY of unstable atoms, with one decay product following another until a stable element is reached. For example, the uranium series begins with uranium-238 and ends with lead-206. The other two naturally occurring radioactive element series are the thorium and actinium series.

radioactive waste Useless residual material and equipment produced by nuclear power generation, nuclear reactor decommissioning, nuclear weapons manufacture, medical applications, and research. Depending on the RADIOACTIVITY and the HALF-LIFE of the materials, they are classified as LOW-LEVEL WASTE or HIGH-LEVEL WASTE.

radioactivity The property exhibited by certain unstable ELEMENTs of spontaneously emitting mass or energy from the nucleus. See RADIOACTIVE DECAY, RADIOISOTOPE.

radiocarbon See CARBON-14.

radiochemical Describing the activities of RADIOCHEMISTRY.

radiochemistry The study of the properties and use of RADIOACTIVE materials.

radiography The production of images by using radiation other than visible light. The term usually refers to the production of images of the human body using X RAYS.

radioisotope An unstable form of an element that undergoes RADIOACTIVE DECAY, emitting energy in the form of GAMMA RAYS or mass in the form of ALPHA PARTICLES or BETA PARTICLES.

radiometer A general term for an instrument that measures ELECTROMAGNETIC RADIATION.

radionuclide See RADIOISOTOPE.

radiopaque Describing a material that does not allow radiation to penetrate or pass through. The property of a shielding agent.

radiopoison A substance, such as boron, that stops a fission CHAIN REACTION by absorbing NEUTRONs and thereby preventing neutrons from striking a nucleus that can undergo fission.

radiosensitivity The susceptibility of living organisms or certain body tissues to the adverse effects of exposure to IONIZING RADIATION. The very young are quite sensitive; sensitivity decreases with increasing age. Sensitive human body tissues include the lining of the gastrointestinal tract and the bone marrow.

radiosonde A device containing meteorological instruments and a radio transmitter that is carried aloft by a balloon. Measurements of air temperature, pressure, and humidity are transmitted to the ground, and wind speed and direction as a function of altitude are determined by ground tracking of the position of the balloon at various times. See RAWINSONDE.

radius of influence For a groundwater well, the horizontal distance from the well to the point at which the WATER TABLE is not influenced (lowered) by the withdrawal of water from the aquifer.

radius of vulnerability zone The distance from the point of an accidental release of a hazardous air pollutant within which the air CONCENTRATION of the chemical could cause acute irreversible health effects: that is, how far one can be from the release of a toxic material and still suffer adverse effects, given the poorest expected air dispersion of the chemical. See LEVEL OF CONCERN, RISK MANAGEMENT PLAN.

radon (Rn) A radioactive element with an atomic number of 86 and an atomic weight of 222. The substance is a gas that is produced directly from radium as part of the uranium decay series. The chemically inert gas enters homes through soil, water, and building materials. The threat is not uniformly distributed across the United States. For example, dwellers in Colorado and Illinois are exposed to about four times the amount of radon residents of Louisiana and South Carolina are. Even in those states where the average exposure is high, specific localities differ markedly from one another. In some areas, a significant source of exposure to radon is tap water obtained from wells. The threat arises from inhalation of the gas released from the water during use: showering, bathing, and cooking. Ingestion of radon-containing water is not a hazard.

radon daughter Element produced by the RADIOACTIVE DECAY of RADON, including isotopes of polonium and bismuth. These compounds, which are also radioactive, contribute significantly to the hazard associated with exposure to radon.

radon decay products See RADON DAUGHTER.

radwaste A term for radioactive waste.

rainbow report The informal term for the USEPA document *Status of Pesticides in Registration, Re-registration, and Special Review,* which reports the REGISTRATION status of pesticides in use, those undergoing SPECIAL REVIEW, and those being reassessed as required by the 1996 FOOD QUALITY PROTECTION ACT (FQPA). The FQPA amended the FEDERAL INSECTICIDE, FUNGICIDE, AND RODENTICIDE ACT, mandating that USEPA review pesticide ingredients with special emphasis on their risks to infants and children and that pesticide ingredients approved before 1984 be reregistered, using the current more restricted review criteria. The report chapters are different (a rainbow of) colors. See TOLERANCES, PESTICIDE; SPECIAL REVIEW.

rain forest See TROPICAL RAIN FOREST.

rain shadow The dry area on the downwind side of a mountain range. Most of the moisture precipitates on the windward or upwind slopes as the air is forced to rise and cool by the topographic features of the mountain. Little water remains in the atmosphere as it descends on the leeward side of the mountain.

ramp method, landfill A method for the placement of solid waste in a LANDFILL in which the waste is handled in the same way as in the AREA METHOD, but part of the COVER MATERIAL is obtained from an excavation at the base of each layer of waste, which forms a ramp.

range 1. For a set of observations, the difference between the lowest and highest values. 2. The thickness of some material needed to absorb a specific type of radiation.

range-finding test An analysis of the toxicity of a pollutant chemical to an aquatic organism during which the concentration range to be used in the more precise DEFINITIVE TEST is determined. The organisms are exposed to a wide range of concentrations of the specific chemical for a relatively short time so that a narrow range for use in future detailed analysis can be determined. See BIOASSAY.

range of tolerance See LIMITS OF TOLERANCE.

rangeland Natural prairie or grassland suitable for grazing cattle. These areas are also useful for agriculture and recreation, generating conflicts in land use choices for federally owned rangeland. See BUREAU OF LAND MANAGEMENT, FEDERAL LAND POLICY AND MANAGEMENT ACT, TAYLOR GRAZING ACT, NATIONAL FOREST MANAGEMENT ACT.

Rankine (R) An ABSOLUTE TEMPERATURE scale. Degrees Rankine equal degrees Fahrenheit plus 460.

Rankine cycle The model for the simple steam power plant. The boiler of the power plant heats the water to a superheated state at a constant pressure, and at the exit of the system, the condenser produces a liquid that allows for the proper operation of the heat pump.

rapid sand filter A water treatment method that removes suspended solids or COLLOIDS as drinking water passes through a sand-filtering medium. Generally used after a SEDIMENTATION treatment. The filters are cleaned periodically by backwashing.

rappers In air pollution control, devices used on an ELECTROSTATIC PRECIPITATOR to dislodge the dust on the collection plates mechanically. Magnetic impulse rappers are raised by an energized coil and drop by gravity, striking rods attached to the collecting plates. Hammer/anvil rappers have hammers attached to a rotating shaft; as the hammers fall, they strike anvils attached to the collecting plates.

raprenox A process used to limit the release of NITROGEN OXIDE (NO_x) produced during combustion. The NO_x is converted to nitrogen gas (N_2) by mixing the hot combustion gas with cyanuric acid.

Raschig rings A type of packing used in PACKED TOWER gas absorption devices.

Rasmussen report A study commissioned by the United States Atomic Energy Agency, named for the study director, Norman Rasmussen, that estimated the probability and consequences of an accident at a nuclear power plant, published in 1975. One conclusion was that the chance of injury to an individual from a serious accident leading to a meltdown was negligible. The report is also referred to as WASH-1400.

rasp A machine that grinds bulky solid waste into manageable smaller pieces.

rate In public health statistics, the frequency of a health-related event, calculated as the number of persons experiencing the event during a period divided by the number of persons in the POPULATION AT RISK. The value is usually expressed in a way that allows the use of whole numbers; for example, if deaths in an area divided by the area population equal 0.008 per year, this is written as 8 deaths per 1000 persons per year.

rate constant (K) The proportionality constant appropriate to the rate of a chemical reaction. In the differential equation $dX/dt = -KX$, the concentration of X is decreasing at a rate proportional to the remaining concentration of X. Such a reaction is called a FIRST-ORDER REACTION, and the K in the equation is the rate constant.

rating curve A graph depicting the DISCHARGE of a river or stream as a function of its water elevation (stage) at a certain point along its bank.

rational method A simple procedure for calculating the direct precipitation peak RUNOFF from a watershed, using the rainfall intensity, the area of the watershed, and a runoff coefficient appropriate for the type of watershed runoff surface.

raw sewage Wastewater that has not undergone any treatment for the removal of pollutants.

raw water Groundwater or surface water before it is treated for use as a public water supply.

rawinsonde A RADIOSONDE that is tracked from the ground to determine wind speed and direction at various altitudes. The term is derived from *ra*dar, *wi*nd, and radio*sonde*.

Rayleigh scattering The scattering of visible radiation by gaseous molecules in an unpolluted, particle-free atmosphere. The degree of scattering is greatest at the shorter wavelengths of visible light. See VISIBLE RANGE, EXTINCTION COEFFICIENT.

reactant An element or compound that takes part in a chemical reaction.

reactive organic gases (ROG) Gases participating in atmospheric reactions that create PHOTOCHEMICAL AIR POLLUTION, or OZONE. These gases include the VOLATILE ORGANIC COMPOUNDS but exclude methane. Also called NON-METHANE HYDROCARBONS and NON-METHANE ORGANIC COMPOUNDS.

reactive waste Solid waste exhibiting the HAZARDOUS WASTE characteristic of interacting chemically with other substances in ways defined by USEPA regulations in the CODE OF FEDERAL REGULATIONS, Title 40, Part 261.23. See CHARACTERISTIC HAZARDOUS WASTE.

reactivity A property of a chemical substance in relation to other substances; the likelihood of a substance's interacting chemically with other substances in the environment or in the body. See REACTIVE WASTE.

reactor-year A unit equaling one nuclear reactor in operation for one year.

Probabilities of serious nuclear accidents are often stated in reactor-years. For example, if a MELTDOWN occurred once in 20,000 reactor-years and if 500 nuclear reactors were in operation, a meltdown would be expected every 40 years.

readily water-soluble substances In water pollution, chemicals that are soluble in water at a concentration equal to or greater than one milligram per liter.

ready biodegradability A property of those substances that produce positive, unequivocal results when used in DECOMPOSITION tests.

reaeration (of streams) The natural process by which flowing stream water is mixed with the atmosphere, resulting in the addition of DISSOLVED OXYGEN to the water.

reagent A laboratory chemical added to a test for the purpose of promoting a specific reaction (for example, sulfuric acid is added to promote the digestion of samples for the measurement of the amount of nitrogen present) or developing a colored compound in a specific procedure (for example, sulfanilamide and N-(1-naphthyl)-ethylenediamine are added to the laboratory test for the presence of nitrites in water; the intensity of the color produced is determined by the amount of nitrite present).

reasonable further progress (RFP) A necessary feature of a STATE IMPLEMENTATION PLAN (SIP) in an area with dirty air. An SIP outlines the emission controls and other methods that a state will use to achieve the NATIONAL AMBIENT AIR QUALITY STANDARDS by a certain date, several years away. The schedule of controls is required to show incremental progress toward the clean air goal during the period before the deadline in a way that prevents "bunching" controls just before the SIP compliance date.

reasonable worst case In risk analysis, a scenario that considers risks of events that are not the very worst that can possibly happen but are roughly in the upper 10%. The assumptions used in estimating the potential dose of a toxic chemical from a contaminated waste site or from an accidental release often use a "worst-case scenario" as an upper bound but also presents a (more) reasonable worst case version.

reasonably available control measure (RACM) The provisions in a STATE IMPLEMENTATION PLAN (SIP) for reducing air emissions of PARTICULATE MATTER, 10-MICROMETER DIAMETER (PM_{10}) from certain sources in an area that does not meet the PM_{10} NATIONAL AMBIENT AIR QUALITY STANDARD.

reasonably available control technology (RACT) The minimal level of emission control technology required for existing air pollutant sources in NONATTAINMENT AREAS. The RACT controls apply to emissions of the pollutant(s) for which the area does not meet the NATIONAL AMBIENT AIR QUALITY STANDARDS.

rebuttable presumption against registration (RPAR) See SPECIAL REVIEW.

recalcitrant Of a substance that is degraded at an extremely slow rate if at all when released into the environment. Consequently, this type of material accumulates in water, soil, or biota.

receiving water Any surface water body into which TREATED or untreated WASTEWATER is discharged.

receptor 1. A substance, compound, or location on the surface of a cell that binds a specific chemical. This binding process can stimulate the cell to respond in certain ways or can take place before the uptake of the substance into the cell. 2. A person, plant, animal, or geographical location that is exposed to a chemical or physical agent released to the environment by human activities.

recessive gene A gene that is not expressed unless paired with a similar recessive gene. When male and female gametes fuse in the reproductive process, individual gametes furnish one member of each pair of chromosomes that will determine the genetic makeup of the offspring. If the gametes are from plants and if plant height is governed by one gene, the possible genes may be *h* for short variety and *H* for tall variety. In the case where *H* is dominant and *h* is recessive, a plant with gene pair *HH* or *Hh* would be tall, since *h* is recessive and is not expressed in the presence of the *H* gene. The gene pair in a short plant would have to be composed of *hh*. Compare DOMINANT GENE.

recessive gene disorder A genetic disorder that is exhibited in an offspring only if that offspring is DOUBLE RECESSIVE for the gene that controls the condition. In that circumstance, each parent would have contributed a RECESSIVE GENE to the child.

recharge basin A synthetic lake or reservoir designed to allow infiltration of water into the ground to recharge an underlying AQUIFER.

recharge zone A land area into which water can infiltrate into an AQUIFER relatively easily. The infiltration replenishes the aquifer. The location is also called a recharge area.

recirculation cooling system A process design that reuses industrial cooling water after heat from the water is transferred to the atmosphere through evaporation of a small amount of it.

RECLAIM An innovative emissions reduction program operated by the South Coast Air Quality Management District in Southern California since 1994. Sources of NITROGEN DIOXIDE and SULFUR DIOXIDE are issued air emission credits equal to their allowable emissions (which decline each year). The source is free to choose the mix of control techniques it uses to meet emission limits. Each source must possess emission credits equal to its actual emissions for the year. This means that if a source controls its emissions to a level below the amount of its emission credits, the credits are surplus. RECLAIM allows a source to sell the credits to another source that needs the allowances to match its emissions. This is the same approach instituted under the 1990 Clean Air Act amendments acid rain control program. See TRADABLE EMISSION ALLOWANCE.

reclamation Restoring land to the natural state after destruction associated with some economic activity such as surface mining. The original contour of the land is restored as much as is feasible, topsoil and fertilizer are applied, and vegetation native to the region is planted.

recombinant bacteria Bacteria in which the deoxyribonucleic acid (DNA) has been modified by technology to introduce new genes. The best known example is *Pseudomonas syringae,* a genetically engineered bacterium usable in the control of frost damage to strawberry and potato crops. See GENETIC ENGINEERING.

recombinant DNA Genetic material modified in the laboratory by combining genetic material from two or more organisms. The technology is used in association with the artificial modification of microorganisms for use in the manufacture of products that are useful to humans. See GENETIC ENGINEERING.

recombinant microorganism See RECOMBINANT BACTERIA.

recommended daily allowance (RDA) A guideline set by the federal Food and Drug Administration for the amounts of nutrients (such as iron, calcium, protein, and vitamins B and C) needed daily for the maintenance of good health. Varying RDAs are set for adults and children over three years old, children under four years, infants to one year old, and pregnant or lactating women.

recommended exposure limits (RELs) Chemical and physical exposure levels determined by the NATIONAL INSTITUTE FOR OCCUPATIONAL SAFETY AND HEALTH

(NIOSH) to be protective against acute and chronic adverse health effects. NIOSH RELs are used by the OCCUPATIONAL SAFETY AND HEALTH ADMINISTRATION in setting its workplace standards, PERMISSIBLE EXPOSURE LIMITS.

recommended maximum contaminant level (RMCL) See MAXIMUM CONTAMINANT LEVEL GOAL.

reconstructed source In air quality management, a facility in which components are replaced to such an extent that the fixed capital cost of the new components exceeds 50% of the capital costs of constructing a totally new facility; the reconstructed source is subject to NEW SOURCE REVIEW. Regulations normally require a new source to meet more stringent pollution control regulations than an old (existing) facility.

recordable injury At a work site, an injury that involves a loss of consciousness, a lost or restricted work day, or requires treatment beyond first aid, regardless of who administers it. The treatment guidelines are defined by the Bureau of Labor Statistics.

recordable injury rate See ACCIDENT RATE.

record of decision (ROD) The document containing the choice of remedial action to be taken at a hazardous waste SUPERFUND site; the ROD is based on the REMEDIAL INVESTIGATION/FEASIBILITY STUDY.

recoverable resource 1. Natural deposits of mineral resources and crude oil that are well defined and accessible with available technology, extraction of which is economically nonfeasible. **2.** Commercial products (e.g., newsprint or bottles) that can be recycled or reused after serving their originally intended purpose.

recovered material Used items, materials, or by-products that have been diverted from the usual discard in a solid waste disposal facility. Items for recycling. This category does not include mill scrap or shavings that are put back into the manufacturing process as a matter of routine.

recovery In industrial operations and waste management, the extraction of useful materials or energy from waste or process operations. One of the goals of the RESOURCE CONSERVATION AND RECOVERY ACT.

recovery rate Percentage of a solid waste stream that is recovered as usable items for recycling.

re-creation Establishment of a biological community to replace one that has been destroyed by the actions of humans. Reestablishment of biota in a severely damaged zone.

recyclable Used to describe a product or material for which the technical ability exists for reformulation within a manufacturing process to make a similar or derivative product. The characterization of a container as recyclable does not mean that it can actually be converted to a new item in every community. For example, if the system for collection, processing, and remanufacture of plastic does not exist in a community, the likelihood that the used container will return to the manufacturing process is remote or nonexistent despite its technical feasibility.

recycling The practice of collecting and reprocessing waste materials for reuse.

recycling economic development advocates Specialists working with states to develop financing, marketing, and permitting options for recycling businesses; part of the USEPA Jobs Through Recycling Initiative. Website: www.epa.gov/epaoswer/non-hw/recycle/jtr/

recycling mill Industrial facility involved in the manufacturing of new

products from items that are collected for recycling.

recycling symbol Identification code stamped on plastic containers. Developed by the Society of the Plastics Industry in association with the recycling industry to facilitate the SORTING of plastic containers into different resin types for subsequent recycling:

Symbol	Plastic
1	Polyethylene terephthalate (PET)
2	High-density polyethylene (HDPE)
3	Vinyl (V)
4	Low-density polyethylene (LDPE)
5	Polypropylene (PP)
6	Polystyrene (Styrofoam™) (PS)
7	Other

red bag waste Infectious waste, the name of which is derived from the red plastic bags used for clear identification.

Red Book **1.** The 1976 publication issued by the USEPA, *Quality Criteria for Water,* that has been used as a basis for ambient water quality standards. **2.** The Food and Drug Administration 1982 guidelines for animal toxicity testing, *Principles for the Safety Assessment of Direct Food Additives and Color Additives Used in Food.* See also GOLD BOOK.

red border review The final stage of a new USEPA regulation before it is signed by the ADMINISTRATOR.

Red Data Book A collection of information on threatened species in a particular country. Red Data Books employ species status classifications defined by the INTERNATIONAL UNION FOR THE CONSERVATION OF NATURE AND NATURAL RESOURCES. See RED LISTS.

Red Lists A collection of the conservation status of animals and plants worldwide compiled by the SPECIES SURVIVAL

COMMISSION of the INTERNATIONAL UNION FOR THE CONSERVATION OF NATURE AND NATURAL RESOURCES (IUCN). Species are classified as extinct, extinct in the wild, critically endangered, endangered, vulnerable, lower risk, data deficient, and not evaluated. The species status report for individual countries is called a RED DATA BOOK. Website: www.iucn.org

red tide An unusual condition associated with a bloom or excessive growth of dinoflagellates in marine waters, resulting in a red, brown, or yellow tint in the water. The event causes the death of marine biota and the accumulation of toxins in mussels or clams. Consumption of the toxin-containing shellfish can cause PARALYTIC SHELLFISH POISONING or severe gastric distress in humans.

redemption program See BOTTLE BILL.

redox potential An expression of the oxidizing or reducing power of a solution.

reduced sulfur compound See TOTAL REDUCED SULFUR, HYDROGEN SULFIDE.

reduced tillage system An agricultural practice that involves the cultivation of crops in fields that are not plowed (or plowed very little) before planting. The objectives are to reduce erosion, conserve energy, and protect water supplies.

reducing agent Any substance that loses electrons when involved in an oxidation/reduction reaction, or a substance that combines with oxygen during a chemical reaction.

reduction A chemical reaction during which electrons are added to an atom or molecule. For organic compounds, the addition of electrons is frequently accompanied by the addition of hydrogen atoms. Also called hydrogenation of organic compounds.

reentry Air that is exhausted from a building but reenters the building via the intake ducts or other openings. This process can contribute to indoor air pollution by the reintroduction of pollutants.

reference daily intake (RDI) The international term equivalent to RECOMMENDED DAILY ALLOWANCE.

reference dose (RfD) The lifetime (chronic) daily exposure level to a noncarcinogen that protects sensitive human populations from adverse effects; usually calculated as the NO-OBSERVED-ADVERSE-EFFECT LEVEL divided by an UNCERTAINTY FACTOR and expressed as milligrams of the chemical per kilogram body weight per day. Developed by the USEPA for exposure evaluations at SUPERFUND sites and applied in agency RISK ASSESSMENTS.

reference method A pollutant- and medium-specific sampling and analysis protocol designated by the USEPA as the officially acceptable technique for determining the CONCENTRATION of that chemical in a certain MEDIUM or the emission rate of that chemical from a source. See EQUIVALENT METHOD.

reflection In air pollution dispersion modeling, the simulated behavior of a PLUME after it strikes the ground. The plume is assumed to bounce off the ground instead of being absorbed or otherwise depleted at the surface.

reformulated gasoline (RFG) Gasoline containing at least 2% oxygen by weight and a maximum of 1% benzene by volume. Also, RFG emissions of VOLATILE ORGANIC COMPOUNDS and selected toxic chemicals must be reduced by certain percentages below established 1990 baselines. Federal regulations require RFG use in areas having excessively high OZONE (SMOG) readings. See also OXYGENATED FUEL, OXYGENATE.

reflux A method that uses SOLVENTs to extract certain classes of chemicals from solid materials. For example, the extraction of methanol-soluble chemicals from soil can be accomplished by placing the soil into a reflux apparatus that allows the methanol in a reservoir to be boiled, the vapor to be condensed, and the condensate to percolate through the soil sample and return to the boiling reservoir. As the methanol cycles through the system several times, the methanol in the reservoir accumulates soil materials that dissolve in the methanol.

reforestation The planting of trees on land from which the forest has been removed.

refraction The bending of light as it passes through transparent material.

refractory 1. A highly heat-resistant material that can be used as a lining for incinerators or furnaces. 2. Describing a material or substance that is difficult to metabolize within biological systems or react chemically during chemical processes. See also WATERWALL INCINERATOR.

refuging In ECOLOGY, the aggregation of large numbers of organisms in a small area with nearby food supplies—bird colonies, for example.

refuse All organic solid waste produced by a community. The combustible fraction of solid waste, including garbage, rubbish, and trash.

refuse-derived fuel (RDF) An energy resource produced by the separation of combustible materials (REFUSE) from MUNICIPAL SOLID WASTE. This material is then combined with coal and burnt as a refuse-derived fuel, which is usually 5% to 25% refuse.

refuse reclamation The conversion of MUNICIPAL SOLID WASTE into useful products by RECYCLING and/or COMPOSTING.

regenerative farming Agricultural techniques and land management practices dedicated to restoring productivity to damaged soil and sustaining productivity on all farmland through crop rotation, use of cover crops, provision of ground cover, retaining of plant residues in the field, and reduction of the use of synthetic pesticides and other chemicals.

regional administrator (RA) The head of one of the 10 multistate regions organized by the USEPA.

Regional Response Center (RRC) Under the NATIONAL CONTINGENCY PLAN, designated location in each of the 10 federal regions that provides communications and coordination for the response to an oil or hazardous substance release. See also REGIONAL RESPONSE TEAM, NATIONAL RESPONSE CENTER.

Regional Response Team (RRT) A group under the joint leadership of the USEPA and the U.S. Coast Guard that serves as the organizational unit to provide for planning and preparedness activities related to spills or discharges of oil and hazardous substances and for coordination and technical advice during such spills or discharges in each of the 10 USEPA multistate regions. See NATIONAL RESPONSE CENTER.

registration In pesticide regulation, the USEPA process leading to approval of a pesticide's use in the United States, following the requirements of the FEDERAL INSECTICIDE, FUNGICIDE, AND RODENTICIDE ACT. The registration defines crops on which it can be applied, timing and amount of its application, requirements for its purchase (over-the-counter or only to licensed individuals), labeling contents, and proper disposal. A pesticide that is "banned" has had its registration revoked or, more commonly, its use severely restricted.

Registry of Toxic Effects of Chemical Substances (RTECS) A publication of the NATIONAL INSTITUTE OF OCCUPATIONAL SAFETY AND HEALTH that is intended to identify all known toxic materials and to list toxicity and reference information on each entry. The *RTECS* is available only through vendors. Website: www.cdc.gov/niosh/rtecs.html

reg-neg A (reg)ulatory (neg)otiation process in which a committee of interested parties (e.g., the USEPA, industry representative, public interest groups) draft a proposed regulation before it is released for public comment in the FEDERAL REGISTER. Normally, the USEPA drafts the proposed regulations alone. The method is designed to speed up the writing of regulations required by statutory law and to minimize litigation.

regulated asbestos-containing material (regulated ACM) Asbestos-containing material that is FRIABLE or is likely to become friable and therefore subject to federal asbestos regulations.

regulated medical waste See MEDICAL WASTE, MEDICAL WASTE TRACKING ACT.

regulations The detailed requirements and standards written by an ADMINISTRATIVE AGENCY such as the USEPA following the dictates of a statute. At the federal level, proposed and final regulations are published in the FEDERAL REGISTER and codified in the CODE OF FEDERAL REGULATIONS. Also called RULES.

regulatory compliance Meeting the requirements imposed by state or federal regulations for planning, construction, operation, or closure of applicable facilities or activities.

Regulatory Flexibility Act (RFA) A federal statute requiring administrative agencies to examine the impact of new regulations and standards on small businesses, small nonprofit organizations, and small governmental units and to explore ways to lessen the regulatory burden.

Regulatory Impact Analysis (RIA) **1.** In the United States, a formal estimate of the benefits and costs of new regulations and standards, as required by EXECUTIVE ORDER 12866, the REGULATORY FLEXIBILITY ACT and, in some cases, the UNFUNDED MANDATES REFORM ACT. **2.** In the member countries of the Organization for Economic Cooperation and Development, a similar system to conduct a formal estimate of the benefits and costs of any new regulations or standards and the communication of the estimated impacts to government decision makers.

rehabilitation When used in an environmental context, the organized replacement or enhancement of physical and biotic components of the natural environment that were degraded by human actions. The complete reestablishment of the original conditions is not necessarily achieved.

reheater A device used to increase the temperature of an exhaust gas after particles have been removed. SCRUBBERS used to remove particles from an exhaust gas require that the temperature of the gas-particle mixture be lowered before cleaning can take place. After the scrubbing process, the gas must be reheated before release in order to prevent damage to downstream equipment, to prevent the release of a visible plume through the smokestack, or to enhance the rising and dispersion of the released gas.

Reid vapor pressure (RVP) The American Society for Testing and Materials method for measuring the VAPOR PRESSURE of VOLATILE ORGANIC COMPOUNDS (VOCs). To control the evaporative losses of VOCs, the USEPA requires refineries to produce gasoline with a lower vapor pressure during the warmer months.

Reilly, William (1940–) American Lawyer, ADMINISTRATOR of the USEPA, 1989–93. Former head of the WORLD WILDLIFE FUND and the Conservation Foundation.

relative biological effectiveness (RBE) A comparison of the effect of some type of IONIZING RADIATION to that caused by 200-keV X rays. The relationship to GAMMA RAYS is about 1, whereas the relationship to FAST NEUTRONS is approximately 10, meaning that the latter type of radiation is about 10 times more effective than X rays in producing biological damage. The RBEs are used as the QUALITY FACTOR to adjust absorbed doses of different types of ionizing radiation to a constant scale. See LINEAR ENERGY TRANSFER.

relative ecological sustainability The ability of a natural environment to maintain and perpetuate itself over an indefinite period.

relative humidity The ratio of the amount or weight of water vapor present in a specified volume of air to the maximal amount that can be held by the same volume of air at a specified temperature and pressure. If the relative humidity is 50% in a room, the air contains one-half the amount of water vapor that it is capable of holding at the temperature and atmospheric pressure in the room. Also expressed as the MIXING RATIO divided by the SATURATION MIXING RATIO times 100.

relative risk In environmental risk assessment or EPIDEMIOLOGY, the ratio of the risk of some adverse condition developing in individuals exposed to pollution to the risk of that same condition developing in an unexposed population. Relative risk can be estimated by using a COHORT STUDY. See ODDS RATIO.

release A spill, leak, escape, or loss of a regulated chemical agent into the environment, including air, water, or land.

release height The height above the ground from which an air pollutant is emitted. A higher release height produces lower ground level air concentration at points downwind of the source.

relict A species or community that is unchanged from some earlier period.

rem See ROENTGEN EQUIVALENT MAN.

remedial action The permanent cleanup or treatment of a contaminated environment. The technology used depends on the nature of the contamination; treatment may use biological, chemical, or thermal technology. Compare REMOVAL ACTION. See REMEDIAL DESIGN/REMEDIAL ACTION, REMEDIAL INVESTIGATION/FEASIBILITY STUDY, REMEDIAL PROJECT MANAGER, REMEDIATION.

remedial design/remedial action (RD/RA) Under the NATIONAL CONTINGENCY PLAN, the cleanup work at the site of the HAZARDOUS SUBSTANCE release, usually a HAZARDOUS WASTE site. The RD/RA follows the RECORD OF DECISION, which itself follows the first step in the cleanup, the REMEDIAL INVESTIGATION/FEASIBILITY STUDY (RI/FS).

remedial investigation/feasibility study (RI/FS) A detailed technical study of the type and extent of contamination at a SUPERFUND site, including the alternatives for its cleanup.

remedial project manager (RPM) The federal official responsible for directing and coordinating federal remedial actions at sites on the NATIONAL PRIORITIES LIST. REMEDIAL ACTION is taken to prevent or minimize the release of a hazardous substance into the environment. REMOVAL ACTION is the responsibility of the ON-SCENE COORDINATOR.

remediation Organized methods applied in response to an excessive release of a HAZARDOUS SUBSTANCE that eliminate or adequately reduce the risk of damage to human health or the environment. Includes PUMP AND TREAT, STABILIZATION, and SOLIDIFICATION. See IN SITU REMEDIATION, REMEDIAL ACTION.

remote sensing 1. The use of imagery generated by high-flying aircraft or Earth-orbiting satellites to study the resources of the Earth or aspects of environmental quality. 2. The use of ground devices to study atmospheric conditions aloft or to determine air pollutant concentrations in the upper urban atmosphere or in a smokestack plume.

removal action Under provisions of the NATIONAL CONTINGENCY PLAN, the selected type and timing of the coordinated response to a chemical release. Actions can be immediate (emergency), time-critical, or non-time-critical. Removal actions are immediate responses to releases or threatened releases, whereas REMEDIAL ACTIONS entail longer-term, permanent cleanups. See TIME-CRITICAL REMOVAL ACTION.

renewable energy A source of energy that is replaced by natural phenomena or natural growth aided by human activity, such as BIOMASS or the water held behind by a dam that is used to generate electricity.

renewable energy production incentive (REPI) A provision of the Energy Policy Act of 1992 intended to encourage the development of renewable power projects. Tax credits are available for electricity produced by wind energy or BIOMASS and for SOLAR ENERGY and GEOTHERMAL investments. See also PUBLIC UTILITIES REGULATORY POLICY ACT (PURPA).

renewable resource A supply of a biological organism that can be replaced after harvesting by regrowth or reproduction of the removed species, such as seafood or timber.

Reorganization Plan Number 3 The executive order issued by President Richard Nixon in 1970 that created the U.S. ENVIRONMENTAL PROTECTION AGENCY.

replicate One or more duplicates of a test, analysis, experiment, exposure, and so on.

reportable quantity (RQ) The size of a spill or release of a hazardous substance or material requiring a report to the U.S. Department of Transportation, the USEPA, or a state or local environmental agency. Title 40, Section 302.4, of the *CODE OF FEDERAL REGULATIONS* provides a list of reportable quantities of USEPA-designated HAZARDOUS SUBSTANCES.

repository A site for the isolation of radioactive wastes from humans or the bio-

logical environment until the waste RADIOACTIVITY decays to near-background levels. Such storage facilities would be located about 2000 feet below the surface in geological strata that are stable, dry environments with good heat-dissipating properties. The site would be designed to isolate HIGH-LEVEL WASTE for about 10,000 years. See NUCLEAR WASTE POLICY ACT.

representativeness Describing how well a sample of air, water, soil, or food represents the whole from which it was taken.

representative sample See REPRESENTATIVENESS, SAMPLE.

reprocessing nuclear fuel See FUEL REPROCESSING.

reproducibility The ability of a test, analysis, or experiment to yield the same results when repeated under the same circumstances. The likelihood that duplicate tests will produce the same results.

Request for Information (RFI) An enforceable request sent to a discharger by the USEPA or a state environmental agency for technical data to be used to establish permit limitations or source control standards, inter alia. RFI provisions are found in the CLEAN AIR ACT, CLEAN WATER ACT, RESOURCE CONSERVATION AND RECOVERY ACT, and COMPREHENSIVE ENVIRONMENTAL RESPONSE, COMPENSATION, AND LIABILITY ACT.

Research and Special Programs Administration (RSPA) The U.S. Department of Transportation agency responsible for regulations covering the transport of hazardous materials in commerce. See HAZARDOUS MATERIALS TRANSPORTATION ACT (HTMA).

Research Triangle Park (RTP) An area in North Carolina containing the offices of many USEPA technical and scientific activities together with a variety of private laboratories and public institutions of higher education.

reservation In public land management, a permanent dedication of land to, usually, a single purpose, such as a national forest, wildlife refuge, or wilderness area.

reserve capacity Normally unused treatment capacity built into a wastewater transport and treatment system to provide for the increased flow that results form excess rainfall or population growth. The capacity to expand operations of a wastewater treatment system without new construction.

reserves Natural resources that can be exploited in an economically feasible manner employing current technology. Compare RESOURCES.

reservoir At the planetary scale, a storage depot for chemical substances; for example, the atmosphere is a reservoir for carbon as carbon dioxide. See STOCKs.

reservoir rock Porous rock containing oil or natural gas.

residence time 1. The time a chemical substance spends in a biological or environmental "container" such as a lake, the human body, or the atmosphere under equilibrium conditions, calculated by the mass of the material in the container divided by the flow rate in (or out). 2. The amount of time a specified volume of liquid or gas spends in a certain device. The residence time for sewage undergoing treatment is the time a specified volume of sewage stays in the treatment facility.

residential use In pesticide management, the application of pesticides in and around homes, apartments, motels, and other living/working areas.

residential waste Solid waste generated in a household, usually discarded consumer goods: yard waste, beverage containers, newsprint, food waste, cardboard, and others.

residual Material that remains. Often used in association with the pollutant load that remains after the operation of some pollution control device, as in nutrients after wastewater treatment or particulate material that passes through a filtering apparatus. The higher the residual, the less efficient the process.

residual risk Under the AIR TOXICS provisions of the 1990 amendments to the CLEAN AIR ACT, the risk to public health or the environment that remains after the application of MAXIMUM ACHIEVABLE CONTROL TECHNOLOGY.

residue Remains after some treatment or process, for example, ash that remains after incineration or salts that remain after the evaporation of seawater.

resilience stability The ability of a biological community or ecosystem to recover its original condition after a severe stress or perturbation. Compare RESISTANCE STABILITY.

resistance stability The ability of a biological community or ecosystem to withstand a stress or perturbation without adverse change to its structure or function. Compare RESILIENCE STABILITY.

resistivity For FLY ASH, the resistance of the ash to the conduction of electricity. Low resistivity reduces the collection efficiency of ELECTROSTATIC PRECIPITATORS by allowing the electrostatic charge attracting the particles of ash to the collection plate to drain away. High-resistivity ash holds a negative charge and can repel other particles (BACK CORONA); resistivity also causes the particles to adhere strongly to the plate, making them difficult to remove. See FLUE GAS CONDITIONING.

resistivity survey A type of noninvasive survey of soil for the detection of buried metal objects and chemical contaminants capable of conducting an electric current. The method is especially useful for the evaluation of groundwater depth and site geological characteristics.

resolving time The time that must elapse between two radiation impulses in order for them to be detected as two discrete events by an electronic radiation detection device.

resource conservation Most commonly used to describe the protection, improvement, and prudent management of natural resources, including land, both agricultural and uncultivated. The protection of soil from erosion or degradation, conservation of water, provision of habitat for wildlife, sustaining of forage and grazing plants, efficient energy use, recycling/reuse, and organization of local, state, and national environmental stewardship efforts are commonly considered to be resource conservation activities.

Resource Conservation and Recovery Act (RCRA) The federal statute providing for the comprehensive management of nonhazardous and hazardous solid wastes. The USEPA, in its implementing regulations, sets minimal standards for all waste disposal facilities and, for HAZARDOUS WASTE, regulates treatment storage and transport. See CORRECTIVE ACTION, MANIFEST SYSTEM, SOLID WASTE, MUNICIPAL SOLID WASTE, SANITARY LANDFILL, SECURE LANDFILL, SUBTITLE C, SUBTITLE D, HAZARDOUS AND SOLID WASTE AMENDMENTS, MEDICAL WASTE TRACKING ACT, BASEL CONVENTION, UNDERGROUND STORAGE TANK.

resource partitioning In ecology, the sharing of different environmental resources (e.g., different food items) by biological organisms that occupy the same general surroundings. The process reduces direct competition among organisms and allows diverse species to occupy the same location.

resource recovery The processing of solid waste to extract recyclable paper, glass, metals, or combustible material. Also, the recovery of energy (steam) by the direct incineration of waste. See RECOVERY, RECYCLING, WASTE-TO-ENERGY.

resources The total amount of any rock, mineral, or fuel in the crust of the Earth. See RESERVE.

respirable particulate See PARTICULATE MATTER, 10-MICROMETER DIAMETER (PM10).

respiration 1. For a person or animal, the act of breathing. 2. The metabolism of an individual cell, tissue, or organism that results in the release of chemical energy derived from organic nutrients. 3. A specific series of reactions in a cell during which electrons removed during the OXIDATION of organic nutrients (substrates) are transferred to a terminal acceptor. When the terminal acceptor is molecular oxygen, the process is termed AEROBIC respiration. When the terminal acceptor is an inorganic substitute for molecular oxygen, such as sulfate or nitrate, the process proceeds without oxygen and is termed ANAEROBIC respiration. In either case, potential energy present in the organic nutrients is converted to a chemical form useful to the cell. Most organisms, including humans, are capable of aerobic respiration, whereas only a few types of bacteria are capable of anaerobic respiration.

respiration/biomass ratio An expression of ecological TURNOVER, the ratio of total community RESPIRATION (often expressed in grams per square meter per day) to community BIOMASS (in grams per square meter). The ratio indicates the rate of energy flowing through the community compared with the mass of the community.

respirator Any device designed to deliver clean air to the wearer. The two general types are an AIR-PURIFYING RESPIRATOR and a SELF-CONTAINED BREATHING APPARATUS.

respirator fit test (RFT) A method used to determine whether a working respirator provides the proper protection when worn by a particular worker. A quick, simple, and low-cost RFT is the qualitative test, which involves the wearer's response to a test atmosphere containing an easily detected material such as banana oil or irritant smoke. If the wearer can detect the material in the test atmosphere, the respirator does not fit properly. Quantitative fit testing is more accurate but more expensive and time-consuming and is not widely used. In the quantitative analysis, trained technicians use instruments to determine the presence and extent of any respirator leaks.

respiratory fibrotic agent See FIBROSIS, SILICOSIS, CRYSTALLINE SILICA, ASBESTOSIS.

response action General term applied to the prescribed resolution of an identified threat to human health and safety or to the environment. The appropriate federal or state agency may direct, for example, the removal of hazardous materials from a disposal site, cleanup of an accidental spill of a dangerous material, containment of a waste on-site, halting of additional groundwater contamination, or removal of asbestos from a public building. See REMOVAL ACTION, REMEDIAL ACTION, NATIONAL CONTINGENCY PLAN.

restoration See RECLAMATION, ECOSYSTEM RESTORATION.

Responsible Care® An initiative of the CHEMICAL MANUFACTURERS ASSOCIATION (currently, the American Chemistry Council) intended to improve the health, safety, environmental performance, and community outreach of member companies. Member companies are required to adopt the initiative as a condition of membership and are expected to improve the safe management of chemicals continuously in a manner that is responsive to the public. The focus of the initiative consists of Codes of Management Practice that address community awareness and emergency response, worker health and safety, pollution prevention, process safety, distribution, and product stewardship. Website: www.chemicalguide.com

restoration ecology See ECOSYSTEM RESTORATION.

restricted-use pesticide (RUP) Pesticide that can be purchased and used only by specially trained and certified people because it constitutes a special hazard to applicators and the environment, even when used as directed.

restriction enzyme Short for *restriction endonuclease*, a class of enzymes that have the ability to cleave both strands of double-stranded deoxyribonucleic acid (DNA). The natural function of these enzymes is to protect a cell from DNA from outside sources. These enzymes are useful in GENETIC ENGINEERING because they cleave DNA on the basis of recognizing a specific sequence of nucleotide bases; consequently, the same location within a DNA molecule is cleaved each time. A large DNA molecule can be cut into a number of small fragments. Furthermore, the ends of the region where the cut occurs are configured in such a way that it possible to attach other DNA units by using laboratory technology. About 2000 different restriction enzymes have been described; the most well known is *Eco*RI from *Escherichia coli.*

resuspended Describing particles that have been remixed with the air or water from which they have settled or collected. For example, sediment particles settle from river water if the water is allowed to stand. Those particles may be remixed with the water if turbulent conditions recur.

retention basin A permanent lake or pond used to slow storm water runoff. See DETENTION BASIN.

retention time See RESIDENCE TIME.

retort A bulb-shaped laboratory apparatus for heating some material in the absence of oxygen.

retrofit The addition of new parts or equipment that were not previously available or required when the facility was built, such as pollution control equipment; a type of upgrading.

retrospective study See CASE-CONTROL STUDY.

reuse The use of an item more than once. As an example, serving the reuse market for washing machines would involve collecting discarded machines, refurbishing old units to put them in working order, then selling the used machines. As a comparison, serving the recycling market for washing machines would entail collecting discarded appliances, shredding to recover ferrous metal, and using the recovered metal to manufacture iron reinforcing bars.

reverse osmosis A process used in water purification in which pressure is applied to the more concentrated (contaminated) solution on one side of a semipermeable membrane. The result is the movement of water, but not contaminants, from the more concentrated side to the more dilute side, thus separating clean water from the contaminated water. See OSMOSIS.

revolving fund As authorized by the Water Quality Act of 1987, a source of loans to be used to construct or upgrade SEWAGE TREATMENT PLANTS (PUBLICLY OWNED TREATMENT WORKS). The USEPA and each state provided the initial funds, 80% federal and 20% state. Previously operated as the CONSTRUCTION GRANTS PROGRAM.

Reynolds number (Re) A dimensionless number that represents the ratio of inertial forces to viscous forces for fluid flow in a pipe or duct or around an obstacle, such as an airborne particle. The expression for fluid flow in a pipe or duct is

$$Re = \frac{\rho D v}{\mu}$$

where ρ is the fluid density, D is the pipe/duct diameter, v is the fluid velocity, and μ is the fluid viscosity. For fluid flow around a particle it takes the form

$$Re = \frac{d_p v_r \rho}{\mu}$$

where d_p is the particle diameter, v_r is the velocity of the particle relative to the fluid, ρ is fluid density, and μ is the fluid viscos-

ity. For fluid flow in a pipe or duct, a Reynolds number below about 2100 (viscous forces dominant) is considered to be streamline, smooth, or LAMINAR FLOW; above 4000 (inertial forces dominant) the flow is turbulent; 2100–4000 is a transition zone. For the flow of fluid around a particle, a Reynolds number less than 1.0 is laminar flow, or in the Stokes regime, and as the value increases above 1.0, turbulence increases. The difference between the conditions for laminar flow around particles and in pipes is explained by the impact of inertial forces as the fluid flows around a particle compared with the straight flow in a pipe or duct.

ribbon sprawl The development of residential areas and businesses along major roadways, such as the Interstate Highway System, that carry traffic into a city.

ribonucleic acid (RNA) A biological macromolecule, common to all organisms, involved in the conversion of the genetic information contained in the deoxyribonucleic acid (DNA), or genes of the cell, into properties or characteristics by which all organisms are described. Two classes of RNA are instrumental in the conversion of the linear sequence of bases in DNA into specific proteins: transfer RNA, which carries specific amino acids for polymerization into proteins, and messenger RNA, which determines the linear sequence of amino acids in specific proteins. The sequence of nucleotides in the genes (DNA) is incorporated into the RNA molecule through a polymerization process termed transcription. Other types of RNA molecules participate in forming the physical structure of parts of a cell.

Richardson number (Ri) A dimensionless number for a fluid (usually air) at a certain place and time relating stability to wind shear. E expressed as follows:

$$Ri = \frac{(g/\Theta)(d\Theta/dz)}{(d\mu/dz)^2}$$

where Θ is the POTENTIAL TEMPERATURE, z is the elevation, μ is the wind velocity, g is the acceleration of gravity. The value of the number indicates the dominant type (low values) of turbulence present, either mechanical or convection (high values). See MECHANICAL TURBULENCE, THERMAL TURBULENCE.

richness 1. The total number of species in an area, usually expressed as the number of species divided by the total number of individuals, or the number of species per unit area. See SPECIES RICHNESS INDEX. 2. The relative abundance of the fissionable form of uranium (^{235}U) in a mixture of uranium containing both of the naturally occurring ISOTOPES of the element (^{235}U and ^{238}U).

rickets A medical condition, normally seen in children, characterized by poor bone structure resulting from a deficiency of vitamin D.

right to know (RTK) 1. Describing statutes, regulations, and policies that require the public disclosure of information about chemical releases and chemical storage, including types and amounts. See TITLE III, SEVESO DIRECTIVE. Website: www.rtk.net 2. Describing regulations that require that employees be given information about workplace chemical hazards. See HAZARD COMMUNICATION STANDARD.

rill erosion A type of damage done to the topsoil of exposed land, commonly pasture or cropland, by rainwater runoff that collects in small channels cutting through the earth. The damage can be corrected with normal tilling.

ring compound Normally used to refer to an organic compound containing an AROMATIC structure typified by BENZENE, in which six carbon atoms are joined in a closed circle or ring.

Ringelmann chart An illustration containing shades of gray that can be used to estimate the density of smoke. The scale varies from 0, clear stack gases, to 5, 100% black smoke. See also SMOKE READER.

Rio Declaration The agreement finalized at the United Nations Conference on Environment and Development held in 1992 at Rio de Janeiro, Brazil. The goal of the accord is the establishment of a new and equitable global partnership through the creation of new levels of cooperation among states. The 27 principles of the declaration commit signatory nations to working toward international agreements that respect the interests of all and protect the integrity of the global environmental and development system. Peace, development, and environmental protection are presented as interdependent and indivisible. See EARTH SUMMIT.

riparian 1. Describing the legal doctrine that a property owner along the banks of a surface water body (lake, river) has the primary right to withdraw water for reasonable use. 2. Related to plant communities located on the banks of rivers.

risk assessment An analytical study of the probabilities and magnitude of harm to human health or the environment associated with a physical or chemical agent, activity, or occurrence. The assessment involves estimates of the types, quantities, and locations of the release of harmful substances or energy; a dose-response evaluation linking exposure to possible harm; a characterization of the exposure of humans, wildlife, or ecosystem components to the release; and a summary using the foregoing analysis to produce the overall risk assessment The final step is also known as RISK CHARACTERIZATION. See HAZARD IDENTIFICATION, SOURCE/RELEASE ASSESSMENT, DOSE-RESPONSE ASSESSMENT, EXPOSURE ASSESSMENT.

risk-based corrective action (RBCA) The use of a RISK ASSESSMENT, along with chosen levels of acceptable risk, to choose cleanup goals at a site contaminated by hazardous substances. RBCA ("Rebecca") differs from other cleanup strategies in that the goal is not necessarily the removal of hazardous substances to levels approximating their BACKGROUND concentrations or to MAXIMUM CONTAMINANT LEVEL GOALS for groundwater but reduction of the actual risk the substances pose to human health, based on exposure. Risk reduction can be accomplished by removal of the hazardous substances (e.g., PUMP AND TREAT) but also by sequestration of the material so that exposure does not occur, POINT-OF-USE/POINT-OF-ENTRY treatments, MONITORED NATURAL ATTENUATION, or other site-specific approaches determined by the risk assessment. The AMERICAN SOCIETY FOR TESTING AND MATERIALS published the first RBCA guidance, which has been adopted by government agencies. RBCA has been extensively applied to leaking UNDERGROUND STORAGE TANK cleanups.

risk-based targeting Concentration of environmental enforcement and compliance efforts in areas and activities that pose relatively greater risk to human health and the environment. See COMPARATIVE RISK ASSESSMENT.

risk characterization An estimation of the nature, magnitude, and likelihood of adverse effects on human health or the environment using the results of the previous analyses in a RISK ASSESSMENT.

risk communication 1. For routine exposures to chemical or physical hazards, the characterization of potential adverse health effects resulting from exposure to toxic materials in such a way that the information is comprehensible to the public. 2. For accidental exposures to chemical or physical agents, the characterization of the probabilities of the release and subsequent range of exposures along with the dose-response relationships potentially leading to acute or chronic health effects in a way that is comprehensible to the public. It is generally agreed that is cannot be fully realized.

risk estimate The probability of an adverse health effect associated with an exposure to a physical or chemical agent. Compare UNIT RISK ESTIMATE.

risk extrapolation model Any of a variety of methods used to apply the measured results of animal toxicology testing, especially those involving a potential CARCINOGEN, to the risk of adverse effects in humans. The EXTRAPOLATION involves the species differences, the estimation of the risk at much lower doses than those used in the testing, a longer human exposure time, and possibly a different route of exposure (human inhalation risk from animal ingestion data). See SCALING FACTOR, SURFACE AREA SCALING FACTOR, LINEARIZED MULTISTAGE MODEL.

risk factor Any exposure, habit, or personal characteristic linked to an increased risk of an adverse health effect or condition.

risk management plan (RMP) Risk analysis and public communications required by provisions of the 1990 amendments to the CLEAN AIR ACT. The USEPA regulations "Section 112(r) regs" require certain industrial, utility, military, water treatment, and small business facilities that either manufacture or use any one of 140 different chemicals to document prevention and emergency response programs, including WORST-CASE SCENARIOS and ALTERNATE CASE SCENARIOS that involve on-site storage of hazardous chemicals. The documentation must include an assessment of the chemical hazards at the facility, description of the steps being taken to reduce the possibility of accidental releases, and plans to mitigate the circumstances of an accident should one occur. The objectives of the program are to establish and maintain communications with the community, emergency responders, police and fire departments, medical services, and neighboring industrial facilities concerning potential chemical emergencies; to assess the "what if" risks associated with potential chemical releases; and, using the analysis, to lower the probability and potential consequences of catastrophic chemical releases. See also BHOPAL; TITLE III; LEVEL OF CONCERN; LOCAL EMERGENCY PLANNING COMMITTEE; TIER I, TIER II REPORTS.

risk quotient In pesticide regulation, exposure divided by toxicity, where toxicity is acute or chronic, using various toxicity END POINTs, and exposure is the ESTIMATED ENVIRONMENTAL CONCENTRATION or an assumed dose. The risk quotient is compared with the LEVEL OF CONCERN to determine the potential risk to NONTARGET ORGANISMS.

risk ratio See RELATIVE RISK.

risk-specific dose (RSD) The daily dose or exposure level that implies a given risk level, for example, the dose of chemical A, in grams per kilogram body weight per day, corresponding to a 1 in 100,000 lifetime excess cancer risk.

R-meter A type of IONIZATION METER designed to measure radiation in ROENTGENs.

Roadless Area Review and Evaluation (RARE) Studies performed by the U.S. Forest Service that included an inventory of roadless areas in the NATIONAL FOREST SYSTEM. Under the provisions of the NATIONAL FOREST MANAGEMENT ACT and the MULTIPLE USE AND SUSTAINED YIELD ACT, the Forest Service may elect to prohibit the construction of roads in part or all of the roughly 40 million acres of now-roadless land (there are 190 million acres of national forests), managing them, in effect, as WILDERNESS AREAS.

rod deck absorber A device used to remove pollutants from an exhaust gas. The exhaust gas is put in contact with a liquid slurry used to remove pollutants from it. Mixing of the liquid and gas is enhanced by decks of cylindrical rods positioned perpendicular to the gas and liquid flows.

rodenticide A PESTICIDE designed to kill, repel, or disrupt the reproductive process of rodents.

roentgen (R) The quantity of radiation (X RAYS or GAMMA RAYS) that produces one electrostatic unit charge (2.08×10^9 ion pairs) in 1 cm^3 of dry air at 0° C and 1 atmosphere of pressure. Roentgens per unit time is the normal expression used when the unit is used to report radiation exposures. One roentgen equals 2.58×10^{-4} coulomb per kilogram of air.

roentgen equivalent man (rem) A non-SI unit of radiation dose that incorporates both the amount of IONIZING RADIATION absorbed by tissue (RAD) and the relative ability of that radiation to produce a particular biological change, called the RELATIVE BIOLOGICAL EFFECTIVENESS (RBE). Expressed as rem = rad × RBE, where rad is an absorbed dose of 100 ergs per gram of tissue (0.01 GRAY). RBE is also called the QUALITY FACTOR. The unit is frequently applied to total body exposure for all types of ionizing radiation. Approximately 0.1 to 0.2 rem per year represents the level of exposure of individuals to natural BACKGROUND RADIATION. Five rem per year is the maximum occupation exposure allowed. An exposure of 100 rems in a single incident is thought to increase the risk of leukemia, and an exposure of 600 rems over 48 hours is a sufficient dose to kill 100% of the individuals exposed. The corresponding SI unit is the SIEVERT (Sv); 1 sievert equals 100 rems.

rollback model A control strategy by which excessive environmental concentrations or pollutants are to be reduced by lowering the emissions of all sources by an equal percentage. The percentage reduction is calculated by using the amount by which the ambient concentration exceeds a standard. For example, if the ambient air standard for pollutant A is 8 and the measured ambient concentration is 10, then a 20% reduction by all sources is required (10 minus 8, divided by 10, times 100).

room-and-pillar mining A method employed in the underground mining of coal in which portions of the coal deposit are left undisturbed as columns that support the roof of the mined cave to prevent collapse.

room constant In noise control, the sum of the surface areas of the various materials in a room, weighted by their respective SABIN ABSORPTION COEFFICIENTS. The room constant includes walls, doors, floor, and ceiling, together with all furniture, cabinets, draperies, and so on. Room constant units are area-sabins, for example, three m²-sabins.

root-mean-square sound pressure (rms sound pressure) An expression of the average sound pressure (P_{rms}) from a series of sound wave AMPLITUDE measurements. The value is represented by

$$P_{rms} = \left[\frac{1}{T} \int_o^T P^2(t)\, dt \right]^{1/2}$$

where P is an instantaneous pressure measurement and T is the averaging time. This method is used because sound wave amplitude values fluctuate around a zero pressure, positive to negative, and squaring the pressure measurements avoids an average pressure of zero. Sound-level meters perform these calculations internally.

rotameter A device used to measure gas flow rates; the meter consists of a tapered tube containing a float that rises as gas flows through the tube and that remains suspended as long as the gas flow is constant. The height of the float is read, using calibrated marks on the tube, and converted to a flow rate.

rotary kiln A long, inclined rotating drum, designed to manufacture cement or lime, used for thermal destruction of HAZARDOUS WASTE. Typical kiln operating temperatures and furnace residence time ensure excellent DESTRUCTION AND REMOVAL EFFICIENCY of the waste.

rotating biological contactor (RBC) See BIODISC.

rotating biological filter See BIODISC.

Rotenone A commercial preparation containing extracts of the tuber roots of plants of the genus *Derris*. The most common use is to paralyze the fish in a pond or other body of water for their removal or collection in ecological studies. The extracts have also been used to control head lice.

rough fish Fish species that are not valued commercially and are not prized in sport fishery. Often these species are more resistant to environmental damage; for example, they would be selected for in water bodies with low DISSOLVED OXYGEN.

roughness A descriptor for the friction produced by a surface; used in AIR QUALITY DISPERSION MODELING to estimate the wind speed at stack height from a measurement of wind speed at 10-meter elevation. Wind speed increases with distance from the ground, but it increases by a larger amount if the surface is rough (friction producing). Roughness is also a factor in MECHANICAL TURBULENCE: the greater the roughness, the stronger the mechanical turbulence and the greater the dispersive power of the atmosphere.

route of exposure The way in which an organism is exposed to a chemical agent, such as inhalation, ingestion, or skin contact.

r-selected A type of reproductive strategy of a species that allocates a relatively high proportion of energy available in nutrients to reproduction and a relatively low proportion to the maintenance and survival of individual members of the species. See R-STRATEGIST. Compare K-SELECTED.

r-strategist A type of animal whose reproductive behavior is characterized by "boom-to-bust" swings in population levels. Typically, animals with this type of reproductive strategy produce large numbers of offspring in a short period to take advantage of favorable environmental conditions. These explosions in numbers of animals are routinely followed by dramatic declines in numbers once the conditions change slightly. Likewise, the dramatic increases in population levels normally exceed the capacity of the habitat to support the species. The result is a precipitous decline in numbers of organisms present. See R-SELECTED.

rubbish Nonkitchen solid waste, such as yard litter, junk, or other material. Also called trash.

rulemaking The public drafting, amending, and finalizing of RULES (regulations) by administrative agencies, such as the USEPA, under the authority of a particular statute and following the process dictated by the ADMINISTRATIVE PROCEDURE ACT. See FEDERAL REGISTER, CODE OF FEDERAL REGULATIONS, NEGOTIATED RULEMAKING, REG-NEG.

rules Standards and regulations issued by administrative agencies to implement statutes. See RULEMAKING.

Rule 702 The federal rule of evidence that states, "If scientific, technical, or other specialized knowledge will assist the trier of fact to understand the evidence or to determine a fact in issue, a witness qualified as an expert by knowledge, skill, experience, training, or education, may testify thereto in the form of an opinion or otherwise." Expert witness testimony in TOXIC TORT and other environmental law cases is subject to Rule 702 and its 1993

interpretation by the U.S. Supreme Court in *Daubert v. Merrell Dow Pharmaceuticals, Inc.*

ruminant animals Cattle, sheep, goats, deer, camels, giraffes, antelopes, buffalo, and similar grazing animals that have a four-chambered stomach. The largest chamber, which receives the ingested hay or grass, is termed the rumen. The rumen is a large fermentation vessel in which a community of bacteria, yeast, and other organisms digest the cellulose-based food to produce nutrients usable by the mammal. Without the community of microbes in the rumen, these animals could not subsist on a diet of hay.

run In environmental sampling and analysis, the period during which an emission sample is taken, as in a sampling run for a smokestack. The term is also used as a slang expression to describe the performance of a laboratory analysis: "The Ames test was run."

runoff That portion of rainwater or snowmelt that enters surface streams rather than infiltrating the ground.

rutherford (Rd) A unit that describes the amount of a radioactive substance that decays at a rate of 10^6 disintegrations per second.

S

sabin The SI unit used to compare the sound absorption ability of materials. A sabin is the unit area of a totally absorbent surface, that is, a surface that does not reflect any sound. One square meter-sabin is one square meter of perfect absorption. Sound-absorbing materials are assigned a SABIN ABSORPTION COEFFICIENT, which is expressed as a fraction of a sabin.

sabin absorption coefficient Fraction of a sabin assigned to different materials indicating their sound absorbance relative to a totally absorbent surface. A sabin absorption coefficient of 0.15 indicates that 15% of the sound energy striking the surface will be absorbed and 85% will be reflected. Sabin absorption coefficients vary with sound frequency. The average absorption coefficient, over several frequencies, for a material is the NOISE REDUCTION COEFFICIENT. See ROOM CONSTANT.

Safe Drinking Water Act (SDWA) A federal statute enacted in 1974, and subsequently amended, that requires the USEPA to set and enforce chemical and radioactivity standards for public drinking water supplies. The act also began an UNDERGROUND INJECTION CONTROL program to protect underground drinking water sources and established a special SOLE SOURCE AQUIFER designation process, WELLHEAD PROTECTION PROGRAM, SOURCE WATER ASSESSMENT PROGRAM, and SOURCE WATER PROTECTION PROGRAM. See MAXIMUM CONTAMINANT LEVEL, SECONDARY MAXIMUM CONTAMINANT LEVEL, MAXIMUM CONTAMINANT LEVEL GOAL, NATIONAL PRIMARY DRINKING WATER REGULATIONS, SECONDARY

DRINKING WATER REGULATIONS, DISINFECTION BY-PRODUCTS.

safe water Water that does not contain dangerous levels of microbes or toxic chemicals. Although the water may be safe, it may have a disagreeable taste, color, or odor. See POTABLE.

safe yield The amount of water that can be removed from an AQUIFER or surface source without threatening the long-term supply available in the resource. If more water is removed from a reservoir during a year than the amount added to the reservoir over the same period, the yield is not safe and the reservoir is being depleted.

safety control rod ax man (SCRAM) The individual responsible for the emergency lowering of control rods into the first nuclear reactor operated by the United States during the World War II Manhattan Project. The control rods were suspended by ropes; in the event of a malfunction, this individual was assigned the responsibility of cutting the ropes with an ax to allow the control rods to fall into the reactor, thereby stopping the FISSION reaction. The acronym SCRAM is used today to signify any sudden shutdown of a nuclear reactor by the emergency insertion of control rods to stop the fission chain reaction.

safety factor In RISK ASSESSMENT, an adjustment used at one or more points in the analysis to account for uncertainty. For example, NO-OBSERVED-ADVERSE-EFFECTS LEVEL for adult males may be reduced by a factor of 10 when applied to the general population. See UNCERTAINTY FACTOR.

sag pipe See INVERTED SIPHON.

Sagebrush Rebellion A political movement that arose in the American West and Alaska in the late 1970s, opposing restrictive rules governing the use of federally owned lands under the FEDERAL LAND POLICY AND MANAGEMENT ACT of 1976 and the administrative agency implementing the act, the BUREAU OF LAND MANAGEMENT. Ninety-three percent of all federally owned land is found in the 12 states that joined the movement to strengthen state authority over public land use. Bureau of Land Management MULTIPLE USE policies helped defuse the movement.

salinity The amount of salts dissolved in water. The term is usually reported in grams per liter (parts per thousand), and the unit symbol $^0/_{00}$ is normally used. Although the measurement takes into account all of the dissolved salts, sodium chloride normally constitutes the primary salt being measured. As a reference, the salinity of seawater is approximately $35^0/_{00}$.

salinization The accumulation of salts in soil to the extent that plant growth is inhibited. This is a common problem when crops are irrigated in arid regions; much of the water evaporates and salts accumulate in the soil.

Salmonella The genus of bacteria that cause typhoid fever and that are associated with some types of food poisonings.

salmonellosis The bacterial disease caused by the presence of bacteria of the genus *Salmonella*. The disease is a type of food poisoning characterized by a sudden onset of gastroenteritis involving abdominal pain, diarrhea, fever, nausea, and vomiting. A variety of foods, such as sweets, meats, sausages, and eggs, can be the mode of infection. Pet turtles and birds can also transmit the bacteria.

salt A chemical class of ionic compounds formed by the combination of an ACID and a BASE. Most salts are the result of a reaction between a metal and one or more nonmetals.

saltwater intrusion The movement of seawater into inland coastal areas normally flooded with freshwater. The term is applied to the flooding of freshwater marshes by seawater, the migration of seawater up rivers and navigation channels, and the movement of seawater into freshwater aquifers along coastal regions.

sample A representative portion withdrawn from a larger whole to determine some characteristic, such as the concentration of some constituent in a body of water, the atmosphere, or a waste stream. For example, if one wishes to investigate the level of lead contamination in a lake, a 500-milliliter portion of water can be withdrawn from the lake and the amount of lead determined from the 500-milliliter sample. The results are taken to be representative of the total lake.

sample size The amount of water, air, waste, and so forth withdrawn from a larger whole in order to measure the level of some constituent. The term is also used to refer to the number of individual portions removed from the whole. See SAMPLE.

sampling plan A description of the procedures to be used for examining a geographic area, the atmosphere in a specified region, or a body of water to survey for the presence of a specific collection of pollutant chemicals or for the presence of pathogenic microorganisms. The document typically includes the type of samples to be collected, the number of samples to be drawn, the correct procedures to be employed, and the handling of samples after collection.

sampling station A defined location from which samples of soil, water, air, or some other medium will be obtained. A location from which a waste stream is monitored.

sampling variability Differences among replicate SAMPLES taken to determine the level of some constituent in a larger whole. For example, five separate 500-milliliter portions may be taken at the same location within a lake to determine the level of lead contamination in the water. The differences among the quantities of lead found in the separate portions are referred to as the sampling variability.

sand filter A device used to remove particles from drinking water prior to distribution to customers. The water is allowed to percolate through a chamber containing sand of various grain sizes, with the finest grain size located on the top. The particles in the water are removed at the surface of the sand and later discarded by reverse flushing.

sanitary waste Wastewater released from households and rest rooms; SEWAGE.

sandstone aquifer The type of aquifer supplying groundwater to the upper Middle West, Appalachia, and Texas. The water-bearing formation is often contained by shale strata, and the water has high levels of iron and magnesium.

sanitary landfill A method of ground disposal of SOLID WASTE in which waste material is spread in relatively thin layers, compacted, and covered with clean earth (COVER MATERIAL) by the end of each working day, as required by current land disposal regulations for MUNICIPAL SOLID WASTE. The daily cover minimizes the odors, insects, rodents, blowing trash, and smoldering fires that often characterized the now-banned open dumps.

sanitary sewer A pipe or network of pipes used to transport municipal, commercial, or industrial wastewater (SEWAGE). See SEWER.

sanitary survey An on-site inspection by a trained public health technician to assess environmental conditions in a community, with special emphasis on communicable diseases. The survey could include sewage treatment and disposal facilities, public or private water supplies and distribution, solid waste collection and disposal methods, pest control activities, and public swimming pools.

sanitized copy A document submitted by a facility to an environmental regulatory agency from which confidential information, such as a trade secret, has been removed.

saprophage An organism that consumes dead organic matter. See DETRITUS, DETRITUS FOOD CHAIN.

saprophytic Describing an organism that derives nutrition from dead organic material, as contrasted with a parasitic organism, which obtains nutrition at the expense of a living organism.

Sanitary landfill

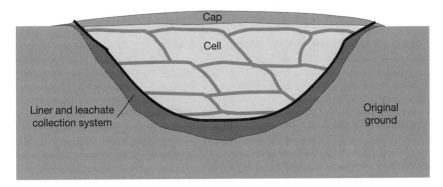

337

sarcoma A malignant tumor, the growth of which begins in connective tissue. See CARCINOGENESIS.

saturated zone The zone in the crust of the Earth, extending from the WATER TABLE downward, in which pore spaces in the soil or rock are filled with water at greater than atmospheric pressure. Also called the zone of saturation. Compare UNSATURATED ZONE.

saturation current The movement, by an applied potential, of all ions in a gas. This is a measure of the ion production rate and can be used as a method of gauging the RADIOACTIVITY of a substance.

saturation mixing ratio The maximal water vapor concentration in the atmosphere for a given air temperature. The higher the air temperature, the higher the saturation mixing ratio.

saturation point 1. The maximal concentration of a chemical (solid, liquid, or gas) that can dissolve in water at a given temperature. 2. Same as SATURATION MIXING RATIO.

savanna One of the major types of natural terrestrial communities. A region where the natural flora is dominated by grasses, sedges, and small shrubs.

Sax hazard ratings A three-point integer scale for chemical hazard used by R. J. Lewis in the multivolume work *Dangerous Properties of Industrial Materials:* 1 represents low hazard; 2, medium hazard; 3, high hazard. The guide also includes rankings for eye and skin irritation, mutagenicity, and carcinogenicity; lists of HAZARDOUS SUBSTANCES and HAZARDOUS MATERIALS; and various environmental and workplace standards.

saxitoxin The primary toxin produced by dinoflagellate protozoan during blooms known as RED TIDE in marine waters. The genus of protozoan involved in the generation of the red color in the water is

Gonyaulax. See PARALYTIC SHELLFISH POISONING.

scale Deposits of solids (usually salts of calcium) that adhere to the inner surfaces of pipes, boilers, scrubbers, and mist control devices. HARD WATER leaves a deposit (scale) in steam irons, coffee makers, and water heaters. Scale in drilling pipes can be RADIOACTIVE. See NATURALLY OCCURRING RADIOACTIVE MATERIAL.

scaling factors Adjustments used in cancer risk assessments to compensate for the differences in size and metabolic rate between humans and experimental animals. The various scaling factors applied to dose-response data are based on relative body weights, relative surface areas, and differences in the mass of food and water ingested per body mass per day. See SURFACE AREA SCALING FACTOR.

scanning electron microscopy (SEM) Use of a microscope that employs a beam of electrons as the source of illumination. The specimen to be examined is usually coated with some metal such as gold, and the beam of electrons moves back and forth over (scans) the surface of the specimen. The electrons that are reflected from the surface form an image on a photographic plate. The instrument can also be used to determine the composition of metal ions in a specimen.

scarify In land restoration activities, to stir the surface of the ground with an implement in preparation for replanting.

scavenger 1. An animal that usually derives its nutrition by consuming refuse or decaying organic matter (dead animals). 2. A person who rummages through discards to recover recyclable items.

scavenging coefficient The exponential constant *(L)* in an EXPONENTIAL DECAY model for the physical removal of particulate from the air by rainfall

$$X_t = X_0 e^{-Lt}$$

where X_0 is the particulate concentration

at time 0, t is the number of time units since the rainfall began, e is the base of the natural logarithm, and X_t is the particulate concentration at time t.

scavenging mechanisms See NATURAL SINK.

scheduled outage A planned shutdown of industrial equipment for periodic inspection and maintenance of machinery.

scheduled wastes A listing of HAZARDOUS WASTES, ranked by their health and environmental hazards and their volume, for which the USEPA was required to set predisposal treatment standards. A waste meeting the treatment standards may be disposed of in a landfill or surface impoundment. The list was divided into thirds and treatment standards were required by the HAZARDOUS AND SOLID WASTE AMENDMENTS to be issued for each third by a certain date, or a restriction on disposal of the waste would take effect. See LAND DISPOSAL BAN, SOFT HAMMER.

schistosomiasis A debilitating disease, common in underdeveloped regions of the world, caused by a small roundworm, *Schistosoma*. The disease is transmitted to humans through contact with water contaminated by FECAL MATERIAL. Infected humans discharge eggs in feces. These eggs hatch in freshwater, producing a small immature form of the parasite that infects snails common in streams, ponds, and lakes. The life cycle of the worm continues in the snail, and a second immature worm reenters the water, where an immature worm later infects humans who have contact with the infected water.

Science Advisory Board (SAB) An independent federal organization established to review the scientific merits of research done by the USEPA and the scientific basis for regulations and standards proposed by the agency. The technical review is accomplished through committees, for example, Drinking Water, Environmental Health, Environmental Engineering, Ecological Processes and Effects,

Integrated Human Exposure, and the CLEAN AIR SCIENTIFIC ADVISORY COMMITTEE. Website: www.epa.gov/science1/

scientific method A systematic method of inquiry that includes the identification of a specific question or problem, the accumulation of the available information relating to that question, the proposal of a tentative answer to the question or problem, the conduct of methodical observations or experiments to test the proposed answer, and the rational interpretation of the results of the observations or experiments. The foundation of the method is the testing of human speculations against the reality of nature.

scientific support coordinator (SSC) Under the NATIONAL CONTINGENCY PLAN for spills or releases of oil or hazardous substances, an official responsible for providing scientific support to the ON-SCENE COORDINATOR or REMEDIAL PROJECT MANAGER. The SSC also serves as the liaison between the federal responders and the scientific community.

scintillation counter An instrument that detects and measures RADIOACTIVITY by counting flashes of light (scintillations) produced when radiation strikes certain chemicals. A sample that is suspected of containing radioactive substances is placed in a vial and mixed with a solution referred to as a cocktail, a mixture of fluors (compounds that emit light when struck by ionizing radiation). The system is useful in quantifying radiation from weak sources used in laboratory experimentation.

scoping An early stage in the environmental impact assessment that is required of all major U.S. government "actions" by the NATIONAL ENVIRONMENTAL POLICY ACT. Scoping identifies the most important environmental issues raised by the proposed action. All public and private organizations that may be affected and thus should be part of the impact assessment are invited to the scoping meetings or are identified and informed as part of the scoping process. See ENVIRONMENTAL IMPACT STATEMENT.

scram 1. See SAFETY CONTROL ROD AX MAN. 2. See SUPPORT CENTER FOR REGULATORY AIR MODELS.

scrap 1. Materials discarded during manufacturing processes, sometimes referred to as mill scrap. These materials are frequently put back into the manufacturing process or used in other economically beneficial ways. Common examples are trimmings produced in the manufacture of paper or cuttings from the manufacture of steel or aluminum products. 2. Postconsumer recyclable steel collected for use in manufacturing of steel.

scrap metal processor Facility involved in the recovery of used metal, commonly steel, for entry into the recycling industry.

scree The stones or debris that accumulate at the base of a hill. Material that has rolled down a hill and come to rest at the bottom.

screening 1. In the testing of chemicals for health effects, any method useful for the preliminary evaluation of their toxic potential. Short-term biological tests can be included, for example, the AMES TEST. 2. In public health studies, any method used to indicate the desirability of further testing of individuals, such as a screening questionnaire that indicates alcohol abuse. 3. In water pollution control, the initial separation of large entrained debris from the wastewater at the entry point of a treatment facility. See BAR RACK.

screening risk assessment A study of the probabilities and magnitude of potential harm to human health or to the environment performed with little specific information concerning a chemical or physical agent or circumstance. The evaluation is done to identify exposures and operations that should be examined more thoroughly. See RISK ASSESSMENT.

scrubber A device designed to remove pollutant particles or gases from exhaust streams produced by combustion or industrial processes.

scrubber, impingement An air particulate control device that passes dirty air through a water spray then removes the particulate-containing water by deposition (impingement) on a solid surface.

scrubber, plate tower An air pollution control device used to remove hydrogen chloride gas from the exhaust gas of an incinerator burning chlorine-containing substances. A series of metal plates containing holes through which an aqueous alkaline solution is forced is used to ensure contact of the exhaust gas with the neutralizing agent. The hydrogen chloride is converted to salt and thus is removed from the exhaust.

scrubber, spray An air pollution control device that removes particulates or gases from an airstream by spraying liquid into the air duct then collecting the pollutant-containing droplets. See SPRAY TOWER.

scrubber, venturi An air pollution control device that operates by the introduction of a liquid into a narrow throat section (venturi) of an air duct that is carrying a contaminant. The high velocity in the venturi, compared with the low initial liquid velocity, produces efficient contact between the injected scrubbing liquid and the contaminant to be removed.

sea breeze The sea-to-land surface wind that typically occurs in coastal areas during the day. The wind is caused by the thermal rising of the air above the land, which warms more readily than the water. See LAND BREEZE.

Secchi depth A crude measurement of the TURBIDITY (cloudiness) of surface water. The depth to which a Secchi disc, which is about 8–12 inches in diameter and has a black-and-white pattern, can no longer be seen.

second law of thermodynamics One of the laws that describe the movement of energy within the environment. Two basic approaches can be used to interpret the second law. One involves the observation that whenever any form of energy is employed to do useful work, the conversion of energy to work is not 100% efficient: a portion of the energy source is always lost from the system as heat. Another way to describe the second law relates to the level of organization within a system. Any system tends to become more disorganized; when energy is converted from one form to another, randomness tends to increase. The measure of randomness is ENTROPY.

second third See LAND DISPOSAL BAN.

Second World Term applied to the socialist countries of Eastern Europe that were part of the former Soviet Union. These countries are industrialized, but the economies of these countries are not well developed.

secondary air pollutant A substance formed in the atmosphere by chemical reactions involving PRIMARY AIR POLLUTANTS. These secondary compounds are not released directly by pollutant sources and therefore can be controlled only indirectly, through controls on the compounds (ingredients) from which they form. For example, OZONE is a secondary air pollutant produced by sunlight-driven reactions involving volatile organic compounds and nitrogen oxides, and emissions of these PRECURSOR materials are controlled to limit ozone formation. See PHOTOCHEMICAL AIR POLLUTION.

secondary association (indirect association) An apparent link between two variables that is actually the result of a CONFOUNDING VARIABLE. When the confounding variable is controlled, the link is no longer seen. For example, a group of workers exposed to chalk dust is seen to have a greater risk of lung disease than a group not exposed to chalk dust, but the exposed group has a much higher portion of smokers than the unexposed group. In this example, the cigarette smoke is the more probable cause of the disease increase in the exposed group than the chalk dust, and the association between chalk dust and lung disease is a secondary association.

secondary clarifier See FINAL CLARIFIER, SECONDARY SETTLING TANK.

secondary consumer An animal that obtains nutrition by eating other animals. Secondary consumers are also referred to as CARNIVORES. An animal that eats only plants is a PRIMARY CONSUMER.

secondary containment Any wall, lining, curbing, DIKE, or other barrier that contains a spill or leak of a container holding a chemical material.

secondary drinking water regulations Guidelines set by the USEPA under the SAFE DRINKING WATER ACT that apply to aesthetic qualities of the drinking water that are important to its public acceptance. See SECONDARY MAXIMUM CONTAMINANT LEVEL. Compare NATIONAL PRIMARY DRINKING WATER REGULATIONS.

secondary electron An electron ejected from an atom or molecule after a collision with a charged particle or photon of light.

secondary irritant A chemical that, upon overexposure, can cause irritation to the skin, eyes, or respiratory system at the site of contact and absorption but that exerts an adverse effect on an organ or organ system elsewhere in the body that is more pronounced than the contact irritation. For example, liquid organic solvents can penetrate the unprotected skin of the hands and forearms and cause irritation in these areas, but absorption of the solvent through the skin and entry into the bloodstream can lead to adverse systemic effects associated with depression of the central nervous system.

secondary maximum contaminant level (SMCL) The maximal concentration or level of certain water contaminants in public water supplies set by the USEPA to protect the public welfare. The secondary levels are written to address aesthetic considerations such as taste, odor, and color rather than health standards. See SAFE DRINKING WATER ACT. Compare MAXIMUM CONTAMINANT LEVEL, MAXIMUM CONTAMINANT LEVEL GOAL.

secondary pollutant A general term applied to pollutants that are formed by chemical reactions in the environment. These reactions form compounds other than the substances released by the primary sources of pollution. See SECONDARY AIR POLLUTANT, PRIMARY POLLUTANT.

secondary productivity The increase in BIOMASS by CONSUMERS at one TROPHIC LEVEL per year, equal to the amount of biomass ingested from lower trophic levels, less predation and respiratory losses. See PRIMARY PRODUCTIVITY.

secondary recovery The injection of water into an underground petroleum deposit to force the remaining oil into recovery wells. This technique is used to recover additional oil from old wells after the removal of the oil that can be easily pumped to the surface (PRIMARY RECOVERY). See ENHANCED OIL RECOVERY.

secondary settling tank A tank used to hold wastewater that has been subjected to SECONDARY TREATMENT. FLOC, or particles of organic matter formed during the secondary processes, is allowed to settle from the suspension for subsequent removal. Also called a secondary clarifier or FINAL CLARIFIER.

secondary standards Allowable amounts of materials in air or water that are set to retain environmental qualities not related to the protection of human health. SECONDARY DRINKING WATER REGULATIONS are set for, among other characteristics, taste and color, and some secondary air standards define concentrations that are not harmful to plant life. Compare PRIMARY STANDARDS.

secondary succession The orderly and predictable changes that occur over time in the plant and animal communities of an area that has been subjected to the removal of naturally occurring plant cover. This type of succession occurs when agricultural fields are taken out of use or when forested areas are subjected to severe fires that destroy all vegetation. In both cases, the topsoil remains for the regrowth of natural plant communities. Compare PRIMARY SUCCESSION.

secondary treatment A phase in the treatment of wastewater. This aspect of treatment usually follows the removal of particulate materials from domestic wastewater (see PRIMARY TREATMENT). In this second phase, conditions are established to maximize the functions of bacteria in destroying or mineralizing dissolved or suspended organic materials that are not removed by the primary process. A BIOCHEMICAL OXYGEN DEMAND (BOD) and TOTAL SUSPENDED SOLIDS (TSS) reduction of 85% or a discharge limit of 30 milligrams/liter, 30-day average, for BOD and TSS is required for secondary treatment. Specific types of units employed in secondary treatment include a TRICKLING FILTER, AERATION, ACTIVATED SLUDGE PROCESS, and the BIODISC.

Section 106 order An ADMINISTRATIVE ORDER issued by the USEPA that directs a POTENTIALLY RESPONSIBLE PARTY to perform a specific remedial or a removal action at a site in response to a release or threatened release of a HAZARDOUS SUBSTANCE that poses "imminent and substantial endangerment to the public health or welfare or the environment." The order is authorized by Section 106 of the COMPREHENSIVE ENVIRONMENTAL RESPONSE, COMPENSATION, AND LIABILITY ACT.

Section 404 permit The WETLAND dredge and fill permit issued under regulations written to conform to Section 404 of the CLEAN WATER ACT. The permit is actually granted by the U.S. Corps of Engineers; the Section 404 regulations are written by the USEPA. See NO NET LOSS,

MITIGATION BANKING, CONSTRUCTED WETLANDS, TAKING.

secular equilibrium The stable relationship established in nature between a RADIOACTIVE substance that has a long HALF-LIFE and a decay product that has a much shorter half-life. For example, radium-226 has a half-life of about 1600 years. As this element decays with the emission of radiation, RADON-222, which has a half-life of about 3.8 days, is produced. An equilibrium between the concentration of these two elements is established such that the activity of each element is equal. The activity is calculated from the equation $A = \lambda N$, where λ is the decay constant for the element and N is the number of atoms of the element present. The decay constant of each element is equal to the natural log of 2 divided by the half-life of the element ($\lambda = [\ln 2]/\text{half-life}$). Consequently, an element with a long half-life demonstrates low ACTIVITY, and the opposite is true of an element with a short half-life. At equilibrium, the activity of radium-226 equals the activity of radon-222 ($\lambda N[\text{radium}] = \lambda N[\text{radon}]$). Since the λ for radium is much smaller than the λ for radon, the concentration of radium is much larger than that of radon when equilibrium is reached.

secular trend A trend over a relatively long period of time. In public health statistics, if a disease is strongly associated with age, as most are, then a long-term rise in the incidence or mortality rate of the disease may be explained by a shift in the age distribution of the population to the older years. Much of the rising secular trend in cancer incidence and mortality rates in the United States during the 20th century is explained by the aging of the population, with the important exception of cancers associated with tobacco consumption.

secure landfill A ground location for the deposit of HAZARDOUS WASTE. The material is placed above natural and synthetic LINERS that prevent or restrict the leaching of dangerous substances into groundwater. A piping network called a LEACHATE COLLECTION SYSTEM is placed beneath the facility to allow the pumped removal of any liquid that accumulates in the landfill. Access to the location is restricted, and wells are used to monitor the leaching of any dangerous materials into GROUNDWATER. A type of TREATMENT, STORAGE, OR DISPOSAL facility.

sediment Soil particles, sand, clay, or other substances that settle to the bottom of a body of water. Wind, water, and glacier erosion are natural sources of sediment. Human causes of sediment accumulation are mainly earth-moving activities, such as agricultural or construction operations. The term is frequently used to signify the mud at the bottom of a stream or lake or the deposits left by receding flood waters.

sediment oxygen demand (SOD) The amount of DISSOLVED OXYGEN removed from the water covering the SEDIMENT in a lake or stream as a result of microbial activity. The greater the contamination of sediment with ORGANIC MATTER, the greater the removal of dissolved oxygen from the water by microorganisms.

sedimentary Describes geological strata comprising deposits of particulate materials that were deposited in some area by mechanical means. Such materials include gravels, muds, clays, and shales. The term also applies to rocks formed through the consolidation of fine-grained materials. For example, limestone can be formed from accumulations of calcareous shells derived from small marine organisms.

sedimentary rock Geological deposits derived from material that has been buried and compacted and undergone fusion into stone.

sedimentation 1. The removal of solid materials from a fluid suspension by gravitational settling, which can be hastened by the addition of ALUM or POLYELECTROLYTES that increase FLOCCULATION and COAGULATION. Solids suspended in water are removed by allowing the water to stand in a tank, pond, or lagoon for a time sufficient to allow them to settle to the bottom. See SETTLING POND, SETTLING CHAMBER, PRIMARY SETTLING TANK, PRIMARY TREATMENT. **2.** The removal of airborne particles from the atmosphere by gravitational settling or their controlled removal by slowing an exhaust stream through a device sufficiently to allow particles to settle out before the air exits the device.

seed bank Organized effort to store under safe conditions seeds from as many different plants as possible. The goal is to protect plant diversity and to provide varieties for future breeding research. The establishment of such repositories is a response to the potential loss of many natural species through habitat destruction and other types of environmental damage.

seepage Percolation of water or other liquid through soil. The material may be from ponds, underground tanks, ditches, or water storage facilities and may represent a threat to underground water supplies.

seeps Groundwater/surface water connections caused by river or stream erosion into a near-surface aquifer.

segregated ballast tank (SBT) Separate compartment in an oceangoing oil tanker for oil and water used as BALLAST. SBTs prevent the discharge of oil when a ship pumps out its ballast tanks before loading.

selection bias A source of error in a scientific study that results when certain people, animals, water samples, or another group is selected at a greater (or lower) frequency than its occurrence in the sampled population. For example, if 50 of the 60 persons selected for a study of the possible health effects of community air pollution are smokers but only 50% of the population smokes, the study may draw incorrect conclusions when comparing health status of one community with that of another (if the sample from the other community is properly drawn) because smoking and the associated health effects are overrepresented in one community sample.

selective harvesting The cutting of mature or diseased trees to encourage the growth of trees of differing ages or species. The removal of trees of only a certain age or species.

selective herbicide A SELECTIVE PESTICIDE that kills certain plant species and is not harmful to others.

selective pesticide A chemical that is acutely toxic to certain pest species but is not significantly harmful to NONTARGET ORGANISMS.

self-absorption In nuclear science, a process in which radioactive material absorbs the radiation released by RADIOACTIVE DECAY events within the material, lowering the external radiation emitted by the material. This phenomenon occurs when the energy of the released radiation is relatively weak.

self-contained breathing apparatus (SCBA) A respirator equipped with an air supply separate from the ambient air. The device is used in confined spaces (which may not contain adequate oxygen) or when actual or threatened air contamination levels make an AIR-PURIFYING RESPIRATOR otherwise inappropriate.

self-purification The removal of ORGANIC material, PLANT NUTRIENTS, or other pollutants from a lake or stream by the activity of the resident biological community. BIODEGRADABLE material added to a body of water is gradually utilized by the microorganisms in the water, lowering the pollution levels. If excessive amounts of additional pollutants are not added downstream, the water undergoes self-cleansing. This process does not apply to pollution by nonbiodegradable organic compounds or metals.

self-quenched counter tube A device for measuring RADIOACTIVITY that includes a substance to control the reactions that occur in the device when radiation is encountered. A Geiger counter may contain such a device.

self-scattering The dispersing of radiation by the material emitting radiation. If the radioactive substance scatters the radiation as the radiation is released, the measured activity of a sample increases.

semicontinuous activated sludge test (SCAS test) A test used by the Soap and Detergent Association (a trade and research organization based in New York) to evaluate the BIODEGRADATION of ALKYLBENZENE SULFONATE. The USEPA used a modified SCAS test to establish the biodegradability of carbon compounds. The test involves the addition of ACTIVATED SLUDGE to an aeration chamber containing the carbon compound to be tested. The modified SCAS test is referenced as Office of Prevention, Pesticides and Toxic Substances (OPPTS) test 835.3210.

semilogarithmic graph A graphic representation of data with one axis having an arithmetic scale and the other axis having a logarithmic scale. The arithmetic axis may be time units, for example, and the axis with the logarithmic units may show the number of individuals in a population. Exponential growth produces a straight line when the number of organisms is plotted against time on a semilogarithmic graph.

semistatic test During measurements of the toxicity of chemical substances to aquatic test organisms, a laboratory procedure in which all of the water in a container holding the test organism is changed by emptying and refilling, frequently on a daily basis. This contrasts with a STATIC TEST, in which the water in the container is not changed for the duration of the procedure, and a FLOW-THROUGH TEST, in which freshwater is constantly flushed through the test container.

semivolatile organic compound A class of organic contaminants consisting of relatively high-molecular-weight hydrocarbons and synthetic derivatives of hydrocarbons. Compounds are placed in this class because of a restricted VOLATILITY at ambient temperatures; they evaporate less readily than VOLATILE ORGANIC COMPOUNDS. Overexposure to many of these agents represents a health hazard. Common examples include a variety of chlorinated derivatives of BENZENE and phenol, aniline and related compounds, toluene, and various POLYCYCLIC AROMATIC HYDROCARBONS.

senescent Describing plants or specific ecosystems that are nearing the end of their normal life span. For example, a lake that is filling with accumulated aquatic vegetation, dead plant material, and sediments can be described as senescent because it is nearing extinction as a productive lake environment.

sensible heat transfer The removal or addition of heat energy that is accompanied by a change in temperature through CONDUCTION and CONVECTION. Compare LATENT HEAT TRANSFER.

sensitivity 1. The ability of a test to identify true cases accurately, such as the number of carcinogens a screening test identifies in a group of materials containing carcinogens. Expressed as the percentage of true cases identified by the test. 2. The relative ability of a laboratory instrument to detect some chemical in a reliable fashion. 3. The characteristic of a group that is susceptible to a physical, chemical, or biological agent, causing adverse effects in them at lower doses than would be required to harm the general population.

sensitivity analysis Testing the components of a MODEL to determine the change in the model's prediction based on the relative change in each component; can be used to decide which component measurements/estimates should be emphasized in the application of the model, that is, which values have most effect on the predicted outcome.

sensitization reaction An allergic response to a chemical substance that is preceded by a sensitizing exposure. Once an individual is sensitized to a substance, quite low doses can elicit an adverse response affecting the skin or respiratory system.

septage Liquids and/or solids pumped out during the periodic cleaning of a domestic SEPTIC TANK.

septic system Apparatus used to treat SANITARY WASTE from an individual residence or business. Such systems range from a simple underground tank with the overflow connected to a system of underground drain pipes (SEPTIC TANK) to sophisticated mechanical units equipped with aeration devices and disinfection capability. See PRIMARY TREATMENT, SECONDARY TREATMENT, PRESSURE SEWER.

septic tank A buried tank used to treat domestic wastes. The SANITARY WASTE from a household is deposited and retained in a covered tank to allow solids to settle to the bottom and to provide an environment for the decomposition of organic components by anaerobic bacteria. The liquid effluent that flows from the tank is allowed to seep into the soil. See LEACHING FIELD.

septic tank effluent filter (STEF) See PRESSURE SEWER.

septic tank effluent pump (STEP) See PRESSURE SEWER.

sequential sampling The collection of a series of separate air or water samples at timed intervals.

seral stages The various transitions in the orderly and predictable changes in a biological community from the PIONEER COMMUNITY stage to the CLIMAX stage. See SUCCESSION.

sere In ecology, a complete sequence of transitions in the orderly and predictable changes that occur in biological communities over time. See SUCCESSION, SERAL STAGES.

set back An energy-saving protocol involving manual or automatic timer setting of a thermostat to a lower temperature for those winter days or hours when a building is not occupied and need not be heated to a comfortable level or at a higher temperature at similar off hours during summer.

settleable solids Bits of debris, sediment, or other solids that are heavy enough to sink when a liquid waste is allowed to stand in a pond or tank. See SETTLING CHAMBER, SETTLING POND, PRIMARY SETTLING TANK, PRIMARY TREATMENT, SEDIMENTATION.

settling chamber An enclosed container into which WASTEWATER contaminated with solid materials is placed and allowed to stand. The solid pollutants suspended in the water sink to the bottom of the container for removal. See SETTLING POND, PRIMARY SETTLING TANK, SECONDARY SETTLING TANK, SEDIMENTATION.

settling pond An open lagoon into which WASTEWATER contaminated with solid pollutants is placed and allowed to stand. The solid pollutants suspended in the water sink to the bottom of the lagoon, and the liquid is allowed to overflow out of the enclosure. See SETTLING CHAMBER, PRIMARY SETTLING TANK, SEDIMENTATION.

settling tank Same as SETTLING CHAMBER.

settling velocity The rate of downward movement of particles through air or water. This gravitational settling removes particles naturally and is also used in pollution control devices, for example, the PRIMARY SETTLING TANKS and SECONDARY SETTLING TANKS in a SEWAGE TREATMENT PLANT and the gravity settling chambers that remove larger-diameter (faster-settling) particulate air contaminants. The settling velocity of a particle is often the same as the TERMINAL SETTLING VELOCITY.

7Q10 The period of lowest stream flow during a seven-day interval that is expected once in every 10 years. During the time of low flow, the amount of DISSOLVED OXYGEN in the water would be expected to be the lowest encountered under normal conditions. Since such conditions are considered to be the worst natural case, the dissolved oxygen levels during such episodes are used to establish ambient water quality standards for the stream.

Seveso The town in northern Italy near which a chemical plant accident occurred in 1976. A trichlorophenol production unit exploded, forming the highly toxic compound DIOXIN in the escaping gases, which contaminated a large downwind area with a population of over 100,000. Seveso received the most severe impact and gives its name to the disaster. See SEVESO DIRECTIVE.

Seveso II See SEVESO DIRECTIVE.

Seveso Directive Industrial safety regulations adopted in 1982 (82/501/EEC) and updated in 1997 (96/82/EEC, called Seveso II) by the 12 nations in the European Economic Community (EEC) in response to the SEVESO disaster. The rules, which mandate additional accident planning and public release of chemical hazard information, are similar to the RIGHT TO KNOW rules in the United States. See TITLE III.

Sevin The trademark for carbaryl, a CARBAMATE insecticide with relatively low acute toxicity to vertebrates.

sewage Wastewater from homes, businesses, or industries; mainly refers to the water transport of cooking, cleaning, kitchen, or bathroom waste.

sewage fungus A thick filamentous growth that develops in water contaminated with sewage. The filamentous material is composed predominantly of the bacterium *Sphaerotilus natans.*

sewage lagoon A shallow pond where natural processes are employed to treat SANITARY WASTE from households or public rest rooms. Solid material settles to the bottom and is degraded by ANAEROBIC microbial communities. The enclosure is open to the atmosphere, which allows for the aerobic MINERALIZATION of organic compounds in the upper layers of the water. The decomposition processes are analogous to those described for PRIMARY TREATMENT and SECONDARY TREATMENT. The effluent from these ponds is usually allowed to flow into nearby streams without further purification.

sewage sludge See SLUDGE.

sewage treatment plant Facility designed to receive the wastewater from domestic sources and to remove materials that damage water quality and threaten public health when discharged into receiving streams. The substances removed are classified into four basic areas: grease and fats, solids from human excrement and other sources, dissolved pollutants from human wastes and decomposition products, and dangerous microorganisms. Most facilities employ a combination of mechanical removal steps and bacterial decomposition to achieve the desired results. Chlorine is often added to discharges from the plants to reduce the danger of spreading disease by the release of PATHOGENS. See PRIMARY TREATMENT, SECONDARY TREATMENT, TERTIARY TREATMENT, CHLORINATION, DISINFECTION BY-PRODUCTS.

sewer The piping system or conduit used to carry runoff water or wastewater. Various types of systems fulfill different functions; for example, a STORM SEWER carries runoff from rainfall, a SANITARY SEWER carries wastewater from a household or public rest room facility, and a COMBINED SEWER transports both RUNOFF and SANITARY WASTE.

sewerage The total system for collecting, transporting, and treating WASTEWATER, including pipes, pumps, and SEWAGE TREATMENT PLANTS.

shale Rock or mineral deposits that contain solid, waxy hydrocarbons termed KEROGEN. The hydrocarbons can be extracted and utilized as source material for petroleum products.

shale oil A thick oil recovered from SHALE rock. The liquid is obtained by heating the pulverized rock to high temperatures to vaporize the kerogen layered within it; condensation of the vapor follows. The recovered oil can be modified and refined to yield useful petroleum products.

shale shaker An oscillating wire mesh that catches cuttings from oil well drilling as drilling fluid passes through.

shallow ecology A term of derision applied to individuals who profess to be interested in the environment but who usually enter into compromises with the industrial community. Those to whom the label is applied contend that they are interested in working within the establishment rather than seeking the overthrow of current societal institutions. Contrast with DEEP ECOLOGY, ENVIRONMENTALISM, ENVIRONMENTAL JUSTICE.

Shannon-Weaver index (H) A SPECIES DIVERSITY expression, equal to
$$-\Sigma P_i \log P_i$$
where P_i is the number of individuals in species (i) divided by the total number of individuals in a community, and the log P_i is the (natural or base 2) logarithm of P_i. The sum is taken for all i species in the community. The Shannon-Weaver index incorporates aspects of SPECIES DIVERSITY, EVENNESS, and RICHNESS.

sharps Implements such as needles and scalpels discarded after medical use. Also includes various glassware items and microscope slides used for medical applications that may pose a danger when broken.

sheen rule The informal term for the minimal reportable oil spill on surface water. A spill large enough to produce a film, discoloration, or sheen must be reported to the NATIONAL RESPONSE CENTER to be included in the EMERGENCY RESPONSE NOTIFICATION SYSTEM database. The oil spill reporting requirements are found in Title 40, Section 110, of the *Code of Federal Regulations*.

sheet erosion The loss of thin layers of soil from unprotected surfaces, commonly agricultural lands. Most typically, the loss of soil that results from the flow of water across the exposed soil surface. Compare RILL EROSION.

sheet piling Material, often concrete or steel, placed vertically in the ground to contain erosion or the lateral movement of groundwater.

sheetflow A broad, shallow overland flow of storm water. Also called a sheetflood.

shelter in place Program to protect people living in proximity to chemical production facilities and oil refineries. In the event of an accidental release of a dangerous agent into the atmosphere, persons in the path of the release as it moves downwind are advised to go indoors, close all windows and doors, turn off airconditioning and heating units, and tune into local radio and television stations for statements from emergency management agencies. See LOCAL EMERGENCY PLANNING COMMITTEE, COMMUNITY AWARENESS AND EMERGENCY RESPONSE, TITLE III, RISK MANAGEMENT PLAN.

shelterwood cutting The harvesting of all mature trees from a forest area through a systematic removal over one or more decades. This method of cutting is followed to encourage the growth of immature trees where their development is inhibited by the shading of older trees.

shielding Material, such as lead, that is used to reduce the passage of RADIATION. See HALF-VALUE LAYER.

Shigella The genus of a bacterium associated with human dysentery, a disease characterized by severe diarrhea with blood and pus in the feces. The disease is transmitted through the consumption of water, food, or beverages contaminated with fecal material.

shock load 1. An extraordinarily large amount of contaminant released to a SEWAGE TREATMENT PLANT during a short period; it may exceed the capacity of the treatment plant, allowing incompletely treated water to be discharged. 2. An industrial wastewater that contains such an unusual amount or class of material that satisfactory treatment is not possible under normal operation. Such an event many occur during an accidental release or a process UPSET. 3. RAW WATER from a source of drinking water containing unusually large amounts of plant matter, suspended solids, color, or other agents that make satisfactory processing difficult.

shock wave A large-amplitude wave created as an object travels through a medium at a faster rate than the typical wave speed for that medium, for example, faster than the speed of sound in air. In such cases, the force of the moving object creates waves that do not travel from the object in all directions, as is the case at slower speeds, but are restricted in movement to the rear and sides of the object. The large amplitude is the result of the great compression at the front of the object and the large degree of rarefaction behind. A SONIC BOOM is an example of a shock wave.

short-term exposure limit (STEL) An occupational air concentration standard using 15-minute AVERAGING TIME within an eight-hour workday. A STEL is set for a material that can cause adverse health effects if workers are exposed to higher concentrations for short periods. Materials not requiring a STEL can have short-term peaks in concentration without adverse effect, and only their overall eight-hour average must be below a certain limit.

short-term sampling The collection of samples over a restricted interval.

shortwave radiation Part of the range of WAVELENGTHs of energy emitted by the Sun, including ULTRAVIOLET, visible, and NEAR INFRARED radiation.

shredding The mechanical division of material into one- or two-inch-diameter pieces. Solid waste is often shredded then compacted to increase the density before disposal in a landfill. See COMPACTING, COMPACTED SOLID WASTE.

SI units The International System of Units (le Système International d'Unités) defined by an international gathering convened to establish agreements on the most frequently used units of measurement. The Conference of Weights and Measures in 1960 adopted standard measures based on the meter/kilogram/second units and radiation quantities based on the becquerel, gray, and sievert. The units adopted by the conference represent the currently preferred measures of length, mass, time, radiation, and so on. See the Appendixes.

sick-building syndrome A medical condition involving acute adverse health effects or discomfort that appear to be related to the accumulation of air pollutants distributed throughout a building or confined to a single room or floor. Some cases are relatively easy to diagnose, as in nausea caused by a solvent used in wax or paint, whereas others defy understanding. The accumulation of mold spores and dust mites leading to allergic responses among sensitive individuals is a common scenario. The drive for energy efficiency by restricted entry of outside air and failure to control moisture levels inside buildings are often contributing factors. See INDOOR AIR POLLUTION.

Sierra Club A large organization in the United States that is concerned with environmental issues. Founded in 1892; 1999 membership 550,000. Based in San Francisco, California, the organization is active in public education, in lobbying of legislative and administrative bodies, and, through the affiliated Sierra Club Legal Defense Fund, in the courts. Website: www.sierraclub.org

sievert (Sv) The SI unit of radiation dose that takes into account both the physical properties of the radiation and the biological properties of the material absorbing it. The unit is equal to the amount of radiation dose actually absorbed by the body (in GRAYS) times a QUALITY FACTOR associated with the energy level and ionizing potential of the particular radiation times modifying factors associated with the tissue being influenced (for example, the distribution of radiation within the tissue being struck). The sievert is the preferred unit of radiation dose for the design of radiation protection devices and for expression of standards limiting human exposure to ionizing radiation. A dose of 0.05 Sv is the maximal annual occupational exposure allowed in the nuclear industry; a dose of 1 Sv in a single incident is thought to increase the probability of leukemia; and a dose of 6 Sv over two days is sufficient to kill 100% of the individuals exposed.

sigmoid growth A population growth pattern that traces out an S-shaped, or sigmoid, curve. In a population, sigmoid growth begins at an exponential rate, but the rate slows as any LIMITING FACTOR is encountered, until a rough equilibrium level of population is reached, the CARRYING CAPACITY of the ecosystem for this

Sigmoid growth

type of population. See EXPONENTIAL GROWTH.

significant biological treatment SECONDARY TREATMENT meeting a discharge limit of 30 milligrams/liter BIOCHEMICAL OXYGEN DEMAND (BOD) and a TOTAL SUSPENDED SOLIDS (TSS) discharge limit of 30 milligrams/liter, as a 30-day average or as a reduction of BOD and TSS of 85%.

significant figures The digits in a number starting with the first nonzero digit to the left of the decimal point (or starting with the first digit to the right of the decimal point if no nonzero digits are to the left of the decimal point) and ending with the rightmost digit. These digits indicate the accuracy of measurements from which the number derives. For example, 78.0 has three significant figures and indicates a known accuracy to the nearest 10th; 0.078 has three significant figures and indicates a known accuracy to the nearest 1000th.

Significant New Alternatives Policy (SNAP) program A USEPA listing of acceptable substitutes for CHLOROFLUOROCARBONS and other chemicals with OZONE-DEPLETING POTENTIAL that are being phased out by provisions of the CLEAN AIR ACT and the MONTREAL PROTOCOL. Website: www.epa.gov/ozone/title6/snap/lists

significant new use rule (SNUR) Under the TOXIC SUBSTANCES CONTROL ACT, the requirement that a producer of a chemical already approved for manufacture notify the USEPA if it will be used in a new and different way. The new use may trigger a regulatory review.

significant potential source of contamination Sources identified as part of the SOURCE WATER ASSESSMENT PROGRAM under the provisions of the SAFE DRINKING WATER ACT. Potential sources of contamination of groundwater and surface water include agriculture, feedlots, atmospheric deposition, erosion, construction activities, dredging, industrial point sources, leakage from UNDERGROUND STORAGE

TANKS, mining, logging, urban runoff, accidental spills, SEPTIC TANKS, municipal landfills, hazardous waste land disposal, oil drilling and production, and waste INJECTION WELLS.

silica Silicon dioxide. The fibrotic form is CRYSTALLINE SILICA, called free silica or quartz. The quartz content of inhaled dust is positively related to the risk of development of SILICOSIS. See FIBROSIS.

silica gel A nontoxic, water-absorbing activated silica used as a drying agent.

silicates Compounds containing silicon and oxygen and sometimes also hydrogen or various metals. CRYSTALLINE SILICA in workplace air is an inhalation hazard. Silicate compounds can occur as deposits on the inside surface of steam and water pipes. See SILICOSIS.

silicic eruption A type of volcanic eruption characterized by highly explosive and violent explosions. The magma is silica-rich and the material is so thick and viscous that the gas does not escape easily; as a result, little lava flow occurs. The buildup of large domes that explode produces large amounts of ASH FLOW. Compare BASALTIC ERUPTION.

silicosis A lung disease caused by prolonged exposure to high levels of small-diameter dust with a high CRYSTALLINE SILICA (free-silica) content. The disease is characterized by scarring of lung tissue (FIBROSIS) and subsequent loss of lung elasticity and gas exchange capacity. Black lung disease is a form of silicosis.

sill Structure produced by the flow of basaltic magma associated with a volcano. As magma rises in a volcano, the pressure associated with the lava flow forces the molten material into the space between geological strata, producing a buried lava flow extending between layers constituting subterranean deposits.

siltation The settling of finely divided particulate (previously suspended solids) on the bottom of a lake, stream, or reservoir. See SEDIMENT.

silviculture A branch of forestry dealing with the cultivation and management of trees in order to produce a crop resource on a continuing basis.

Simon, Julian (1932–98) American Economist, author of *The Ultimate Resource* (1981; 1996) and, with Herman Kahn, *The Resourceful Earth: A Response to Global 2000* (1984), two of his many publications in which he argues strongly against those worrying about overpopulation or resource depletion. Won a famous bet with PAUL EHRLICH over raw material price trends (EHRLICH thought that scarcity would force prices up; they went down).

simple asphyxiant A gas present at a concentration high enough to reduce the oxygen concentration in the air to a level below the minimum required for proper human respiration, about 18% by volume. The gases that can act as simple asphyxiants may be nontoxic to humans, but they become dangerous if their presence is at oxygen-excluding concentrations. They are used as purge or blanketing gases in closed tanks and vessels and pose a simple asphyxiation hazard to workers entering these confined spaces for inspection or cleaning. Examples of simple asphyxiant gases are nitrogen, carbon dioxide, helium, ethane, and methane. See CHEMICAL ASPHYXIANT.

Sinclair, Upton (1878–1968) American Author of *The Jungle* (1906), an exposé of unsanitary conditions in the meatpacking industry that strongly influenced the passage of the Pure Food and Drug Act (1906) and the Meat Inspection Act (1906) in the early wave of sanitary reform in the United States. See WILEY, HARVEY; FOOD, DRUG, AND COSMETIC ACT.

sink See NATURAL SINK.

sinkhole Collapse of the ground surface as a result of the loss of underground support provided by groundwater. These areas can be quite large, involving loss of roads, trees, and homes.

sinking agent Chemical additive that, when applied to a floating oil discharge, causes oil to sink below the surface of the water. The application of sinking agents is generally not permitted by regulations of the USEPA governing the treatment of oil discharges.

sinusoidal wave A wave varying in amplitude with the sine of an independent variable. Sound waves are oscillations around atmospheric pressure and are sinusoidal.

SIP call A regulatory notice issued by the USEPA requiring a state to revise a particular feature of its STATE IMPLEMENTATION PLAN (SIP). The requested revision will, according to the USEPA, work toward or achieve compliance with the NATIONAL AMBIENT AIR QUALITY STANDARDS.

site A general term used to describe a location of interest, such as a waste disposal site, industrial site, site inspection, or sampling site.

site assessment program See PRELIMINARY ASSESSMENT AND SITE INSPECTION (PA/SI).

site inspection See PRELIMINARY ASSESSMENT AND SITE INSPECTION (PA/SI).

site safety and health plan 1. A general workplace plan that identifies potential safety and health hazards, outlines the controls necessary to minimize the hazards, provides employee training on the basic features of the plan, and, to be effective, has senior management's commitment to the plan's execution. 2. A special plan for workers at sites containing HAZARDOUS WASTE or HAZARDOUS SUBSTANCES. See HAZARDOUS WASTE OPERATIONS AND EMERGENCY RESPONSE (HAZWOPER).

six nines The incineration destruction requirement for waste containing POLY-CHLORINATED BIPHENYLS (PCBS) at a concentration greater than 500 parts per million: 99.9999%. See FOUR NINES, DESTRUCTION AND REMOVAL EFFICIENCY, TRIAL BURN.

60-day letter See NOTICE LETTER 2.

skimmer See OIL SKIMMER.

skin dose The amount of RADIATION absorbed by the skin; the sum of the amount of radiation dose in the air above the skin and that scattered from below by bone and other body parts.

skin sample Sampled water that is not representative of the cross-sectional flow in a pipe or conduit. A skin sample can result if the tap opens on the inside wall of the pipe and can be prevented by using sampling taps that extend toward the center of the pipe, away from the wall.

skyshine The gamma radiation released through the roof of a nuclear source and reflected back to the ground by the atmosphere.

slag The nonmetallic glasslike material formed in a blast furnace or smelter when impurities react with a flux. Slag floats as a liquid on top of the process and after cooling and solidification is used in cement.

slaker A mechanical device in which dry lime (calcium oxide) or magnesium oxide is powdered and mixed with water to produce alkali consisting of calcium or magnesium hydroxide.

SLAPPs See STRATEGIC LAWSUITS AGAINST PUBLIC PARTICIPATION.

slash and burn An agricultural practice involving the rapid destruction of natural forest for limited farming activity. The natural forest is cleared, the residue from the clearing process is burned, and crops are planted for a few years. When the fertility of the soil is depleted, the process is repeated in a new area, and the initial land is abandoned.

slope factor See CANCER POTENCY FACTOR.

slow neutron Neutron (subatomic particle) possessing a kinetic energy of about 100 electron volts or less. This class of neutrons is needed to produce nuclear FISSION reactions. Compare FAST NEUTRON.

slow sand filtration (SSF) A gravity filtration system to render surface water safe to drink. RAW WATER from a RIVER or LAKE is allowed to percolate slowly through a fine sand filter. A rich biological community, a biofilm, consisting of bacteria and protozoa, develops in the top few inches of the filter. As the water moves, the chemical and biological quality of the water improves through a complex of biological, biochemical, and physical processes. The technology is widely used to process drinking water in major European cities and is one of the methods of choice in developing countries. Slow sand filtration is gaining popularity in those communities in the United States that rely on surface water as a source of drinking water because the technology decreases contamination by DISINFECTION BY-PRODUCTS such as the TRIHALOMETHANES, reduces the regrowth of microorganisms in the distribution system, and removes natural organic material contaminating the raw water.

sludge A general term used to designate a thick suspension of waste products having the consistency of paste or soft mud, such as the particulate waste collected during the treatment of sewage or the muddy sediment that collects in a boiler used to produce steam. Sludge high in organic matter is produced by municipal SEWAGE TREATMENT PLANTS, food processors, refineries, and paper mills. Sludge low in organic matter is produced by some chemical and power plants. See ACTIVATED SLUDGE, ACTIVATED SLUDGE PROCESS, SLUDGE DIGESTION, SLUDGE DISPOSAL, SLUDGE VOLUME INDEX.

sludge digestion The biological decomposition of solids collected during the operation of a facility designed to remove organic wastes from domestic or industrial sources. The total volume of solids is reduced by the mineralizing activity of bacteria, and the sludge remaining is rendered less reactive because the easily degraded compounds have been removed.

sludge disposal The removal and discarding of thick watery suspensions of particulate waste matter. Final disposal may involve the removal of excess water and the subsequent burning of the solids or placement of the dewatered material in a LANDFILL.

sludge volume index (SVI) A laboratory test result used to indicate the rate at which sludge is to be returned from the discharge end of an AERATION TANK to the inflow (upstream) end. The SVI is calculated as $SVI = (SV/MLSS) \times 1000$ milligrams/gram, where SV is the sludge volume (solids settled in a one-liter graduated cylinder after 30 minutes, in milliliters/liter) and $MLSS$ is the level of MIXED LIQUOR SUSPENDED SOLIDS, in milligrams/liter. The SVI employs units of milliliters/gram.

slug A short rod containing fissionable material (fuel) for a nuclear reactor.

slurry A watery mixture of insoluble matter such as lime. The mixture is pourable and can be transported by pipe. The form in which some raw material is added to an industrial process. Compare LIQUOR.

slurry wall Material placed in a trench in the ground to prevent the lateral movement of groundwater. Often made of a cement-bentonite (clay) mixture, a slurry wall is used to retard the movement of LEACHATE out of and groundwater into a hazardous waste disposal landfill and to contain the spread of a contaminated groundwater PLUME.

small-quantity generator (SQG) A facility that produces small amounts of hazardous waste and is therefore subject

Slurry wall

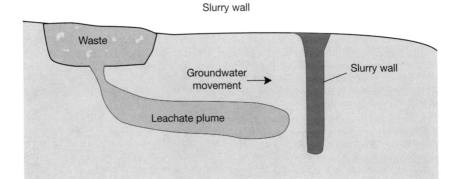

to less stringent management requirements by federal regulations. SQGs (often pronounced "squeegees") are currently defined as those producing more than 100 kilograms of hazardous waste but less than 1000 kilograms per month. State regulatory agencies may require registration of SQGs. See CONDITIONALLY EXEMPT SMALL-QUANTITY GENERATOR.

smelter A facility that melts metal ores as part of the process used to separate metals from other compounds present in them. Smelters are potential sources of SULFUR DIOXIDE and PARTICULATE MATTER air pollutants, and air and water discharges must be controlled for metal contamination. Smelters are also large generators of waste material, including HAZARDOUS WASTE.

smog A general term that is currently used to describe (1) any condition in which visibility is reduced by air pollution or (2) elevated levels of PHOTOCHEMICAL OXIDANTS in the atmosphere. See PHOTOCHEMICAL AIR POLLUTION.

smoke An AEROSOL produced by the incomplete burning of carbonaceous materials (i.e., wood or fossil fuels). The aerosol consists of a mixture of gases and visible particles.

smoke reader A person trained to quantify the opacity (darkness) of emissions from a smokestack or flare. The reader classifies the smoke with a 0–5 scale, from white to deep black. Many pollution control agencies have regulations limiting smoke emissions to a certain number on the opacity scale. See RINGELMANN CHART.

snail darter A small fish in the perch family found in the Little Tennessee River system and listed as an ENDANGERED SPECIES, thus potentially blocking the completion of the Tellico Dam being constructed by the Tennessee Valley Authority. The first test case of the 1973 ENDANGERED SPECIES ACT (ESA), *Tennessee Valley Authority v. Hiram Hill, et al.*, was decided by the U.S. Supreme Court in favor of the snail darter. In 1979, a review board, also known as the GOD COMMITTEE, created by an amendment to the ESA to rule on exemptions to the act, again found for the snail darter, but later that year Congress voted to complete the dam. The main darter population was lost when the river flow was disrupted, but the fish has been found in other area streams and has been downgraded to THREATENED SPECIES status.

social ecology A radical humanist outlook based on the rejection of the centrality of humans, simple living, intimate contact with nature, decentralization of power, support for biological diversity, direct personal action to protect nature, and fostering of societal change based on ecological principles.

sodium tripolyphosphate (STPP) A builder, or water softener, contained in many detergents. Water softeners bind calcium and magnesium ions in water, preventing the reaction of the ions with the surfactant (soap) and the loss of cleaning effectiveness. The phosphates in STPP may not be removed by municipal water treatment plants and are contributors to EUTROPHICATION in the surface waters receiving municipal wastewater.

soft detergent A synthetic cleaning agent that is degradable by microorganisms. These chemicals do not accumulate in the environment. See BIODEGRADABLE.

soft hammer The provision in the HAZ-ARDOUS AND SOLID WASTE AMENDMENTS allowing SCHEDULED WASTES to avoid an automatic LAND DISPOSAL BAN if the USEPA failed to issue treatment standard regulations for a waste on time. Land disposal is permitted only if the disposal site is equipped with an approved composite LINER and a LEACHATE COLLECTION SYSTEM and the disposer has attempted to treat the waste in an acceptable manner in the absence of a treatment standard.

soft pesticide A PESTICIDE that is readily BIODEGRADABLE and is decomposed a short time after distribution into the environment.

soft water Water that contains low concentrations of metal ions such as calcium and magnesium. This type of water does not precipitate soaps and detergents. Compare HARD WATER.

softening The removal of metal ions such as calcium and magnesium from water supplies. The conversion of HARD WATER to SOFT WATER. See SODIUM TRIPOLYPHOSPHATE.

softwoods Wood products from conifers such as pine, spruce, and fur, which dominate the forests of the temper-

ate regions. The term does not relate to the hardness of the wood. Compare HARDWOODS.

soil absorption field See LEACHING FIELD.

soil air Below-ground air in the spaces between soil particles. DECOMPOSITION of organic matter in the soil can cause high concentrations of CARBON DIOXIDE in the soil air. The carbon dioxide combines with water to form CARBONIC ACID, thereby increasing the acidity of the groundwater.

soil amendment Any material added to soil that enhances plant growth.

soil and sediment adsorption isotherm test A test method that measures the ADSORPTION of a chemical substance to soil or SEDIMENT and thus indicates the likely distribution pathways in the environment. A readily adsorbed substance increases in concentration in a soil or sediment layer upon addition to the soil, and a substance that is not adsorbed moves freely through the soil into GROUNDWATER or runs off into surface water.

soil conditioner Any bulky, complex natural organic material, such as plant litter, PEAT, COMPOST, or other organic material derived from the partial DECOMPOSITION of plant or animal remains, that is mixed with soil to restrict compaction, improve aeration, control acidity, and provide for mineral balance and beneficial microbial growth. Soil conditioner also enhances the slow release of plant nutrients, such as nitrogen and phosphorus compounds. See HUMUS.

Soil Conservation Service (SCS) The federal agency, created by the Soil Conservation Act of 1935, that administers erosion control projects and research. The SCS works through local Soil Conservation Districts.

soil core A sample of soil taken by forcing a cylindrical device into the ground perpendicular to the horizontal. The resulting sample contains a circular section of each layer of sediment.

soil flushing A treatment technique for cleaning soil contaminated with inorganic or organic HAZARDOUS WASTE; the process involves the flooding of the soil with a flushing solution, which may be acidic or basic or may a contain SURFACTANT, and the subsequent removal of the LEACHATE via shallow wells or subsurface drains. The recovered leachate is then purified.

soil horizon Horizontal layers of soil, each of which has a different characteristic. A collection of soil horizons is a SOIL PROFILE.

soil moisture regime The changes in the moisture content of soil during a year.

soil profile The layers of soil revealed when one digs through the Earth or removes a core sample: The three basic horizontal layers that may be observed in a soil profile are the A, B, and C SOIL HORIZONS. The A horizon, or topsoil, is the main source of plant nutrients. Soluble materials leach from the A horizon down to the SUBSOIL, or B horizon. This is the zone of clay accumulation. The deepest layer, the C horizon, is composed of partially weathered bedrock. See SOIL CORE.

soil screening levels (SSLs) Risk-based soil concentrations developed by the USEPA for use at sites on the NATIONAL PRIORITIES LIST, that is, sites being cleaned up after the NATIONAL CONTINGENCY PLAN, as provided for in the COMPREHENSIVE ENVIRONMENTAL RESPONSE, COMPENSATION, AND LIABILITY ACT (SUPERFUND). The SSL concentrations were developed by using exposure assumptions and dose-response data. They are not intended to be acceptably low levels for cleanups, nor do all concentrations above the SSLs necessarily trigger soil remediation for that chemical. The USEPA intends the screening levels to be used to eliminate parts of cleanup sites from further evaluation and to eliminate chemicals or chemical expo-

sure pathways during the REMEDIAL INVESTIGATION/FEASIBILITY STUDY (RI/FS) required of all Superfund cleanups. USEPA guidance notes that the SSLs can also be applied at sites undergoing voluntary remediation or at site cleanups conducted as part of a CORRECTIVE ACTION/ORDER.

soil sorption coefficient (K_d) A parameter relating the partitioning of a chemical between soil and water in a soil-water mixture. The coefficient is computed by

$$K_d = C_s/C_w$$

where C_s is the amount of a specific chemical bonded to the soil (micrograms chemical per gram of soil) and C_w is the concentration of the same substance dissolved in the water (micrograms of the chemical per gram of water).

soil structure The physical properties of different soils. Sand has little structure because sand particles do not tend to stick together, whereas clay has a firm structure because clay particles pack tightly together to form a solid material.

soil texture A classification of soils based on the size distribution of mineral grains composing the soil. The relative proportions of silt, sand, clay, and gravel are normally given.

soil thin-layer chromatography A method used to estimate the potential for LEACHING a chemical from soil by measuring the mobility of the chemical in soil under controlled conditions.

soil vapor extraction The removal of VOLATILE ORGANIC COMPOUNDS from a defined soil volume by applying a vacuum via a set of wells, drawing organic vapors to the surface. See IN SITU STRIPPING.

soil vapor survey A noninvasive method for the detection of VOLATILE ORGANIC COMPOUNDS or SEMIVOLATILE ORGANIC COMPOUNDS in shallow subsurface soil. The technology is especially useful in the analysis of soils that have high sand content and of shallow groundwater.

A probe is driven into the ground, and samples of vapors in the soil are drawn to the surface for analysis on-site or in the laboratory.

solar backscatter ultraviolet (SBUV) data Vertical profile stratospheric ozone measurements by NATIONAL OCEANIC AND ATMOSPHERIC ADMINISTRATION satellite instruments. Used in OZONE LAYER DEPLETION studies.

solar cell An electrical device that converts sunlight directly into an electric current. Also referred to as a PHOTOVOLTAIC cell, it consists of a thin wafer of silicon containing a small amount of a metal that emits electrons when struck by sunlight.

solar constant A quantity used to identify the amount of solar radiation striking the upper portion of the atmosphere. Specifically, the amount of radiant energy striking a surface positioned at the top of the atmosphere and lying perpendicular to the solar radiation when the Earth is an average distance from the Sun. The quantity is about two calories per square centimeter per minute, or two LANGLEYs per minute. The quantity is used in calculations relating to the input of solar energy into the atmosphere.

solar energy The term is normally applied to the conversion of direct sunlight into usable forms of energy. PASSIVE SOLAR SYSTEMS are designed to maximize the warming of a residence by the Sun, and ACTIVE SOLAR SYSTEMS involve the storage of water heated by the Sun in tanks with subsequent pumping of the hot water to heat a home or other structure. Sunlight can also be converted to electricity, either directly, by using SOLAR CELLS, or indirectly, by focusing sunlight on a boiler, producing electricity with a steam turbine. The most active use being made of solar energy by individual homeowners is the heating of water. Several experimental programs are under way to generate commercial quantities of electricity by employing the energy in sunlight to generate steam. Solar cells are used to power handheld calculators as well as to provide electricity to operate electrical devices in remote areas.

solar irradiance in water An expression of the amount of light flux from a clear sky, over a certain wavelength range, that is available to cause photochemical transformations (PHOTOLYSIS) in chemicals present in surface water bodies at shallow depths. Solar irradiance varies with latitude and the time of year. Values are used with the MOLAR ABSORPTIVITY of the waterborne chemical substance to calculate the HALF-LIFE of the specific chemical in the environment due to photolysis.

sole source aquifer An aquifer designated by a provision of the SAFE DRINKING WATER ACT (the Gonzales Amendment) as the principal or only source of drinking water for a geographical area. This designation bars the use of federal funds for projects in the RECHARGE ZONE that may lead to a significant hazard to the public health by degrading groundwater quality in the aquifer. See CRITICAL AQUIFER PROTECTION AREA, SOURCE WATER ASSESSMENT PROGRAM, SOURCE WATER PROTECTION PROGRAM, SOURCE WATER PROTECTION AREA.

solid waste Waste material not discarded into surface waters via water treatment systems or directly into the atmosphere. Under federal regulations, the term can include waste in solid or liquid form as well as gaseous material. MUNICIPAL SOLID WASTE and all HAZARDOUS WASTE are by definition solid waste.

solid waste management A unified approach to the handling and disposal of discards from the point of generation to the point of final disposition or disposal. The primary elements include generation, storage, collection, source reduction, recycling, processing, and final disposal. See RESOURCE CONSERVATION AND RECOVERY ACT.

solid waste management plan Documentation specifying the key elements of waste disposal for an industrial facility, city, region, or state. See SOLID WASTE MANAGEMENT.

solid waste management unit (SWMU) The property on which hazardous waste management occurs, such as a surface IMPOUNDMENT, LANDFILL, incinerator, waste pile, or tank and the adjacent land used for storage, transfer, or preliminary treatment of the waste.

solidification The addition of agents that solidify liquid or semiliquid HAZARDOUS WASTE before burial to reduce the LEACHING of the waste material and the possible migration of the waste or constituents of the waste from the facility. Usually accompanied by stabilization. See STABILIZATION, WASTE.

solubility The relative capacity of a substance to serve as a SOLUTE. Sugar has high solubility in water, whereas gold has low solubility in water.

solubility product constant (K_s) The product of the molar concentrations of the ions in solution that result from partial dissolution of a solid chemical compound in water. Each compound has its own equilibrium dissolved concentration, and the solubility product is a constant for that compound. For example, the K_s of an equilibrium between solid ferric hydroxide [$Fe(OH)_3$] and dissolved Fe^{+3} and OH^- ions is 1×10^{-38}, which means that the product of the molar concentrations of Fe^{+3} and OH^- is 1×10^{-38}. Usually expressed as pK_s, which is the negative logarithm of K_s. Therefore, the pK_s for ferric hydroxide is 38.

solum The top two soil layers, composed of the topsoil (A horizon) and the SUBSOIL (B horizon, or layer of leached material deposition). The solum excludes the parent material layer (C horizon). See SOIL PROFILE.

solute The substance that is dissolved in a SOLUTION.

solution A homogeneous mixture of a SOLUTE in a SOLVENT. When sugar (the solute) is dissolved in water (the solvent), the molecules that the sugar crystal comprises are separated from one another and dispersed throughout the liquid medium.

solution mining The removal of a mineral deposit that is soluble in water. Water is injected into the geological strata containing the mineral, and the dissolved material is recovered by wells. The method is applied to mine salt (sodium chloride) and potash (potassium chloride), among others.

solvent The dissolving medium, or liquid portion, of a solution. Water is frequently referred to as the universal solvent. The term *solvent* is also applied to organic materials (e.g., benzene, acetone, or gasoline) used to clean (dissolve) oils or grease from machinery, fabrics, and other surfaces or to extract hydrocarbons from some source material. Many organic solvents are flammable and/or toxic. See SOLVENT REFINING.

solvent recovery A method to minimize hazardous waste by recovering process solvents for reuse. Common techniques involve DISTILLATION of the solvent from a solvent-containing mixture.

solvent refining A process used to remove sulfur and other contaminants from coal. Powdered coal is mixed with an organic SOLVENT, such as anthracene, which puts about 95% of the carbon compounds in the coal into solution. The coal recovered from the solution contains less than 1% of the sulfur and ash-forming material present in the original coal.

somatic cell Every cell of the body except the GERM CELLS, which produce gametes (sperm and egg).

sone A unit used to express the perceived LOUDNESS of a sound. The unit is equal to the loudness of a 1000-hertz tone with a sound pressure of 40 decibels; if listeners judge that a sound is three times as loud, for example, as this 1000-hertz tone, the sound has a loudness of 3 sones.

sonic boom The intense sound waves created by objects traveling faster than the speed of sound (about 1100 feet per second or 750 miles per hour). The SHOCK WAVE pressure wave be a large as 2000–3000 newtons per square meter, which is strong enough to shake houses. Normal speech sound levels are about 0.01 newton per square meter. Concern over sonic booms restricts some faster-than-sound aircraft to certain flight paths. For example, supersonic speeds by the *Concorde* are restricted to areas over open water.

sorbent A material that absorbs or adsorbs solids, liquids, gases, or vapors, such as the material in a workplace RESPIRATOR that selectively removes gases or vapors as air passes through. Sorbents must be matched to the type of substance(s) to be removed.

sorbing agent A material placed on an oil spill to absorb or adsorb the oil; the oil and the material then can be removed together.

sorption The physical or chemical linkage of substances, either by ABSORPTION or by ADSORPTION.

sorting The separation of municipal solid waste into recyclable categories of material: paper, metals, and glass, for example. Sorting can be performed by households or at centralized facilities. See MATERIALS RECOVERY FACILITY.

sound absorption coefficient See SABIN ABSORPTION COEFFICIENT.

sound intensity The average sound power passing through a unit area perpendicular to the direction that the sound is traveling. Common units for sound intensity are watts per square meter.

sound intensity level The expression of SOUND INTENSITY in decibel units. The sound intensity level (L_I), in decibels, is computed as

$$L_I = 10 \log (I/I_o)$$

where I is the measured sound intensity and I_o is the reference intensity $(1 \times 10^{-12}$ watt per square meter).

sound level See SOUND PRESSURE LEVEL.

sound power The sound energy emitted by a source per unit time, usually expressed in watts. Sound power causes SOUND PRESSURE.

sound power level The sound energy emitted by a sound source per unit time and expressed in decibels. SOUND POWER (W), in watts, is converted to sound power level (L_W), in decibels, by

$$L_W = 10 \log (W/W_0)$$

where W_o is the reference power $(1 \times 10^{-12}$ watt).

sound pressure The periodic fluctuation above and below atmospheric pressure created by an oscillating body. The pressure differences are often expressed in newtons per square meter.

sound pressure level The expression of SOUND PRESSURE in DECIBEL units. Because the average of the sound pressure fluctuations above and below atmospheric pressure (equal positive and negative values) would be zero, the measured pressures are squared, summed, then averaged. The square root of the average, termed the root mean square, is then converted to decibels by

$$L_P = 20 \log_{10} (p/p_0)$$

where L_P is the sound level, p is the root mean square sound pressure, and p_o is the reference sound pressure (commonly 2×10^{-5} newton per square meter). Noise meters perform these conversions internally and display sound pressure level in decibels.

sound wave A sinusoidal variation around atmospheric pressure caused by a vibrating body.

sour gas Natural gas that contains a high level of HYDROGEN SULFIDE, which has a very foul odor at the least and poses a severe health hazard at the worst.

source material Any material that is not a SPECIAL NUCLEAR MATERIAL but contains at least 0.05% uranium, thorium, or any combination of the two.

source measurement The noise level produced by a tool, machine, or process. See AMBIENT NOISE.

source reduction Solid waste management program designed to reduce the amount of solid waste from households and industrial facilities through alteration of practices so that solid waste is not produced. Therefore, the amount of material to be used as landfill, incinerated, or recycled is lowered. See POLLUTION PREVENTION.

source/release assessment An estimate of the types, quantities, and locations of the release of harmful substances or energy into the indoor or outdoor environment. The results are used in a RISK ASSESSMENT.

source separation A system intended to facilitate the RECYCLING of items in normal household discards. Newsprint, white paper, glass, plastics, aluminum, and steel (cans) items (to mention some of the possible categories) are placed in separate containers for collection at curbside or for deposit at drop-off locations. The system facilitates the cleaning and preparation of the recyclable items for sale and conversion into new commercial products. See MATERIALS RECOVERY FACILITY.

source water Groundwater or surface water used for a public drinking water supply.

Source Water Assessment Program (SWAP) A SAFE DRINKING WATER ACT program that requires states to identify and describe the water sources for the drinking water supplies in the state, including the land above groundwater supplies and the watershed areas that feed surface water supplies; to identify signifi-

cant potential sources of contamination of the water sources; and to investigate the susceptibility of the water sources to that contamination. See SOURCE WATER PROTECTION AREA, SOURCE WATER PROTECTION PROGRAM, WELLHEAD PROTECTION AREA.

source water protection area Land identified under the SOURCE WATER ASSESSMENT PROGRAM required by the SAFE DRINKING WATER ACT that contains sources of potential contamination of the SOURCE WATER for an area's drinking water. The protection areas include land with potential sources of GROUNDWATER contamination and any WATERSHED that contributes RUNOFF that may contaminate surface waters used for drinking water supplies.

Source Water Protection Program (SWPP) Under provisions of the SAFE DRINKING WATER ACT, after a state completes its SOURCE WATER ASSESSMENT PROGRAM, the next step is the implementation of protective measures for the SOURCE WATER serving as drinking water supplies. The SWPP is to be operated by state and local representatives, using regulations or voluntary measures to reduce or eliminate the potential threat to drinking water supplies within the state. See WELLHEAD PROTECTION PROGRAM, SIGNIFICANT POTENTIAL SOURCE OF CONTAMINATION, SOURCE WATER PROTECTION AREA.

southern oscillation The changing sea level atmospheric pressures over the eastern Indian Ocean (measured at Darwin, Australia) and the pressure over the eastern Pacific Ocean (measured at Tahiti). When the pressure is high at Darwin and low over Tahiti, El Niño conditions occur.

Spaceship Earth Term coined in separate 1966 works by Kenneth Boulding, "The Economics of the Coming Spaceship Earth," and BARBARA WARD, *Spaceship Earth*. Usually used to emphasize the precarious aspect of life and the environmental factors that sustain it.

span drift A gradual change in the instrument response to a SPAN GAS. This source of measurement error is reduced by frequent CALIBRATION.

span gas A gas with a known concentration of a chemical substance introduced into an analytical instrument as a CALIBRATION step. The initial concentration of a span gas is typically about 80% of the operating range of an instrument. Other, more dilute gas concentrations are then used to create a CALIBRATION CURVE.

span value For measurements of environmental pollutant concentrations, the upper limit of the scale on the measuring instrument. See SPAN GAS, SPAN DRIFT.

sparging Bubbling air through water to remove volatile chemicals.

spatter cone Feature associated with lava flows. Areas around fissures or cracks in the surface where globs of lava splash out and form little conical mounds.

special nuclear material Elements capable of undergoing nuclear FISSION, including PLUTONIUM-239, uranium-233, URANIUM-235, or URANIUM enriched with uranium-233 or uranium-235. See ENRICHMENT.

Special Review A declaration of the USEPA issued under the authority of the FEDERAL INSECTICIDE, FUNGICIDE, AND RODENTICIDE ACT that the REGISTRATION (use permit) of a pesticide is in question because new data indicate that the agent poses an "unreasonable" risk to human health or the environment. The declaration does not imply CANCELLATION OR SUSPENSION of the chemical; the agency can still weigh the risks and benefits of the pesticide and allow a continuation of the registration. Formerly called a rebuttable presumption against registration (RBAR).

special waste Any solid waste requiring special handling beyond the usual and customary procedures; examples are used tires, house furniture and appliances, crankcase oil, hospital waste, household hazardous waste, and some industrial solid waste.

speciation The formation of two or more genetically distinct groups of organisms after a division within a single group or SPECIES. A group of organisms capable of interbreeding is segregated into two or more populations, which gradually develop barriers to reproduction. These barriers can be extrinsic— that is, the separated populations may reproduce at different locations or at different times of the year—or intrinsic, characterized by some genetic barrier that makes attempts to crossbreed unsuccessful. When the reproductive isolation is maintained long enough, the separated groups may develop into separate, distinct, identifiable species.

species The members of a group of organisms that can successfully interbreed with each other under natural conditions. The members of the group generally resemble each other and occupy a specific geographic region. The term is also applied to taxonomic classifications into which individual specimens are placed. These taxonomic groupings usually, but not always, correspond to the natural biological groups.

species composition The types and abundance of organisms inhabiting a specified locality.

Species Conservation Commission (SCC) Over 7000 volunteer scientists and conservationists from 179 countries working with the INTERNATIONAL UNION FOR CONSERVATION OF NATURE AND NATURAL RESOURCES (the WORLD CONSERVATION UNION) to collect data on animal and plant species and their habitat toward the goal of species conservation and the maintenance of BIODIVERSITY. The SSC, operating through 110 specialist groups, publishes the RED LISTS, publications covering the conservation status of animals and plants worldwide. Website: http://www.iucn.org/themes/ssc

species density The total numbers of individuals of a SPECIES found in a defined area of a habitat for a particular period. Such calculations are of value in that they provide information on the magnitude of the population of some species occupying a given area. The formula is

$$D = (n/a)/t$$

where D is the density, n in the number of individuals, a is the area studied, and t is the period during which the study was conducted.

species diversity A measurement that incorporates both the number of different SPECIES, or individual types of organisms, that inhabit a given location and the number of individuals of each type present. Generally, undisturbed locations have a higher species diversity than that found in similar habitats that have undergone extensive environmental alteration. See SHANNON-WEAVER INDEX.

species frequency The percentage of the sampling areas in which a representative of a specific SPECIES is found. The frequency is independent of the number of individuals of each species located in each sampling area.

species richness index A mathematical expression that indicates the number of SPECIES in a COMMUNITY relative to the total number of all individuals in that community. The number of organisms constituting each species is not considered in the index.

specific activity The total radioactivity of a specified substance per gram of that compound or element.

specific collection area (SCA) The ratio of the total COLLECTING SURFACE area of an ELECTROSTATIC PRECIPITATOR to the VOLUMETRIC FLOW RATE of the airstream. Other variables being equal, higher specific collection areas are more efficient at PARTICULATE MATTER removal.

specific conductance A measure of the ability of a solution to conduct an electric current, mainly dependent on the concentrations and types of ION in the solution, expressed in units of conductance/length, such as micromhos per centimeter. Used in groundwater monitoring as an indication of the presence of ions of chemical substances that may have been released by a leaking landfill or other waste disposal facility. A higher specific conductance in water drawn from DOWNGRADIENT WELLS than in that from UPGRADIENT WELLS indicates possible contamination from the facility.

specific gravity A comparison of the density of a substance (frequently a liquid) to the density of a reference substance (normally water). Specific gravity (SG) is given by

$$SG = r/r_w$$

where r is the density (weight per unit volume) of the unknown substance and r_w is the density of water. The parameter has no units and is frequently used to determine the concentration of a SOLUTION.

specific heat The ratio of the amount of heat (calories) required to raise the temperature of one gram of a substance 1° C to the amount of heat required to raise the temperature of one gram of water 1° C (which is one calorie). Every substance has a characteristic specific heat. For example, the value is 1.000 for water and 0.092 for copper, meaning that more heat is required to raise the temperature of one gram of water than is needed to produce the same temperature increase in one gram of copper.

specific molal volume The volume occupied by one MOLE of a liquid. The volume can be calculated by dividing the MOLECULAR WEIGHT of the substance by the liquid DENSITY. Common units are cubic centimeters per gram-mole.

specific rate In public health statistics, a RATE that is limited to a certain age, sex, ethnic group, or year category, such as the 1987 death rate for 70-year-old African-American females. See CRUDE RATE.

specific weight (γ)   The weight (force) per unit volume of a material, expressed as $\gamma = \rho g$, where ρ is the material DENSITY and g is the acceleration of gravity.

specific yield   The volume of water available per unit volume of aquifer if drawn by gravity. Specific yield is expressed as a percentage; for example, if 0.2 cubic meter of water drains from 1 cubic meter of aquifer sand, the specific yield is 20%.

specifications standards   Environmental or workplace health and safety standards that stipulate in detail what actions must be taken to be in compliance: how a control device will be configured, where labels will be placed on a container and what size print they will have, and so on. Compare PERFORMANCE STANDARDS.

spectrometer   An instrument used to identify and measure the WAVELENGTH of ELECTROMAGNETIC RADIATION emitted by elements when heated to high temperatures. Since each chemical element emits a unique spectrum, the technique can be used to detect small quantities of substances. The elemental composition of stars can be deduced from the spectrum of radiation emitted by the star.

spectrophotometry   An analytical method that uses the intensity of radiation absorbed at certain wavelengths to detect the presence of specific chemical elements.

spent fuel   Fuel, from a nuclear power reactor, that can no longer efficiently sustain a continuous FISSION reaction. The rods containing the nuclear fuel used to power a nuclear power plant become depleted of fission fuel (i.e., certain forms of uranium), and waste products from the fission reactions accumulate within the fuel elements. As a result, the efficiency of the nuclear reaction decreases and the rods are said to be spent. The rods are usually removed after 1.5 years of use. The used rods emit a significant amount of heat and are very RADIOACTIVE. Spent

FUEL RODS removed from commercial power reactors are stored on site and constitute a significant disposal or recycling problem. See NUCLEAR WASTE POLICY ACT.

spent fuel pool   A storage facility where FUEL RODS that have been removed from a nuclear reactor are held underwater for cooling, then moved to dry storage on site. See SPENT FUEL.

spike   1. A known amount of a chemical added to an environmental sample to detect the accuracy of an analytical instrument. 2. To add a known amount of a chemical to an environmental sample.

Spill Prevention Control and Countermeasure Plan (SPCC Plan)   A written description of prevention, detection, and containment measures a facility has or will use to prevent, minimize, or respond to oil spills into water. Regulations written by the USEPA under the authority of the CLEAN WATER ACT require certain "non-transportation-related onshore and offshore facilities" to have a SPCC plan. Transportation-related facilities that have the potential to spill oil in harmful quantities are regulated by the U.S. Department of Transportation. See NATIONAL CONTINGENCY PLAN, OIL POLLUTION ACT OF 1990.

spirometer   An instrument for determining lung flow and volume capacities. The lung function test results are used to detect lung disease or adverse effects on the respiratory system of the inhalation of occupational or community air contaminants.

spirometry   The testing of lung function through the use of a SPIROMETER.

split sample   Material collected from an environmental medium (for example, sediment or water) that has been divided into two or more portions so that each portion can be a different analytical laboratory. Split samples are used in comparison of the results from different laboratories or different technologies and in quality assurance procedures.

spoil 1. The refuse or rubble that accumulates when soil, rock, or sand is removed to allow access to mineral deposits. 2. The sediment removed from a channel by dredging operations.

spongy parenchyma Chlorophyll-containing cells in a leaf that lie under the palisade parenchyma. If excessive air contaminant exposure to vegetation damages the spongy parenchyma or palisade cells to the point of tissue collapse, the leaf becomes discolored and the damaged areas may fall away, leaving holes in the leaf.

spotted owl The northern spotted owl (*Strix occidentalis caurina*) was listed as a THREATENED SPECIES under the provisions of the ENDANGERED SPECIES ACT (ESA) in 1990. The key features of species protection by the ESA are the delineation and protection of critical HABITAT, which for the spotted owl has been defined as OLD GROWTH forests from northern California to British Columbia. Around 80% of the old growth forests in this area have been logged, diminishing the spotted owl population, leading to its threatened status. (Definition of the owl's habitat has been the subject of much debate since the finding that the spotted owl lives in a variety of forest types, not solely in old growth forest.) Around 90% of the remaining old growth forest in the spotted owl's range is on federal land managed by the U.S. FOREST SERVICE and the BUREAU OF LAND MANAGEMENT, and logging on this land has been drastically curtailed by the habitat protection provisions of the ESA. The decline in the logging industry in the Pacific Northwest became a national issue of logging jobs versus owls.

spray-back system A treatment technique for LEACHATE in which the leachate is collected then directed back through the landfill. The aerated liquid enhances biological DECOMPOSITION of the leachate

and the waste in the landfill. See LEACHATE COLLECTION SYSTEM.

spray chamber A device that removes certain organic compounds from an airstream by condensation. A cooling material, usually water, is sprayed into a chamber, and the condensed organics exit with the water. In addition to the removal of the condensable contaminants, the condensation greatly reduces the volume of the waste exhaust. Also called a contact condenser.

spray tower A device used for the removal of gases or particles from an exhaust gas. The dirty gas stream typically is directed through the bottom of a tower and flows past a finely divided spray that removes the pollutants. The cleaned gas

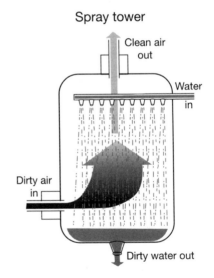

Spray tower

Clean air out

Water in

Dirty air in

Dirty water out

exits the top of the tower, and the water drains to the bottom. See also PACKED TOWER, TRAY TOWER.

spring Natural flow of groundwater from a subsurface AQUIFER to the surface.

spring turnover The mixing of water in a lake during the spring of the year. This process takes place most frequently in lakes located in temperate zones where the winter temperatures are low enough during the winter to result in freezing of the lake surface. As the ice melts in the early spring, the water temperature is the surface gradually warms to that of the underlying water. Since the water temperature is then uniform from top to bottom, the force exerted by the winds is sufficient to promote mixing of the surface and deeper waters. Compare FALL TURNOVER.

spring water Under the FOOD AND DRUG ADMINISTRATION regulatory definition, BOTTLED WATER derived from an AQUIFER from which water flows naturally to the surface. Spring water can be collected through a well that taps an underground formation that feeds a natural spring, but it cannot be pumped to the surface. The water collected by the use of a well must be of the same chemical composition as that of the water flowing from the natural spring.

spurious count A term used to describe the recording of false RADIATION reading caused by malfunctions of, interference with, or improper use of radiation detection devices.

stability See ATMOSPHERIC STABILITY; STABILITY CLASS, ATMOSPHERIC.

stability class, atmospheric A classification of atmospheric TURBULENCE, or air contaminant dispersive capability, developed by F. Pasquill and F. A. Gifford. Atmospheric turbulence is categorized by the letters *A* through *F* with *A* being the most dispersive and *F* the least. The most important variables determining the stability class at a given time are wind speed and amount of sunshine or solar insolation.

stability index See LANGELIER INDEX.

stabilization A broad expression used to denote a process that is intended to lessen the damage that a pollutant or discharge causes in the environment. For example, the stabilization of sewage involves allowing microorganisms to degrade those components that can be decomposed. The water that is subsequently released into the environment does less damage than the release of RAW SEWAGE. See STABILIZATION, WASTE; SOLIDIFICATION.

stabilization pond A shallow diked area used to allow adequate mixing of treated or untreated WASTEWATER before discharge into a surface water body. The mixing prevents excessive swings in the ACIDITY or ALKALINITY of a discharge, for example.

stabilization/solidification (S/S) See STABILIZATION, WASTE; SOLIDIFICATION.

stabilization, waste A HAZARDOUS WASTE treatment process that decreases the mobility or solubility of waste constituents by means other than SOLIDIFICATION. Techniques include chemical precipitation or pH alteration to limit solubility and mixing of the waste with SORBENTS such as FLY ASH to remove FREE LIQUIDS. Liquid wastes can be solidified into a material that resembles concrete, and the block of material can then be safely disposed of in a hazardous waste landfill.

stabilized grade The grade (slope) of a water channel at which no erosion or deposition occurs.

stable 1. Describing an ELEMENT or ISOTOPE that does not undergo RADIOACTIVE DECAY. A stable material may be the end of a RADIOACTIVE SERIES. 2. Describing air that does not readily disperse pollutants. See STABILITY CLASS, ATMOSPHERIC.

stable isotope A form of an ELEMENT that is not radioactive: that is, an ISOTOPE that does not undergo RADIOACTIVE DECAY.

stack downwash The movement of a smokestack PLUME toward the ground, instead of, more commonly, upward. Stack downwash (also called stack-tip downwash) can occur if the exit velocity of the stack gas is significantly less than the wind speed at the top of the stack. Certain AIR QUALITY DISPERSION MODELS have a stack downwash option, which simulates this phenomenon. Model results under the stack downwash assumption, all other variables being equal, predict higher ground-level concentrations.

stack flue The innermost channel in a smokestack through which exhaust gas is carried.

stack gas The gaseous product of a process (usually combustion) that exits through a stack or flue.

stack sampling The collection of representative portions of the gases and particulate matter that are being discharged through a smokestack or duct. This type of sampling allows direct estimation of the amount and types of air pollutants being released.

stage II control System designed to limit the release of gasoline vapors (VOLATILE ORGANIC COMPOUNDS) during the refueling of automobiles. The technology, which captures and returns vapors, is applied to the pumps at gasoline outlets. Compare ONBOARD REFUELING VAPOR RECOVERY.

stagnation Persistent atmospheric conditions characterized by limited vertical and horizontal air motion, resulting in an increase in the concentration of air contaminants. See TEMPERATURE INVERSION.

stakeholder Any individual or any organization that has an interest in the outcome or operation of, for example, a regulation, industrial operation, or legislative action. An interest in this context transcends financial interest in the outcome; for example, residents near an industrial facility are stakeholders in the facility even if they or their family do not own stock in the company, do not work in the facility, or do not have any financial interest in a supplier to or customer of the facility. The drafting and enforcement of environmental regulations and permits by local, state, and federal administrative agencies allow participation by STAKEHOLDERS.

standard addition technique In analytical chemistry, a method in which a known amount of a standard solution is added to the sample being measured. The process can be used if a constituent in the sample other than the one being measured interferes with the response of the analytical instrument, thus giving incorrect results because the pure calibration standards are not subject to this interference. The standard addition technique corrects for the interference.

standard air Air under defined standard temperature and pressure conditions. The USEPA defines 25° C and one atmosphere as STANDARD CONDITIONS. Many occupational health calculations use 21° C and one atmosphere, and most other applications use 0° C and one atmosphere.

standard air density The density of dry air at the chosen STANDARD CONDITIONS. At 0° C and one atmosphere of pressure, the density of dry air is 1.293 kilograms per cubic meter; at 25° C, one atmosphere, the density is 1.184 kilograms per cubic meter.

standard conditions Arbitrary reference conditions established for comparing reactions that involve gases. The density and other properties of gases depend on the temperature and pressure under which the gas exists. Standard conditions represent a way that gases can be compared under different circumstances. They are frequently set at 0° C and one atmosphere, but environmental and occupational calculations assume different air temperatures. See STANDARD AIR.

standard cubic feet per minute (SCFM) Common VOLUMETRIC FLOW RATE units for air. The STANDARD CONDITIONS used determine the density and therefore the standard flow rate expressed.

standard curve See CALIBRATION CURVE.

standard deviation (s) The square root of the VARIANCE of a data set, an indication of the "spread" of the data set around the mean, or average. The variance for a large data set is calculated as follows:

$$\sigma = \sqrt{\frac{\sum (x_i - \bar{x})^2}{n}}$$

where x_i is an individual observation, $\bar{x}$ is the mean, and n is the number of observations. For smaller data sets (less than 50) the sample standard deviation (σ) is calculated by replacing n with $n - 1$ in the equation.

Standard Industrial Classification (SIC) A numbering system used by the federal government to group industrial facilities by category. The full SIC number has four digits. The first two digits (range 01–99) indicate the major industrial category, and the last two digits classify the company further. For example, major category codes 20–39 identify manufacturing facilities; code 28 within this range is the major category number for chemicals and allied products, and SIC number 2869 is the classification for industrial organic chemicals not elsewhere classified. Environmental regulations are often applied to facilities within certain SIC codes. The SIC codes are being phased out by the NORTH AMERICAN INDUSTRY CLASSIFICATION SYSTEM (NAICS).

Standard Methods A short form for *Standard Methods for the Examination of Water and Wastewater,* which is prepared and published jointly by the AMERICAN PUBLIC HEALTH ASSOCIATION, AMERICAN WATER WORKS ASSOCIATION, and WATER ENVIRONMENT FEDERATION. The book serves as the primary reference for analytical methods employed in investigations and monitoring of water purification, sewage treatment and disposal, water pollution, sanitary quality, and other functions. A new edition is published every five years.

Standard Metropolitan Statistical Area (SMSA) See METROPOLITAN STATISTICAL AREA.

standard population A group used to compare the MORTALITY RATES or MORBIDITY RATES of two different populations. The distribution of some characteristic within the standard population as related to mortality or morbidity (especially age) is used as the standard, and any differences in the distribution of the characteristic in the two groups is statistically taken into account in the comparison. See AGE ADJUSTMENT, DIRECT METHOD; AGE ADJUSTMENT, INDIRECT METHOD.

standard mortality ratio (SMR) The ratio of the occurrence of a specific cause of death in a given test population to the MORTALITY RATE of the same cause in a STANDARD POPULATION. The numerical value is based on 100 incidents in the standard population; therefore, a ratio of 500 means that the test population had a mortality rate five times that of a standard population due to the same specific cause.

standards, environmental Allowable conditions or actions that protect against unwanted effects on human health or welfare, wildlife, or natural processes. Such standards can be defined as ambient chemical concentrations, chemical or radioactive material emission or effluent rates, emission or effluent chemical concentrations, sound levels, or radioactivity levels as well as harder-to-measure taste, odor, and appearance criteria. Many standards are set to a level, concentration, or emission rate that equates to an acceptably low risk. See AMBIENT STANDARD, EMISSION STANDARD, EFFLUENT STANDARD.

standing crop The quantity of plant BIOMASS in a given area. For example, the amount of plant material per acre of forest or swamp or per cubic meter of water. Usually expressed as mass (dry) per unit area or energy content per unit area.

standing stock See STANDING CROP.

standing, legal In environmental law, the legal position required to seek judicial review of the actions of a government agency. Standing is gained by a plaintiff who is experiencing an adverse personal impact (legal or bodily injury), called injury in fact. Once thought to be a barrier to third-party (concerned citizen) intervention in environmental agency decisions, the definition of standing has been much relaxed by court decisions, giving environmental interest groups ready access to the courts. Also, most major environmental statutes have specific provisions for private citizen legal action that do not require injury in fact to a plaintiff. See CITIZEN SUIT PROVISION.

stannosis The presence of tin oxide dust in the lungs, resulting from occupational exposure. Stannosis is detected by a chest X ray. Few if any of the 200 reported cases has been accompanied by the chronic lung damage, such as FIBROSIS, that is associated with dusts containing FREE SILICA or ASBESTOS.

State and Local Air Monitoring System (SLAMS) Nonfederal air quality monitoring stations meeting siting and quality assurance requirements established by the USEPA. Some stations are used to track compliance with AIR QUALITY STANDARDS, some to measure the air in areas of expected high concentrations for certain pollutants, and others to determine the BACKGROUND CONCENTRATION for air contaminants.

State and Territorial Air Pollution Program Administrators/Association of Local Air Pollution Control Officials (STAPPA/ALAPCO) A group of air pollution control agency representatives from 50 states, 4 territories, and over 150 municipalities who share information on air pollution issues and regulatory compliance. Headquarters in Washington, D.C. Website: www.4cleanair.org

state emergency response commission (SERC) A group in each state appointed by the governor in accordance with provisions of the Emergency Planning and Community Right-to-Know Act, which is TITLE III of the SUPERFUND AMENDMENTS AND REAUTHORIZATION ACT OF 1986. The commission is responsible for forming and working with LOCAL EMERGENCY PLANNING COMMITTEES (at the county level) to collect information on chemicals stored and released by affected facilities. The SERC also reviews emergency plans written by the Local Emergency Planning Committees and responds to public requests for the inventory and chemical release information.

state hazardous waste plan Documentation detailing how a state intends to manage hazardous waste generated within it or hazardous waste stored, treated, or transported through it.

state implementation plan (SIP) The written agreement between an individual state and the USEPA describing how the state will comply with the provisions of the CLEAN AIR ACT, especially the attainment and maintenance of the NATIONAL AMBIENT AIR QUALITY STANDARDS. See SIP CALL.

state management plan (SMP) A program to restrict the use of certain pesticides that threaten groundwater quality under the provisions of the FEDERAL INSECTICIDE, FUNGICIDE, AND RODENTICIDE ACT. Each state SMP is to assess the potential for groundwater resources to be contaminated by pesticides, describe measures to be used to lessen the threat of contamination, implement a monitoring program to detect contamination, inform pesticide users of the SMP and its goals, and provide for public awareness and participation in the groundwater protection program.

state variables The components that make up a system in mathematical models used to describe the environment. The state variables have certain states, or conditions, at a given time. For example, in a simple BOX MODEL, the amounts of phosphorus in aquatic vegetation, sediments, and water at a given time could be chosen as the state variables.

static pressure The pressure exerted by a fluid at rest. Static pressure in air is due to the weight of the atmosphere and is usually equivalent to BAROMETRIC PRESSURE. See HYDROSTATIC PRESSURE.

static reserve index A resource use model that assumes that there is a fixed reserve size and that resource use rates will remain constant. Under this assumption, the depletion time for the resource is calculated by dividing the current RESERVES by the use rate. The model does not account for the dynamic effects of price increases resulting from resource depletion. Higher resource prices both dampen demand (use rate) and increase discovery and production of the resource. Technological improvements that increase resource use efficiency and lower discovery and production costs are also omitted.

static test A laboratory test in which the water is not changed and there is no flow in or out of the test vessel when the toxicity of chemical substances to aquatic test organisms is being measured. See SEMISTATIC TEST, FLOW-THROUGH TEST.

stationary growth phase A particular growth phase in the cultivation of bacteria in which there is no increase in the number of cells over time. Any cell division that takes place is balanced by cell death. This phase of growth follows the EXPONENTIAL GROWTH phase.

stationary source A fixed (nonmoving) source of air emissions, such as an oil refinery or power plant. See MOBILE SOURCE, AREA SOURCE.

statistical tests of significance Mathematical methods of stating the probability that two data sets are not from the same population: in other words, there is an actual difference in some characteristic between the two groups. These tests are commonly used in cases in which one group has been exposed to a treatment (or pollutant) and the other has not. After the treatment, some characteristic of the two groups is measured. A difference between the treated and untreated groups may be due to SAMPLING VARIABILITY or to the exposure. A statistical test can determine the probability that the difference is not due to sampling variability. If the observed difference between the exposed group and the nonexposed group has only a small probability of being the result of sampling variability, then the observer may conclude that the difference observed is the result of the exposure. See STATISTICALLY SIGNIFICANT.

statistically significant Describing a difference between two groups, one of which functions as the control and is not subjected to manipulation in the laboratory or environment and one that is subjected to some alteration in the physical, chemical, or biological surroundings (termed the experimental or test group). For example, an experimental group of animals may be subjected to increased noise levels while the control group is held under the same conditions except for the noise. If subsequent measurements of the hearing ability of the animals in the two groups demonstrate a difference between the two groups, STATISTICAL TESTS OF SIGNIFICANCE can be applied to determine the likelihood that the observed differences are due to the exposure to the sound or are the result of natural SAMPLING VARIABILITY. See P VALUE.

statutory law Written law passed by state or federal legislatures, such as the CLEAN WATER ACT. The common law is not written by legislative bodies but is based on precedent. Environmental law is mainly concerned with statutory law and the regulations issued in support of the legislative acts. See ADMINISTRATIVE LAW.

stay time The time that personnel may remain in a restricted area before accumulating the MAXIMUM PERMISSIBLE DOSE of radiation.

steady state In a system with a flow-through of material or energy, the EQUILIBRIUM condition in which flow in equals flow out. See BOX MODEL.

steady-state or apparent plateau In testing chemical substances for their BIOCONCENTRATION potential in fish, the situation in which the amount of chemical substance taken into the test fish from the water is equal to the amount being eliminated from the test fish. If the substance bioconcentrates in the fish, the steady-state condition follows an uptake phase in which the ABSORPTION rate exceeds the elimination rate.

steam stripping The removal of VOLATILE ORGANIC COMPOUNDS from contaminated GROUNDWATER or WASTEWATER by forcing steam through the liquid. The higher wastewater temperature increases the evaporation rate of the volatile contaminants. Steam stripping can remove more contaminants than AIR STRIPPING. See PACKED TOWER AERATION.

Stefan-Boltzmann law The equation describing the rate of radiation emitted by an object, expressed as $H = Ae\sigma T^4$, where A is the surface area, e is an EMISSIVITY factor (between 0 and 1), s is the Stefan-Boltzmann constant, and T is the ABSOLUTE TEMPERATURE of the object, assumed to be a BLACKBODY.

steno- A prefix meaning "narrow," used in ecology to describe the LIMITS OF TOLERANCE for organisms. For example, a fish species may be described as stenothermal, meaning the organism can grow and reproduce well within only a narrow range of water temperatures. See EURY-.

sterile 1. A condition in which a quantity of water, soil, or other substance does not contain viable organisms such as viruses and bacteria. The term, however, is not synonymous with *clean*. 2. Animals or humans that are not capable of reproduction because of the absence of GAMETES.

sterilization 1. The process of killing, inactivating, or removing microorganisms (bacteria, viruses, or fungi) from a quantity of water, soil, or other liquid or solid material. The normal methods of sterilization involve agents such as heat, chemicals, or radiation. 2. The process of rendering an animal incapable of reproduction.

stewardship 1. An environmental philosophy holding that humans have responsibility to protect and care for nature. 2. Under the RESPONSIBLE CARE initiative of the Chemical Manufacturers Association, a code holding chemical companies answerable for their products from the time the idea for the product is developed until the ultimate disposal of the product after use.

sticky tape sampler A device designed to collect airborne PARTICULATE MATTER by surface adhesion.

stippling The spotty injury to a leaf surface caused by excessive exposure to air contaminants. Chronic exposure to OZONE, for example, causes stippling in broad-leaved vegetation.

stochastic Describing an event or process that involves random chance or probability.

stock In ecological cycles and models, the amount of a material in a certain MEDIUM or RESERVOIR, for example, the stock of the carbon dioxide in the atmosphere, which equals about 2.9×10^{15} kilograms.

stock solution A concentrated solution of a chemical used as a reagent in a laboratory procedure. For example, a laboratory test may require 100 milliliters (ml) of a 0.01% solution of sodium chloride. To eliminate the requirement of weighing 0.01 gram (g) of salt and adding that amount to 100 ml of distilled water every time that one is to perform the test, a stock solution of sodium chloride containing 1 g salt per 100 ml can be prepared. When the test is to be performed, 1 ml of the concentrated solution can be added to 99 ml of distilled water, and the test performed much more quickly.

Stockholm Conference See UNITED NATIONS CONFERENCE ON THE HUMAN ENVIRONMENT.

Stoddard solvent A solvent used for dry cleaning, spot removal, and degreasing that has relatively low human toxicity, as evidenced by its THRESHOLD LIMIT VALUE of 100 ppm(v).

stoichiometric condition See STOICHIOMETRIC RATIO.

stoichiometric ratio In air pollution control, the air-fuel mixture containing just enough air to oxidize the fuel completely. Under these conditions, all air oxygen would be consumed and the STACK GAS would contain no oxygen. Combustion of fuels actually requires more oxygen than the amount indicated by the theoretical stoichiometric ratio, and EXCESS AIR is introduced to the combustion chamber.

stoker Mechanical equipment (or person, in the past) for feeding waste materials or coal into an incinerator or furnace.

Stokes diameter An equivalent diameter for a (usually airborne) particle, defined as the diameter of a perfect sphere with the same SETTLING VELOCITY and density as the particle. For example, a nonspherical particle that is six micrometers across with density of two grams per cubic centimeter may have a Stokes diameter of five micrometers, meaning that the nonspherical particle has the same settling velocity as a five-micrometer sphere with a density of two grams per cubic centimeter. See AERODYNAMIC DIAMETER.

Stokes's law The relationship that defines, for low velocities, the frictional force resisting the movement of a spherical body through a fluid (air or water) as 6π times the product of the radius of the sphere, the velocity of the sphere, and the VISCOSITY of the fluid. Stokes's law is used

to determine the rate at which an object will fall through air or water, which is important in the design of pollution control systems. The TERMINAL SETTLING VELOCITY is reached when the gravitational force downward is equal to the frictional force (upward).

stomata Small openings in the surface of plant leaves that allow gas transfer. Air pollutants can cause direct internal damage to vegetation by entering the stomata.

stopping power The rate of energy loss by IONIZING RADIATION per unit thickness or per unit mass of radiation SHIELDING.

storage In reference to groundwater reservoirs, the water naturally retained in an AQUIFER plus any water artificially recharged to the aquifer by pumping.

Storage and Retrieval of Aerometric Data (SAROAD) A database of the USEPA containing air quality measurements from federal, state, and local monitoring stations. Each station has a unique SAROAD number. The SAROAD database is now part of the AEROMETRIC INFORMATION RETRIEVAL SYSTEM (AIRS).

storage facility In hazardous waste management, a facility that stores HAZARDOUS WASTE before treatment or disposal on the site or before transport to another treatment or disposal facility. Industries or businesses that generate hazardous waste are not usually classified as storage facilities. Regulations for large hazardous waste generators state that, with certain exceptions, hazardous waste may accumulate on site for at most 90 days before the generator is subject to the rigorous permit requirements of a storage facility. Longer storage times are allowed for SMALL-QUANTITY GENERATORS.

storm sewer A network of pipes and conduits buried underground that drain rainwater from city streets.

storm water discharge See STORM WATER RUNOFF.

storm water runoff Rainwater that is potentially contaminated by flowing over ground in industrial facilities, municipal roadways, or vehicle parking areas. Under the 1987 amendments to the CLEAN WATER ACT, many sites with storm water runoff are required to obtain a NATIONAL POLLUTION DISCHARGE ELIMINATION SYSTEM (NPDES) permit.

straggling The variation in energy content of particles released in nuclear events. Although they may all be of the same energy level upon release, over time they may vary in their range of energy level because of interactions within the material being traversed.

straight-chain hydrocarbons Compounds of carbon and hydrogen in which multiple carbon atoms are bonded to each other in a straight line. Other hydrocarbons can be branched, as when carbons are bonded at right angles to the primary chain. Some hydrocarbons can have two or more ISOMERS, one straight chain and the other(s) branched. Straight-chain hydrocarbons are ALIPHATIC compounds. Compare AROMATIC.

strata Distinct horizontal layers in geological deposits. Each layer may differ from adjacent layers in terms of texture, grain size, chemical composition, or other geological criteria. The term is also applied to layering of other material such as the atmosphere.

strategic lawsuits against public participation (SLAPPs) Countersuits filed by industries, developers, or other defendants against environmental activist groups in response to the activists' suing to stop land development, curb pollution, or conserve natural resources. The SLAPPs accuse the activist-plaintiffs of libel, slander, or defamation, inter alia. Although a SLAPP may have a poor chance of proving the allegations, the burden of defending the SLAPP may cause a citizen group to abandon the original claim.

stratification The layering of a body of water caused by temperature or salinity differences among the layers. Also, the layering of materials in SEDIMENTARY ROCK.

stratopause The boundary in the atmosphere between the STRATOSPHERE and the next highest layer, the MESOSPHERE.

stratosphere The second layer of the atmosphere above the Earth. The air surrounding the Earth consists of distinct zones, distinguished by the temperature gradient within each layer. The stratosphere begins at about 7 miles of altitude and extends up to about 30 miles. The air temperature in this layer generally increases as altitude increases. The OZONE LAYER of the atmosphere is located within the stratosphere. See ATMOSPHERE.

Straight-chain hydrocarbons

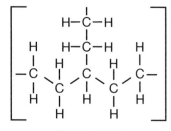

Straight chain   Branched chain

Stratospheric Aerosol and Gas Experiment (SAGE) data Vertical profile stratospheric ozone measurements by satellite instruments. Used in OZONE LAYER DEPLETION studies.

stratospheric ozone See OZONE LAYER.

streamline A line parallel to the non-turbulent flow (LAMINAR FLOW) of a fluid, indicating the path that the fluid takes around objects, such as airflow around a building.

Streeter-Phelps equation An equation representing the change in DISSOLVED OXYGEN content with distance downstream from a source of BIOCHEMICAL OXYGEN DEMAND. See OXYGEN SAG CURVE.

stressed waters As defined by the CLEAN WATER ACT, ocean waters that lack a "balanced indigenous population of shellfish, fish, and wildlife," whose poor SPECIES DIVERSITY is caused solely by human actions. PUBLICLY OWNED TREATMENT WORKS are not eligible for an exemption from the SECONDARY TREATMENT requirements if they discharge into stressed waters.

strict liability A legal standard applying to certain abnormal uses of land (often hazardous activities) that treats any injuries to persons resulting from the activity as the fault of the activity, without proof of negligence. HAZARDOUS WASTE management is held to a strict liability standard.

strip cropping An agricultural practice involving the planting of crops in bands that serve as barriers to erosion by wind and water.

strip cutting Forestry method for the cutting of trees in such a way that the integrity of the forest system is maintained. The land is laid out in parallel strips about 30 to 40 yards wide. Trees are removed from alternating strips, leaving the forest undisturbed between harvested

areas. During the next cutting cycle, the untouched areas are harvested. Judicious scheduling of cutting within the strips allows the forest to be maintained as a continuing resource. The technique permits natural reseeding and regrowth to occur and has less impact on wildlife than does CLEAR CUTTING.

strip farming An agricultural technique involving the planting of different crops in alternating strips within a field. The alternation of crops in strips within the same field helps prevent the spread of insect pests and can be used to limit soil erosion if one crop remains in the field while the crop in the alternate row is removed.

strip mining See SURFACE MINING.

stripper well A low-output oil well that is only marginally profitable to operate; a well producing fewer than 10 barrels of oil per day.

stripping Method for the removal of unwanted dissolved gases from water. Stripping techniques involve increasing the surface area of the water to be stripped and maintaining the atmospheric PARTIAL PRESSURE of the gas(es) to be removed at a low level relative to the partial pressure of the gas dissolved in the water. Oxygen, ammonia, hydrogen sulfide, volatile organic compounds, and carbon dioxide are commonly stripped from water. See AIR STRIPPING, STEAM STRIPPING.

strontium-90 (^{90}Sr) A RADIOACTIVE isotope released by the detonation of nuclear weapons. With a HALF-LIFE of about 28 years, the isotope can BIOACCUMULATE in the bones, where strontium replaces calcium in the human body.

structure-activity relationship (SAR) The connection between the molecular structure of a chemical and its toxicity. Although SAR is only poorly predictive, an untested chemical similar in composition and spatial arrangement to a material known to be a human toxicant would be suspected of exhibiting similar toxic effects on the basis of a SAR analysis.

stuff and burn The practice of introducing waste to an incinerator at a rate greater than the waste combustion rate in the incinerator.

subbituminous A grade of coal that has a heat content higher than that of LIGNITE but lower than that of BITUMINOUS. See COAL.

subchronic exposure In toxicology, doses that extend to approximately $1/10$ the lifetime of an organism. For humans, 5–10 years; for rats, 2–4 months.

subcutaneous Beneath the skin; a route of exposure of chemical substances to test organisms.

subduction zone According to the theory of PLATE TECTONICS, a region at the boundary of two of the plates that form the CRUST of the Earth where one of the plates is forced downward into the mantle. The deep ocean trenches off the Pacific coasts of Mexico and South America are examples of these regions. The zones are also referred to as convergent plate boundaries.

sublimation The conversion of a substance in the solid phase directly to the gas phase without an intervening liquid phase, as in the changing of snow directly into water vapor without melting.

submerged plant Aquatic vegetation that has roots, stems, and leaves. The plant is rooted in the bottom of a watercourse, but the leaves remain submerged below the surface of the water. Sea grasses are examples of this type of vegetation.

submicron Synonym for SUBMICROMETER.

submicrometer A distance less than one micrometer, or less than one-millionth of a meter. Often used to describe the diameter of very small particles.

subsidence The sinking of the land. The process can be natural, as in the sinking of the land relative to sea level in coastal areas, or artificial, as in the collapse of the ground as a result of the removal of water, coal, or mineral deposits from the underlying strata.

subsidence inversion A TEMPERATURE INVERSION. The sinking air in the center of a high-pressure weather system compresses the air below and raises the temperature of this upper-level air above that of the air below. Inversion conditions suppress vertical motion of air contaminants and thus allow their air concentrations to increase. The high-pressure systems associated with subsidence inversions can remain in an area for several days or more. See RADIATION INVERSION.

subsidiarity The principle directing that decisions and actions should be made and undertaken at the lowest level of governance that is feasible and effective. In environmental management, following the admonition, Think Globally, Act Locally.

subsoil A SOIL HORIZON located between the topsoil and bedrock. The subsoil generally lacks organic content and consists of materials reflecting the weathering of bedrock and the washing of material from the overlying horizons.

substituted ring compound A chemical compound consisting of one or more closed rings of carbon atoms and their attached hydrogen atoms but with one or more of the hydrogen atoms replaced by another chemical substance. For example, polychlorinated biphenyl is composed of two benzene rings with a varying number of hydrogens replaced by chlorine atoms. See RING COMPOUND.

substrate 1. The solid surface on which an organism moves about or attaches, such as the sediments, rock, or sand on the bottom of a water body. 2. In biochemical reactions, the chemical material on which ENZYMES act. Enzyme action is typically specific to a particular substrate.

The combination of enzyme and substrate is called the enzyme-substrate complex.

Subtitle C   The section of the federal RESOURCE CONSERVATION AND RECOVERY ACT that applies to HAZARDOUS WASTE management.

Subtitle D   The section of the federal RESOURCE CONSERVATION AND RECOVERY ACT in which standards for the construction and operation of MUNICIPAL SOLID WASTE disposal facilities are outlined. The standards apply primarily to the management of nonhazardous waste.

succession   The orderly and predictable changes in the biological communities within a specific locality over time from an initial or PIONEER COMMUNITY to a final stage, or CLIMAX. The climax community is stable and self-sustaining, with little change in SPECIES COMPOSITION in the absence of large-scale perturbations such as extreme weather events, catastrophic fire, or extensive human activity. Also referred to as ecological succession.

suction lysimeter   A sampling device for the collection of groundwater from the UNSATURATED ZONE; a sample is drawn by applying a negative pressure to a porous ceramic cup embedded in the soil layer.

sulfate   In the atmosphere, a small-diameter AEROSOL composed of oxidized SULFUR DIOXIDE in the form of sulfate salt (e.g., ammonium sulfate) or sulfuric acid. Sulfates are an inhalation hazard and an important contributor to ACID DEPOSITION. In the soil, inorganic molecules (SO_4^{3-}) serve as plant nutrients when present in appropriate concentrations. In the salt marsh, precursors for the microbial production of HYDROGEN SULFIDE.

sulfide ore   A natural metal deposit chemically combined with sulfur. Copper, zinc, mercury, and lead are commonly found in nature as sulfides. The removal of sulfur from the ore in smelters, called roasting, is an important (localized) source of atmospheric SULFUR DIOXIDE.

sulfur cycle   A series of integrated biological, geological, and chemical reactions (termed BIOGEOCHEMICAL CYCLING) involved in the conversion of sulfur from an inorganic state in soil, water, air, and rocks to a soluble form; to organic BIOMASS; and back to the inorganic state once again. Some aspects of the cycling of sulfur are (1) the ANAEROBIC DECOMPOSITION of biomass, producing the many REDUCED SULFUR COMPOUNDS implicated in the odors associated with the decomposition process; (2) the anaerobic reduction of sulfate in seawater sediments; (3) the oxidation of iron sulfide compounds in abandoned mines, leading to the development of extremely acidic water that drains from the mine (ACID MINE DRAINAGE); and (4) the production of SULFUR DIOXIDE when hydrogen sulfide or high-sulfur coal is burned.

sulfur dioxide (SO_2)   A compound of sulfur containing one sulfur atom and two oxygen atoms; a major air pollutant produced by the burning of sulfur-containing compounds. See CRITERIA POLLUTANTS, ACID RAIN, ACID AEROSOL, TRADABLE EMISSION ALLOWANCE, SULFUR OXIDE, SULFURIC ACID.

sulfur oxides (SO_x)   A mixture of compounds of sulfur containing an indeterminate number of oxygen atoms. When sulfur-containing compounds are burned, a variety of compounds containing sulfur and oxygen can be produced (SO_2, SO_3). When the exact composition is not known, the notation SO_x is used to signify the unknown mixture. Sulfur oxides are a major category of air pollution.

sulfuric acid   A strong, highly corrosive acid. The acid is a colorless, oily liquid that is a strong dehydrating and oxidizing agent. Sulfuric acid is a widely used industrial chemical and is formed in the atmosphere as a result of reactions produced by the emission of SULFUR DIOXIDE. See ACID RAIN.

sump   A tank or drainage pit that collects water or other fluids for eventual removal.

sunscreen Lotion containing chemical agents that absorb ULTRAVIOLET LIGHT. These preparations can protect individuals from the damaging effects of sunlight, including sunburn and skin cancer. Sunscreens are rated with a sun protection factor (SPF) ranging from about 4 to 50 or better. The numerical value relates to the time that an individual can stay in the Sun without burning. If a person can safely stay in the Sun for one hour, a sunscreen with an SPF of 5 extends the time to 5 × 1 hour or 5 hours. The SPF time to sunburn applies to ULTRAVIOLET RADIATION-B RANGE (UV-B) because UV-B is the primary cause of sunburn and skin cancer risk. Some sunscreens also protect against ULTRAVIOLET RADIATION-A RANGE (UV-A), which promotes tanning but also contributes to skin aging. UV-A protection does not have an approved, standard protection rating yet. In order to be effective, the lotion must be applied in the amount and at the intervals suggested on the instructions.

sunset provision The stipulation that a regulation or an administrative agency automatically loses effect or ceases operations on a specific date unless positive measures are taken to justify continuance.

sunspot A relatively cooler area on the Sun's surface formed by strong magnetic fields. Sunspot activity oscillates from a maximum to a minimum on a cycle that average 11 years. The changes in sunspots have been linked to changes in the weather and climate of the Earth.

sunspot minima Relatively long periods of very low SUNSPOT activity; observed during 1100–1250, 1460–1550, and 1645–1715 A.D. The 70-year period starting in 1645 is called the Maunder minimum.

superchlorination The addition of CHLORINE to water at levels that oxidize DISSOLVED ORGANIC CARBON compounds that may otherwise cause taste or odor problems. This chlorine treatment is beyond that necessary to kill PATHOGENS.

Superfund See HAZARDOUS SUBSTANCES SUPERFUND; COMPREHENSIVE ENVIRONMENTAL RESPONSE, COMPENSATION, AND LIABILITY ACT.

Superfund Amendments and Reauthorization Act (SARA) The 1986 amendments to the COMPREHENSIVE ENVIRONMENTAL RESPONSE, COMPENSATION, AND LIABILITY ACT (the Superfund law). SARA included provisions that increased the size of the HAZARDOUS SUBSTANCES SUPERFUND, required new cleanup standards, and started the SUPERFUND INNOVATIVE TECHNOLOGY EVALUATION (SITE) program. TITLE III of SARA, the Emergency Planning/Community Right-to-Know Act of 1986, requires that for certain chemicals, facilities make public annually the amounts they routinely (or accidentally) release into the air, water, or ground. Title III also requires, for certain chemicals, that facilities make public annually the amounts stored and their locations within the facility. The chemical storage provision was accompanied by the creation of STATE EMERGENCY RESPONSE COMMISSIONS and LOCAL EMERGENCY PLANNING COMMITTEES, which compile this information and plan for public responses to the possible accidental release of the stored chemicals.

Superfund Comprehensive Accomplishments Plan (SCAP) Part of the COMPREHENSIVE ENVIRONMENTAL RESPONSE, COMPENSATION, AND LIABILITY INFORMATION SYSTEM (CERCLIS), a listing of the status of SUPERFUND SITES that are included and those that are not included on the NATIONAL PRIORITIES LIST, including completed and planned removal/remediation actions. Website: www.epa.gov/superfund/sites/

Superfund Innovative Technology Evaluation program (SITE) A cooperative arrangement between private companies and the USEPA to demonstrate, evaluate, and encourage the commercial development of improved technologies for the permanent cleanup of SUPERFUND SITES.

Superfund site A HAZARDOUS WASTE landfill or other area that represents a threat to public health or the environment because of the presence of HAZARDOUS SUBSTANCES. Locations are placed on the NATIONAL PRIORITIES LIST and are cleaned up by the responsible parties or by use of proceeds from the HAZARDOUS SUBSTANCE SUPERFUND. Cleanups follow the NATIONAL CONTINGENCY PLAN. See COMPREHENSIVE ENVIRONMENTAL RESPONSE, COMPENSATION, AND LIABILITY ACT; PRELIMINARY ASSESSMENT AND SITE INSPECTION; LISTING SITE INSPECTION; HAZARD RANKING SYSTEM.

supernatant The clear fluid that is removed from the top of tanks or ponds used to allow solids to settle from suspension. The liquid is also referred to as overflow.

supersonic Faster than sound. See SHOCK WAVE, SONIC BOOM.

supplemental environmental project (SEP) An activity that is directly beneficial to the environment performed by a company that has been convicted of violating pollution standards. The cost of the project is considered partial fulfillment of the company's required penalty payment.

supplied-air respirator A respirator delivering clean air to the wearer from an external source through a hose. In contrast, an AIR-PURIFYING RESPIRATOR cleans the air by filtering or chemical sorption as the wearer inhales.

Support Center for Regulatory Air Models (SCRAM) The USEPA website that supplies technical documentation and executable computer code for the AIR QUALITY DISPERSION MODELS approved for use in air pollution regulatory programs (called GUIDELINE MODELS). Website: www.epa.gov/ttn/scram/

surface area scaling factor In cancer risk assessments, a factor used to convert animal dose rates to equivalent human dose rates on the basis of their body surface areas. This method relies on the proportionality between body surface area and BASAL METABOLISM. The dose rate units are typically milligrams of a chemical per square meter of surface area per day. Of the several adjustment methods available, surface area scaling is the method most frequently used by the USEPA to calculate cancer risks from laboratory data. See SCALING FACTOR.

surface casing The well pipe inserted as a lining nearest to the surface of the ground to protect the well from near-surface sources of contamination.

surface collecting agents Chemical additives spread on oil spills in an aquatic environment to control the thickness of the oil layer.

surface compaction Increasing soil density by applying force at the surface; the process is used in the installation of a clay LINER. In contrast, solid waste typically undergoes COMPACTION into bales or compaction in collection vehicles before it is buried.

surface impoundment See IMPOUNDMENT.

surface mining The process of removing mineral deposits that are found close enough to the surface so that the construction of tunnels is not necessary. The soil and strata that cover the deposit are removed to gain access to the mineral deposit. The primary environmental concerns related to this technique are the disposition of spoils removed to gain access to the deposit and the need to restore the landscape that remains after the complete removal of the deposit. Water pollution is also a concern because runoff from the mining area is frequently rich in SEDIMENT and minerals. Also called strip mining. Compare UNDERGROUND MINING. See SURFACE MINING CONTROL AND RECLAMATION ACT.

Surface Mining Control and Reclamation Act (SMCRA) A federal law passed in 1977 requiring all surface coal mining operators to meet detailed performance standards, including the restoration of the surface-mined land to its original condition. The act also imposed a fee on each ton of coal removed, to be used for the reclamation of land subject to unregulated or poorly regulated strip mining in the past.

surface roughness See ROUGHNESS.

surface tension The force that causes the surface of a liquid to contract, causing it to behave as though it were a stretched membrane or elastic skin. The behavior is the result of the intermolecular forces between the molecules constituting the liquid. Each liquid molecule (for example, liquid water) is pulled in all directions by the forces exerted by surrounding molecules. The molecules at the air-liquid interface are attracted only by those molecules below and on the sides, as a result, the molecules are pulled back into the body of the liquid. For water, the result is that needles and some insects sit on top of the surface, and water beads on waxed surfaces. Detergents reduce surface tension, allowing water to wet surfaces by decreasing the attraction between water molecules.

surface water Water that occupies rivers, streams, lakes, reservoirs, and wetlands. Also, water that falls to the ground as rain or snow and does not evaporate or percolate into the ground.

surface water rule See SURFACE WATER TREATMENT RULE.

Surface Water Treatment Rule (SWTR) USEPA regulations issued under the provisions of the SAFE DRINKING WATER ACT that require public water systems that use surface water supplies or groundwater influenced by surface water to filter and disinfect the SOURCE WATER before distribution to the public. The basis of the regulation is to control the levels of GIARDIA, LEGIONELLA, and CRYPTOSPORIDIUM species and of viruses in the source water. The *Cryptosporidium* sp.

control regulation is also called the Enhanced Surface Water Treatment Rule.

surfactant An agent that is used to decrease the surface tension of water, useful for removing or dispersing oils or oily residues. Most DETERGENTS are surfactants; the term is derived from *surface active agent.*

surrogate standard An organic compound used in gas CHROMATOGRAPHY as a standard for accuracy and precision estimates. The compound is chosen to be of similar chemical makeup to the actual organic compounds being measured and therefore behaves similarly in the analytical procedure. A known amount of the surrogate is introduced along with the sample containing other organic compounds as an internal control of methods and instruments.

Surveillance, Epidemiology, and End Results (SEER) A summary of information from each CANCER REGISTRY in the United States. The registries collect data from 10% of the U.S. population and include a cross section of urban, rural, and regional populations. A program of the NATIONAL CANCER INSTITUTE, the SEER data are considered to be national average cancer incidence rates. Local and state cancer incidence rates are often compared with SEER rates. Website: www.seer.ims.nci.nih.gov/

survival curve A graph obtained by plotting the fraction of organisms surviving an increasing dose of some dangerous agent such as radiation or toxic chemical.

survivorship curve A graph of the number of individuals born in the same year who are alive at the beginning of a series of succeeding periods. For example, if in year one 100 individuals are born, at the beginning of year two perhaps 90 are living, at the start of year three 87 survive, and so on.

suspect material Roofing, flooring, siding shingles, ceiling tiles, insulation, and other building materials that may contain ASBESTOS.

suspended particulate matter 1. In water quality management, a sample drawn from natural water or from a wastewater stream consists of a mixture of both dissolved and suspended matter. Those solid materials that are retained on a filter prescribed by the specific technique being followed are referred to as particulate matter. The suspended particulate matter can by subdivided into two fractions: volatile and fixed. The volatile particulates are those that are lost when the filter is heated to about 550° C, and the fixed particulates are those that are not lost upon heating to 550° C. The volatile substances are generally considered to be of biological origin, and the fixed solids are considered to be minerals. 2. In air quality management, solid particles or liquid droplets suspended in air or carried by a stream of air or other gases through a duct, for example. See PARTICULATE MATTER; PARTICULATE MATTER, 10-MICROMETER DIAMETER.

suspended solid Same as SUSPENDED PARTICULATE MATTER.

suspension 1. The dispersion of small particles of a solid or liquid in a gaseous or aqueous medium. 2. In pesticide regulation, the immediate ban of a pesticide deemed by the USEPA to pose an IMMINENT HAZARD to human health or the environment. If a high probability of serious harm cannot be demonstrated, the agency may begin CANCELLATION proceedings.

sustainable agriculture Practices employed in the production of food and fiber that take a long-term view of the requirements for maintaining the productivity of the land; CONSERVATION and STEWARDSHIP of the land. Some of the aspects are maintenance and regeneration of the soil, water, and essential biological resources through the prevention of erosion, nutrient deprivation, and salt accumulation. Also included are the production of food and

fiber in sufficient quantity and quality to provide for the needs of the public and for the financial stability of the farmer.

sustainable development Describes efforts to guide economic growth, especially in less developed countries, in an environmentally sound manner, with emphasis on natural resource CONSERVATION. Also described as development that proceeds in a way that extends the lifetime of natural resources as long as possible.

sustained nuclear reaction A continuous series of FISSION events. Nuclear fission involves the splitting of the nucleus of an atom such as URANIUM 235. As a result of the fission process, heat energy, radiation, large FISSION PRODUCTS (which are actually atoms of other elements produced by the fragmentation of the uranium atom), and NEUTRONs are released. The process continues as individual uranium atoms undergo sequential fission reactions if certain conditions are met: the energy level of neutrons that are released must be lowered or moderated, and some of the moderated neutrons must hit another atom of uranium 235. When these two conditions are satisfied, the chain reaction perpetuates itself. See MODERATOR.

sustained-yield harvesting The removing of RENEWABLE RESOURCES, such as trees, at a rate that allows sufficient regrowth to maintain a continuous supply for cutting in the future.

swamp A tract of land that is saturated with water and that is covered intermittently with standing water. The area is usually overgrown with thick vegetation dominated by shrubs and trees.

swampbuster provision A WETLAND conservation measure in federal farming legislation excluding farmers from participation in USDA farm support programs if they plant on land meeting the definition of a wetland.

sweetening The removal of odorous sulfur contaminants from petroleum products.

SW-846 The common term for the USEPA publication *Test Methods for Evaluating Solid Waste, Physical/Chemical Methods,* which contains sampling and analytical protocols for testing waste regulated under the RESOURCE CONSERVATION AND RECOVERY ACT. The publication is available in paperback, CD-ROM, and on-line at www.epa.gov/epaoswer/hazwaste/test/main.htm. See CHARACTERISTIC HAZARDOUS WASTE, TOXICITY CHARACTERISTIC LEACHING PROCEDURE, IGNITABILITY, CORROSIVITY, REACTIVITY.

swill A thick liquid waste material consisting of food scraps and water.

symbiosis An association between two different organisms so that both profit from the relationship. See MUTUALISM.

symbiotic Describing a relationship between either bacteria and animals or plants, an animal and plant, animals of different species, or plants of different species such that both members of the pair benefit from the association. The association is so strong that neither member can exist or carry out certain activities alone. For example, the presence of bacteria of the genus *Rhizobium* in root structures of legumes (plants that produce seeds in a pod) results in the fixation of atmospheric nitrogen for use as a nutrient. Neither the bacterium nor the plant can carry out the function alone.

sympatric Describing two or more species occupying the same geographical area.

synapse The point at which a nerve impulse travels from one nerve cell to another. This transfer requires the release and activity of a variety of chemicals. Some toxic agents, such as the ORGANOPHOSPHATE and CARBAMATE insecticides, affect organisms by suppressing the functioning of these chemicals. See CHOLINESTERASE INHIBITORS.

syncrude See SYNFUEL.

synecology The study of the interactions among different POPULATIONS within an ecosystem. Compare AUTECOLOGY.

synergistic effect An effect that occurs when two or more agents act in such a way that the total effect is greater than the predicted sum of the individual agents acting alone. For example, the two air pollutants SULFUR DIOXIDE and PARTICULATE MATTER have a greater adverse effect on human health than would be expected from the sum of their individual toxicities.

synfuel A gas or liquid hydrocarbon fuel produced from coal or shale. The use of coal to produce these alternative fuels presents a variety of environmental problems, ranging from damage done at the mine to emissions of SULFUR OXIDE, POLYCYCLIC AROMATIC HYDROCARBONS, and HEAVY METALS. The problems related to the use of shale center around the high demand for water during processing and the substantial amount of solid waste that results.

synoptic scale Large-scale weather patterns, as used on a typical weather map, with horizontal units of several hundred to several thousand kilometers and time units of one day to one week. This scale of analysis combines atmospheric data from hundreds of weather stations to produce patterns such as low-pressure systems and fronts.

synthetic fuel Same as SYNFUEL.

synthetic minor Under the TITLE V PERMIT regulations of the 1990 amendments to the CLEAN AIR ACT, a source that has the POTENTIAL TO EMIT pollutants at a rate that would classify it as a major source but has actual emissions that put it in the minor source category. The actual emissions take into account enforceable limitations (control devices) imposed on the source. See TRUE MINOR.

synthetic natural gas (SNG) A gaseous hydrocarbon fuel produced by the processing of coal, primarily composed of methane. See SYNFUEL.

synthetic organic chemical manufacturing industry (SOCMI) About 400 major sources regulated as a group by certain air and water pollution control regulations (e.g., the HAZARDOUS ORGANIC NESHAP ([HON]) rules).

synthetic organic chemicals (SOCs) General term used to describe carbon-containing molecules made by human beings. Hydrocarbon feedstocks, crude oil, and natural gas are common raw materials that are converted into synthetic materials.

synthetic seawater An artificial product of the approximate ionic composition of water in the oceans.

systemic In toxicology, affecting the whole body or portions of the body other than the site of entry of a chemical substance. For example, a material may contact the skin and cause a localized irritant effect at the site of contact. If it also penetrates the skin and gains access to the blood, it can have a systemic effect.

systems audit A comprehensive examination of an environmental sampling and analysis project, including sampling techniques, calibration methods, and an appraisal of the QUALITY ASSURANCE plan.

systems ecology The use of mathematical analyses to study the operations, factors, and processes that influence the association of organisms and their surroundings.

T

tacking The binding of mulch fibers by mixing them with an adhesive chemical compound during land restoration projects.

taggant chips Microscopic multilayered and multicolored particles used for identification. The chips are manufactured with a unique sequence of colored layers that positively identifies a registered user. Fluorescent and magnetic layers are included to locate the chips, and the color sequence is read with a 100× microscope. If mixed with explosives or toxic wastes, the particles can be used to solve crimes that involve terrorism or illegal waste dumping. Also called microtaggants.

tagged molecule An atom of a radioactive element used within some molecule for the purpose of studying the behavior of that molecule. For example, carbon dioxide (CO_2) containing a radioactive isotope of carbon (^{14}C) might be used to study carbon dioxide fixation by plants through determining the inclusion of the radioactive form of carbon in the carbohydrates produced by the plant during photosynthesis. See CARBON 14.

taiga See BOREAL FOREST.

tail water The runoff from land that is being irrigated. The water may be high in fertilizer and pesticide concentrations and may contain an accumulation of inorganic salts from the washing of salts from the soil coupled with the evaporation of the water as it covers the field. See SALINIZATION.

tailings The remaining waste material after metal is extracted from ore.

tailpipe standards Allowable vehicle emission rates set by USEPA regulations under the provisions of the CLEAN AIR ACT. The standards are written as the mass of pollutant that may be emitted per mile traveled; for example, a light-duty vehicle may not exceed 3.4 grams of CARBON MONOXIDE per mile driven. See CATALYTIC CONVERTER, INSPECTION AND MAINTENANCE, LOW-EMISSION VEHICLE, ZERO EMISSION VEHICLE.

take Under the ENDANGERED SPECIES ACT, no person may "take" an ENDANGERED SPECIES, which is defined in the act as "to harass, harm, pursue, hunt, shoot, wound, kill, trap, capture, or collect, or to attempt to engage in such conduct," 16 USC 1532 (19). Focusing on the word *harm* in the *take* statutory definition, the courts have held that HABITAT modification, including activities on private land, may be stopped by the ESA if the activities will *take* a protected species. See INCIDENTAL TAKE for a partial exemption to the take prohibition.

taking In law, the Fifth Amendment to the U.S. Constitution prohibits the taking of private property for public use without just compensation. An issue in environmental law is whether some regulations prevent a landowner from using property to an extent that is the equivalent of a taking of the property, for example, for a public highway. Takings questions have arisen most often in WETLANDS or ENDANGERED SPECIES protection regulations that prevent land development or other use.

talc A soft mineral, usually light tan or white in color, composed of magnesium silicate. The main ingredient in talcum powder. Chronic exposure to this agent of workers in the rubber and cosmetics industries results in pulmonary FIBROSIS.

tar balls Nonvolatile hydrocarbon clumps remaining in water after the VOLATILE fractions have evaporated from CRUDE OIL that has been discharged or spilled into the marine environment. When washed ashore, these residues, which range from marble size to beachball size, spoil beaches.

tar sands Sandy deposits containing BITUMEN, a viscous petroleum-like material that has a high sulfur content. Bitumen can be thermally removed after surface mining of the sands and upgraded to a synthetic crude oil. Large tar sand deposits are in Alberta, Canada. The smaller amounts in the United States are almost all in Utah. See HEAVY OIL.

tare The determination of the empty weight of a container or vessel in order to allow for the future quantification of contents by weighing the full or partially full container and computing the difference.

target organ The body organ (e.g., the liver) or organ system (respiratory, nervous) that is most likely to be adversely affected by overexposure to a chemical or physical agent.

target organism An undesirable or harmful insect, fungus, rodent, or other pest that a PESTICIDE is designed to kill.

target theory A concept used to explain the interaction of IONIZING RADIATION and biological specimens in which the ionization that is produced in the cell by the radiation damages a specific target or location within that cell. One or more ionizing events may be required to take place within the cell before a specific physiological condition is observed.

Taylor Grazing Act A 1934 federal statute that, with the FEDERAL LAND POLICY AND MANAGEMENT ACT, governs the management and preservation of federal public land, excluding national forests and national parks. Establishes grazing districts and issues grazing permits.

technical adviser See TECHNICAL ASSISTANCE GRANT.

technical assistance grant (TAG) Under the provisions of the COMPREHENSIVE ENVIRONMENTAL, RESPONSE, COMPENSATION, AND LIABILITY ACT, funds provided to communities directly affected by the cleanup of a hazardous waste site on the NATIONAL PRIORITIES LIST (Superfund site). The grant is awarded to a community group to hire a technical adviser to interpret scientific and engineering data developed as part of the cleanup. The goal is for the community to understand the site's risk to human health and the environment sufficiently to participate in the decisions concerning the goals and methods of the cleanup work.

technical-grade active ingredient (TGA) The pure form of a chemical used as a PESTICIDE that is delivered to a processor for formulation into a commercial product.

technological optimist One who believes that technology and human enterprise will overcome the adverse effects of industrialism and its accompanying pollution problems. See CORNUCOPIAN.

technologically enhanced naturally radioactive material (TENR material) A naturally occurring source of RADIOACTIVE elements to which humans are exposed as a result of some human activity, such as underground mining, ore processing (e.g., PHOSPHOGYPSUM PILES), or drilling of wells. Also called technologically enhanced naturally occurring radioactive material. See NATURALLY OCCURRING RADIOACTIVE MATERIAL, PRODUCED WATER.

technology based Describing emission or effluent limitations that are not defined in terms of allowable releases that achieve a desirably low ambient pollutant concentration (AMBIENT STANDARD) but instead are based on the pollutant control efficiency that is achievable using current technology, for example, the BEST AVAILABLE CONTROL TECHNOLOGY standard for air emissions and the BEST AVAILABLE TECHNOLOGY ECONOMICALLY ACHIEVABLE for water discharges of toxic chemicals.

technology-based effluent limitation (TBEL) In water quality management under the CLEAN WATER ACT, the TECHNOLOGY-BASED standards applicable to various types of pollutants and pollutant sources. See BEST AVAILABLE TECHNOLOGY ECONOMICALLY ACHIEVABLE, BEST CONVENTIONAL CONTROL TECHNOLOGY, BEST DEMONSTRATED AVAILABLE TECHNOLOGY.

technology forcing Describing standards or levels of control called for in environmental statutes or regulations for which existing technologies are inadequate and therefore require technical advancements to achieve.

technopolis 1. Name applied to a model of futuristic city development proposing the vertical city. Under this conception, the growth in a city would be vertical rather than outward in a horizontal manner. 2. An urban area attractive to and the home of high-technology businesses. 3. An urban area with a high-technology telecommunications and transportation infrastructure.

temperate deciduous forest A geographic region characterized by distinct seasons, moderate temperatures, and rainfall from 30 to 60 inches per year. These forests are found in eastern North America; eastern Australia; western, central, and eastern Europe; and parts of China and Japan. Typical trees in the North American deciduous forest are oak, hickory, maple, ash, and beech.

temperature A measure of the average energy of the molecular motion in a body or substance at a certain point.

temperature inversion In the atmosphere, the condition in which air tempera-

Temperature inversion

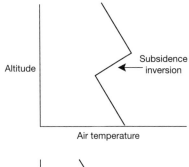

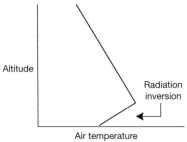

ture increases with increasing altitude over a certain altitude range. The inversion layer can be at ground level or aloft. The condition results in a layer of warmer air above cooler air, a circumstance that inhibits atmospheric mixing and dispersion of pollutants. The two types are RADIATION INVERSION and SUBSIDENCE INVERSION.

temporary hardness Water HARDNESS that can be reduced or removed by heating. Heating drives off carbon dioxide, shifting the CARBONATE BUFFER SYSTEM equilibrium so that CARBONATE ions combine with dissolved calcium or magnesium ions, form solids, and precipitate. This process lowers the calcium/magnesium ion water concentration, lowering the hardness. Also called carbonate hardness.

temporary threshold shift (TTS) The short-lived reduction in hearing ability for certain frequencies caused by noise overexposure. Exhibited by an increase in the HEARING THRESHOLD LEVEL. Recovery is usually complete within 16 to 24 hours. See PERMANENT THRESHOLD SHIFT.

ten percent rule In ecology, the maxim stating that about 10% of the energy available at one TROPHIC LEVEL is passed upward and stored in the bodies of organisms at the next-higher trophic level. The relatively small transfer of energy is due in part to RESPIRATION requirements of the energy CONSUMERS. Also, parts of ingested organisms are excreted, and not all available organisms are harvested. Approximately the same relationship can be demonstrated when the transfer of carbon from one trophic level to the next is examined. See ECOLOGICAL PYRAMID.

tephra Dust, ash, and pumice ejected by the eruption of a volcano and carried downwind before settling to Earth.

teratogen A chemical substance or physical agent that significantly increases the risk of malformation or adverse development of a fetus.

terminal settling velocity For a particle falling in a nonturbulent fluid (liquid or gas), the maximal possible velocity reached when the drag, or frictional resistance, on the particle equals the gravitational force on the particle. The measure is used in the design of chambers in which particles are removed from air or from water by gravitational settling. The horizontal flow rate through the chamber must allow time for the particles to reach the bottom of the settling chamber. See STOKES'S LAW.

terpenes Natural products consisting of 10 carbon atoms found in the oils extracted from many types of plants. This family of compounds provides plants and flowers with their fragrance, and some of the terpenes are used commercially in perfumes and flavoring agents. Some, such as camphor from the camphor tree, have been used as medicines since antiquity. Terpenes can contribute to the BIOGENIC VOLATILE ORGANIC COMPOUNDS in the atmosphere during certain times of the year.

terracing A series of levels on a hillside, one above another. Hillside farming on terraces greatly reduces water erosion of soil.

terrestrial radiation The INFRARED RADIATION emitted by the surface of the Earth and the atmosphere. Solar radiation is absorbed in the atmosphere, by surface waters, and by the ground; the energy is reradiated as heat (longer-wavelength infrared radiation) and, after atmospheric absorption and transfers, eventually lost to space. See WIEN'S LAW, GREENHOUSE EFFECT.

territoriality Animal behavior trait in which animals define an area around their nest, den, or home and then defend the area with gusto, most commonly against members of the same species.

tertiary treatment Method for the purification of wastewater beyond the conventional methods of PRIMARY TREATMENT (physical) and SECONDARY TREATMENT (biological). Facilities may apply one or more of the tertiary treatment methods, which remove inorganic nitrogen and phosphorus or dissolved organic compounds. Methods range from chemical techniques, such as the addition of lime to remove phosphates, to biological techniques, such as the application of wastewater to land to allow the growth of plants to remove plant nutrients. See ADVANCED WASTEWATER TREATMENT, AMMONIA STRIPPING, CARBON POLISHING, PHYTOTREATMENT.

tetrachlorodibenzofuran (TCDF or TCDBF) A polyhalogenated hydrocarbon found as a contaminant in commercial preparations of POLYCHLORINATED BIPHENYLS (PCBs). This contaminant may be responsible for some of the adverse effects of PCBs. The molecule is very similar in chemical structure to TETRACHLORODIBENZO-*PARA*-DIOXIN (TCDD) and is frequently referred to as DIBENZODIOXIN. The toxicity, potency, and biological effects are very similar to those of TCDD.

Tetrachlorodibenzo-*para*-dioxin

tetrachlorodibenzo-*para*-dioxin (TCDD) An AROMATIC HALOGENATED hydrocarbon that is one of the most feared chlorinated compounds. Dioxin is produced during the synthesis of precursors used in the manufacture of TRICHLOROPHENOXYACETIC ACID (2,4,5-T). The compound is toxic to liver and kidney function and has been shown to induce a variety of tumors in animal models. Adverse effects on the immune system of mammals have also been noted. CHLORACNE is the most common symptom resulting from human exposure. TCDD has been involved in a number of well-publicized environmental cases, the most famous of which is contamination of the herbicide mixture known as AGENT ORANGE used as a defoliant during the Vietnam War. Also referred to as dioxin and dibenzo-*para*-dioxin. See SEVESO, TIMES BEACH.

tetraethyl lead An ANTIKNOCK ADDITIVE in gasoline that was the single largest source of lead emissions into the atmosphere until the phaseout of the additive in the United States completed in 1989.

theoretical maximum residue contribution (TMRC) In pesticide regulation of a proposed new pesticide or a new use of an existing pesticide, an assumed maximal exposure used in the REGISTRATION review by the USEPA. The TMRC is used for pesticides that do not pose a cancer risk. The hypothetical maximal amount of a PESTICIDE RESIDUE received by a person is estimated by using the tolerance (maximal concentration of the pesticide residue on food) and a person's assumed daily intake of *all* foods that may contain the residue, all containing the maximal tolerance (concentration) allowed. The TMRC is then compared with the allowable daily intake of the chemical (REFERENCE DOSE). If the TMRC exceeds the reference dose, the USEPA denies a REGISTRATION for the new pesticide. See TOLERANCES, PESTICIDE; FEDERAL INSECTICIDE, FUNGICIDE, AND RODENTICIDE ACT.

theoretical oxygen demand (TOD) The amount of oxygen that is calculated to be required for complete DECOMPOSITION of a substance. The calculation is based on the empirical formula of specific substances.

theory of diminishing returns David Ricardo's economic theory that as factors of production (input) were added, the return (output) per additional input declined. As labor inputs rose, therefore, the return (output) declined per unit of labor added. Thomas Malthus extended Ricardo's theory to predict that as the human population rose, more labor would be added to food production, but, given the fixed amount of available land and the theory of diminishing returns, food production would not keep pace with the growing population. As a result, this would regulate population at a starvation level. The predictions of Malthus were based on extremely pessimistic assumptions. See MALTHUSIAN.

therapeutic index (TI) The ratio of the dose of a drug that causes adverse effects to the dose producing therapeutic effects. The higher the TI, the safer the drug.

therm A unit of heat equal to 100,000 BRITISH THERMAL UNITS (BTUS).

thermal inversion See TEMPERATURE INVERSION.

thermal NO_x NITROGEN OXIDE formed by the heating of combustion air to the point at which atmospheric nitrogen and atmospheric oxygen combine. Fuel NO_x is the primary source of the nitrogen oxide emissions from human activities. Compare FUEL NO$_x$.

thermal plume The hot water discharged from a power generation facility or other industrial plant. When the water at elevated temperature enters a receiving stream, the hot water is not immediately dispersed and mixed with the cooler waters of the lake or river. The warmer water moves as a single mass (PLUME) downstream from the discharge point until it cools and gradually mixes with that of the receiving stream. See THERMAL POLLUTION.

thermal pollution The addition of excessive waste heat to a water body, usually by the discharge of cooling water from an electric power plant. The shift to a warmer aquatic environment can cause a change in SPECIES COMPOSITION and lower the DISSOLVED OXYGEN content of the water.

thermal power plant A facility that generates electricity by using some energy source to convert water to steam. The steam is then used to turn turbine generators.

thermal stratification The formation or condition of well-defined water temperature zones with depth in a lake or pond. See HYPOLIMNION, EPILIMNION, FALL TURNOVER, SPRING TURNOVER.

thermal system insulation (TSI) Material applied to the exterior of pipes, ducts, tanks, or similar items to prevent heat loss or gain.

thermal treatment of hazardous waste Any treatment of HAZARDOUS WASTE that involves exposure of the material to elevated temperatures in an effort to change the characteristics of the waste, for example, INCINERATION, PYROLYSIS, plasma processes, and microwave discharge. See DESTRUCTION AND REMOVAL EFFICIENCY, FOUR NINES, SIX NINES, TRIAL BURN.

thermal turbulence Randomly fluctuating air motion caused by ground-level air being heated at the surface and rising past and through the upper air. The erratically moving eddies thus produced are typically larger than those produced by MECHANICAL TURBULENCE and are more effective in the dilution of air pollutants by mixing cleaner air with contaminated air. See EDDY DIFFUSION.

thermocline The sharp boundary between the EPILIMNION and the HYPOLIMNION in certain lakes and ponds in the temperate zone during the summer months. Characterized by a rapid change in water temperature over a short distance, from the warmer epilimnion to cooler hypolimnion. Little mixing occurs across the thermocline.

thermodynamics The study of the involvement of heat energy in chemical or physical reactions and the conversion of energy from one form to another. See FIRST LAW OF THERMODYNAMICS, SECOND LAW OF THERMODYNAMICS.

thermonuclear A nuclear reaction that requires extremely high temperature (10^7–10^8 K) as the activation energy to initiate the FUSION process.

thermoplastic The common plastics, such as POLYETHYLENES and POLYETHYLENE TEREPHTHALATE, composed of long, linear POLYMERS. Termed thermoplastic because they become soft when heated; consequently, they can be molded into various shapes while soft.

thermoset polymer The hard plastic formed from cross-linked molecules. Examples include acrylics and epoxys. These cannot be melted with heat.

thickener A settling pond or tank where the concentration of solids is increased by allowing settling and removal of clarified liquid. The solids that are pumped from the bottom of the pond or tank are much thicker than the incoming fluid.

Thiobacillus An aquatic or terrestrial genus of bacteria that is capable of oxidizing elemental sulfur, sulfide ions, thiosulfates, and other forms of inorganic sulfur to derive the energy needed in metabolism. Bacteria belonging to this genus fix carbon dioxide (are AUTOTROPHIC) and produce sulfuric acid as an end product. They can increase the acidity of soil or water to levels that result in the destruction of natural environments. Bacteria belonging to this genus are termed CHEMOAUTOTROPHS.

third third   See LAND DISPOSAL BAN.

Third World   Less developed countries that have not created advanced industrial economies.

Thoreau, Henry David (1817–62) American Writer who argued the virtues of a simple life. He is best known for *Walden; Or Life in the Woods* (1854), in which he described his time living in a small cabin near Concord, Massachusetts. Also having environmental themes are *A Week on the Concord and Merrimac Rivers* (1849) and *The Maine Woods* (1864, after his death).

thorium   A naturally occurring element with an atomic number of 90 and an atomic weight of about 232. The element can be converted to uranium-233, a fissionable isotope of uranium, by exposure to neutron irradiation.

threatened species   A specific plant or animal species whose population level in some sections of the natural range is very low and that is likely to become an ENDANGERED SPECIES without intervention.

Three Mile Island   Location of the most publicized accident associated with the operation of a nuclear reactor in the United States. The island is in the Susquehanna River near Harrisburg, Pennsylvania. Beginning at 4:00 A.M. on March 28, 1979, the Three Mile Island Unit 2 power plant suffered a partial meltdown resulting from the loss of cooling water caused by a combination of equipment malfunc-

tion and human error. The accident caused significant damage to the reactor and contamination by radioactive materials within the containment structure; however, little, if any, damage was done in the surrounding community. In the wake of the potential disaster, President Carter appointed the Kemeny Commission to recommend improvements in reactor safeguards, which included improved operator training, additional emergency planning, and renewed emphasis on PROBABILISTIC RISK ASSESSMENT to identify potential failures. See LOSS-OF-COOLANT ACCIDENT.

three-way catalyst   A device that controls all three of the top automobile emissions. An oxidizing catalyst converts hydrocarbons (unburned fuel) to carbon dioxide and water and oxidizes CARBON MONOXIDE to carbon dioxide; a reducing catalyst converts NITROGEN OXIDES to atmospheric nitrogen.

threshold   The lowest dose of a chemical or other agent that results in an observable effect, either adverse or beneficial. Dose levels lower than the threshold do not have a discernible effect.

threshold dose   The smallest physical or chemical exposure that results in an observable adverse biological effect. Compare NONTHRESHOLD POLLUTANT.

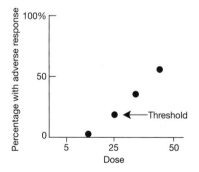

Threshold dose

threshold effect   The adverse effect caused by the THRESHOLD DOSE for a particular physical or chemical agent and

route of exposure; the dose required to cause the threshold effect varies by individual. See END POINT.

threshold hypothesis The concept that holds that no damage is done to a cell or organism when that unit is exposed to chemical agents or radiation below certain concentrations or intensities. Compare NONTHRESHOLD POLLUTANT.

threshold level See THRESHOLD DOSE.

threshold limit value (TLV) The airborne concentration of a gas or particle to which most workers can be exposed on a daily basis for a working lifetime without adverse effects. The TLVs are set by the AMERICAN CONFERENCE OF GOVERNMENTAL INDUSTRIAL HYGIENISTS. See PERMISSIBLE EXPOSURE LIMIT.

threshold planning quantity (TPQ) For a chemical designated as an EXTREMELY HAZARDOUS SUBSTANCE by TITLE III of the SUPERFUND AMENDMENTS AND REAUTHORIZATION ACT, the inventory amount at a facility that necessitates that the material be reported by the facility and be included in the emergency response plan for chemical spills or releases. TPQs are found at Title 40, Part 355, of the *CODE OF FEDERAL REGULATIONS* (40 CFR 355).

threshold shift A change in the ability of an individual to hear minimal sounds at a given frequency. The change can be either permanent or temporary. See TEMPORARY THRESHOLD SHIFT, PERMANENT THRESHOLD SHIFT.

Ti plasmid From *tumor-inducing plasmid*. A small, extrachromosomal piece of deoxyribonucleic acid (DNA) characteristically found in the bacterium *Agrobacterium tumefaciens*. This PLASMID induces crown gall tumor (a knotty growth) in plants that are infected with it. The T-DNA segment of this plasmid can be used as a carrier to introduce foreign genes into plants.

tidal energy The mechanical energy associated with the rising and falling of water level during the movement of the tides. See TIDAL POWER.

tidal marsh Coastal property that is alternately flooded during high tide and exposed at low tide. This margin around the continents contains plant and animal communities that are like neither those in the sea nor the uplands away from the coast. These areas are unusually productive in terms of BIOMASS accumulation (PRODUCTIVITY) and serve as nursery areas for many marine species constituting fishery resources. The marsh acts as a buffer area protecting the uplands from storm damage because it absorbs a significant part of the energy of a storm coming in from the sea. They also protect the oceanic system because they absorb many pollutants washed down from the uplands and thereby prevent the pollutants from entering the ocean.

tidal power The conversion of the mechanical energy associated with the rising and falling of water level caused by the tides into electric energy. Rising and falling waters are forced into large turbines, which then turn and activate generators.

tidal station 1. A recording location for measuring the fluctuations in tidal height over time. 2. A facility designed to generate electricity by capturing some of the energy available as a result of the ebb and flood of the tides. See TIDAL POWER.

tidal volume 1. The volume of air in one normal breath, about 0.5 liter in adult humans. 2. The volume of water entering and leaving a bay or salt marsh as the water level fluctuates as a result of the tides.

Tier I, Tier II reports As required by the EMERGENCY PLANNING AND COMMUNITY RIGHT-TO-KNOW ACT, chemical inventory forms that must be submitted to the STATE EMERGENCY RESPONSE COMMISSION (SERC), the LOCAL EMERGENCY PLANNING COMMITTEE (LEPC), and the local fire department. Tier I reports are annual public disclosures of the estimated ranges and the maximal amounts of certain hazardous chemicals stored at a facility, the average daily amounts, and the general locations of the stored materials. Tier II forms give more detailed information on the same chemicals covered by Tier I, such as the storage conditions and specific location within the facility. Tier II reports may be submitted in place of Tier I forms voluntarily, and they must be submitted on request by the SERC, LEPC, or local fire department. See TITLE III, RISK MANAGEMENT PLAN.

tiering The preparation of an ENVIRONMENTAL IMPACT STATEMENT (EIS) for a broad action by a federal agency (PROGRAM IMPACT STATEMENT) and the subsequent preparation of short, detailed statements for site-specific projects.

tilth 1. The general physical condition of soil in reference to agricultural use. 2. Land used for agriculture, as opposed to pasture or forest.

timberline See TREE LINE.

time constant The time required for a physical quantity to rise from 0 to 1 − $1/e$, or 63.2 percent of the final steady-state value, when the quantity varies with time (t) as $1 - e^{-kt}$, where $e = 2.71828$. Also, the time required for a physical unit to fall to $1/e$ (36.8%) of the initial value when the physical unit varies with time as e^{-kt}.

time-critical removal action (TCRA) Under the provisions of the NATIONAL CONTINGENCY PLAN, a chemical release that does not require an emergency (immediate) response; response must begin within six months. See REMOVAL ACTION, NON-TIME-CRITICAL REMOVAL ACTION (NTCRA).

Times Beach The city of 2800 in Missouri that, in 1972–73, hired a waste contractor to spray oil on its unpaved roads as a dust control measure. Ten years later, in 1982, the waste oil was discovered to be contaminated with the toxic chemical DIOXIN. In late 1982, after the USEPA had begun sampling for dioxin along the roadways, the nearby Meramec River flooded the city, spreading the dioxin-contaminated soil throughout business and residential areas. The CENTERS FOR DISEASE CONTROL issued a health advisory for Times Beach residents to relocate permanently. Federal SUPERFUND money was allocated to purchase property from all the residents. After a massive cleanup effort, including the incineration of over 250,000 tons of contaminated soil, what was formerly the town of Times Beach is now a 500-acre state park.

time-weighted average (TWA) A method used to calculate the daily exposure of a worker to an airborne chemical. The duration of exposure to various airborne concentrations, usually during an eight-hour day, is summed, then averaged. Expressed as

$$TWA = \frac{\sum_{i=1}^{N} C_i T_i}{N}$$

where C is the air concentration measured during period i, T_i is the length of period i, and N is the number of periods. For example, the TWA for an exposure to a chemical level of 20 parts per million (ppm) for 2 hours and 100 ppm for 6 hours is 80 ppm (2 × 20 + 6 × 100, divided by 8).

tipping At a solid waste disposal facility, the dumping of the contents of a waste truck, often by hydraulic lifting of one end of the load. See TIPPING FEE.

tipping fee The per truck or per ton monetary charge to dispose of solid waste at a SANITARY LANDFILL.

tire processor Business or facility engaged in the processing of used tires for disposal or recycling. For example, the tires may be shredded into small pieces.

Title III The section of the SUPERFUND AMENDMENTS AND REAUTHORIZATION ACT that required nationwide chemical emergency planning and annual public reporting of industrial releases of hazardous substances. Also known as the Emergency Planning and Community Right-to-Know Act. (1) The emergency planning is for chemical accidents but also includes the public release of information about chemicals used and stored in a community. Emergency planning began with the private COMMUNITY AWARENESS AND EMERGENCY RESPONSE program established by the chemical industry and continued with the chemical emergency planning under Title III. Similar analyses were repeated by the RISK MANAGEMENT PLANS required by the CLEAN AIR ACT. (2) Under the right-to-know provisions, Title III requires facilities to make annual public reports of the amounts of routine or accidental releases of any of over 600 toxic chemicals and chemical categories to the air, water, or land. See EXTREMELY HAZARDOUS SUBSTANCE; HAZARDS ANALYSIS; CAMEO®; TIER I, TIER II REPORTS; THRESHOLD PLANNING QUANTITY; TOXICS RELEASE INVENTORY.

Title V permit Under the 1990 amendments to the CLEAN AIR ACT, an integrated PERMIT program for a facility's emissions of air pollutants, similar to the NATIONAL POLLUTANT DISCHARGE ELIMINATION SYSTEM (NPDES) of the CLEAN WATER ACT. The permit specifies emission limits, reporting requirements, an annual compliance certification, public notice rules, and, as applicable, CONTINUOUS EMISSION MONITORING (CEM).

tolerance limits See LIMITS OF TOLERANCE.

tolerances, pesticide An extensive listing (prepared by the USEPA and published in Title 40, Parts 180, 185, and 186, of the CODE OF FEDERAL REGULATIONS) of the amounts of PESTICIDE RESIDUES that are allowable in or on agricultural products as a result of pesticide application prior to harvest or slaughter. The amounts are expressed in parts by weight of each specific chemical per 1 million parts by weight of the commodity (ppm). See FED-

ERAL INSECTICIDE, FUNGICIDE, AND RODENTICIDE ACT; FOOD QUALITY PROTECTION ACT; THEORETICAL MAXIMUM RESIDUE CONTRIBUTION.

tons per day (TPD) A mass per time unit often used to describe MUNICIPAL SOLID WASTE generation rates. When describing disposal at a municipal solid waste facility, the annual tonnage divided by 260 days (52 five-day weeks).

top-down Describing the emission control approach that, when setting BEST-AVAILABLE CONTROL TECHNOLOGY for an individual source PERMIT, begins with the most stringent control technology available for facilities of that industry type. If the particular source can demonstrate that, for economic or technical reasons, the most stringent technology is not appropriate, then the next most stringent technology is examined, and so on, until a technology cannot be eliminated. This approach puts the burden on the permit applicant to demonstrate that it cannot use the most stringent technology.

topsoil The topmost layer of soil, the A HORIZON. The composition of the material normally is a mixture of substances ranging from clay and silt particles to organic matter derived from the decomposition of plant and animal remains. The texture and color are distinct from those of the underlying layers. Without this layer of soil, the land is of little use in agriculture. See HUMUS.

total carbon (TC) A measure of the amount of carbon-containing compounds in water. The measure includes both organic and inorganic forms of carbon as well as compounds that are soluble and insoluble. The typical laboratory analysis involves the conversion of all forms of carbon to carbon dioxide and the subsequent measurement of the carbon dioxide produced. The TC value is an estimate of the potential damage that an effluent can cause in a receiving stream as a result of the removal of DISSOLVED OXYGEN from the water. The measurement of total carbon requires less sample, is quicker, and yields more reproducible results than the measurement of either CHEMICAL OXYGEN DEMAND or BIOCHEMICAL OXYGEN DEMAND. See TOTAL ORGANIC CARBON.

total coliforms All COLIFORM BACTERIA, regardless of their source. FECAL COLIFORMS are bacteria from the digestive tract of humans and warm-blooded animals. See TOTAL COLIFORM RULE.

Total Coliform Rule The USEPA regulatory program under provisions of the SAFE DRINKING WATER ACT that sets limits on TOTAL COLIFORMS in drinking water. Coliform microorganisms are present in the digestive tract of humans and some animals. If they are found in water, their presence indicates that the water treatment is inadequate or the distribution system is being contaminated, and follow-up tests for FECAL COLIFORMS are required. Health problems associated with coliform contamination include GASTROENTERITIS or diseases caused by other PATHOGENS possibly present in water contaminated with feces. See INDICATOR ORGANISMS, MAXIMUM CONTAMINANT LEVEL, MAXIMUM CONTAMINANT LEVEL GOAL.

total dissolved phosphorus That form of phosphorus-containing chemical that passes through a filter. Most commonly, this class consists of inorganic forms of phosphorus and represents that form of phosphorus that is available for uptake by plants and use as a plant nutrient. Contrast PARTICULATE PHOSPHATE. See PHOSPHATES.

total dissolved solids (TDS) A measure of the amount of material dissolved in water (mostly inorganic salts). The inorganic salts are measured by filtering a water sample to remove any SUSPENDED PARTICULATE MATTER, evaporating the water, and weighing the solids that remain. An important use of the measure involves the examination of drinking water. Water that has a high content of inorganic material frequently has taste problems and/or water HARDNESS prob-

lems. As an example, water that contains an excessive amount of dissolved salt (sodium chloride, a dissolved solid that is frequently encountered) is not suitable for drinking. See TOTAL SUSPENDED SOLIDS.

total fertility rate Within the human population, the number of children, on average, born to women during their lifetime, calculated by summing, for a particular year, the age-specific birth rates for women during their childbearing years. The total shows the results of current birth rates extended through women's childbearing years and is therefore a projection. A rate equal to 2 for a particular country is considered to be a replacement rate, assuming no infant deaths or immigration. Rates greater than 2 indicate that the population of the country is increasing, and a rate of less than 2 indicates that the population of a country in on the decline.

total growth rate The net change in the population of a country resulting from all positive and negative influences on the number of people in it: births and movements of people into the country versus (minus) deaths and movements of people out of the country. See IMMIGRATION, EMIGRATION.

total injury incident rate See ACCIDENT RATE.

total inorganic carbon (TIC) The total amount of inorganic salts of carbonates and bicarbonates present in water without regard to whether they are in suspended particulate form or dissolved. Water that contains an excessive amount of these salts is considered to be HARD WATER. The dissolved materials interfere with the functioning of soaps and detergents and can form adherent SCALE in boilers, pipes, and steam equipment.

total Kjeldahl nitrogen (TKN) See KJELDAHL NITROGEN.

total maximum daily load (TMDL) The maximal quantity of a particular water pollutant that can be discharged into a water body without violating a water quality standard. The amount of pollutant is set by the USEPA, under provisions of the CLEAN WATER ACT, when the agency determines that existing TECHNOL-OGY-BASED EFFLUENT LIMITATIONS on the water pollution sources in the area will not achieve one or more ambient water quality standards. (The polluted lake or stream in this case is termed an *impaired* water body.) The process results in the allocation of the TMDL to the various sources in the area; the amount from POINT SOURCES is the WASTELOAD ALLOCA-TION, and the amount from NONPOINT SOURCES is the LOAD ALLOCATION.

total organic carbon (TOC) A measure of the amount of organic materials suspended or dissolved in water. The measure is very similar to the assay of the TOTAL CARBON content; however, samples are acidified prior to analysis to remove the inorganic salts of CARBONATES and BICARBONATES. The assay of total organic carbon is an estimate of the potential damage that an effluent can cause in a receiving stream as a result of the removal of DISSOLVED OXYGEN from the water. The measurement of total organic carbon requires less sample, is quicker, and yields more reproducible results than the measurement of either CHEMICAL OXYGEN DEMAND or BIOCHEMICAL OXYGEN DEMAND. As a pollution indicator, this method is more reliable than the assay of TOTAL CARBON when the wastewater contains high amounts of TOTAL INORGANIC CARBON.

total ozone mapping spectrometer (TOMS) data Stratospheric ozone measurements by *Nimbus 7* satellite instruments for 1978–93; used in the analysis of trends in OZONE LAYER DEPLE-TION.

total petroleum hydrocarbons (TPH) A general term used to describe the amount of hydrocarbons (molecules containing primarily carbon and hydrogen as part of a relatively large molecular struc-

ture) derived from CRUDE OIL or hydrocarbons manufactured from CRUDE OIL FRAC-TIONS that can be extracted and quantified from contaminated soil or water. The absolute "total" is operationally difficult to determine because of the great array of material in crude oil and the limitations of extraction technology and instrumentation.

total reduced sulfur (TRS) The sum of the nonoxidized sulfur compounds emitted to the air from a facility. These compounds are responsible for some of the disagreeable odors associated with industrial sources. Sulfur emissions from a KRAFT PAPER mill using the KRAFT PROCESS might include HYDROGEN SULFIDE, methyl mercaptan, dimethyl sulfide, and dimethyl disulfide. TRS from a sulfur recovery process in an oil refinery might include hydrogen sulfide, carbonyl sulfide, and carbon disulfide.

total solids (TS) A measure of the amount of material that is either dissolved or suspended in a water sample, obtained by allowing a known volume to evaporate then weighing the remaining residue. Total solids equals the sum of the measurements of TOTAL DISSOLVED SOLIDS and TOTAL SUSPENDED SOLIDS.

total suspended particulate (TSP) The measured concentration of airborne PARTICULATE MATTER, often expressed as micrograms of particulate per cubic meter of sampled air.

total suspended solids (TSS) A measure of the amount of PARTICULATE MAT-TER that is suspended in a water sample. The measure is obtained by filtering a water sample of known volume. The particulate material retained on the filter is then dried and weighed. Compare TOTAL DISSOLVED SOLIDS.

total trihalomethanes (TTHMs) The sum of the concentrations of individual members of a family of halogenated derivatives of methane in drinking water. The concentrations of the following are employed to compute the sum in milligrams per liter: chloroform, dibromochloromethane, bromodichloromethane, and bromoform. This class of water pollutants are generated when water containing dissolved organic material, much from natural sources, is disinfected with chlorine. See TRIHALOMETHANES, DISINFECTION BY-PRODUCTS.

totally enclosed treatment facility A building containing pollution-control equipment that is directly connected to an industrial process. The arrangement is constituted to prevent the release of any hazardous material into the environment.

toxaphene A commercial CHLORINATED HYDROCARBON insecticide, prepared as a mixture of around 170 different compounds. Toxaphene was used extensively on cotton crops. The INTERNATIONAL AGENCY FOR RESEARCH ON CANCER (IARC) has rated toxaphene 2B (probable human carcinogen). In 1982, the USEPA canceled the REGISTRATION for toxaphene for almost all uses in the United States.

toxemia A pathological condition in a person or animal caused by the presence of a toxic substance of chemical or biological origin in the body. The toxic substance can be natural or synthetic.

toxic Describing a material that can cause acute or chronic damage to biological tissue after excessive physical contact or absorption.

toxic chemical A term that lacks precision. Generally, a compound that has an adverse effect on the normal biological functions of an organism when the organism is exposed to a relatively small dose. The compound can be either natural (commonly called a TOXIN) or synthetic (commonly called a TOXICANT), and the adverse biological outcome can vary from the inhibition of a simple enzymatic reaction, which would be unnoticed by a person, to a dramatic outcome such as death. Complicating the issue further are the

observations that some food items (salt, for instance) can cause death if consumed in sufficient quantities and some beneficial medications owe their therapeutic usefulness to an inherent toxicity. See XENOBIOTICS.

Toxic Chemical Release Inventory System (TRIS) A USEPA database of chemicals released and chemicals stored on-site submitted to the USEPA under provisions of the Emergency Planning and Community Right-To-Know Act; the annual data are available from 1987 on. See TITLE III. Query USEPA data at www.epa.gov/enviro/.

toxic cloud Airborne plume of a gas, vapor, fume, radioactive material particles, or aerosol that is capable of causing damage to the human population or to the environment. See TOXIC CHEMICAL.

toxic colonialism A term given to the practice of shipping waste from developed countries to less developed countries for the purpose of ultimate disposal. Several programs control waste exports. The PRIOR INFORMED CONSENT (PIC) PROCEDURE, managed by the United Nations, limits shipments of certain hazardous chemicals to those expressly approved by the receiving country. Under the provisions of the RESOURCE CONSERVATION AND RECOVERY ACT, shipments of HAZARDOUS WASTE from the United States to other countries must have the informed consent from the importing country. The terms of the BASEL CONVENTION, ratified by over 120 countries, ban the export of hazardous waste from developed to lesser-developed nations.

toxic dose The amount of some material (TOXIC CHEMICAL) to which a person must be exposed before a measurable negative outcome is noted. See DOSE, DOSE RATE, DOSE RESPONSE, DOSE-RESPONSE RELATIONSHIP, END POINT.

toxic pollutant A synthetic substance that, when added to the environment in amounts and under conditions that result in excessive exposure, causes damage to the natural environment or to human health. See END POINT, TOXIC CHEMICAL, TOXIC DOSE.

Toxic Substances Control Act (TSCA) The 1976 federal law that authorizes the USEPA to regulate or prohibit the manufacture, sale, or use of any new or existing chemical substance if the agency determines that the material poses an unreasonable risk to human health or the environment. The main focus of the law is on the review of new chemicals or proposed new uses of existing materials, which starts with a PREMANUFACTURING NOTICE (PMN) sent by the chemical producer to the agency. The PMN may include the results of toxicity testing of the chemical and other pertinent information required for the "unreasonable risk" determination. The INDOOR RADON ABATEMENT ACT and the ASBESTOS HAZARD AND EMERGENCY RESPONSE ACT are actually part of the TSCA. See SIGNIFICANT NEW USE RULE.

toxic tort The basis of a lawsuit alleging harm to health or property caused by exposure to emissions of chemicals from an industrial site. Toxic tort litigation seeks monetary damages and/or a court injunction against the chemical releases.

toxic waste Superfluous chemical compounds or mixtures that have the potential to cause adverse effects on human health or the environment. See HAZARDOUS WASTE; compare HAZARDOUS SUBSTANCE.

toxicant Any chemical that has the potential of causing acute or chronic adverse effects in animals, plants, or humans.

toxicity The ability of a chemical substance to cause acute or chronic adverse health effects in animals, plants, or humans.

toxicity assessment A compilation of research and observations of a chemical's potential TOXICITY for different DOSES and DOSE RATES via different exposure routes in different species. See DOSE RESPONSE, DOSE-RESPONSE RELATIONSHIP, END POINT, TOXIC CHEMICAL.

toxicity characteristic leaching procedure (TCLP) A test that measures the mobility of organic and inorganic chemical contaminants in wastes. The test, designed by the USEPA, produces an estimate of the potential for LEACHATE formation by a waste if placed in the ground. If the TCLP is applied to a solid waste sample and the extract leached from the waste contains concentrations of specified materials exceeding allowable levels, the waste is defined as a HAZARDOUS WASTE, meeting the toxicity characteristic. The TCLP replaced the EXTRACTION PROCEDURE TOXICITY TEST. See CONSTITUENT CONCENTRATIONS IN WASTE EXTRACT TABLE, CONSTITUENT CONCENTRATIONS IN WASTES TABLE, CHARACTERISTIC HAZARDOUS WASTE, SW-846.

Toxicity Reduction Evaluation (TRE) As part of a facility's NATIONAL POLLUTANT DISCHARGE ELIMINATION SYSTEM water permit, a study to determine the chemical compounds that must be limited to comply with water quality–based standards or to reduce acute or chronic toxicity of the facility's discharge water as determined by WHOLE EFFLUENT TOXICITY TESTING.

toxicity testing Use of fish, invertebrates, or mammals to determine the adverse effects of a potential toxic substance. See BIOASSAY.

toxicological profile A fact sheet containing a characterization of a potentially dangerous compound in terms of properties, effects on biota, and DOSE-RESPONSE RELATIONSHIP.

toxicology The study of chemical agents that cause diminished health and death in organisms, including humans. The study involves the chemical properties, recognition, identification, measurement, distribution, and metabolism of hazardous substances to which organisms are exposed. The science also includes the prediction of potential adverse effects on organisms, including humans, associated with different doses of chemicals.

Toxicology Data Network See TOXNET.

Toxicology Information on Line (TOXLINE) A database operated by the National Library of Medicine in Bethesda, Maryland, containing information on adverse effects of chemical compounds. The National Library of Medicine is part of the National Institutes of Health, Department of Health and Human Services. See TOXNET. Website: http://toxnet.nlm.nih.gov/

Toxics Release Inventory (TRI) Information supplied by companies to the USEPA and subsequently made available to the public covering the toxic emissions from certain manufacturing facilities. The report is mandated by the EMERGENCY PLANNING AND COMMUNITY RIGHT-TO-KNOW ACT (EPCRA) as part of the SUPERFUND AMENDMENTS AND REAUTHORIZATION ACT (SARA) and also by provisions of the POLLUTION PREVENTION ACT. Certain industrial operations that produce, prepare, use, or import above-threshold amounts of 600+ chemicals and chemical categories are required to report. In general, the chemicals are chosen for the list if, at excessive doses, they increase the risk of cancer, birth defects, nervous system disorders, or other chronic health effects or are expected to cause serious adverse effects in the environment. The report includes the amounts of these chemicals released into the air and surface water, disposed of by deepwell injection, sent to PUBLICLY OWNED TREATMENT WORKS, transferred off-site, recycled off-site, and used for energy production during each calendar year. The basic purpose behind requiring the inventory is to inform citizens about the chemical releases in their community. The TRI is a right-to-know law, not a pollution control statute; how-

ever, the public reporting requirement has had a significant positive effect in reducing the amounts of material released. Website: www.epa.gov/tri/

toxin A substance produced by plants or microorganisms that has the ability to cause adverse health effects in animals or humans upon excessive exposure.

TOXNET The Toxicology Data Network, a suite of toxicology databases available from the National Library of Medicine. The following environmental health and toxicology databases are available: (1) Chemical Carcinogenesis Research Information System (CCRIS), maintained by the NATIONAL CANCER INSTITUTE, contains 8000 chemical records on carcinogenicity and mutagenicity. (2) ChemID®, maintained by the National Library of Medicine, has over 340,000 chemical records and has the capacity to search lists of chemicals in various regulatory programs. (3) ChemID-*plus*® has over 50,000 structure and nomenclature files for over 349,000 chemical records in National Library of Medicine databases. (4) Developmental and Reproductive Toxicology/Environmental Teratology Information Center (DART®/ETIC) Database, maintained by the National Library of Medicine, contains more than 90,000 references on reproductive and development toxicology. (5) Environmental Mutagen Information Center (EMIC) Database, maintained by the National Library of Medicine, has over 100,000 references on genotoxicity by physical, chemical, or biological agents. (6) GENE-TOX, created by the USEPA, includes mutagenicity data on over 3000 chemicals. (7) Hazardous Substances Data Bank (HSDB)®, maintained by the National Library of Medicine, has human exposure data, industrial hygiene data, and regulatory information on 4500 chemicals. (8) INTEGRATED RISK INFORMATION SYSTEM (IRIS), from the USEPA, has data on over 500 chemicals for use in human health risk assessments. (9) TOXLINE® is the National Library of Medicine database of 2.5 million bibliographic entries on chemical toxicology. (10) The TOXICS RELEASE INVENTORY, compiled by the USEPA, lists annual releases of selected chemicals into the air, water, or land. Website: http://toxnet.nlm.nih.gov/

trace elements Elements essential to plant or animal life but required in only small amounts, such as the trace amounts of manganese, zinc, iron, molybdenum, cobalt, and copper.

trace metals Metals present in low concentrations in air, water, soil, or food chains.

tracer A stable, easily detected substance or a RADIOISOTOPE added to a material to follow the location of the substance in an organism or in the environment or to detect any physical or chemical changes it undergoes. Tracer applications include study of atmospheric dispersion of pollutants, following the uptake of fertilizers by plants, and study of the metabolism and excretion of compounds introduced to the human body. See CARBON-14, TAGGED MOLECULE.

tradable emission allowance A permit giving the holder the right to emit one unit of a pollutant per given period, for example, the right to emit one ton of SULFUR DIOXIDE per year. Tradable allowances are limited by an administrative agency to the total level that they determine will not cause any significant adverse effects. The allowances are initially allocated by past emission patterns, lot, auction, or set price. If the program is part of a plan involving emission reductions over several years, the number of permits in circulation drops with time. Facilities with lower marginal costs for emission control may reduce their emissions below the number of permits they hold and sell those they do not need to a facility with higher control costs, for which the purchase of allowances is cheaper than implementation of further controls. This market-efficient approach is being used as part of the ACID RAIN control program begun under the CLEAN AIR ACT amendments of 1990 and in Southern California's RECLAIM program. See also EMISSIONS TRADING.

Tragedy of the Commons Idea adapted from the title of an essay by Garrett Hardin originally published in *Science* in 1968. Hardin likened the human effect on the planetary environment to overgrazing of a commons (shared grazing land): individual decisions based on incremental personal benefit (adding another animal to the pasture, driving an automobile), when added together, ruin the common environment (overgrazing, pollution). His chief concern in the essay was world overpopulation, but the analogy has been widely applied to pollution of common resources such as air and water.

Train, Russell (1920–) American Conservationist, lawyer. Russell Train served as the second USEPA ADMINISTRATOR, 1973–77, and first chairman of the COUNCIL ON ENVIRONMENTAL QUALITY, 1970–73. He is the former head of the WORLD WILDLIFE FUND and the CONSERVATION FOUNDATION.

trammel net A net used in fish sampling, which contains a layer of large-mesh netting on each side of a smaller-mesh gill net.

Trans-Amazon Highway Begun in 1970, a 3000-mile road running across northern Brazil from the Atlantic Ocean to Peru. The highway was intended to open the Brazilian interior to economic development but has been severely criticized as producing extensive DEFORESTATION, loss of SPECIES DIVERSITY, and species EXTINCTION in the Amazon Basin by farming, ranching, logging, and mining activities. See TROPICAL RAIN FOREST, LATERITIC SOIL, EARTH SUMMIT, GLOBAL WARMING, AGENDA 21, SUSTAINABLE DEVELOPMENT.

transfer station In solid waste management, a facility at which waste collected by smaller trucks is reloaded into larger trucks for more efficient transport to a distant disposal site.

transformation The conversion of a normal animal or human cell to a tumor cell.

transformed cell Cell that has been converted from a normal cell to a tumor cell. The term is frequently used in association with cells grown in tissue culture in the laboratory and converted by exposure to a chemical or radiation.

transgene See TRANSGENIC.

transgenic From one species, a gene that is cloned and transferred to an organism of a different species. The process results in the conferring of a new trait on the recipient organism. See GENETIC ENGINEERING.

transgenic plant Plant, usually of agricultural importance, that contains genes from other species. These added genes confer certain desirable traits, such as resistance to insects, plant diseases, or particular herbicides. The best-known transgenic plant is the FlavrSavr tomato, which has been engineered genetically to contain an artificial gene that prevents overripening for a longer period than is possible in traditional tomatoes. Therefore, these tomatoes can be allowed to ripen on the vine instead of being picked green.

translocation A disorder of cellular CHROMOSOMES caused by breakage of a chromosome followed by a combining of the broken fragment with the wrong chromosome as the cell attempts to repair the damage caused by the initial breakage. Such damage can be caused by exposure to an excessive amount of X rays.

translocation factor In models of the distribution of chemical materials in the environment, a factor, different and specific for each chemical, used to estimate the daily movement of a chemical from soil to plants, from animal feed to animal tissues, or from animal feed to milk.

transmission electron microscopy (TEM) A microscopy technique in which illumination is provided by a beam of electrons passed through electron-transparent items. Cells, even bacteria, must be sliced very thinly and treated with specific reagents to allow observation of their interior. Compare SCANNING ELECTRON MICROSCOPY.

transmissivity (T) The rate at which water can flow through a unit width of an AQUIFER under a unit HYDRAULIC GRADIENT. Calculated as $T = KD$, where K is the HYDRAULIC CONDUCTIVITY and D is the aquifer thickness. Transmissivity is expressed in units of area per time.

transmutation The process of RADIOACTIVE DECAY during which one element is converted to another through the loss of radiation of some type.

transparency The portion of light that passes through water without distortion or absorption. A measure of the TURBIDITY of water or other liquid.

transpiration The direct transfer of water as a gas from plant leaves to the atmosphere. Transpiration combined with evaporation from the soil is called EVAPOTRANSPIRATION.

transportation control measure (TCM) Lowering of air pollution in a community through reductions in the use of automobiles (reducing the VEHICLE MILES TRAVELED). Auto-related pollution is reduced through such steps as increased use of public transportation, requirements for car pooling, park-and-ride lots, special lanes to encourage high occupancy rates in vehicles, parking restrictions, restriction or elimination of vehicular traffic in some areas, telecommuting, and encouragement of the use of bicycles. Compare INSPECTION AND MAINTENANCE, INDIRECT SOURCE REVIEW.

transporter Commercial enterprise involved in picking up properly prepared HAZARDOUS WASTE from those generating the waste and conveying the material to a TREATMENT, STORAGE, OR DISPOSAL facility. See MANIFEST SYSTEM, POTENTIALLY RESPONSIBLE PARTY, ARRANGER LIABILITY.

transuranic waste (TRU waste) A type of HIGH-LEVEL WASTE containing uranium-233, DAUGHTER products of uranium-233, and radioactive elements with ATOMIC NUMBER greater than 92 (the atomic number of uranium) produced during the operation of a NUCLEAR REACTOR and nuclear weapon fabrication. The ALPHA PARTICLE emissions from TRU waste have an ACTIVITY greater than 100 nanocuries per gram. Part of the TRU waste now at Department of Energy facilities around the United States will be transported to the WASTE ISOLATION PILOT PLANT in New Mexico.

transuranium Those elements with an ATOMIC NUMBER greater than that of uranium, which is 92.

trapping Describing the behavior of a PLUME being emitted from a smokestack beneath a TEMPERATURE INVERSION. The inversion prevents vertical dispersion of the stack gases, causing higher ground-level pollutant concentrations. Compare CONING, FANNING, FUMIGATION, LOFTING, LOOPING.

trash The fraction of solid waste that is not food waste; includes paper, glass, wood, and aluminum cans. Some components are BIODEGRADABLE (such as paper and wood), and others potentially can be recycled (such as aluminum cans and glass). Also called rubbish.

trash-to-energy See WASTE-TO-ENERGY.

tray tower An air pollution control device that removes pollutants from an exhaust by forcing the gas up through a tower containing trays above which the scrubbing liquid is introduced. Holes in the trays increase the gas-liquid contact. See also PACKED TOWER, SPRAY TOWER.

treatability study A test of a hazardous waste cleanup technology either in a laboratory (BENCH-SCALE) or as a pilot test at an actual waste cleanup site. Part of the REMEDIAL INVESTIGATION/FEASIBILITY STUDY at a SUPERFUND site.

treated Describing WASTEWATER that has undergone one or more forms of pollutant removal.

treated regulated medical waste Discarded paraphernalia associated with medical practice (e.g., bandages, syringes, tubing, gloves) that has been disinfected to reduce the threat of infection associated with processing; however, the discarded items have not been destroyed by incineration or shredding. See MEDICAL WASTE.

treatment The removal, mitigation, or control of chemical, biological, or radiation hazards from water, air, soil, or food. For water, see PRIMARY TREATMENT, SECONDARY TREATMENT, TERTIARY TREATMENT, PRETREATMENT, PUBLICLY OWNED TREATMENT WORKS, SEWAGE TREATMENT PLANT, PRESSURE SEWERS, CHLORINATION. For waste, see TREATMENT, STORAGE, AND DISPOSAL FACILITY; SOLIDIFICATION; STABILIZATION, WASTE; PYROLYSIS; INCINERATION. For soil and groundwater, see BIOREMEDIATION, IN SITU OXIDATION, IN SITU STRIPPING, IN SITU REMEDIATION, AIR STRIPPING, ION EXCHANGE, PRESSURE FILTER, PUMP AND TREAT, SOIL VAPOR EXTRACTION. For air, see SCRUBBER, BAGHOUSE, CYCLONE COLLECTOR, ELECTROSTATIC PRECIPITATOR, ACTIVATED CHARCOAL, CATALYTIC CONVERTER, PARTICULATE CONTROL DEVICE.

treatment, storage, or disposal (TSD) Describing a facility where hazardous waste is treated, stored, and/or disposed of as defined by the RESOURCE CONSERVATION AND RECOVERY ACT. Such facilities include LANDFILLS, SURFACE IMPOUNDMENTS, waste piles, and INCINERATORS.

treatment, storage, or disposal facility (TSDF) See TREATMENT, STORAGE, OR DISPOSAL.

treatment technology See TREATMENT.

tree line The dividing point, caused by changing latitudes or altitudes, between areas with environmental conditions that support trees and those that do not. The tree line in North America defined by latitude runs across northern Canada. Tree lines also exist at lower latitudes because of the altitudes in such places as the Rocky Mountains.

tremolite A fibrous mineral of the ASBESTOS group consisting of hydrated silicates of calcium and magnesium. A small amount of this mineral is marketed as asbestos.

trench method, landfill A technique for the placement of MUNICIPAL SOLID WASTE in a LANDFILL. The waste is spread and compacted in 2-foot layers within trenches 100–400 feet long, 15–25 feet wide, and 3–6 feet deep. The COVER MATERIAL for a full trench is obtained from the excavation of an adjacent trench. See also AREA METHOD, LANDFILL.

trial burn An extensive controlled test of the DESTRUCTION AND REMOVAL EFFICIENCY of an incinerator. The results are used to establish the operating conditions specified in the permit for a facility. See THERMAL TREATMENT OF HAZARDOUS WASTE.

trichloroethylene A two-carbon ALIPHATIC compound (C_2HCl_3) in the general class of chemicals referred to as HALOGENATED or CHLORINATED HYDROCARBONS. The chemical is used as a solvent and industrial chemical. The compound has been shown to cause liver and kidney damage and to cause tumors in some animal models. Listed as a HAZARDOUS AIR POLLUTANT by the CLEAN AIR ACT amendments of 1990. See AIR TOXICS.

trichlorophenoxyacetic acid (2,4,5-T) A chlorophenoxy HERBICIDE used to control broadleaf weeds and woody plants. An ingredient, along with dichlorophenoxyacetic acid (2,4-D), in AGENT ORANGE, the toxic defoliant used during the Vietnam War. Synthesis of the chemical results in a slight contamination with TETRACHLORODIBENZO-*PARA*-DIOXIN.

trickle irrigation See DRIP IRRIGATION.

trickling filter A wastewater treatment apparatus used to remove soluble or colloidal organic compounds. The filter consists of a large open tank filled with small stones. Clarified wastewater (from which the particulate material has been removed) is sprayed over the surface of the stone bed and allowed to trickle through. Microbial communities that coat the stones mineralize the organic compounds by aerobic metabolic processes as the water trickles downward. See CLARIFICATION, SECONDARY TREATMENT.

trihalomethanes (THMs) A group of low-molecular-weight HALOGENATED hydrocarbons including chloroform, bromodichloromethane, dibromochloromethane, and bromoform. The group includes suspect human CARCINOGENS. Small amounts of THMs have been detected in RAW WATER collected from surface sources used as a public water supply, and concentrations have been shown to be increased during the CHLORINATION phase of the water purification process. The most marked increase during chlorination has been recorded in water containing SUSPENDED PARTICULATE MATTER and/or HUMIC substances. See DISINFECTION BY-PRODUCTS.

trip blank A sample container to which a clean sample MATRIX (water, soil, or other) is added, which is carried to and from the location where environmental samples are collected. The trip blank is analyzed along with the actual samples and serves as a check for sample or sample container contamination during transit.

triple rinse 1. A laboratory procedure associated with the processing of laboratory glassware and other implements to remove any vestige of contaminants. After cleaning, the items are washed three times with deionized or distilled water. **2.** In pesticide management, the method used to remove pesticide residue from spent containers, which involves three rinses of freshwater.

tritium   A synthetic ISOTOPE of hydrogen containing one proton and two neutrons in the nucleus.

trommeling   The removal of small-diameter dense solids with a TROMMEL SCREEN. Shredded solid waste is placed in a rotating screened drum, and smaller particles fall through the screen.

trommel screen   A cylindrical screen with a shaft through the center, around which the screen rotates. The apparatus is used for waste separation. Smaller-sized, denser materials, such as glass fragments and grit, fall out and larger components remain behind.

trophic level   A feeding level within a FOOD WEB. The nature of the food of an organism determines the trophic level of the organism. PRIMARY PRODUCERS, which constitute the first trophic level, are plants that utilize inorganic materials as nutrients. Organisms at the second trophic level, animals that obtain nutrients from the consumption of plants, feed on organisms in the first trophic level. Some animals that feed only on other animals are described as occupying the third trophic level. See ECOLOGICAL PYRAMID, FOOD CHAIN.

trophic structure   The FOOD CHAIN or feeding relationships within a specific biological community.

tropical rain forest   The richest natural terrestrial environment in terms of the amount of BIOMASS and the number of species present. These forests occupy the equatorial regions of South and Central America, Central Africa, and Southeast Asia. Such forests include up to five times the number of species of trees found in temperate regions and an unrivaled number of species of associated plants and animals. Paradoxically, the soil of the typical rain forest is very thin and low in plant nutrients; many rain forests have LATERITIC SOIL. The temperatures usually remain between 70° F and 95° F year-round, and rainfall typically varies between 50 and 200 inches per year. The destruction and disturbance of tropical rain forests are of environmental concern because of the loss of many species (some yet to be described), conversion of the land to a semidesert condition, disruption of the climactic patterns of the region, and worsening of the global balance in carbon dioxide production and utilization, among other factors. See RICHNESS, GLOBAL WARMING, CONVENTION ON INTERNATIONAL TRADE IN ENDANGERED SPECIES OF WILD FAUNA AND FLORA, AGENDA 21, CONVENTION ON BIODIVERSITY, DEBT FOR NATURE SWAP, EARTH SUMMIT.

tropopause   The boundary in the atmosphere between the layer next to the surface of the Earth (troposphere) and the next highest layer (stratosphere). See ATMOSPHERE.

troposphere   The layer of the atmosphere closest to the surface of the Earth extending up to about 9 to 16 km. See ATMOSPHERE.

true minor   Under the TITLE V PERMIT regulations of the 1990 amendments to the CLEAN AIR ACT, a source that does not have the POTENTIAL TO EMIT pollutants at a rate that would classify it as a MAJOR SOURCE. See SYNTHETIC MINOR.

trustee   1. Under the COMPREHENSIVE ENVIRONMENTAL RESPONSE, COMPENSATION, AND LIABILITY ACT, a federal or state agency that may sue for NATURAL RESOURCES DAMAGES. **2.** Under the PUBLIC TRUST DOCTRINE, persons, organizations, companies, or government agencies that are stewards of natural resources for present and future generations' use.

tsunami   (pronounced "sue-nah-me"). A Japanese word roughly meaning "large waves in harbors." These waves are commonly termed "tidal waves," which they are not, since they have no relation to tides. A tsunami is an oceanic wave, also termed a seismic sea wave, generated most commonly by slippage (vertical displacement) of the ocean floor along a fault in association with an earthquake or a rock slide. The waves are characterized by long wavelengths (200 km) and small wave height (1 m). Consequently, such waves would pass under a vessel unnoticed. As the wave approaches a shoreline, the bottom of the wave is affected by the bottom topographical features, dragging across the bottom and slowing. The wave length decreases, and the height increases to tens of meters (10 to 30 feet is not uncommon). On the coast, one would experience the beaching of the wave as first a falling of water level then a rising of sea level in a relatively short period, 10 to 20 minutes. The damage done to coastal areas by these waves is normally associated with the strong currents associated with the flood and ebb currents as the wave washes over the land then recedes. Those areas that have building within a few meters of the normal high tide line are most severely impacted.

tuberculation The accumulation of knots or small mounds of iron and manganese oxide inside iron pipes, causing a restriction of flow. The deposits are caused by corrosion and bacterial metabolism of the iron pipe, forming an oxide PRECIPITATE.

tuff Rock formed from the accumulation of ash resulting from a volcanic eruption. The ashlike materials that are ejected from the volcano settle and fuse into a rock formation that is similar to that of SEDIMENTARY ROCK.

tumor See NEOPLASM, BENIGN NEOPLASM, MALIGNANT NEOPLASM.

tumor registry See CANCER REGISTRY.

tumorigenicity The ability of cells from a tissue culture to grow and produce tumors when inoculated into a host organism.

tundra A BIOME located in the northern regions of the continents of North America, Europe, and Asia, characterized by the absence of trees and the presence of permafrost (permanently frozen subsoil). The natural vegetation consists of small shrubs, mosses, and lichens.

turbidity A measure of the amount of suspended matter in water or some other fluid as determined by the relative light transmission of the suspension. The typical scales used are percentage transmission or transmittance, which varies from 0% to 100% of the light passing through the sample, and optical density or absorbance, which is a logarithmic scale varying from 2 to 0, in which 2 is the most turbid and 0 is the least turbid. About 0.3 on the absorbance scale corresponds to 50% transmittance.

turbine A device with blades connected to a central shaft that converts the energy in a moving fluid or gas into rotational mechanical energy. In steam and gas turbines, hot gases expand through and rotate the turbine blades. Water and wind turbines capture and convert the energy of falling water or blowing wind into rotation. The mechanical energy from turbines is used in pumps and electric generators.

turbulence See MECHANICAL TURBULENCE, THERMAL TURBULENCE.

turbulent flow Fluid flow exhibiting random fluctuations in speed and direction. Compare LAMINAR FLOW; see REYNOLDS NUMBER.

turnaround The period during which an industrial facility or generating station is shut down for planned maintenance.

turnover 1. In ecology, the rate of PRODUCTIVITY divided by the STANDING CROP, or BIOMASS. Expressed as

$$T = P/B$$

where P is productivity (in units of mass per area-time) and B is biomass (in units of mass per area). Turnover is expressed in units of 1/time. **2.** See FALL TURNOVER, SPRING TURNOVER.

turnover rate The rate (for example, in milligrams per cubic meter of water per day) at which some material is metabolized or decomposed in the environment.

turnover time **1.** In ecology, the average time (t) required for the BIOMASS in an ecosystem to replace itself, the inverse of TURNOVER, in time units. Expressed as

$$t = B/P$$

where P is the PRODUCTIVITY (in units of mass per area-time) and B is the biomass with units of mass per area. **2.** In microbiology, the time required to metabolize a specific substance in a body of water or in soil.

turtle excluder device (TED) An addition to a shrimper's trawl net that allows shrimp into the net but blocks entrance of larger organisms. Required by federal and state regulations to prevent the accidental netting and drowning of sea turtles, especially the Kemp's Ridley *(Lepidochelys kempii)*, which is on the ENDANGERED SPECIES list.

type I error When judging the results of a scientific study, the rejection of the NULL HYPOTHESIS although the null hypothesis is true. See TYPE II ERROR.

type II error When judging the results of a scientific study, the acceptance of the NULL HYPOTHESIS although the null hypothesis is false. See TYPE I ERROR.

type A/type B packaging Containers designed to meet federal regulations for the transport of RADIOACTIVE materials. The containers must maintain their integrity without leakage of radioactive material or loss of SHIELDING under certain instances of heat, cold, free drop, and other factors. Type B packaging must meet higher standards in terms of radiation shielding and structural integrity.

type A/type B quantities Types and maximal amounts of radioactive materials that can be transported in TYPE A and TYPE B PACKAGING, respectively.

U

ultimate analysis A chemical analysis that measures the concentrations of certain individual elements or compounds in a material. When it is performed for combustible waste, the concentrations of carbon, oxygen, sulfur, nitrogen, hydrogen, and noncombustibles (ASH) are typically determined. Compare PROXIMATE ANALYSIS.

ultraclean fuels Fossil fuels that have had impurities removed before burning so that little or no pollutant removal from the exhaust gas is required. For example, an ultraclean coal would have sulfur impurities removed and a low ash content, producing low SULFUR DIOXIDE and PARTICULATE MATTER emissions. The U.S. DEPARTMENT OF ENERGY has a program to encourage the development of ultraclean fuels. See COAL GASIFICATION, COAL LIQUEFACTION. The Clean Coal Centre is based in London. Website: www.iea-coal.org.uk/

ultrasonic Describing acoustic waves with frequencies above 20 kilohertz and therefore not audible by the human ear.

ultraviolet (UV) That portion of the ELECTROMAGNETIC SPECTRUM that extends from the violet range of visible light (wavelength equal to about 400 nanometers) to the X RAYS (wavelength equal to about 10 nanometers). Radiation falling between the wavelengths of X rays and 100 nanometers is referred to as extreme or vacuum ultraviolet and is of little biological interest since radiation in this range of wavelengths is not transmitted through air. The remaining portion of the spectrum (100 to 400 nanometers) is divided into A, B, and C ranges for the purpose of differentiating the biological

effects of UV light of different wavelengths. Ultraviolet radiation within the range of 130 to 200 nanometers is responsible for the generation of OZONE within the stratosphere. See ULTRAVIOLET RADIATION—ACTINIC RANGE, ULTRAVIOLET RADIATION-A RANGE, ULTRAVIOLET RADIATION-B RANGE, ULTRAVIOLET RADIATION-C RANGE.

ultraviolet photometry An analytical method that employs the selective molecular absorption of particular ultraviolet wavelengths to identify and measure the concentrations of certain chemicals in air or water. The method is also called ultraviolet spectrophotometry.

ultraviolet radiation—actinic range That range of the spectrum of ultraviolet (UV) radiation that has the most pronounced biological effect, such as mutations and skin cancer; the wavelengths extend from 280 to 320 nanometers (all of the UV-B and part of the UV-C range).

ultraviolet radiation-A range (UV-A) That part of the spectrum of ultraviolet radiation that encompasses the wavelengths from 320 to 400 nanometers. This portion is also referred to as near-ultraviolet since the range is nearest to that of visible light (wavelengths longer than 400 nanometers). Radiation within this range is transmitted through air and to some extent through glass. A portion of the radiation within this class causes tanning of the skin, especially when the radiation is of high intensity. UV-A also contributes to skin aging and cataract risk. Most (95% to 99.9%) of the ultraviolet light used in tanning salons is within the A range. See SUNSCREEN.

ultraviolet radiation-B range (UV-B)
That part of the spectrum of ultraviolet (UV) radiation that encompasses the wavelengths from 280 to 320 nanometers (middle-ultraviolet). Radiation within this range is not well transmitted through glass. UV-B is the primary cause of sunburn, skin aging, and skin cancer risk. See SUNSCREEN, OZONE LAYER DEPLETION.

ultraviolet radiation-C range (UV-C)
That part of the spectrum of ultraviolet (UV) radiation that encompasses the wavelengths from 100 to 280 nanometers. This portion is also referred to as far-ultraviolet since the range is most distant from the wavelengths of visible light. The germicidal effect of UV light is strongest within this region: a wavelength of 260 nanometers is the most effective at killing of microorganisms.

ultraviolet spectrophotometry See ULTRAVIOLET PHOTOMETRY.

ultraviolet-visible absorption spectrum (UV-VIS) The absorption pattern of certain chemical compounds of electromagnetic energy at wavelengths within those of visible and ultraviolet radiation. Various molecules absorb radiation at specific wavelengths and consequently can be analyzed in both a qualitative and a quantitative manner. The technique is especially useful in the analysis of organic molecules that contain double bonds between carbon atoms.

uncertainty factor (UF) An adjustment applied to experimental toxicity data to set acceptable human dose levels to protect against noncancer health effects. The UF is intended to account for the uncertainties introduced by the EXTRAPOLATION of animal data to humans, the variation in susceptibility within the human population, the use of ACUTE EXPOSURE data to predict safe CHRONIC EXPOSURE levels, and/or the use of data from an oral ROUTE OF EXPOSURE to set inhalation standards. The USEPA uses the following uncertainty factors as guidelines; their application is specific to each risk assessment. The higher the UF, the lower the acceptable human dose:

Factor	Applied to
10	Valid data on human exposures
100	Valid chronic animal studies
1000	Animal studies with less than chronic exposure
1–10	An additional factor when using a LOWEST-OBSERVED-ADVERSE-EFFECT LEVEL instead of a NO-OBSERVED-ADVERSE-EFFECT LEVEL

A REFERENCE DOSE, a lifetime daily exposure set to protect sensitive human populations, is calculated by using a NO-OBSERVED-ADVERSE EFFECT LEVEL or a LOWEST-OBSERVED-ADVERSE-EFFECT LEVEL divided by an appropriate uncertainty factor. The FOOD QUALITY PROTECTION ACT of 1996 requires that the USEPA apply an additional factor of 10 for protection against PESTICIDE RESIDUE in foods that have a THRESHOLD DOSE (i.e., to protect against noncancer health effects). The additional factor is to protect infants and children. The USEPA may apply a factor lower than 10× only if the agency can show that it is protective to infants and children.

unconfined aquifer An aquifer with no low-PERMEABILITY zones between the SATURATED ZONE and the surface; an aquifer with a WATER TABLE. Also called a water table aquifer.

unconventional oil Hydrocarbon contained within such geological deposits as SHALE or TAR SANDS. A hydrocarbon product can be recovered from these materials and refined into fuels. See SHALE OIL, BITUMEN, SYNFUEL.

underflow The slurry of concentrated solids or SLUDGE that is removed from the bottom of a SETTLING CHAMBER, CLARIFIER, or THICKENER.

Underground Injection Control (UIC)
A program required in each state by a provision of the SAFE DRINKING WATER ACT for the regulation of INJECTION WELLS,

including a permit system. An applicant must demonstrate that the well has no reasonable chance of adversely affecting the quality of an underground source of drinking water before a permit is issued. See CLASS I, II, III, IV, AND V INJECTION WELLS.

underground mining A technique of removing coal and mineral ores from the Earth that involves cutting tunnels below the surface to gain access to the deposit(s). The environmental problems that can be associated with underground mining include the runoff from the SPOIL carried to the surface and ACID MINE DRAINAGE. Compare SURFACE MINING.

underground source of drinking water (USDW) An aquifer that (1) supplies a public water system or (2) can potentially supply a public water system and is currently used for human consumption or has a TOTAL DISSOLVED SOLIDS concentration of less than 10,000 milligrams per liter and is not exempted. UNDERGROUND INJECTION CONTROL rules protect underground sources of drinking water. The regulatory definition, including the list of exemptions, is found at 40 CFR 144.3.

underground storage tank (UST) Under RESOURCE CONSERVATION AND RECOVERY ACT regulations, a tank with at least 10% of the volume beneath the ground, including attached pipes, with some exemptions. Underground storage tanks must meet certain performance standards, have spill and overfill controls, and be monitored regularly for leaks. See LEAKING UNDERGROUND STORAGE TANK TRUST FUND. For regulatory details from the USEPA Office of Underground Storage Tanks see www.epa.gov/swerust1/.

undiscovered resource Usually designates oil or mineral deposits that are suspected to be present on the basis of preliminary evaluation of geological data but whose presence has not been confirmed by exploration.

Unified Air Toxics Website An Internet clearinghouse for resources on the measurement, risk assessment, sources, health effects, and control of routine or accidental releases of HAZARDOUS AIR POLLUTANTS, also called AIR TOXICS. Although the stated purpose of the site is sharing of information by and for regulatory agencies, this is a valuable source for students, industry, and the public. Website: www.epa.gov/ttn/uatw/

unfunded mandate Any action required of state and local governments by federal statutes and regulations but without appropriations to support the new mandatory program. Many federal environmental protection statutes have been guilty of this offense. The Unfunded Mandate Reform Act of 1995 attempts to lessen this problem by requiring that proposed major federal legislation include an estimate, by the Congressional Budget Office, of the burden the law will impose on state and local governments and the private sector and that the cost estimate will be available to Congress before the members vote on the legislation.

unit density A density of one gram per cubic centimeter or one gram per milliliter; the density of water at 4° C.

unit risk The lifetime risk of cancer per unit dose of exposure. See UNIT RISK ESTIMATE, CANCER POTENCY FACTOR.

unit risk estimate The CANCER POTENCY FACTOR expression for air exposures; expressed as an inverse concentration, usually per micrograms per cubic meter. The actual or estimated air concentration of a chemical, in micrograms per cubic meter, multiplied by the unit risk estimate, gives the lifetime risk of contracting cancer as a result of inhaling the chemical.

$$\text{Lifetime risk} \atop \text{of cancer} = \frac{1}{\frac{\mu g}{m^3}} \times \frac{\mu g}{m^3}$$

unit risk units	air concentration units

The exposure is assumed to be 24 hours per day for 70 years. See CANCER POTENCY FACTOR.

United Nations Conference on Environment and Development See EARTH SUMMIT.

United Nations Conference on the Human Environment The 1972 meeting of 113 nations, many other intergovernmental agencies, and NONGOVERNMENTAL ORGANIZATIONS held in Stockholm, Sweden, to discuss global environmental issues. Established action plans for over 100 environmental problems and became the model for subsequent global conferences on population, energy, water, climate, human settlements, food, and outer space. The Stockholm conference began the UNITED NATIONS ENVIRONMENT PROGRAM. See EARTH SUMMIT, CONVENTION ON BIODIVERSITY.

United Nations Environment Program (UNEP) An agency of the United Nations, headquartered in Nairobi, Kenya, responsible for coordinating intergovernmental efforts to monitor and protect the environment. See EARTHWATCH, BASEL CONVENTION, CONVENTION ON BIODIVERSITY, CONVENTION ON INTERNATIONAL TRADE IN ENDANGERED SPECIES OF WILD FAUNA AND FLORA, MONTREAL PROTOCOL, INTERGOVERNMENTAL PANEL ON CLIMATE CHANGE, GLOBAL ENVIRONMENT FACILITY, GLOBAL ENVIRONMENT MONITORING SYSTEM. Website: http://www.unep.ch/

United Nations Framework Convention on Climate Change See FRAMEWORK CONVENTION ON CLIMATE CHANGE, KYOTO PROTOCOL.

United Nations/North America number (UN/NA number) A four-digit number used internationally to identify a hazardous material; for example, UN/NA 1203 is gasoline.

United States Code (U.S.C.) A multivolume compilation of all federal statutes, arranged by 50 subject titles. A new edition is published every six years and supplements issued during the intervening period. The code is on-line at www.access.gpo.gov/congress/cong013.html or www4.law.cornell.edu/uscode/. See UNITED STATES CODE ANNOTATED, UNITED STATES CODE SERVICE.

United States Code Annotated (U.S.C.A.) A multivolume collection of the entire UNITED STATES CODE plus state and federal court decisions applicable to specific code sections and other reference material.

United States Code Service (U.S.C.S.) A cumulative supplement to the UNITED STATES CODE published during the six-year interval between releases of new editions of the complete code.

United States Geological Survey (U.S.G.S.) An agency of the U.S. Department of the Interior based in Reston, Virginia; responsible for assembling technical geographical and geological information about public lands and waterways to improve management of wildlife, water, air, energy, and mineral resources. Website: www.usgs.gov

universal gas constant (R) The constant of proportionality (R) in the EQUATION OF STATE and the IDEAL GAS LAW. For one gram-mole of a gas, $PV = RT$, where P is the pressure, V is the volume, and T is the ABSOLUTE TEMPERATURE. The constant is expressed in many different units to match the units for P, V, and T. A common form is 82.06 atm-cm^3 gram-mole^{-1} K^{-1}, where atm is the pressure in ATMOSPHERES and K is the Kelvin temperature.

Universal Transverse Mercator coordinates (UTM coordinates) A map coordinate system covering the world from 80 degrees north to 80 degrees south with 60 north-south zones, each covering 6 degrees of longitude and divided into 8-degree latitude sections. The zones overlap 0.5 degree on each side. Each zone has an individual origin, and the coordinates are read in meters east and meters north of the origin. The coordinates often are used to define emission locations in AIR QUALITY DISPERSION MODELING.

unreasonable risk Part of the definition of *unreasonable adverse effects* in the provisions of the FEDERAL INSECTICIDE, FUNGICIDE, AND RODENTICIDE ACT (FIFRA). The REGISTRATION of a pesticide can be changed (restricted) if unreasonable adverse effects are being caused by its current use. The USEPA must determine that the pesticide poses an "unreasonable risk to man or the environment, taking into account the economic, social, and environmental costs and benefits of the use of the pesticide," to conclude that unreasonable adverse effects are occurring. See CANCELLATION, SUSPENSION.

unsaturated zone The upper layers of soil in which pore spaces or rock are filled with air or water at less than atmospheric pressure; the area of the ground from the surface down to the WATER TABLE. The region is also called the zone of aeration or vadose zone. Compare SATURATED ZONE.

unstable Describes elements or compounds that react easily or spontaneously to form other elements or compounds. For example, OZONE (O_3) is an extremely unstable gas because it reacts readily with many materials. Likewise, radioactive materials form new elements as they undergo RADIOACTIVE DECAY: URANIUM 238 is converted to THORIUM as a result of the release of radiation.

upgradient well A groundwater MONITORING WELL, such as those required at facilities that treat, store, or dispose of HAZARDOUS WASTE using a SURFACE IMPOUNDMENT or LANDFILL, that allows sampling and analysis of groundwater that is upstream from the facility before it can be affected by any escaping contaminants (LEACHATE). The results of the analyses are used for comparison to the results of groundwater sampled from DOWNGRADIENT WELLS.

upper-bound risk level A risk level derived from a cancer RISK ASSESSMENT. The upper bound is derived by using conservative (on the side of increased risk) assumptions in the risk assessment and the upper CONFIDENCE LIMIT of the statistically estimated CANCER POTENCY FACTOR. The level of risk is reported as the 95% upper-bound risk, meaning that there is only a 5% chance that the true risk is greater than the upper-bound risk.

upper detection limit The highest concentration of a chemical that can be reliably measured with available instrumentation technology. When the environmental concentration exceeds this upper threshold, the samples must be diluted to assure the recovery of meaningful data.

upper explosive limit (UEL) The highest concentration of a substance in air that burns or explodes when ignited. At concentrations higher than this level, the amount of oxygen in the mixture is not sufficient to support combustion. Compare LOWER EXPLOSIVE LIMIT.

upper flammable limit (UFL) See UPPER EXPLOSIVE LIMIT.

upper respiratory tract (URT) The mouth, nasal passages, pharynx, and larynx.

upset An exceptional incident, beyond the reasonable control of a company holding a water discharge or air emission PERMIT, in which there is an accidental release of pollutants greatly in excess of the discharges or emissions allowed by the permit. An equipment failure is a common cause of an upset.

uptake The incorporation of a pollutant into an exposed organism.

upwelling The appearance of water from the deep ocean at the surface. This usually occurs along the coasts of continents (such as the coast of Peru along the west coast of South America) where the prevailing winds push the surface waters away from the land area, allowing waters from the deep ocean to rise to the surface. The deep waters carry a significant input of plant nutrients to the surface, resulting in an elevated level of PRIMARY PRODUCTIVITY and abundant fish populations.

uranium (U)   A rare heavy metal with an ATOMIC NUMBER of 92. The most common ATOMIC WEIGHT in naturally occurring uranium is 238. Other ISOTOPES range from 234 to 239. The element was little used until about 1942, when uranium was identified as a potential FISSION fuel useful in the operation of NUCLEAR REACTORS and as a fuel in nuclear weapons. Uranium is found in natural deposits combined with other elements, principally oxygen.

uranium enrichment   A process that results in an increase in the amount of the fissionable isotope of uranium (URANIUM 235) in a given mass of uranium. As recovered from natural deposits, uranium consists of about 0.7% of the fissionable isotope (^{235}U) and 99.3% of the nonfissionable isotope URANIUM-238 (^{238}U). The relative abundance of the fissionable uranium 235 must be increased to about 3% before the uranium can be used as a fuel in a NUCLEAR REACTOR. See GRAHAM'S LAW, FUEL ENRICHMENT.

uranium fuel cycle   The mining, refining, fabrication, transport, and recycling of URANIUM along with the disposal of wastes from both the purification processes and the reprocessing of used FUEL RODS. See FUEL REPROCESSING.

uranium tailings   The residue of the mining of URANIUM ores and the purification of uranium from those ores. This residue presents a disposal problem because of the release of radioactive elements such as RADON gas.

uranium 235 (U-235 or ^{235}U)   A fissionable ISOTOPE of the element URANIUM found in naturally occurring uranium deposits at a concentration of about 0.7% of the total uranium present in the geological deposit. See URANIUM ENRICHMENT.

uranium 238 (U-238 or ^{238}U)   A nonfissionable ISOTOPE of the element URANIUM. U-238 contains three more neutrons than U-235 and is the most commonly occurring form of the element (99.3%). See URANIUM ENRICHMENT.

urban heat island   The elevated release and retention of energy characteristics of an urban area in which larger quantities of FOSSIL FUEL are burned than in the surrounding countryside and that contains asphalt, brick, concrete, and other materials that absorb and retain higher amounts of heat energy than does vegetative cover. This phenomenon is illustrated by the higher average annual temperatures measured in urban areas than in adjacent rural areas.

urban plume   The downwind impact zone of air contaminants or products of atmospheric reactions involving compounds released within the confines of an urban area.

urban runoff   Storm water from city streets and associated property. Potentially, this class of wastewater can contribute significant amounts of pollutants to receiving streams because of materials that accumulate on covered surfaces. SEDIMENT, sand, DETRITUS, oil and grease, PLANT NUTRIENTS, and PESTICIDES that accumulate during dry periods are flushed from the surface and swept into the receiving stream. Because the soil surface is covered with streets, homes, businesses, parking lots, and so forth, both the amount and the velocity of urban runoff are much higher than those from the same extent of rural property receiving the same amount of rainfall. Some urban runoff sources now require a STORM WATER RUNOFF permit under the CLEAN WATER ACT. See FIRST FLUSH.

use cluster Chemicals, processes, and technologies that can be substituted for each other in performing a particular function. If the use of a specific chemical results in the release of a hazardous waste that presents a significant pollution problem, perhaps some alternate material within the use cluster can be substituted, maintaining the process while reducing the pollution potential.

used oil Oil drained from the crankcase of automobiles, vans, trucks, and other vehicles. This oil presents a potential pollution problem, especially when individuals change oil in their personal vehicles. The oil that is drained from the vehicle is often used to control weeds on the lawn and in ditches around the home, poured down the drain into a sewer system serving a municipal sewer system, or put into an old container and thrown into the trash. Such practices are not advisable because the oil is rich in hydrocarbon oxidation products and suspended metals.

Users' Network for Applied Modeling of Air Pollution (UNAMAP) See SUPPORT CENTER FOR REGULATORY AIR MODELS (SCRAM).

utilitarian conservation A natural resource management philosophy based on the proposition that resources should be used for the greatest good, for the greatest number of people, and for the longest time. Careful scientific management preserves resources for future generations without denying their use and enjoyment by the current generation. This was the basis of the pragmatic progressive programs followed by President Theodore Roosevelt and GIFFORD PINCHOT in the establishment of the National Park System in the United States. Compare ALTRUISTIC PRESERVATION; CONSERVATION; PRESERVATION; MUIR, JOHN.

UV index A daily forecast of the amount of ULTRAVIOLET (UV) radiation reaching the surface of the Earth in a specific location during the peak hours of sunlight, 11:30–12:30 standard time. The index is available in 58 U.S. cities, for at least 1 city in each state, and is based on methodology and data from the NATIONAL OCEANIC AND ATMOSPHERIC ADMINISTRATION (National Weather Service) and the USEPA. The amount of surface UV radiation at the peak period is dependent on the height of the Sun in the sky (varying by latitude and time of year), the elevation above sea level, the amount of ozone in the stratosphere (OZONE LAYER), and the expected cloud cover for the day. The index is not based on measurements but is calculated as a forecast. Next-day stratospheric ozone is predicted by using SOLAR BACKSCATTER ULTRAVIOLET DATA or *TIROS* operational vertical sounder data from satellite measurements. Cloud cover is predicted by using standard weather forecasts. The ultraviolet irradiance is adjusted for latitude, season, and altitude, then weighted by the sensitivity of human skin to different ultraviolet wavelengths. The adjusted irradiance is divided by a standard 25-milliwatt-per-square-meter scaling factor to produce the UV index value. The numerical scale varies from 0 to about

Index	Category	Protection
0–2	Minimal	Sunscreen SPF 15
3–4	Low	Above, plus protective clothing
5–6	Moderate	Above, plus UV-A/UV-B sunglasses
7–9	High	Above, plus attempt to avoid the sun, 10 A.M. to 4 P.M.
10+	Very high	Above, plus avoid sun, 10 A.M. to 4 P.M.

15; it can be interpreted by using a table from the USEPA.

See ULTRAVIOLET RADIATION-A RANGE (UV-A), ULTRAVIOLET RADIATION-B RANGE (UV-B), SUNSCREEN, OZONE LAYER DEPLETION.

U waste Toxic, but not acutely hazardous chemicals defined as LISTED HAZARDOUS WASTE by the USEPA in the *Code of Federal Regulations,* Title 40, Part 261.33. See also F WASTE, K WASTE, P WASTE, HAZARDOUS WASTE.

V

vacuum filtration Drawing air or a liquid through a filter medium by creating reduced air pressure on the opposite side of the filter medium from the air or liquid material that will pass through the filter.

vacuum sewer A type of wastewater collection system used in rural areas as a less expensive alternative to the GRAVITY SEWER used in urban areas. The wastewater is collected through small-diameter pipes by a central vacuum pump. The costs of smaller pipes and shallow pipe burial are much lower than the costs of clay or concrete pipe systems in urban areas needing a downward gravity flow between LIFT STATIONS. However, vacuum sewers do not separate solids before collection, and the sewage is treated at a conventional wastewater treatment facility; that treatment raises the costs associated with this system when compared with PRESSURE SEWERS, another system of rural wastewater collection. Vacuum sewers have not been used extensively in the United States.

vadose zone The area of the ground below the surface and above the region occupied by groundwater. See UNSATURATED ZONE.

valence The number of electrons an atom contributes or receives when a chemical bond is formed or when ions are formed from the elements. Atoms contributing electrons in reactions or during ionization, such as metals, have positive valences. Nonmetals, such as chlorine, have negative valences: that is, they receive electrons during bonding.

Valley of the Drums The A. L. Taylor site in Brooks, Kentucky, that served as a chemical dump and drum recycling center from 1967 to 1977. The operator would empty the contents of the drums into a shallow pit then recycle the container. The chemical contamination on the 13-acre site, which was largely waste from paints and coatings, included metals and a variety of volatile organic compounds. A USEPA emergency response began in 1979 under provisions of the CLEAN WATER ACT, and the notorious site gained SUPERFUND status in 1981, having been instrumental in the passage, in 1980, of the COMPREHENSIVE ENVIRONMENTAL RESPONSE, COMPENSATION, AND LIABILITY ACT (SUPERFUND). The cleanup removed over 8000 drums and installed controls to prevent off-site contamination. The site was removed from the NATIONAL PRIORITIES LIST in 1996.

valued environmental component Those characteristics or attributes of a natural system considered to be sufficiently important to justify special consideration when threatened by human activity or natural hazards.

van Dorn sampler See EKMAN WATER BOTTLE.

vapor The gaseous form of a material that is normally found in the solid or liquid state at room temperatures, for example, water vapor, which is water in the gas phase. Vaporization occurs when molecules in the solid or liquid gain enough energy to escape the material, especially with an increase in temperature.

vapor capture system Any arrangement of hoods, ducts, hoses, piping, or a ventilation system designed to recover the organic vapors released in a process or activity. Usually the captured vapors are routed into the process for recovery. See STAGE II CONTROLS, ONBOARD REFUELING VAPOR RECOVERY, CARBON ADSORBER, ACTIVATED CHARCOAL.

vapor density The density of a pure gas or vapor compared with the density of hydrogen or the density of air. When used in chemical hazard analysis, the comparison is with air and the ratio is based on an assumed air density of 1.0; therefore, lighter-than-air gases have a vapor density less than 1.0, and those with values greater than 1.0 are heavier than air. Note that a gas or vapor released into the environment mixes with the surrounding air, causing the vapor density of the mixture to approach 1.0 in many cases, and behavior of the vapor from a spilled liquid may not be accurately predicted by using a vapor density based on the pure material. See HEAVIER-THAN-AIR GAS.

vapor incinerator See AFTERBURNER.

vapor pressure The pressure exerted by a gas or vapor. These pressures are experimentally determined by establishing an equilibrium between the gas and liquid phases of the substance in a closed vessel at a specific temperature. Vapor pressures increase with an increase in liquid temperature. The higher the vapor pressure, the greater the tendency of the liquid to evaporate. Common units are millimeters of mercury (mm Hg) and POUNDS PER SQUARE INCH (ABSOLUTE).

vaporize To change liquid (or solid) into a VAPOR; to evaporate.

variance 1. A statistical expression of the spread of a data set about the mean, or average. See STANDARD DEVIATION. 2. A waiver of certain environmental regulatory requirements for a facility, process, or unit.

vascular bundle Tissue in a plant responsible for transport of materials from the roots up and from the leaves down.

vector An organism, such as a mosquito, flea, or tick, that carries a pathogenic microorganism from one host to another.

vector-borne Describing a disease transmitted by a VECTOR.

vegetative control Pollution prevention design involving the use of plant cover and grass BUFFER STRIPS to reduce erosion. Used in STORM WATER RUNOFF control.

vehicle miles traveled (VMT) A statistical measure describing the vehicle use in a defined geographical area as calculated from the sum of the miles traveled by all vehicles in the area over a prescribed period. VMT can rise with an increase in the number of vehicles in use or with an increase in the average miles driven per vehicle, or with both.

velocity (v) The distance moved in a given direction per unit time, such as meters per second (m s^{-1}).

velocity head (V$_H$) The kinetic energy in a hydraulic system. Velocity head is given by

$$v_H = \frac{v^2}{2g}$$

where v is the fluid flow velocity and g is the acceleration due to gravity, expressed in length units. The sum of the ELEVATION HEAD, PRESSURE HEAD, and velocity head equals the total energy of a hydraulic system. See HEAD, TOTAL.

velocity pressure (VP) For air flowing in a duct, the air pressure attributable to the impaction of the moving air molecules. Expressed in the same way as velocity head,

$$VP = \frac{v^2}{2g}$$

where VP is the velocity pressure, v is air velocity, and g is gravitational acceleration. The total pressure in a duct is the sum of the STATIC PRESSURE and the velocity pressure.

velometer A device for measuring air velocity, often used in studies and analysis of workplace airflow and in the design and operation of local exhaust ventilation systems. The portable instrument indicates air velocity by using a spring-loaded vane (plate) that is deflected by the moving air. Also called a swinging vane anemometer. See ANEMOMETER.

ventilation In exposure studies, air intake to the lungs.

ventilation rate (respiration) The volumetric breathing rate: for adult males, about 23 cubic meters/day; for adult females, 21 cubic meters/day; and for children, 15 cubic meters/day.

venturi effect The increase in the velocity of a fluid stream as the fluid passes through a constriction in a channel, pipe, or duct. Calculated by the CONTINUITY EQUATION $Q = VA$, where Q is the volumetric flow rate, A is the area of flow, and V is the fluid velocity. Because Q does not change, if A gets smaller, then V must increase.

venturi scrubber See SCRUBBER, VENTURI.

vertical dispersion coefficient See DISPERSION COEFFICIENT.

Vienna Convention for the Protection of the Ozone Layer A 1985 international agreement reached in Vienna committing signatory nations to research the causes and consequences of OZONE LAYER DEPLETION; no protective measures were included. Subsequent meetings introduced a worldwide phaseout of the CHLOROFLUOROCARBONS and other compounds with OZONE-DEPLETING POTENTIAL. The first supplement to the Vienna Convention that provided for controls to protect the ozone layer was the MONTREAL PROTOCOL. Amendments to the Montreal Protocol were agreed on at meetings in London,

Copenhagen, Montreal again, and Beijing. For the text of the Vienna Convention and its amendments see Website: www.unep.org/ozone/vienna.htm.

vinyl chloride A gaseous organic compound composed of carbon, hydrogen, and chlorine; vinyl chloride monomers (individual units) are used to make the polymer (a long chain), POLYVINYL CHLORIDE. Occupational exposure to vinyl chloride is linked to a rare liver cancer, angiosarcoma. Workplace exposure to vinyl chloride is regulated by the OCCUPATIONAL SAFETY AND HEALTH ACT. The USEPA has set a community air emission standard for vinyl chloride under the NATIONAL EMISSION STANDARDS FOR HAZARDOUS AIR POLLUTANTS provision of the CLEAN AIR ACT. See HAZARDOUS AIR POLLUTANTS.

virtually safe dose (VSD) A dose or exposure level for a CARCINOGEN corresponding to an individual lifetime cancer risk considered to be essentially zero; sometimes suggested as a LIFETIME RISK of 1 in 1 million.

viscosity (h) A measure of the resistance of a fluid to flow. For liquids, viscosity increases with decreasing temperature. For gases, viscosity increases with increasing temperature. Expressed as mass per length-time (e.g., kilograms per meter-second). A common viscosity unit is the poise, named for the French physician Jean Louis Poiseuille (1799–1869): 1 poise equals 1.0 gram per centimeter-second. The viscosity of water at 20° C is 0.01002 poise; therefore the centipoise (one-hundredth of a poise) is often used; this makes the viscosity of water equal to 1.002 (rounded, 1.0) centipoise. Also called dynamic viscosity. Compare KINEMATIC VISCOSITY.

visible range That part of the electromagnetic spectrum that can be seen by the human eye. The wavelengths range from about 400 nanometers (violet light) to about 710 nanometers (red light). Also called visible light.

visibility   In the atmosphere, the distance to which an observer can distinguish objects from their background. The determinants of visibility include the characteristics of the target object (shape, size, color, pattern), the angle and intensity of sunlight, the observer's eyesight, and the extent of light absorption and scattering caused by air contaminants. See KOSCHMIEDER RELATIONSHIP, EXTINCTION COEFFICIENT, AIR QUALITY–RELATED VALUE.

visibility protection   The CLEAN AIR ACT requirement that AIR QUALITY–RELATED VALUES, including visibility, be protected in CLASS I AREAS as part of the PREVENTION OF SIGNIFICANT DETERIORATION control program. The 156 Class I areas include national parks and WILDERNESS AREAS. The STATE IMPLEMENTATION PLANS for states with CLASS I AREAS must include visibility protection measures, such as controls on small-diameter PARTICULATE MATTER and SULFUR DIOXIDE emissions (to control SULFATE formation). The USEPA has added a Regional Haze Rule to its existing visibility regulations. See VISIBILITY.

vitrification   A process forming a highly stable noncrystalline material; glassification. Proposed as a treatment for HIGH-LEVEL RADIOACTIVE WASTE. The waste is encased in a glasslike medium, which resists high temperatures and LEACHATE formation indefinitely. See HIGH-LEVEL NUCLEAR WASTE FACILITY.

volatile   Describes a substance that evaporates or vaporizes rapidly at room temperatures.

volatile hydrocarbons   Organic compounds composed of carbon and hydrogen that evaporate rapidly at room temperatures, for example, gasoline, methanol, and benzene.

volatile organic analysis (VOA)   The assay of carbon compounds with a high vapor pressure, commonly hydrocarbons and hydrocarbon derivatives, in the atmosphere or wastewater. The technology commonly involves GAS CHROMATOGRAPHY coupled with use of a MASS SPECTROMETER. See VOLATILE ORGANIC COMPOUNDS.

volatile organic carbon (VOC)   A measure of the amount of particulate material in a water sample that is lost upon heating. The measure is obtained by passing a given quantity of water through a glass fiber filter, then drying and weighing the solids retained on the filter. The preweighed filter is then heated to about 500°–600° C, and a second weight is obtained. The amount lost during the heating process is termed VOC.

volatile organic compounds (VOCs)   A category of organic compounds with relatively high VAPOR PRESSURE, a major category of air contaminants. Most VOCs are carbon-hydrogen compounds (hydrocarbons), but they may also be ALDEHYDES, ketones, CHLORINATED HYDROCARBONS, and others. Thousands of individual compounds exist, including the unburned hydrocarbon compounds emitted from automobiles or industrial processes and the ORGANIC SOLVENTs lost to evaporation from household, commercial, or industrial cleaning and painting operations and other activities. Some VOCs participate in the atmospheric reactions that lead to PHOTOCHEMICAL AIR POLLUTION, and excessive exposure to certain individual compounds is associated with skin irritation, central nervous system depression, and/or an increased risk of cancer. Large quantities of VOCs are introduced to the air by vegetation; these are termed BIOGENIC VOLATILE ORGANIC COMPOUNDS. Control techniques for VOC include ADSORPTION on ACTIVATED CHARCOAL, use of the CATALYTIC CONVERTER, STAGE II CONTROL, ONBOARD REFUELING VAPOR RECOVERY, use of the CARBON ADSORBER, lowering of REID VAPOR PRESSURE, and use of REFORMULATED GASOLINE. See HAZARDOUS AIR POLLUTANTS, AIR TOXICS.

volatile organic sampling train (VOST)   The air-sampling apparatus specified by the USEPA for the collection of organic material that is not completely destroyed in a waste incinerator. Analysis of the collected organics determines incinerator efficiency. See also DESTRUCTION AND REMOVAL EFFICIENCY, PRODUCTS OF INCOMPLETE COMBUSTION, PRINCIPAL ORGANIC HAZARDOUS CONSTITUENTS.

volatile solids   See VOLATILE SUSPENDED SOLIDS.

volatile suspended solids (VSS)   An indirect measure of the amount or weight of particles composed of organic material suspended in a given amount of water. A specified amount of water or wastewater is filtered. The filtrate, the liquid passing through the filter medium, is discarded. The material remaining on the filter is dried and weighed. The filtered material is then heated to 550° C and weighed again. The difference between the two measures constitutes the weight loss on ignition and is considered to be the amount of all animal matter, plant matter, and detritus to have been suspended in the water passed through the filter.

volatility   A measure of the tendency of a solvent or other material to evaporate at normal temperatures.

volatilization   The process of evaporation.

volcanic ash   Rock formed as lava droplets and globs are blown out of a volcanic vent and cool before reaching the ground, forming particles of varying size. Small particles of ash can travel great distances from the volcano.

volcanic glass   Rock with a glassy texture that forms when molten rock cools so rapidly that the formation of crystal grains is inhibited. Glassy texture forms on the crust of lava flows.

volcanic neck   A vent of a volcano through which gases, ash, and larger particles exit to the outside.

volume reduction   COMPACTION, SHREDDING, INCINERATION, or COMPOSTING of SOLID WASTE for the purpose of reducing the volume occupied in a SANITARY LANDFILL. The processes are intended to reduce the expenses associated with landfill disposal or to extend the life of a landfill.

volumetric flow rate   For a liquid or gas, the volume moving past a point per unit time. For example, a smokestack exhaust of 10 cubic meters per second would result if gas exited a stack 2 meters square at 5 meters per second. Actual flow rate (Q) is expressed as $Q = AV$, where A is the cross-sectional area of the pipe, duct, or stack and V is the velocity of the liquid or gas.

volumetric water content   That portion of the volume of a soil sample that is occupied by water, expressed as percentage by volume.

vulnerability analysis   As part of a HAZARDS ANALYSIS, an assessment of the locations and sizes of the areas that might be harmed by an accidental release of a HAZARDOUS SUBSTANCE. The RADIUS OF VULNERABLE ZONES are estimated by using mathematical models that simulate the behavior of a chemical moving downwind or the spatial effects of an explosion or fire. The vulnerable zone is examined for the presence of sensitive populations, such as schools, hospitals, and nursing homes. Hazards analyses under TITLE III (of the Emergency Planning and Community Right-to-Know Act) and RISK MANAGEMENT PLANS required by the CLEAN AIR ACT both include a vulnerability analysis. See BHOPAL; LEVEL OF CONCERN; LOCAL EMERGENCY PLANNING COMMITTEE; TIER I, TIER II REPORTS; EXTREMELY HAZARDOUS SUBSTANCE.

vulnerable species   A category used in the RED LISTS published by the SPECIES SURVIVAL COMMISSION of the INTERNATIONAL UNION FOR THE CONSERVATION OF NATURE AND NATURAL RESOURCES (IUCN). A vulnerable species has "a high risk of extinction in the wild in the medium-term future" if it meets any of the IUCN detailed criteria, including a declining trend in population size, a shrinking of the species range, a sufficiently low breeding stock, or a forecast probability of 10% that the species will become extinct in the wild within 100 years.

vulnerable zone See VULNERABILITY ANALYSIS.

W

Ward, Barbara (1914–81) English Economist, conservationist, and prolific author. Works on the environment include *Spaceship Earth* (1966), *Progress for a Small Planet* (1980), and *Only One Earth: The Care & Maintenance of a Small Planet* (1972), with RENE DUBOS.

waste broker An individual or company serving the waste management industry through matching buyers and sellers of waste material that has commercial potential. A material recovery facility involved in the separation of recyclable items from municipal solid waste may use such a broker to locate and make arrangements with a buyer of the discarded material. See WASTE EXCHANGE.

waste characterization 1. An estimation of the types and quantities of materials in a waste stream; for instance, the waste can be sampled, sorted by type, then weighed to determine the relative sizes of the waste categories by weight. An alternate method of characterization is the materials flow methodology, which uses data on the production of goods (by weight) combined with product life to estimate the relative sizes of the waste categories. Materials flow makes adjustments for food and yard wastes, imports and exports, and recycling, none of which is captured by production data. 2. Determination of the physical, chemical, and microbiological properties of discards.

waste diversion credit A financial reward system to provide incentives to municipalities and private recycling companies and paid on the basis of the total tonnage of recyclable resources that were diverted from deposit in a landfill.

waste exchange A clearinghouse that matches disposers of waste materials to facilities that can use the waste as a fuel or raw material. For an example, see www.recycle.net/exchange/.

waste-heat recovery Transfer of excess heat produced in one process to provide heat to another system.

waste immobilization See IMMOBILIZATION.

waste ink Fluid remaining from the blending or use of inks in the printing industry. Such fluids are considered hazardous because of the likely presence of solvents, organic compounds, oils, and metals used as pigments.

Waste Isolation Pilot Plant (WIPP) An underground facility near Carlsbad, New Mexico, used for the disposal of TRANSURANIC WASTE arising from nuclear weapon activities. Wastes are buried 2000+ feet below the surface in rooms formed inside geologically stable salt beds that are 2000 feet thick. WIPP is run by the U.S. DEPARTMENT OF ENERGY and government contractors. Website: www. wipp.carlsbad.nm.us/wipp.htm

waste minimization A reduction in the types and quantities of industrial waste, a reduction in waste toxicity, or an increase in RECYCLING and REUSE. For HAZARDOUS WASTE, reducing quantity is not as important as minimizing the wastes that are PERSISTENT, undergo BIOACCUMULATION, or are toxic (referred to as PBT chemicals). Waste minimization focuses on the waste produced at the site of production or manufacturing. See POLLUTION PREVENTION,

SOURCE REDUCTION, WASTE REDUCTION, RESOURCE CONSERVATION AND RECOVERY ACT.

waste reduction A decrease in the amount of MUNICIPAL SOLID WASTE requiring disposal; also called SOURCE REDUCTION. Some methods and management initiatives include restricting or otherwise lowering consumption, increasing the expected lifetime of a product, reducing the amount of packaging needed, reducing the weight of products, and enhancing the ability of a product to be recycled. Waste reduction focuses on the waste produced by consumers. See POLLUTION PREVENTION, WASTE MINIMIZATION, RESOURCE CONSERVATION AND RECOVERY ACT.

waste separation See SOURCE SEPARATION, MATERIALS RECOVERY FACILITY.

waste shipment record (WSR) A required document describing the generation, transportation, and ultimate disposition of a waste containing ASBESTOS.

waste sources Refers to the different sectors that produce solid waste: commercial buildings, households, agricultural operations, mining, construction projects, demolition operations, medical facilities, wastewater treatment facilities, and others.

waste stabilization See STABILIZATION/SOLIDIFICATION.

waste stream The total quantity of solid discards from a city, company, or institution. The discards may be recycled, burned, or landfilled.

waste-to-energy (WTE) The incineration of MUNICIPAL SOLID WASTE with recovery of energy, usually in the form of hot water, steam, or, via further conversion, electricity.

waste treatment lagoon Pond operated to facilitate the biological treatment of wastewater; often used for animal waste. When maintained in an AEROBIC condition, such ponds represent a low-cost and efficient treatment, reducing BIOCHEMICAL OXYGEN DEMAND and TOTAL SOLIDS. However, when these ponds are overloaded with organic material, ANAEROBIC conditions develop, causing odor problems. In-ground systems for industrial wastewaters are subject to LAND DISPOSAL BAN and LAND DISPOSAL RESTRICTIONS and must have LINERS, LEACHATE COLLECTION SYSTEMS, and groundwater MONITORING WELLS. See IMPOUNDMENT, SEWAGE LAGOON, SETTLING POND.

waste treatment plant **1.** For wastewater, a facility designed and operated so that the water can be released into the environment without causing damage to human health or to the environment. See PRIMARY TREATMENT, SECONDARY TREATMENT, TERTIARY TREATMENT, PUBLICLY OWNED TREATMENT WORKS, SEWAGE TREATMENT PLANT. **2.** For a public drinking water supply, a facility operated to assure that the piped water delivered to the public is safe and pleasing to use. See SAFE DRINKING WATER ACT, GROUNDWATER RULE, SURFACE WATER TREATMENT RULE, TOTAL COLIFORM RULE, CHLORINATION, DISINFECTION BY-PRODUCTS, MAXIMUM CONTAMINANT LEVEL, SECONDARY MAXIMUM CONTAMINANT LEVEL.

wasteload allocation (WLA) A system designed to limit the total discharge of pollutant materials into a receiving stream because the stream does not achieve one or more ambient water quality standards specified by TECHNOLOGY-BASED EFFLUENT LIMITATIONS. In this case, a TOTAL MAXIMUM DAILY LOAD (TMDL), which is the maximum amount of the pollutant(s) that can be released without violating water quality standards, is calculated. Each source is allocated part of the TMDL; the amount allotted to POINT SOURCES is the wasteload allocation; the amount allotted to NONPOINT SOURCES is termed the LOAD ALLOCATION. See WATER QUALITY–BASED EFFLUENT LIMITATION.

wastewater Water discharged from homes, businesses, and industries that contains dissolved, suspended, and particulate inorganic or organic material. The term is also used as a synonym for SEWAGE. Also called domestic wastewater.

water-air ratio (K_w) An expression for the partitioning of a substance present in a dilute solution between the water and the overlying air. The ratio is computed by

$$K_w = C_w/C_A$$

where C_w is the water concentration of the chemical (micrograms of the chemical per cubic centimeter of water) and C_A is the air concentration of the chemical (micrograms of the chemical per cubic centimeter of air).

water balance A measure of the amount of water entering and the amount leaving a system.

water column A hypothetical cylinder of water from the surface to the bottom of a stream, lake, or ocean within which physical, chemical, and/or biological properties can be measured.

Water column

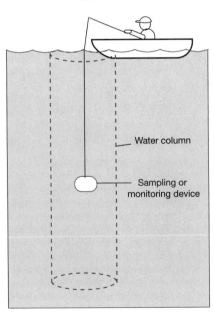

Water column

Sampling or monitoring device

water-cooled reactor A NUCLEAR REACTOR that employs water to cool the reactor CORE. A nuclear reactor is a device designed to promote the FISSION of an appropriate fuel (such as URANIUM 235) in a controlled manner. The heat produced during the fission event must be removed from the device to prevent excessive buildup. Water is usually used as the heat transfer agent. Other coolants used in nuclear reactors of other designs are liquid sodium and inert gases. See LIGHT-WATER REACTOR.

water cycle The progression of water through several phases and processes as it moves within and between major compartments of the Earth. As an example, trace one of many different pathways in the cycle: Water is converted from the liquid phase to the gas phase as it evaporates from the sea. Water vapor CONDENSATION forms liquid droplets, leading to precipitation over the land and subsequent runoff into rivers and then to the sea, illustrating the cycling of water from the ocean, into the atmosphere, onto or into the ground, and back to the sea.

water dilution volume (WDV) The volume of water required to dilute radioactive waste to a concentration meeting drinking water standards. Typically expressed in cubic meters of water per metric ton of radioactive waste.

water droplet coalescence Merging of small particles of water in clouds into drops that are sufficiently large to fall as precipitation.

Water Environment Federation A private, nonprofit technical and educational group dedicated to wastewater treatment and water quality protection, including surface water, groundwater, and both POINT SOURCE and NONPOINT SOURCE controls. Membership in 1999, 40,000. Based in Alexandria, Virginia. Website: www.wef.org

Water Environment Research Federation A private research group funded by municipal sewage treatment plants and industrial wastewater treatment facilities; affiliated with the WATER ENVIRONMENT FEDERATION. Supports ongoing research in collection and treatment systems, human heath and environmental effects, BIOSOLIDS management, and watershed management. Some research is done in cooperation with USEPA water programs. Website: www.werf.org

water hyacinth A floating freshwater plant belonging to the genus *Eichhornia*. Introduced into the United States in the late 19th century as an ornamental plant, water hyacinth has become a prolific nuisance weed that clogs waterways in the southern United States.

water-quality–based effluent limitation (WQBEL) The allowable discharge of any water pollutant by a POINT SOURCE into a WATER QUALITY LIMITED SEGMENT: that is, a stream that does not meet the ambient water quality standard for that pollutant even though TECHNOLOGY-BASED EFFLUENT LIMITATIONS have been applied. In this circumstance, the regulatory agency sets a TOTAL MAXIMUM DAILY LOAD for that stream segment and allocates to the POINT SOURCES for that pollutant part of what is termed the WASTELOAD ALLOCATION. The TMDL is the maximal amount of the pollutant that can be released into the stream without violating the ambient water quality standard for the pollutant. A point source permit would reflect the source wasteload allocation for the offending pollutant as a WQBEL, for example, an allowable phosphorus discharge of five pounds per day.

water quality criteria The aqueous concentration limits for pollutants in water that is to be used for specific purposes. The criteria are set for individual pollutants and are based on different water uses, such as a public water supply, an aquatic habitat, an industrial supply, or recreational facility.

water quality limited segment A portion of a stream where the condition of the water does not meet water quality standards and/or where standards are not expected to be achieved after TECHNOLOGY-BASED EFFLUENT LIMITATIONS on all POINT SOURCES are applied. Therefore, controls beyond the technology-based discharge limits are required for the stream segment to meet the ambient standards. See TOTAL MAXIMUM DAILY LOAD, WATER-QUALITY-BASED EFFLUENT LIMITATION, WASTELOAD ALLOCATION, LOAD ALLOCATION.

water quality standards Regulations specifying the intended use of a body of water and establishing the criteria to be used to protect the designated use. The standards are prepared by each state and are subject to the approval of the USEPA.

water reactive Describing any substance that reacts spontaneously with water to release a flammable or toxic gas, such as sodium metal.

water softener An apparatus designed to remove divalent metal ions (the most important of which are calcium, magnesium, and iron) from water, often replacing the divalent or trivalent ions with the monovalent sodium ion. See ION EXCHANGE.

water soluble Describes a material that dissolves in water, for example, table sugar.

water-soluble fraction (WSF) The portion of crude oil or oil product that is soluble in water.

water-source heat pump Heating unit designed to transfer heat from groundwater to the inside of a building (heating) or from a building into the ground (cooling). Open well systems pump groundwater up then inject it back into the ground in a different well. Ground loop (closed) systems recirculate the same water or water/coolant mixture into the ground and back. Also called ground source heat pump, geothermal heat pump, and earth-coupled heat pump.

water table The uppermost level of the below-ground, geological formation that is saturated with water. Water pressure in the pores of the soil or rock is equal to atmospheric pressure.

water treatment The processing of SOURCE WATER (well water or surface water) for distribution in a public drinking water system. See WATER TREATMENT PLANT.

waterborne Of or related to something that is carried by water, for example, a disease transmitted by water contaminated by a disease-causing microorganism.

waterborne disease Diseases of humans, commonly involving the digestive tract, that are transmitted from one individual to another by water. Typhoid fever, dysentery, cholera, and giardiasis are common examples.

waterlogging Filling of the spaces among soil grains with water, thereby displacing air. The process deprives plant roots of needed oxygen and promotes the development of ANAEROBIC conditions as the soil microbes remove the remaining oxygen. Most agricultural crops cannot grow in such soils. The event characteristically occurs when fields are overirrigated, rainfall is excessive, fields are flooded, and drainage is inhibited.

watershed That area of land that drains into a lake or stream.

waterwall furnace See WATERWALL INCINERATOR.

waterwall incinerator An energy recovery system used in some municipal waste incinerators. The combustion chamber of the incinerator is lined with steel tubes containing circulating water. The heat from the combustion boils the water, and the steam can be sold or used to turn turbines in an electric generator.

watt (W) The SI unit of power equal to one JOULE per second. Expressed as 1 W = 1 J s^{-1}.

Watt, James (1938–) American U.S. secretary of the interior, 1981–83. Controversial lightning rod for environmental activist groups. Watt advocated exploration for and extraction of fossil fuels, minerals, and timber on federal lands. Supporter of the SAGEBRUSH REBELLION.

wavelength For ELECTROMAGNETIC RADIATION, the distance between corresponding points of a wave cycle. Wavelength and FREQUENCY are inversely proportional. See ELECTROMAGNETIC SPECTRUM.

weak acid A compound that releases small amounts of hydrogen ions when dissolved in water. As a result, low concentrations of hydronium ions are formed relative to the volume of water employed. An example of a weak acid is acetic acid, which is used as vinegar. See ACID.

weathering The breakdown of rock through a combination of chemical, physical, geological, and biological processes. The ultimate outcome is the generation of soil.

Weibull model A DOSE-RESPONSE RELATIONSHIP represented by $P(e) = \gamma + (1 - \gamma)(1 - e^{(-a\ D^b)})$, where $P(e)$ is the probability of an adverse effect given a continuous dose D, γ is the background response rate, and a and b are fitted constants. Used for EXTRAPOLATION from high-dose observations to low-dose exposures.

weighted average For a series of recorded observations, the sum of the products of the frequency of certain values and the value of the observation divided by the total number of observations. For example, for one measurement of 5 grams, three measurements of 7 grams, and two measurements of 2 grams, the weighted average is $[1(5) + 3(7) + 2(2)]/6 = 5$ grams. See TIME-WEIGHTED AVERAGE.

weight fraction An expression of concentration of materials in solutions or mixtures. The weight fraction of a certain material is the weight of the material in question divided by the total weight of the solution or mixture. For example, in a mixture of A, B, and C,

weight fraction of A
= (weight of A)/(weight of A + B + C)

weighting networks A frequency-specific adjustment made by a sound level meter to the measured DECIBEL levels to account for the increased sensitivity of the human ear to higher-frequency sounds and lower sensitivity to lower frequencies. Three networks are available, called A, B, and C. The A-weighted scale is by far the most frequently used because it best matches the sound frequency sensitivity of the human ear at the sound levels most commonly encountered. See FLETCHER-MUNSON CONTOURS; DECIBELS, A-WEIGHTING NETWORK.

weir An underwater dam or barrier in a channel or ditch placed to limit or control water flow; water flows over the top of the weir.

welded tuff Structure formed after the accumulation ash, rock, hot crystal fragments, and volcanic glass ejected from a volcano. The hot materials fuse together, producing a coherent mass.

well field A relatively small area of land containing several wells producing water or oil.

well plug Any watertight or gastight seal installed in a well to prevent the flow of fluids or gases.

well purging The removal of water that has accumulated in a well so that a fresh quantity of groundwater can enter the well. The water that enters the well after purging is more reflective of the chemical status of water in the sands of the aquifer than the water that has been standing in the well piping for some time.

well stimulation Cleaning, enlarging, or increasing the pore space of a well used

for the injection of fluids into subsurface geological strata.

well water Water produced from a well drilled or bored into the ground; usually pumped but sometimes free-flowing (ARTESIAN WATER). Often a source of RAW WATER for a public water supply. See AQUIFER, GROUNDWATER.

wellhead protection area A prescribed area around a well producing water for distribution to the public. Depending on the circumstances, the area consists of a circle with the well at the center and a radius of one to two miles. Potential sources of contaminants that may infiltrate into the water well through the BOREHOLE, such as abandoned wells, UNDERGROUND STORAGE TANKS, hazardous materials storage locations, and businesses that have a potential for contaminating groundwater, are identified, then removed or controlled. Local government controls include zoning restrictions, inspections, public education, and monitoring. See SOURCE WATER ASSESSMENT PROGRAM, SOURCE WATER PROTECTION AREA, SOURCE WATER PROTECTION PROGRAM, SIGNIFICANT POTENTIAL SOURCE OF CONTAMINATION, UNDERGROUND INJECTION CONTROL.

West Antarctic ice sheet That part of the ice covering Antarctica about which most concern is expressed in discussions of GLOBAL WARMING. The concern is based on past episodes of rapid sea-level rise. The West Antarctica ice, separated from the eastern sheet by the Transantarctic mountain range, is the most prominent remaining ice-filled marine basin on Earth. The ice sheet on the western side of the continent has lost about two-thirds of its mass since the last glacial maximum about 20,000 years ago.

wet adiabatic lapse rate The rate of temperature decrease as a parcel of air saturated with water rises and the pressure decreases, given by

$$\gamma_s = -\frac{dT}{dz}$$

where dT is the temperature change, dz is

the change in altitude, and γ_s is the saturated (wet) ADIABATIC LAPSE RATE. Because moisture is condensing in the rising parcel of air and releasing latent heat, the temperature drop with increasing altitude is less than the (dry) adiabatic lapse rate, or about 0.6° C per 100 meters. The rate assumes that there is no exchange of heat between the parcel and the surrounding air by conduction or mixing. See DRY ADIABATIC LAPSE RATE.

wet-bulb temperature The temperature reading from a thermometer with a wetted wick surrounding the bulb. The evaporative loss of LATENT HEAT from the wick lowers the temperature reading. Used with the DRY-BULB TEMPERATURE and a table to compute RELATIVE HUMIDITY. See PSYCHROMETER.

wet deposition The introduction of acidic material to the ground or to surface waters by sulfuric and nitric acids dissolved in rainfall or snow. Compare DRY DEPOSITION. See ACID RAIN.

wet scrubbing A process that removes particles, gases, or vapors from an exhaust gas by passing the exhaust through a shower of water or water that contains an agent to react with the material to be removed. See SCRUBBER, IMPINGEMENT; SCRUBBER, SPRAY; SCRUBBER, VENTURI; SCRUBBER, PLATE TOWER.

wet stack A stack that is capable of handling moisture that condenses from the exhaust gas exiting a SCRUBBER.

wet test meter A laboratory instrument used mainly to calibrate the volume of airflow in other instruments. The meter consists of a set of compartments of equal volume that can rotate inside an outer casing. The compartments are partially submerged in water or other liquid. Air enters the meter and forces the liquid out of a compartment, causing the compartments to rotate. As the open end of each compartment rotates past the top of the meter, the air flows out of the meter. The measurement of the liquid level in the meter indicates the amount of air released by each compartment, and a counter determines the number of compartments that were filled with air. This information is used to compute the airflow.

wetlands The regulatory definition used by the USEPA and the U.S. Army Corps of Engineers reads, "Wetlands are areas that are inundated or saturated by surface or ground water at a frequency and duration sufficient to support, and that under normal circumstances do support a prevalence of vegetation typically adapted for life in saturated soil conditions. Wetlands generally include swamps, marshes, bogs, and similar areas." The three parts of the definition address soil HYDROLOGY, soil types, and vegetation. The *Wetlands Delineation Manual* published by the Army Corps of Engineers in 1987 is the official guide used for wetlands identification. In 1989, a new wetlands manual, the *Federal Manual for Identifying and Delineating Jurisdictional Wetlands,* was released, and a 1991 revision was also proposed, but Congress voted to continue with the 1987 guidance pending further study. A new manual incorporating the improvements is pending. The CLEAN WATER ACT requires anyone who discharges dredged material or fill into the waters of the United States (which include wetlands) to get a permit (SECTION 404 PERMIT) from the Army Corps of Engineers, with approval from the USEPA. This means that any development or clearing of land classified as a wetland, even if far from an estuary, stream, or lake, must have a permit to proceed. The wetlands permit requires the applicant to show that alternatives to the wetlands destruction are not available, that the project will incorporate reasonable controls to minimize loss, and that unpreventable damage will be replaced by wetlands addition or conservation elsewhere. See WETLAND LOSS, NO NET LOSS, MITIGATION BANKING, CONSTRUCTED WETLANDS, SWAMPBUSTER PROVISION. Website: www.epa.gov/OWOW/wetlands/

wetland loss The conversion of land that is intermittently covered with water to uses that degrade the unique biological community characteristic of such areas. The loss of coastal wetlands and the subsequent conversion of wetland areas into open water in Louisiana are major environmental problems in that state.

Wetlands Reserve Program (WRP) A voluntary program, directed by the U.S. Department of Agriculture, organized to preserve and restore WETLANDS. A landowner can sell a CONSERVATION EASEMENT (agreement to manage the land as wetland habitat only) to the Department of Agriculture that will last permanently or for 30 years. The WRP will also share the cost of restoring wetland with the landowner, including restorations without an easement agreement.

wheeling The transfer of electricity from one system through another to users; occurs when one system buys power from another or when retail customers purchase power from a company that does not have electric lines connecting it to the customer. Can also refer to the transfer of natural gas or drinking water.

whey The clear fluid that separates from the solid curd when milk is allowed to coagulate, or sour. The curd contains most of the protein solids from the milk (casein), and the whey contains most of the small, soluble compounds. Whey represents a waste liquid produced in the manufacture of some cheeses and has a high BIOCHEMICAL OXYGEN DEMAND.

white damp Carbon monoxide present in an underground mine after blasting or remaining after a mine fire or explosion. From the German *damf,* "vapor." See DAMP.

white goods Refrigerators, stoves, clothes washers and dryers, and other appliances contained in MUNICIPAL SOLID WASTE.

whole-body dose The exposure of the entire human body to radiation, for example, BACKGROUND RADIATION. This type of exposure is also of importance when a RADIOISOTOPE is inhaled or ingested then uniformly distributed throughout the body.

whole effluent toxicity testing (WET testing) The exposure of living organisms to the discharge released from an industrial or wastewater treatment facility for the purpose of determining the effects of the effluent on natural biological organisms. The method replaces the measurement of the concentrations of specific chemicals in the discharged water. Some of the organisms tested are fish, shrimp, water fleas, oysters, and sea urchins. Also called biomonitoring. See BIOASSAY.

Wien's law The physical law relating the peak wavelength of electromagnetic radiation emitted by a radiating body to the surface temperature of that body. The higher the temperature, the shorter the peak wavelength of the energy emitted. For example, the Sun emits an energy spectrum with a peak wavelength of about 480 nanometers, which is within the visible part of the electromagnetic spectrum; since the surface temperature of the Earth is much lower than that of the Sun, the energy emission spectrum of the Earth has a much longer peak wavelength, about 10,000 nanometers, which is in the INFRARED, or radiant heat, range. The emission can be expressed as λT = constant, where λ is the peak wavelength in nanometers, T is the absolute temperature of the body in Kelvin, and the constant is equal to 2.9×10^6 nanometers-K. The Earth's outgoing infrared radiation is absorbed by GREENHOUSE GASES. See GREENHOUSE EFFECT, GLOBAL WARMING.

Wild and Scenic Rivers System Over 100 designated stream segments throughout the United States chosen for their scenic, historic, aesthetic, or scientific characteristics to be protected from overuse and economic development. Managed by the National Park Service, U.S. Department of Interior. Website: www.nps.gov/rivers/

Wilderness Act   See WILDERNESS AREA.

wilderness area   Large tracts of federal land, most over 5000 acres, that are set aside and allowed to remain in a natural state. Such activities as construction of roads, development of recreational facilities, removal of trees, and hunting are prohibited. In some cases, even the fighting of fires started by natural means is limited. The 1964 Wilderness Act allows the U.S. government to set aside sections within the NATIONAL FOREST SYSTEM, national parks, and national wildlife refuges as wilderness areas, which are administered by the National Park Service, Forest Service, FISH AND WILDLIFE SERVICE, and BUREAU OF LAND MANAGEMENT. There are about 650 areas in the National Wilderness Area Preservation System, totaling over 100 million acres, about 50% of which is in Alaska. Website: www.wilderness.net/nwps/

Wilderness Society, The   An American environmental organization concerned with protecting wildlife habitat and wildlife refuges as well as preserving public lands. In 1999, 255,000 members. Headquarters in Washington, D.C. Website: www.wilderness.org

wildlife   Most commonly used to refer to the vertebrate animal population in natural environments. May also be used to describe all biota that have not been domesticated.

Wiley, Harvey (1844–1930)   American Head of the U.S. Department of Agriculture Bureau of Chemistry. Led "Poison Squad" tests, which demonstrated the toxicity of many early FOOD ADDITIVES, leading to the Pure Food and Drug Act of 1906, a predecessor of the FOOD, DRUG, AND COSMETIC ACT. See SINCLAIR, UPTON.

wind farm   Area occupied by a multitude of wind turbines installed to generate electricity from the wind power produced by strong prevailing winds. Altamont Pass, California, is a typical example.

wind profile power law   The expression used to estimate the (higher) wind speed at the top of a smokestack by using a measure of wind speed at ground level; the law is applied in AIR QUALITY DISPERSION MODELING. It is expressed as
$$(u_2/u_1) = (z_2/z_1)^p$$
where u_2 is the wind speed at the higher altitude z_2, u_1 is the wind speed at the lower altitude z_1, and the exponent p varies with atmospheric TURBULENCE.

wind rose   A diagram depicting the strength and direction of the winds as measured over time at a specific station.

windbreak   Rows of trees or shrubbery established to protect or shield homes, land, or crops from the damage caused by high winds.

window (infrared)   The wavelength range of the infrared emissions from the surface of the Earth, between about 8 and 12 micrometers, that is poorly absorbed in the atmosphere by the GREENHOUSE GASES, water vapor, and carbon dioxide. This characteristic allows a portion of the heat radiated by the Earth to escape directly to space.

windrow   A long, narrow COMPOST pile. In large-scale operations, the design allows convenient access of machines, which turn (mix) the material periodically.

Winkler method   A standard procedure for measuring the level of DISSOLVED OXYGEN in water. This laboratory analysis, also called the iodometric method, is a reliable titrametric procedure. A solution of divalent manganese and a strong alkali are added to a water sample. The oxygen dissolved in the water oxidizes the manganous hydroxide to hydroxides of higher valences. In the presence of iodide ions and acidification, the oxidation of the metal is reversed with the release of iodine in amounts equivalent to the amount of oxygen dissolved in the original water sample. The released iodine is titrated with a standard solution of thiosulfate, using starch as the indicator.

Wind rose

N

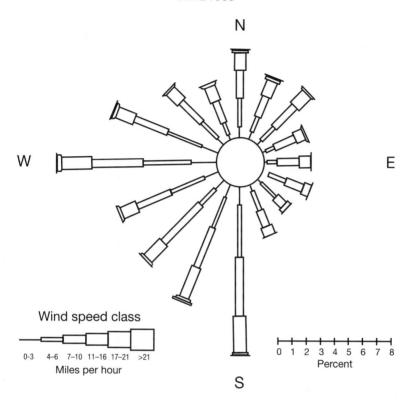

W

E

Wind speed class

0-3 4–6 7–10 11–16 17–21 >21
Miles per hour

0 1 2 3 4 5 6 7 8
Percent

S

Winogradsky column A laboratory microenvironment containing soil, water, and microorganisms. The column was instrumental in the discovery of microbial involvement in the process of NITRIFICATION and the description of chemoautotrophic bacteria. The apparatus can be used to observe the ecological changes caused by the metered addition of test chemicals. See CHEMOAUTOTROPH.

wipe test A sampling method used to determine the presence of hazardous or radioactive substances that can be removed from surfaces. For the detection of radioactive contamination, for example, a small piece of absorbent material (filter paper) is wiped across a table, sink, or other surface, and the absorbent material is then assayed for radioactivity.

wise use movement A diverse collection of interest groups, emerging in the late 1980s, that takes its name from GIFFORD PINCHOT, who wrote in his biography, "Conservation means the wise use of the earth and its resources for the lasting good of men" (*Breaking New Ground*, 1947). Their common enemy is the agenda of radical environmentalist groups, especially restrictive federal land use policies and the controls on economic development implemented by the ENDANGERED SPECIES ACT and WETLANDS regulation. The movement represents a strong backlash against the tactics of radical environmental groups, although wise use activists employ many of the same methods used successfully by the environmentalists. Some wise use advocates have reacted violently to what they consider tree worshiping at the expense of human needs. Also used to identify those who advocate unrestricted access to natural resources. See SNAIL DARTER; SPOTTED OWL; WILDERNESS AREA; WILD AND SCENIC RIVERS SYSTEM; WATT, JAMES; CONSERVATION; PRESERVATION; TAKING; FEDERAL LAND POLICY AND MANAGEMENT ACT; MULTIPLE USE; ROADLESS AREA REVIEW AND EVALUATION; SAGEBRUSH REBELLION; LEOPOLD, ALDO; ABBEY, EDWARD.

withdrawal of water The removal of water from a surface or underground source without regard for any quality change that occurs as a result of withdrawal and use. In contrast, if the water quality is degraded such that reuse is limited, the water is said to be consumed. See CONSUMPTION OF WATER.

withdrawal, public land In the management of public lands, the temporary or permanent suspension of all or some of the laws that allow public land to be used for certain purposes, such as mineral extraction. Allowed by the FEDERAL LAND POLICY AND MANAGEMENT ACT. Withdrawal followed by RESERVATION for a single use, such as a wilderness area, has been a method of setting aside public lands for environmental preservation.

work A force acting over a distance. The work W done by a constant force F on a body that undergoes a displacement s is expressed as $W = (F \cos q)s$, where q is the angle between the force and the displacement. The SI work unit is the joule, or one newton-meter. The work unit in the U.S. system is the FOOT-POUND.

working face The location within a solid waste disposal site (LANDFILL) where waste is discharged and compacted before burial with cover material.

working level (WL) A measure of the RADIOACTIVITY of gases released by the RADIOACTIVE DECAY of RADON. The unit is used as an expression of radon exposure level for individuals. One WL is the combined radioactivity of radon and its short-lived DAUGHTER products in one liter of air that emits a total alpha energy of 1.3×10^5 million electron volts upon complete decay to ^{210}Pb. USEPA guidelines for residential radon levels recommend taking actions to lower exposure if indoor WL measurements are greater than about 0.02 WL, which corresponds to about 4 picocuries per liter of air. See RADON DAUGHTERS.

working-level month (WLM) A unit for individual exposure to radioactivity resulting from the inhalation of air contaminated with a sufficient amount of the RADIOACTIVE DECAY products of RADON to equal one WORKING LEVEL of radioactivity for 170 working hours (one month of working in a contaminated environment).

World Climate Applications and Services Programme (WCASP) Operated by the United Nations, a subprogram of the WORLD CLIMATE PROGRAMME intended to promote the use of climate information and other data in activities relating to food, water, energy, urban planning, and construction. Website: www.wmo.ch/

World Climate Data and Monitoring Programme (WCDMP) Operated by the United Nations, a subprogram of the WORLD CLIMATE PROGRAMME intended to promote the improved collection, digitization, quality control, storage, retrieval, and use of climate-related information. Website: www.wmo.ch/

World Climate Programme (WCP) A United Nations activity operating under the WORLD METEOROLOGICAL ORGANIZATION. Started in 1979 to coordinate and sponsor data collection and research on the world's climate. Website: www.wmo.ch/

World Commission on Environment and Development (WCED) The United Nations panel that published *Our Common Future* in 1987, in which it promoted SUSTAINABLE DEVELOPMENT, drawing international attention to the concept. See BRUNDTLAND COMMISSION REPORT.

World Conservation Union See INTER-NATIONAL UNION FOR THE CONSERVATION OF NATURE AND NATURAL RESOURCES.

World Health Organization (WHO) A multinational agency dedicated to improving and maintaining public health; founded in 1948. The organization works through regional offices with national governments to support programs in food and water sanitation, epidemiological surveillance, communicable disease control, and many other areas. Headquarters are in Geneva, Switzerland. Website: www. who.int/

World Meteorological Organization (WMO) A United Nations agency started in 1950 to promote and coordinate meteorological data collection and research among member nations. The organization, based in Geneva, Switzerland, provides a scientific voice on the state and behavior of the atmosphere and world climate. Some of the activities are weather prediction, climate change monitoring and research, ozone depletion monitoring, and forecasting of tropical storms. The WMO established the WORLD CLIMATE PROGRAMME and its related research activities. Website: www.wmo.ch/

World Wildlife Fund International private organization dedicated to natural resource conservation and wildlife protec-tion, especially of ENDANGERED SPECIES. Founded in 1961, with a U.S. membership in 1999 of 1.2 million. Website: www. worldwildlife.org

Worldwatch Institute A private research and policy organization emphasizing global environmental issues. Publishes *State of the World* and *Vital Signs*, annual reports of trends in population, energy use, food resources, pollution emissions, and species conservation. Headquarters in Washington, D.C. Website: www.worldwatch.org

worst-case scenario Part of the RISK MANAGEMENT PLANS required of certain industrial, utility, military, water treatment, and small business facilities that either manufacture or use any of 140 different chemicals. The worst case is an accident at a facility involving the theoretical release of the largest amount of a hazardous chemical or a theoretical catastrophic fire or explosion. Documents must be developed to illustrate steps that are being taken to reduce the likelihood of the theoretical worst-case occurrence and to manage the emergency response in the event of an accidental release that affects the community outside the facility. Worse-case scenario planning is also performed under the Emergency Planning and Community Right-to-Know Act (TITLE III). Compare ALTERNATE-CASE SCENARIO.

X

xenobiotics A general term for chemicals foreign to their surroundings, for example, chemicals present in organisms or the environment that are not naturally found there. The presence of xenobiotics almost always is an indicator of human activity. Examples include most pesticides, SYNTHETIC ORGANIC CHEMICALS, and most FOOD ADDITIVES.

xenon (Xe) A heavy, relatively inert gas found in very minute quantities in the atmosphere. The gas is employed in flashbulbs used for photographic purposes and may be produced as a FISSION product in a NUCLEAR REACTOR.

xeric Describing an organism that requires little moisture or a habitat containing little moisture. Dry.

xerophyte A plant, such as cactus, adapted to a dry environment. Compare HYDROPHYTE, MESOPHYTE.

X ray ELECTROMAGNETIC RADIATION having a wavelength shorter than that of ultraviolet light and usually longer than that of gamma rays. Essentially the same as a GAMMA RAY except that X rays originate outside the atomic nucleus and gamma rays originate inside the nucleus. X rays are a form of IONIZING RADIATION and have excellent penetrating ability; they are used in a variety of medical treatments (cancer therapy, for example) and diagnostic imaging technologies. Excessive exposure to X rays has resulted in mutations, abnormal fetal development, and cancer development.

X-ray diffraction An analytical method that involves exposing samples to X rays. The radiation is reflected from the sample in response to the structure of the crystals that compose the sample. Consequently, the method is useful in the identification of elements in materials and coatings. Because of the structure of the asbestos crystal, it is especially useful in detecting this element in old construction material.

X-ray fluorescence (XRF) A technique that can be used to visualize the presence of X rays. Certain substances absorb the electromagnetic energy of X rays, then emit that energy as an altered form of electromagnetic radiation, usually as visible light. The glowing of a material when struck by X rays. The technology is employed to produce X-ray images on video screens (fluoroscopy) that allow instant viewing or viewing of a system such as the gastrointestinal tract in motion rather than the still photographic images produced on standard X-ray films. The technology is also employed to lessen the medical exposure of patients to X rays. Intensifying screens made of fluorescent materials can produce X-ray images by using radiation at a much lower dose than that required by standard film techniques.

yard waste   That part of solid waste produced at a household or business consisting of grass clippings, leaf litter, plant residues, landscaping wastes, and similar items. Some communities require the separation of such material from the other household discards. Alternatives to LAND-FILL disposal include INCINERATION and COMPOSTING.

years of life lost   **1.** The expected shortening of a lifetime due to exposure to a hazardous material or to radiation. Calculated from the expected lifetime for a non-exposed person minus the expected lifetime of a person exposed to the dangerous agent. **2.** The estimated years of life lost from premature death caused by a disease or trauma, such as the years of life lost caused by automobile accidents or prostate cancer. Also called years of potential life lost.

Yellow Book   A USEPA publication, *The Yellow Book: Guide to Environmental Enforcement and Compliance at Federal Facilities,* written for managers at the roughly 15,0000 federal facilities to explain and outline the requirements of federal and state environmental regulations. The 350+ page manual is a useful compendium of environmental laws for students, industry representatives, and interested citizens. Website: http://es.epa.gov/oeca/fedfac/yellowbk/

yellow boy   Deposit on the bottom of streams contaminated with acidic drainage from coal mines and some mines from the removal of other minerals. The deposits vary in color from yellow to brownish red and consist of various oxides of iron. The most common cause is ACID MINE DRAINAGE or runoff from piles of mine tailings. The yellow boy deposits indicate a strongly acidified water.

yellow cake   Uranium oxide (U_3O_8) that results from the refining of uranium ore. The purified material contains 99.3% URANIUM 238 and 0.7% URANIUM 235. The term is applied because of the color and texture of the material.

yes, in my backyard, for a price (Y, IMBY, FAP)   The slogan adopted by certain communities willing to site a waste disposal facility nearby in exchange for an acceptable economic benefit.

Yucca Mountain   The site in Nevada proposed as the repository for SPENT FUEL from NUCLEAR REACTORS and HIGH-LEVEL RADIOACTIVE WASTE. Approximately 90% of the waste to be deposited in this location will be from commercial nuclear power plants, with the remainder to come from defense programs. See NUCLEAR WASTE POLICY ACT. Website: www.epa.gov/radiation/yucca/

Z

Z tables Listings of workplace exposure limits for toxic and hazardous substances published by the Occupational Safety and Health Administration. The tables are found in the CODE OF FEDERAL REGULATIONS, Subpart Z, Title 29, Part 1910. Website: www.osha-slc.gov/OshStd_toc/OSHA_Std_toc_1910_SUBPART_Z.html

zero discharge 1. The goal, in the preamble to the CLEAN WATER ACT, of zero pollutants in water discharges. 2. Describing a facility that does not release any wastewater to the environment but recycles and reuses it internally. Also called zero wastewater discharge systems or closed-circuit systems. 3. Describing a regulatory requirement that certain (not all) pollutants be undetectable in a waste stream, for example, a zero discharge standard for certain persistent bioaccumulative toxic chemicals (PBT chemicals) into Lake Superior. Zero discharge is sometimes softened to "virtually eliminate."

zero drift For pollution-monitoring equipment, the change in the monitor response from the zero calibration value and the monitor response to a zero calibration input at a later time, such as 24 hours later; how the instrument's "zero" reading changes on the instrument readout, above or below zero, with time.

zero emission vehicle (ZEV) Automobile or light truck that operates without the release of air pollutants from the vehicle itself, typically an electric car or one that uses hydrogen as the fuel (FUEL CELL technology). The CALIFORNIA AIR RESOURCES BOARD has mandated that in 2003 at least 10% of new vehicles in Cali-

fornia be ZEVs. Of course, a bicycle is a zero emission vehicle.

zero-infinity dilemma In a risk analysis that ranks environmental hazards by the product of their magnitude and probability, the policy problem posed by a catastrophic harmful scenario that has a very low probability of occurrence. For example, the introduction of an extremely potent human carcinogen as an additive to aspirin might cause tens of thousands of excess cancers over several decades, but the likelihood of such a chemical evading the normal toxicity screenings for additives is very low.

zero-order reaction A chemical reaction in which the rate of reaction is independent of the concentration of a reactant or the concentration of any other chemicals present. Compare FIRST-ORDER REACTION.

zero population growth (ZPG) A condition in which a population in a given location neither increases nor decreases over time. The increases due to births and immigration are balanced with decreases caused by deaths and emigration.

zero pressure A complete vacuum; the zero reference point on the ABSOLUTE PRESSURE scale. Note that zero GAUGE PRESSURE equals ATMOSPHERIC PRESSURE.

zero tolerance In pesticide regulation, a requirement that no amount of pesticide may remain on an agricultural commodity when shipped; in effect, the TOLERANCE for these pesticides is the accepted analytical limit of detection. Tolerance levels are established under the authority of the FED-

ERAL INSECTICIDE, FUNGICIDE, AND RODENTICIDE ACT and the FOOD, DRUG, AND COSMETICS ACT. See PESTICIDE RESIDUE, REGISTRATION, DELANEY CLAUSE, FOOD QUALITY PROTECTION ACT.

zone of aeration See UNSATURATED ZONE, VADOSE ZONE.

zone of engineering control The area occupied by a hazardous waste TREATMENT, STORAGE, OR DISPOSAL facility that the owner or operator can readily decontaminate if a leak is detected, thus preventing hazardous waste or hazardous constituents from entering groundwater or surface water.

zone of initial dilution (ZID) **1.** That area where the discharge from an outfall first mixes with the receiving water; much smaller than the MIXING ZONE. Water quality standards for acutely toxic pollutants apply outside the zone of initial dilution but inside the (larger) MIXING ZONE. Water quality standards for chronic effects apply only outside the MIXING ZONE. **2.** For PUBLICLY OWNED TREATMENT WORKS that discharge into marine waters and are allowed to discharge wastewater that has

not undergone SECONDARY TREATMENT, that area where the discharge from an outfall first mixes with the receiving water. Beyond this area, WATER QUALITY STANDARDS apply. This volume is dependent upon water density gradients near the outfall and the design of the diffuser (end of the outflow pipe). See MIXING ZONE.

zone of leaching A layer of soil, also termed the E horizon, usually consisting of a thin layer of soil sandwiched between the topsoil or A horizon and the subsoil or B horizon. This zone represents a transition layer that is distinct chemically and physically from the layers above or below. The zone of leaching is a layer of soil that has been modified by the accumulation of dissolved and suspended materials from the layer above and the removal of soluble materials by percolation of the water. See SOIL HORIZON, SOIL PROFILE.

zone of saturation See SATURATED ZONE.

zoonotic Describing a pathogen that normally infects wild animals but can infect humans if the carrier and humans have contact. For example, Lyme disease is caused by

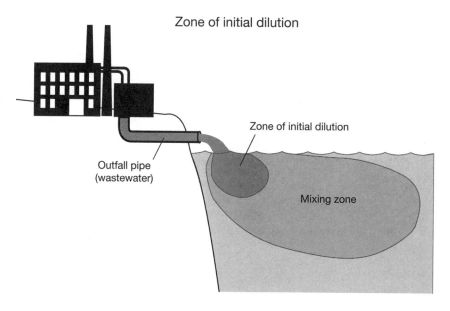

Zone of initial dilution

a pathogenic bacterium normally carried by ticks that infect deer. The bacterium can be transmitted to humans by the bite of deer ticks, given a close association of deer and humans in some locations.

zooplankton The small, often microscopic, animals in aquatic environments that possess little or no means of propulsion. Consequently, animals belonging to this class drift along with the currents.

APPENDIXES

I. Acronyms (and Other Abbreviations)

a	acceleration
Å	angstrom
AA	atomic absorption spectrophotometer
ABC lines	adiabatic cooling lines
ABS	alkylbenzene sulfonate
A/C	air-to-cloth ratio
ACFM	actual cubic feet per minute
ACGIH	American Conference of Governmental Industrial Hygienists
ACH	air changes per hour
ACLs	alternate concentration limits
ACM	asbestos-containing material
ADI	acceptable daily intake
ADP	AHERA designated person
ADR	alternate dispute resolution
AES	atomic emission spectroscopy
AFR	air/fuel ratio
AHERA	Asbestos Hazard Emergency Response Act
AI test	adsorption isotherm test
AIRS	Aerometric Information Retrieval System
AL	action level
ALARA	as low as reasonably achievable
amu	atomic mass unit
ANOVA	analysis of variance
ANPR	advance notice of proposed rulemaking
ANSI	American National Standards Institute
APA	Administrative Procedure Act
APCA	Air Pollution Control Association
APHA	American Public Health Association
API	American Petroleum Institute
APR	air-purifying respirator
APWA	American Public Works Association
AQCR	air quality control region
AQMA	air quality maintenance area
AQRV	air quality related value
AQS	air quality standard
AQS	ambient quality standard
ARAR	applicable or relevant and appropriate requirement
As	arsenic
ASCE	American Society of Civil Engineers
ASTM	American Society for Testing and Materials
ASTSWMO	Association of State and Territorial Solid Waste Management Officials
ATP	adenosine triphosphate
ATSDR	Agency for Toxic Substances and Disease Registry
AWMA	Air and Waste Management Association
AWWA	American Water Works Association
BACM	best available control measures
BACT	best available control technology

BADT	best available demonstrated technology
BANANA	build absolutely nothing anywhere near anyone
BaP	benzo(a)pyrene
BAPMon	Background Air Pollution Monitoring Network
BART	best available retrofit technology
BAT	best available technology economically achievable
BCF	bioconcentration factor
BCT	best conventional control technology
BDAT	best demonstrated available technology
Be	beryllium
BEI	Biological Exposure Index
BEIR report	Biological Effects of Ionizing Radiation report
BEJ	best engineering judgment
BHC	benzene hexachloride
BIRP	Beverage Industry Recycling Program
BLM	Bureau of Land Management
BMP	best management practices
BNA	Bureau of National Affairs
BOD	biochemical oxygen demand
BPJ	best professional judgment
BPT	best practicable control technology
Bq	becquerel
BTEX	benzene, toluene, ethylbenzene, and xylene
BTU	British thermal unit
BWR	boiling-water reactor
C	carbon
C	coulomb
^{14}C	carbon 14
CAA	Clean Air Act
CAER	Community Awareness and Emergency Response
CAFE standard	corporate average fuel economy standard
CAG	Carcinogen Assessment Group
CAI	carcinogenic activity indicator
CAIR	comprehensive assessment information rule
cal	calorie
CAM	compliance assurance monitoring
CAMEO®	Computer-Aided Management of Emergency Operations
CAMU	corrective action management unit
CANDU	Canadian deuterium-uranium reactor
CAPA	critical aquifer protection area
CARB	California Air Resources Board
CASAC	Clean Air Scientific Advisory Committee
CAS number	Chemical Abstracts Service Registry number
CBA	cost-benefit analysis
CBI	confidential business information
CCR	Consumer Confidence Report
CCW table	constituent concentrations in wastes table
CCWE table	constituent concentrations in waste extract table
CDBFs	chlorinated dibenzofurans
CDC	Centers for Disease Control
CEA	cost-effectiveness analysis
CEC	cation exchange capacity

CEC	Commission for Environmental Cooperation
CEM	continuous emission monitoring
CEQ	Council on Environmental Quality
CERCLA	Comprehensive Environmental Response, Compensation, and Liability Act
CERCLIS	Comprehensive Environmental Response, Compensation, and Liability Information System
CERES	Coalition for Environmentally Responsible Economies
CFCs	chlorofluorocarbons
CFR	*Code of Federal Regulations*
CGI	combustible gas indicator
CGL policy	comprehensive general liability policy
CHEMTREC	Chemical Transportation Emergency Center
CHIPs	Chemical Hazard Information Profiles
CHRIS	Chemical Hazard Response Information System
Ci	curie
CI pipe	cast-iron pipe
CITES	Convention on International Trade in Endangered Species of Wild Fauna and Flora
Cl_2	chlorine
CLP	contract laboratory program
CMA	Chemical Manufacturers Association
CNG	compressed natural gas
CO	carbon monoxide
CO_2	carbon dioxide
COD	chemical oxygen demand
COE	Corps of Engineers
Coh	coefficient of haze
COHb	carboxyhemoglobin
COLIWASA	composite liquid waste sampler
CPF	cancer potency factor
CPF	carcinogenic potency factor
CPSC	Consumer Product Safety Commission
CPT	cone penetrometer testing
CRP	child-resistant packaging
^{137}Cs	cesium-137
CSF	confidential statement of formula
CTGs	Control Techniques Guidelines
CWA	Clean Water Act
CWLM	cumulative working-level months
CWP	coal workers' pneumoconiosis
CZMA	Coastal Zone Management Act
2,4-D	dichlorophenoxyacetic acid
DAL	defect action level
dB	decibel
dBA	decibels, A-weighting network
DBPs	disinfection by-products
DDD	dichlorodiphenyldichloroethane
DDE	dichlorodiphenyldichloroethene
DDT	dichlorodiphenyltrichloroethane
DE	dose equivalent
DEIS	draft environmental impact statement

DES	diethylstilbestrol
DfE	design for the environment
DMR	discharge monitoring report
DNA	deoxyribonucleic acid
DNA	Designated National Authority
DNAPL	dense nonaqueous phase liquid
DO	dissolved oxygen
DOC	dissolved organic carbon
DOE	Department of Energy
DOM	dissolved organic matter
DRE	destruction and removal efficiency
DU	Dobson unit
ECCS	emergency core cooling system
ECD	electron capture detector
ECOS	Environmental Council of the States
ECx	experimental concentration—percentage
EDB	ethylene dibromide
EDC	ethylene dichloride
EDD	enforcement decision document
EDF	Environmental Defense Fund
EEC	estimated environmental concentration
EGR	exhaust gas recirculation
EHS	extremely hazardous substance
EIL policy	environmental impairment liability policy
EIQ	emission inventory questionnaire
EIS	environmental impact statement
ELP	Environmental Leadership Program
EMAP	Environmental Monitoring and Assessment Program
EMMI	environmental monitoring methods index
EMPACT	Environmental Monitoring for Public Access and Community Tracking
EOP	end-of-pipe technology/treatment
EOX	extractable organic halogens
EPA	Environmental Protection Agency
EPCRA	Emergency Planning and Community Right-to-Know Act
EPRI	Electric Power Research Institute
EP toxicity test	extraction procedure toxicity test
ERA	expedited removal action
ERCs	emission reduction credits
ERDA	Energy Research and Development Administration
ERNS	Emergency Response Notification System
ERT	environmental response team
ESA	Endangered Species Act
ESP	electrostatic precipitator
η	viscosity
ETS	environmental tobacco smoke
eV	electron volt
EW	equivalent weight
f/cc	fibers per cubic centimeter
FAO	Food and Agricultural Organization
FD	forced draft

FDA	Food and Drug Administration
FDCA	Food, Drug, and Cosmetic Act
FDFs	fundamentally different factors
FEPCA	Federal Environmental Pesticide Control Act
FERC	Federal Energy Regulatory Commission
FEV_1	forced expiratory volume
FGD	flue gas desulfurization
FHSA	Federal Hazardous Substances Act
FID	flame ionization detector
FIFRA	Federal Insecticide, Fungicide, and Rodenticide Act
FIP	Federal Implementation Plan
FLPMA	Federal Land Policy and Management Act
FMVCP	Federal Motor Vehicle Control Program
F/M ratio	food-to-microorganism ratio
FOE	Friends of the Earth
FONSI	finding of no significant impact
FPC	Federal Power Commission
FPC	fish protein concentrate
FQPA	Food Quality Protection Act
FR	*Federal Register*
FRRRPA	Forest and Rangeland Renewable Resources Planning Act
FTP	Federal Test Procedure
FVC	forced vital capacity
FWPCA	Federal Water Pollution Control Act
g	gram
GAC	granular activated carbon
GAO	General Accounting Office
GATT	General Agreement on Tariffs and Trade
G/C	gas-to-cloth ratio
GC	gas chromatography
GCM	general circulation model
GC/MS	gas chromatography/mass spectrometry
GCR	gas-cooled reactor
GEMI	Global Environmental Management Initiative
GEMS	Global Environment Monitoring System
GEMS	Graphical Exposure Modeling System
GEP stack height	good engineering practice stack height
GIS	geographic information system
GLC	ground-level concentration
GLP standards	Good Laboratory Practice standards
GMW	gram molecular weight
GPCD	gallons per capita per day
GPO	Government Printing Office
GPS	global positioning system
GRAS	generally recognized as safe
GW	gigawatt
GWP	global warming potential
Gy	gray
h	Planck's constant
H	Shannon-Weaver index
HAA5	haloacetic acids

HAAs	hormonally active agents
HACCP	Hazard Analysis Critical Control Point
HACS	Hazard Assessment Computer System
HAPs	hazardous air pollutants
HAZMAT	hazardous material
HAZOP	hazard and operability study
HAZWOPER	hazardous waste operations and emergency response
HCB	hexachlorobenzene
HCP	habitat conservation plan
HCS	hazard communication standard
HDD	halogenated dibenzo-p-dioxin
HDF	halogenated dibenzofuran
HDPE	high-density polyethylene
HEPA filter	high-efficiency particulate air filter
HHW	household hazardous waste
HIT	Hazard Information Transmission
hi-vol	high-volume air sampler
HLW	high-level waste
HMTA	Hazardous Materials Transportation Act
HOCs	halogenated organic compounds
HON	hazardous organic NESHAP
HQ	hazard quotient
HRS	Hazard Ranking System
HSI	heat stress index
HSL	hazardous substance list
HSWA	Hazardous and Solid Waste Amendments
HTGR	high-temperature gas reactor
HVAC system	heating, ventilation, and air-conditioning system
HVL	half-value layer
HWM facility	hazardous waste management facility
HWR	heavy-water reactor
HYVs	high-yielding varieties
Hz	hertz
IAEA	International Atomic Energy Agency
IAQ	indoor air quality
IARC	International Agency for Research on Cancer
ICR	information collection request
ICRP	International Commission on Radiological Protection
ICS	intermittent control system
ICS	incident command system
ICSU	International Council of Scientific Unions
ID	induced draft
IDL	instrument detection limit
IDLH	immediately dangerous to life and health
I&I	infiltration and inflow
IIR	injury incident rate
IJC	International Joint Commission
I&M	inspection and maintenance
INFOTERRA	International Environmental Information System
[I]/[O]	indoor/outdoor concentration ratio
IPCC	Intergovernmental Panel on Climate Change
IPM	integrated pest management

IR	incident rate
IR	infrared radiation
IRIS	Integrated Risk Information System
IRLG	Interagency Regulatory Liaison Group
IRPTC	International Register of Potentially Toxic Chemicals
ISC model	Industrial Source Complex model
ISO	International Organization for Standardization
ISR	indirect source review
IUCN	International Union for Conservation of Nature and Natural Resources
J	joule
JTU	Jackson turbidity unit
k	kinetic energy
K	carrying capacity
K	rate constant
K	Kelvin
K_a	acid dissociation constant
K_b	base dissociation constant
K_d	soil sorption coefficient
K_{oc}	organic carbon partition coefficient
K_{ow}	octanol-water partition coefficient
K_s	solubility product constant
K_w	water-air ratio
kcal	kilocalorie
kg	kilogram
kW	kilowatt
kWh	kilowatt-hour
l	liter
L_{dn}	day-night sound level
LA	load allocation
LAER	lowest achievable emission rate
LAS	linear alkyl sulfonate
LC_{50}	lethal concentration—50%
LC_{50}	median lethal concentration
LC_{LO}	lethal concentration, low
LD	lethal dose
LD_{50}	lethal dose—50%
LD_{LO}	lethal dose, low
LDPE	low-density polyethylene
LEA	local education agency
LEL	lower explosive limit
LEPC	local emergency planning committee
LET	linear energy transfer
LEV	low-emission vehicle
LFL	lower flammable limit
LIDAR	light detection and ranging
LLRW	low-level radioactive waste
LLW	low-level waste
LMFBR	liquid-metal fast breeder reactor
ln	natural logarithm

LNAPL	light nonaqueous phase liquid
LNG	liquefied natural gas
LOAEL	lowest-observed-adverse-effect level
LOC	level of concern
LOCA	loss-of-coolant accident
LOD	limit of detection
LOEL	lowest-observed-effect level
log	logarithm
low-E	low emissivity
LPG	liquefied petroleum gas
LSI	listing site inspection
LSS	life span study
LULU	locally undesirable land use
LUST trust fund	leaking underground storage tank trust fund
LWR	light-water reactor
ly	langley
m^3	cubic meter
MAB	Man and Biosphere program
MAB reserve	Man and Biosphere program reserve
MAC	maximum allowable concentration
MACT	maximum achievable control technology
MATC	maximum acceptable toxicant concentration
MCF	thousand cubic feet
MCL	maximum contaminant level
MCLG	maximum contaminant level goal
MCS	multiple chemical sensitivity syndrome
MDC	more-developed country
MDL	method detection limit
MEDLINE	Medical Literature Analysis and Retrieval System
MEI	most exposed individual
MEI	maximum exposed individual
meq/l	milliequivalents per liter
mg	milligram
Mg	megagram
mg/l	milligrams per liter
MHz	megahertz
MIR	maximum individual risk
ml	milliliter
MLD	median lethal dose
MLSS	mixed liquor suspended solids
MLVSS	mixed liquor volatile suspended solids
MMCF	million cubic feet
MMCFD	million cubic feet per day
MMMFs	man-made mineral fibers
MOE	margin of exposure
MOS	margin of safety
MPC	maximum permissible concentration
MPD	maximum permissible dose
MPN	most probable number
MPRSA	Marine Protection, Research, and Sanctuaries Act
mrem	millirem
MRF	materials recovery facility

MS	mass spectrometer
MSA	Metropolitan Statistical Area
MSDs	musculoskeletal disorders
MSDS	material safety data sheet
MSR	mammalian selectivity ratio
MSW	municipal solid waste
MTBE	methyl-tertiary-butyl ether
MTD	maximum tolerated dose
MTTP	maximum total trihalomethane potential
μg	microgram
$\mu g/m^3$	micrograms per cubic meter
μl	microliter
μm	micrometer
MWTA	Medical Waste Tracking Act
N	newton
N_2O	nitrous oxide
N_A	Avogadro's number
NAA	nonattainment area
NAAEC	North American Agreement for Environmental Cooperation
NAAQS	national ambient air quality standards
NAFTA	North American Free Trade Agreement
NAICS	North American Industry Classification System
NAMS	National Monitoring System
NAPAP	National Acidic Precipitation Assessment Program
NAPL	nonaqueous phase liquid
NASQAN	National Air Stream Quality Accounting Network
NAWAPA	North American Water and Power Alliance
NBAR	nonbinding allocation of responsibility
NCI	National Cancer Institute
NCP	National Contingency Plan
NCP	National Oil and Hazardous Substances Pollution Contingency Plan
NCP	net community productivity
NCRPM	National Council on Radiation Protection and Measurements
NDIR	nondispersive infrared analysis
NDT	nondestructive testing
NEP	National Estuary Program
NEPA	National Environmental Policy Act
NEPPS	National Environmental Performance Partnership System
NESHAP	National Emission Standards for Hazardous Air Pollutants
NFMA	National Forest Management Act
NFPA	National Fire Protection Association
NFRAP	no further remedial action planned
NGOs	nongovernmental organizations
NIEHS	National Institute for Environmental Health Sciences
NIHL	noise-induced hearing loss
NIMBY	not-in-my-backyard syndrome
NIMTOO	not in my term of office
NIOSH	National Institute for Occupational Safety and Health
NIST	National Institute of Standards and Technology

nm	nanometer
NMHC	nonmethane hydrocarbon
NMOC	nonmethane organic compound
NO	nitric oxide
NO_2	nitrogen dioxide
NO_x	nitrogen oxides
NO_x	oxides of nitrogen
NOAA	National Oceanic and Atmospheric Administration
NOAEL	no-observed-adverse-effect level
NOEL	no-observed-effect level
NOM	natural organic matter
NORM	naturally occurring radioactive material
NOV	notice of violation
NOW	nonhazardous oil field waste
NPCA	National Parks and Conservation Association
NPDES	National Pollutant Discharge Elimination System
NPDWR	National Primary Drinking Water Regulations
NPL	National Priorities List
NPP	net primary productivity
NRC	National Research Council
NRC	National Response Center
NRC	Nuclear Regulatory Commission
NRDC	Natural Resources Defense Council
NRT	National Response Team
NSDWR	National Secondary Drinking Water Regulations
NSF	National Science Foundation
NSF	National Strike Force
NSPS	new source performance standards
NSR	new source review
NTA	nitrilotriacetic acid
NTIS	National Technical Information Service
NTP	National Toxicology Program
NTU	nephelometric turbidity unit
NVPP	notice of violation—potential penalty
NWF	National Wildlife Fund
NWPA	Nuclear Waste Policy Act
OCRWM	Office of Civilian Radioactive Waste Management
ODP	ozone-depleting potential
OH^-	hydroxide ion
OHMTADS	Oil and Hazardous Materials Technical Assistance Data System
OIRA	Office of Information and Regulatory Affairs
OMB	Office of Management and Budget
ONRW	Outstanding Natural Resource Waters
OPA90	Oil Pollution Act of 1990
OSC	on-scene coordinator
OSHA	Occupational Safety and Health Administration
OSHAct	Occupational Safety and Health Act
OSWER	Office of Solid Waste and Emergency Response
OTA	Office of Technology Assessment
OTAG	Ozone Transport Assessment Group
OTEC	ocean thermal energy conversion

p	pico-
p	pressure
P	phosphorus
P	power
P	poise
P_2	pollution prevention
Pa	pascal
PAH	polycyclic aromatic hydrocarbon
PAH	polynuclear aromatic hydrocarbon
PAIR	preliminary assessment information rule
PAMS	Photochemical Assessment Monitoring Stations
PAN	peroxyacetyl nitrate
PAN	peroxyacyl nitrate
PAR	population at risk
PA/SI	preliminary assessment and site inspection
Pb	lead
PCBs	polychlorinated biphenyls
PCDFs	polychlorinated dibenzofurans
pCi/l	picocuries per liter
PCP	pentachlorophenol
PCV	positive crankcase ventilation
PE	polyethylene
PE	population equivalent
PEL	permissible exposure limit
PET	polyethylene terephthalate
PF	protective factor
PFLT	paint filter liquids test
PHA	process hazards analysis
PIC procedure	prior informed consent procedure
PICs	products of incomplete combustion
PID	photoionization detector
PM_{10}	particulate matter, 10-micrometer diameter
$PM_{2.5}$	particulate matter, 2.5-micrometer diameter
PMN	premanufacturing notice
PMR	proportionate mortality ratio
POC	point of compliance
POC	purgeable organic carbon
POHCs	principal organic hazardous constituents
POM	particulate organic matter
POM	polynuclear organic matter
POTW	publicly owned treatment works
POU/POE	point-of-use/point-of-entry
POX	purgeable organic halogens
ppb	parts per billion
PPE	personal protective equipment
ppm	parts per million
PRA	probabilistic risk assessment
PRGs	preliminary remediation goals
PRP	potentially responsible party
PSD	prevention of significant deterioration
PSES	pretreatment standards for existing sources
PSI	Pollutant Standards Index
psia	pounds per square inch (absolute)

PSNS	pretreatment standards for new sources
PSP	paralytic shellfish poisoning
PTS	permanent threshold shift
^{239}Pu	plutonium-239
PURPA	Public Utility Regulatory Policies Act of 1978
PVC	polyvinyl chloride
PWR	pressurized-water reactor
PYAR	person-years at risk
q	quad
Q	quint
Q	directivity factor
QA/QC	quality assurance/quality control
QF	quality factor
R	rankine
R	roentgen
R	universal gas constant
R_0	net reproductive rate
RA	regional administrator
RACM	reasonably available control measure
RACT	reasonably available control technology
rad	radiation absorbed dose
RARE	Roadless Area Review and Evaluation
Raw	airway resistance
RBC	rotating biological contactor
RBCA	risk-based corrective action
RBE	relative biological effectiveness
RCRA	Resource Conservation and Recovery Act
Rd	rutherford
RDA	recommended daily allowance
RDF	refuse-derived fuel
RDI	reference daily intake
RD/RA	remedial design/remedial action
Re	Reynolds number
regulated ACM	regulated asbestos-containing material
RELs	recommended exposure limits
rem	roentgen equivalent man
REPI	renewable energy production incentive
RFA	Regulatory Flexibility Act
RfD	reference dose
RFG	reformulated gasoline
RFI	Request for Information
RFP	reasonable further progress
RFT	respirator fit test
RGR	mean relative growth rate
Ri	Richardson number
RIA	Regulatory Impact Analysis
RI/FS	remedial investigation/feasibility study
RMCL	recommended maximum contaminant level
RMP	risk management plan
RMS sound pressure	root-mean-square sound pressure
Rn	radon

RNA	ribonucleic acid
ROD	record of decision
ROG	reactive organic gas
RPAR	rebuttable presumption against registration
RPM	remedial project manager
RQ	reportable quantity
RRC	Regional Response Center
RRT	Regional Response Team
RSD	risk-specific dose
RSPA	Research and Special Programs Administration
RTECS	Registry of Toxic Effects of Chemical Substances
ρ	density
RTK	right to know
RTP	Research Triangle Park
RUP	restricted-use pesticide
RVP	Reid vapor pressure
S	entropy
SAB	Science Advisory Board
SAGE data	Stratospheric Aerosol and Gas Experiment data
SAR	structure-activity relationship
SARA	Superfund Amendments and Reauthorization Act
SAROAD	Storage and Retrieval of Aerometric Data
SBT	segregated ballast tank
SBUV data	solar backscatter ultraviolet data
SCA	specific collection area
SCAP	Superfund Compliance Accomplishments Plan
SCAS test	semicontinuous activated sludge test
SCBA	self-contained breathing apparatus
SCC	Species Conservation Commission
SCFM	standard cubic feet per minute
SCRAM	safety control rod ax man
SCRAM	Support Center for Regulatory Air Models
SCS	Soil Conservation Service
SDWA	Safe Drinking Water Act
SEER	Surveillance, Epidemiology, and End Results
SEM	scanning electron microscope
SEP	supplemental environmental project
SERC	state emergency response commission
SIC	Standard Industrial Classification
SIP	state implementation plan
SITE	Superfund Innovative Technology Evaluation
SLAMS	State and Local Air Monitoring System
SLAPPs	strategic lawsuits against public participation
SMCL	secondary maximum contaminant level
SMCRA	Surface Mining Control and Reclamation Act
SMP	state management plan
SMR	standardized mortality ratio
SMSA	Standard Metropolitan Statistical Area
SNAP program	Significant New Alternatives Policy program
SNG	synthetic natural gas
SNUR	significant new use rule
SO_2	sulfur dioxide

SO$_x$	sulfur oxides
SO$_x$	oxides of sulfur
SOCMI	synthetic organic chemical manufacturing industry
SOCs	synthetic organic chemicals
SOD	sediment oxygen demand
SPCC Plan	Spill Prevention Control and Countermeasure Plan
SQG	small-quantity generator
^{90}Sr	strontium-90
S/S	stabilization/solidification
SSC	scientific support coordinator
SSF	slow sand filtration
STAPPA	State and Territorial Air Pollution Program Administrators
STEF	septic tank effluent filter
STEL	short-term exposure limit
STPP	sodium tripolyphosphate
Sv	sievert
SVI	sludge volume index
SW-486	*Test Methods for Evaluating Solid Waste, Physical/Chemical Methods*
SWAP	Source Water Assessment Program
SWMU	solid waste management unit
SWPP	Source Water Protection Program
SWTR	Surface Water Treatment Rule
t	metric ton
T	absolute temperature
T	transmissivity
2,4,5-T	trichlorophenoxyacetic acid
TAG	technical assistance grant
TBEL	technology-based effluent limitation
TC	total carbon
TCDBF	tetrachlorodibenzofuran
TCDD	tetrachlorodibenzo-*para*-dioxin
TCDF	tetrachlorodibenzofuran
TCLP	toxicity characteristic leaching procedure
TCM	transportation control measure
TCRA	time-critical removal action
TDS	total dissolved solids
TED	turtle excluder device
TEM	transmission electron microscopy
TENR	technologically enhanced natural radioactivity
TGA	technical-grade active ingredient
THMs	trihalomethanes
TI	therapeutic index
TIC	total inorganic carbon
TKN	total Kjeldahl nitrogen
TLV	threshold limit value
TMDL	total maximum daily load
TMRC	theoretical maximum residue contribution
TOC	total organic carbon
TOD	theoretical oxygen demand
TOMS data	Total Ozone Mapping Spectrometer data
TOXLINE	Toxicology Information on Line

TOXNET	Toxicology Data Network
TPD	tons per day
TPH	total petroleum hydrocarbons
TPQ	threshold planning quantity
TRE	Toxicity Reduction Evaluation
TRI	Toxics Release Inventory
TRIS	Toxic Chemical Release Inventory System
TRS	total reduced sulfur
TRU waste	transuranic waste
TS	total solids
TSI	thermal system insulation
TSCA	Toxic Substances Control Act
TSD	treatment, storage, or disposal
TSDF	treatment, storage, or disposal facility
TSP	total suspended particulate
TSS	total suspended solids
TTHMs	total trihalomethanes
TTS	temporary threshold shift
TWA	time-weighted average
U	uranium
^{235}U	uranium 235
^{238}U	uranium 238
UEL	upper explosive limit
UF	uncertainty factor
UFL	upper flammable limit
UIC	Underground Injection Control
UNAMAP	Users' Network for Applied Modeling of Air Pollution
UNEP	United Nations Environment Program
UN/NA number	United Nations/North America number
URT	upper respiratory tract
USDW	underground source of drinking water
USEPA	United States Environmental Protection Agency
U.S.C.	*United States Code*
U.S.C.A.	*United States Code Annotated*
U.S.C.S.	*United States Code Service*
USGS	United States Geological Survey
UST	underground storage tank
UTM coordinates	Universal Transverse Mercator coordinates
UV	ultraviolet
UV-A	ultraviolet radiation-A range
UV-B	ultraviolet radiation-B range
UV-C	ultraviolet radiation-C range
UV-VIS	ultraviolet-visible absorption spectrum
v	velocity
V_H	velocity head
VMT	vehicle miles traveled
VOA	volatile organic analysis
VOC	volatile organic carbon
VOCs	volatile organic compounds
VOST	volatile organic sampling train
VP	velocity pressure

VSD	virtually safe dose	
VSS	volatile suspended solids	

W	watt
WCASP	World Climate Applications and Services Programme
WCDMP	World Climate Data and Monitoring Programme
WCED	World Commission on Environment and Development
WCP	World Climate Programme
WDV	water dilution volume
WET testing	whole effluent toxicity testing
WHO	World Health Organization
WIPP	Waste Isolation Pilot Plant
WL	working level
WLA	wasteload allocation
WLM	working-level month
WMO	World Meteorological Organization
WQBEL	water quality–based effluent limitation
WRP	Wetlands Reserve Program
WSF	water-soluble fraction
WSR	waste shipment record
WTE	waste-to-energy

Xe	xenon
XRF	X-ray fluorescence

Y, IMBY, FAP	yes, in my backyard, for a price

ZEV	zero emission vehicle
ZID	zone of initial dilution
ZPG	zero population growth

II. Unit Prefixes

tera (T)	1×10^{12}	trillion
giga (G)	1×10^{9}	billion
mega (M)	1×10^{6}	million
kilo (k)	1×10^{3}	thousand
hecto (h)	1×10^{2}	hundred
deka (da)	1×10^{1}	ten
deci (d)	1×10^{-1}	one-tenth
centi (c)	1×10^{-2}	one-hundredth
milli (m)	1×10^{-3}	one-thousandth
micro (μ)	1×10^{-6}	one-millionth
nano (n)	1×10^{-9}	one-billionth
pico (p)	1×10^{-12}	one-trillionth

III. Approximate Unit Equivalents

1. Mass

> 1 kilogram = 1000 grams = 2.205 pounds
> 1 metric ton = 1000 kilograms
> 1 pound = 453.6 grams = 7000 grains = 16 ounces = 0.4536 kilogram
> 1 ton = 2000 pounds = 907.2 kilograms

2. Length

> 1 meter = 100 centimeters = 3.281 feet = 39.37 inches
> 1 kilometer = 1000 meters = 0.6214 mile = 3281 feet
> 1 mile = 5280 feet = 1760 yards = 1609 meters
> 1 inch = 2.54 centimeters
> 1 micrometer = 1×10^{-6} meter
> 1 nanometer = 1×10^{-9} meter
> 1 angstrom = 1×10^{-10} meter

3. Area

> 1 square meter = 1×10^4 square centimeters
> = 10.764 square feet
> = 1550 square inches
> = 1.196 square yards
> 1 square foot = 929.03 square centimeters
> = 0.0929 square meter
> = 144 square inches
> = 0.1111 square yard
> 1 hectare = 10,000 square meters = 2.47 acres
> 1 acre = 0.405 hectare = 43,560 square feet
> 1 square kilometer = 0.386 square mile
> 1 square mile = 2.59 square kilometers = 640 acres

4. Volume

> 1 cubic centimeter = 0.001 liter = 0.061 cubic inch
> 1 liter = 1000 cubic centimeters
> = 1.057 quarts (U.S.)
> = 61.02 cubic inches
> = 0.0353 cubic foot
> 1 cubic meter = 1000 liters = 1 stere
> = 1×10^6 cubic centimeters
> = 35.31 cubic feet
> = 264.2 gallons (U.S.)
> 1 gallon (U.S.) = 4 quarts (U.S.) = 3.785 liters
> 1 quart (U.S.) = 2 pints (U.S.) = 0.946 liter
> 1 pint (U.S.) = 16 ounces = 0.473 liter
> 1 cubic mile = 4.17 cubic kilometers
> 1 acre-foot = 43,560 cubic feet = 1233.5 cubic meters
> 1 barrel of petroleum = 42 gallons (U.S.)

5. Energy

> 1 joule = 1 kilogram-meter^2/second^2 = 1 newton-meter = 1 watt-second
> = 0.239 calorie = 9.48×10^{-4} British thermal units
> = 0.7376 foot-pound

$$= 2.78 \times 10^{-7} \text{ kilowatt-hour}$$
$$= 1 \times 10^7 \text{ ergs} = 6.24 \times 10^{18} \text{ electron volts}$$
1 calorie = 4.184 joules
1 calorie (food) = 1 kilocalorie = 1000 calories
1 British thermal unit (BTU) = 1055 joules = 252 calories
1 kilowatt-hour = 3.6×10^6 joules = 3412 British thermal units
1 quad = 1×10^{15} British thermal units = 1.05×10^{18} joules

6. Power
   1 watt = 1 joule/second = 3.412 British thermal units/hour
   1 kilowatt = 1000 watts = 1.34 horsepower
   1 horsepower = 550 foot-pounds/second = 0.746 kilowatt

7. Force
   1 newton = 1 kilogram-meter/second^2 = 1×10^5 dynes

8. Pressure
   1 pascal = 1 newton/square meter
             = 1.45×10^{-4} pound per square inch
             = 9.87×10^{-6} atmosphere
   1 atmosphere = 101,325 pascals = 101.33 kilopascals
             = 14.7 pounds per square inch
             = 760 millimeters (mm) Hg = 29.92 inches Hg
             = 406.8 inches of water = 33.9 feet of water
             = 1013 millibars = 1.013 bars
             = 760 torrs
   1 pound per square inch = 0.068 atmosphere = 6895 pascals
             = 51.7 mm Hg = 27.68 inches of water
   1 torr = 133.32 newtons/square meter = 1/760 atmosphere
   1 bar = 1×10^5 pascals = 0.9869 atmosphere

9. Speed
   1 meter/second = 3.281 feet/second = 2.237 miles per hour
   1 mile/hour = 0.447 meter/second = 1.609 kilometers/hour
             = 1.467 feet/second = 88 feet/minute

10. Viscosity (dynamic)
    1 poise = 100 centipoise = 0.1 kilogram/(meter-second)
             = 1 dyne-second/square centimeter
    1 centipoise = 1×10^{-3} kilogram/(meter-second)
             = 0.01 gram/(centimeter-second)
             = 3.6 kilogram/(meter-hour)
             = 2.42 pounds/(foot-hour)

11. Radioactivity and radiation dose
    1 curie = 3.7×10^{10} nuclear disintegrations per second
    1 becquerel = 1 nuclear disintegration per second
    1 roentgen = an exposure to X rays or gamma rays causing an electric charge
             of 2.58×10^{-4} coulomb per kilogram of dry air = approximately
             10×10^{-3} sievert dose
    1 rad = 100 ergs of absorbed radiation per gram of absorbing medium
             = 0.01 joule per kilogram of medium
    1 gray = 100 rads = 1 joule per kilogram

1 sievert = 100 rems
1 rem = an equivalent radiation dose
= rad times a quality factor (*QF*), which varies by type of radiation
for gamma, beta, and X rays, *QF* = 1
for fast neutrons and protons, *QF* = 10
for alpha particles, *QF* = 20

12. Temperature
degrees Celsius = 5/9 × (degrees Fahrenheit − 32)
degrees Fahrenheit = (9/5 × degrees Celsius) + 32
degrees Kelvin = degrees Celsius + 273.15

IV. Concentrations

Concentration is an expression of how much of a material is in a given amount of another material or environmental medium (air, water, soil, food). Environmental contaminant concentrations are usually described as the mass of a chemical in a given mass (or volume) of a medium or as the volume of a material in a given volume of a medium.

Concentrations of contaminants in water are expressed as the mass of a contaminant per given volume (typically one liter) of water or as the mass of a contaminant per given mass of water. An example of a mass/volume concentration is the number of milligrams of chemical X in one liter of water. Now, one liter of water (at 4° C) has a mass of 1000 grams. Therefore, the mass of chemical X, in milligrams/liter, is equivalent to the mass of chemical X in milligrams/1000 grams, a mass/mass expression. There are 1 million milligrams in 1000 grams, or one milligram is one-millionth of 1000 grams. Therefore, a concentration of one milligram per liter is equal to one part per million. It follows that one microgram per liter is equal to one part per billion, and so on. The following units are used interchangeably in expressions of water pollution concentrations.

mass/volume	*mass/mass*	*dimensionless*
milligrams/liter	= milligrams/1000 grams	= parts per million (ppm)
micrograms/liter	= micrograms/1000 grams	= parts per billion (ppb)
nanograms/liter	= nanograms/1000 grams	= parts per trillion (ppt)

The concentrations of aerosols (airborne solids or liquids) in air are expressed as the mass of a substance in a given volume (typically one cubic meter) of air. Typical expressions are milligrams of chemical X per cubic meter (m^3) of air or micrograms of chemical X per cubic meter of air. These aerosol concentrations are not equivalent to parts per million, parts per billion, and so forth.

Concentrations of gases or vapors in air are expressed as the volume of a gaseous material in a given volume of air (volume/volume) or as the mass of the material per given volume of air (mass/volume). A concentration of one liter of chemical X per 1 million liters of air is equivalent to a concentration of one part per million of X; one liter of chemical Y in 1 billion liters of air is a concentration of one part per billion. The alternative expression of the air concentration of a gaseous chemical is a mass per volume, for example, the number of milligrams of X in a given volume (typically one cubic meter) of air.

Volume/volume concentrations and mass/volume concentrations of gaseous air contaminants are convertible, using the molecular weight of the gas or vapor and the density of air. For air at 25° C and a pressure of one atmosphere (USEPA standard conditions), the following conversions can be used.

micrograms of a substance A per cubic meter of air =
(parts per million of A) × (molecular weight of A) × 40.9

parts per million of A = micrograms of substance A per cubic meter of air
× (1/molecular weight of A) × (1/40.9)

V. Standard (Average) Human Factors

Factor	Man	Woman	Child (3–12)
Body mass (in kilograms)	70	60	15–40
Skin surface area (in square meters)			
no clothing	1.8	1.6	0.9
normal clothing	0.1–0.3	0.1–0.3	0.05–0.15
Respiration (in liters per minute)			
resting	7.5	6.0	5.0
light activity	20	19	13
Air volume breathed			
(in cubic meters per day)	23	21	15
Fluid intake			
(in liters per day of water, milk, beverages)	2	1.4	1.0

See the three-volume *Exposure Factors Handbook,* from the USEPA National Center for Environmental Assessment, for an extensive compilation of standard factors used to estimate human exposure to toxic chemicals (www.epa.gov/ncea/exposfac.htm).

VI. Plastic Recycling Codes

Code	Resin	Name	Primary Use	Recycled Product
1	PETE	polyethylene terephthalate	soft drink bottles water bottles	fiberfill products carpet backing
2	HDPE	high-density polyethylene	milk jugs food containers	plastic lumber detergent bottles
3	V	polyvinyl chloride	construction plastics	drainage pipe
4	LDPE	low-density polyethylene	plastic bags	garbage bags
5	PP	polypropylene	frozen food containers	automotive products
6	PS	polystyrene	Styrofoam items	insulation
7	Other	mixed resins	containers	construction

VII. The Chemical Elements

Element	Symbol	Proton No.	Relative atomic mass
actinium	Ac	89	[227]
aluminum	Al	13	26.9815
americium	Am	95	[243]
antimony	Sb	51	121.75
argon	Ar	18	39.948
arsenic	As	33	74.9216
astatine	At	85	[210]
barium	Ba	56	137.34
berkelium	Bk	97	[247]
beryllium	Be	4	9.0122
bismuth	Bi	83	208.98
boron	B	5	10.81
bromine	Br	35	79.904
cadmium	Cd	48	112.40
caesium	Cs	55	132.905
calcium	Ca	20	40.08
californium	Cf	98	[251]
carbon	C	6	12.011
cerium	Ce	58	140.12
chlorine	Cl	17	35.453
chromium	Cr	24	51.996
cobalt	Co	27	58.9332
copper	Cu	29	63.546
curium	Cm	96	[247]
dysprosium	Dy	66	162.50
einsteinium	Es	99	151.96
erbium	Er	68	167.26
europium	Eu	63	151.96
fermium	Fm	100	[257]
fluorine	F	9	18.9984
francium	Fr	87	[223]
gadolinium	Gd	64	157.25
gallium	Ga	31	69.72
germanium	Ge	32	72.59
gold	Au	79	196.967
hafnium	Hf	72	178.49
helium	He	2	4.0026
holmium	Ho	67	164.930
hydrogen	H	1	1.00797
indium	In	49	114.82
iodine	I	53	126.9044
iridium	Ir	77	192.2
iron	Fe	26	55.847
krypton	Kr	36	83.80
lanthanum	La	57	138.91
lawrencium	Lr	103	[257]
lead	Pb	82	207.19
lithium	Li	3	6.939
lutetium	Lu	71	174.97

Element	Symbol	Proton No.	Relative atomic mass
magnesium	M	12	24.305
manganese	Mn	25	54.938
mendelevium	Md	101	[258]
mercury	Hg	80	200.59
molybdenum	M	42	95.94
neodymium	Nd	60	144.24
neon	Ne	10	20.179
neptunium	Np	93	[237]
nickel	Ni	28	58.71
niobium	Nb	41	92.906
nitrogen	N	7	14.0067
nobelium	No	102	[255]
osmium	Os	76	190.2
oxygen	O	8	15.9994
palladium	P	46	106.4
phosphorus	P	15	30.9738
platinum	Pt	78	195.09
plutonium	Pu	94	[244]
polonium	Po	84	[209]
potassium	K	19	39.102
praseodymium	Pr	59	140.907
promethium	Pm	61	[145]
protactinium	Pa	91	[231]
radium	Ra	88	[226]
radon	Rn	86	[222]
rhenium	Re	75	186.20
rhodium	Rh	45	102.905
rubidium	Rb	37	85.47
ruthenium	Ru	44	101.07
samarium	Sm	62	150.35
scandium	Sc	21	44.956
selenium	Se	34	78.96
silicon	Si	14	28.086
silver	Ag	47	107.868
sodium	Na	11	22.9898
strontium	Sr	38	87.62
sulfur	S	16	32.064
tantalum	Ta	73	180.948
technetium	Tc	43	[97]
tellurium	Te	52	127.60
terbium	Tb	65	158.924
thallium	Tl	81	204.37
thorium	Th	90	232.038
thulium	Tm	69	168.934
tin	Sn	50	118.69
titanium	Ti	22	47.90
tungsten	W	74	183.85
unnilhexium	Unh	106	263
unnilpentium	Unp	105	262.11
unnilquadium	Unq	104	261
uranium	U	92	238.03

Element	Symbol	Proton No.	Relative atomic mass
vanadium	V	23	50.942
wolfram (tungsten)	W	74	183.85
xenon	Xe	54	131.30
ytterbium	Yb	70	173.04
yttrium	Y	39	88.905
zinc	Zn	30	65.37
zirconium	Zr	40	91.22

VIII. The Greek Alphabet

A	α	alpha
B	β	beta
Γ	γ	gamma
Δ	δ	delta
E	ε	epsilon
Z	ζ	zeta
H	η	eta
Θ	θ	theta
I	ι	iota
K	κ	kappa
Λ	λ	lambda
M	μ	mu
N	ν	nu
Ξ	ξ	xi
O	o	omicron
Π	π	pi
P	ρ	rho
Σ	σ	sigma
T	τ	tau
Y	υ	upsilon
Φ	φ	phi
X	χ	chi
Ψ	ψ	psi
Ω	ω	omega